T0311894

NITRIC OXIDE DONORS

NITRIC OXIDE DONORS

Novel Biomedical Applications and Perspectives

Edited by

AMEDEA BAROZZI SEABRA

Center for Natural and Human Sciences,
Federal University of ABC (UFABC),
Santo André, São Paulo, Brazil

ACADEMIC PRESS

An imprint of Elsevier
elsevier.com

Academic Press is an imprint of Elsevier
125 London Wall, London EC2Y 5AS, United Kingdom
525 B Street, Suite 1800, San Diego, CA 92101-4495, United States
50 Hampshire Street, 5th Floor, Cambridge, MA 02139, United States
The Boulevard, Langford Lane, Kidlington, Oxford OX5 1GB, United Kingdom

British Library Cataloguing-in-Publication Data
A catalogue record for this book is available from the British Library

Library of Congress Cataloging-in-Publication Data
A catalog record for this book is available from the Library of Congress

ISBN: 978-0-12-809275-0

For Information on all Academic Press publications
visit our website at https://www.elsevier.com/books-and-journals

Working together
to grow libraries in
developing countries

www.elsevier.com • www.bookaid.org

Publisher: Mica Haley
Acquisition Editor: Rafael Teixeira
Editorial Project Manager: Mariana Kuhl
Production Project Manager: Edward Taylor
Designer: Alan Studholme

Typeset by MPS Limited, Chennai, India

CONTENTS

List of Contributors   *xi*

1. Nitric Oxide Derivative Ruthenium Compounds as NO-Based Chemotherapeutic and Phototherapeutic Agents   **1**

Laísa Bonafim Negri, Tassia Joi Martins, Loyanne C.B. Ramos and Roberto S. da Silva

Introduction   1
Nitrosyl Ruthenium Complexes as NO Delivery Agents   2
Ruthenium-Nitrogen Oxide Derivatives as NO Sources   5
Nitrosyl Ruthenium Complexes as NO Delivery Agent by Reduction Process   5
Photo-Nitric Oxide Release Promoted by Ruthenium Complexes   11
Image: A Powerful Tool to Elucidate Cytotoxicity Mechanisms of No Donor Agents
    Based on Ruthenium Complexes   17
Conclusion   21
References   22

2. Nitric Oxide Donors for Treating Neglected Diseases   **25**

Amedea Barozzi Seabra and Nelson Durán

Neglected Diseases   25
Why Using NO Donors to Treat Neglected Diseases?   26
Malaria   27
Leishmaniasis   32
Trypanosomiasis   35
Schistosomiasis   40
Tuberculosis   41
Dengue Fever   42
Blastocystis Hominis Infection   44
NO Donors and Lymphatic Filariasis   45
NO Donors and Other Neglected Diseases   45
Mechanisms of Action of NO Against Pathogens That Cause Neglected Diseases   46
Final Remarks: Challenges and Perspectives   47
Acknowledgments   48
References   48

3. Nitric Oxide-Donating Devices for Topical Applications   **55**

Kathleen Fontana and Bulent Mutus

Introduction   55
NO's Role in the Body   57

Topical Applicators   67
Conclusion   71
References   72

4. Nitric Oxide Donors and Therapeutic Applications in Cancer   75

Khosrow Kashfi and Pascale L. Duvalsaint

Introduction   75
Cellular Actions of NO   76
NSAIDs Protect Against Cancer, Proof of Principle   79
Side Effects Associated with NSAID Use   79
Nitric Oxide-Releasing NSAIDs (NO-NSAIDs) and the Rationale for Their Development 80
Structural Features of Nitrate NO-NSAIDs   81
NO-Releasing Coxibs   84
Cancer Prevention with NO-NSAIDs   86
Biological Actions of the Spacer in NO-ASA   93
Diazeniumdiolate-Based NO-Releasing Compounds   94
HNO-NSAIDs   97
NOSH-NSAIDs   100
RRx-001 an Aerospace Compound for Cancer Treatment   102
Concluding Remarks and Future Directions   105
Acknowledgment   106
References   106

5. Nitric Oxide Donors and Penile Erectile Function   121

Serap Gur, Allen L. Chen and Philip J. Kadowitz

Introduction   121
Nitric Oxide Synthase (NOS) Isoforms   122
The Role of NO in Erectile Physiology   123
ED and Its Prevalence   126
NO Donors   126
NO Donors for Erectile Function   127
S-nitroacetylpenicillamine   129
Sodium Nitroprusside   129
Organic Nitrates   130
Sydnonimines   130
Oximes   131
Spermine NONOate   131
S-nitrosothiols   131
Nitrosamines   131
PDE-5 Inhibitors   132
sGC Stimulators and Activators   133
NO-Donating Statins   134
Conclusions and Perspectives   134
References   136

6. Nitric Oxide Donors in Nerve Regeneration **141**

Vinod B. Damodaran, Divya Bhatnagar, Heather Rubin and Melissa M. Reynolds

Introduction	141
NO in Neuroregeneration	141
Neuronal Growth and Synaptic Plasticity	143
Neuroprotectant	143
NO in Neurodegeneration	145
Alzheimer's Disease	145
Parkinson's Disease	147
Huntington's Disease	147
Examples of NO Donors Evaluated for Nerve Regeneration	148
Locust Studies on NO-cGMP with NOC-18 and SNP	151
Goldfish Optic Nerve Regeneration Studied with NOR-2 and SNAP	152
Regenerating RGCs of Cats with Nipradilol	155
Reinnervation After Penile Nerve Crush Studies with SNP in Rats	157
Exogenous SNAC on Motor Functional Recovery in Rats	158
S-Nitrosoglutathione (GSNO) Effects After I/R injury in Rats	159
GSNO Influences Angiogenesis in Rats	162
Conclusion	164
References	165

7. Nitric Oxide Donors as Antimicrobial Agents **169**

Samuel K. Kutty, Kitty Ka Kit Ho and Naresh Kumar

Nitric Oxide	169
Biosynthesis of NO	169
NO Inactivation Chemistry	170
Biological Effects of NO	170
NO Regulation in Microbes	171
NO as an Antimicrobial and Antibiofilm Agent	173
NO Gas as an Antimicrobial Agent	175
NO Donors as Antimicrobials	176
Dual Action NO Donors as Antimicrobials	181
Conclusion	184
References	185

8. Improving the Performance of Implantable Sensors with Nitric Oxide Release **191**

Megan C. Frost

Introduction	191
Biological Response to Implanted Sensors	192
The Role NO Could Play	196
NO Donors and Generators Applied to Sensors	197
Conclusion and Future Directions	215
References	216

9. Strategies to Deliver Nitric Oxide Donors to Control Biofilms of Clinical and Industrial Interest   **221**

Massimiliano Marvasi and Tania Henríquez

Introduction   221
Dual Function of NO: Antibacterial Agent and Genetic Regulator in Bacteria   223
NO Donor Platforms to Control Biofilm of Clinical Interest   226
Delivery of NO via Small Molecular Weight Donors   234
Conclusion   236
References   237

10. Developing New Organic Nitrates for Treating Hypertension   **243**

Camille M. Balarini, Josiane C. Cruz, José L.B. Alves, Maria S. França-Silva and Valdir A. Braga

Introduction   243
Historical Perspective   245
Important Organic Nitrates Used in Clinic   246
Organic Nitrates and Arterial Hypertension   248
Nitrate Tolerance   251
New Insights on Drug Development   253
Acknowledgments   257
References   258

11. Nitric Oxide Donors in Brain Inflammation   **263**

Marisol Godínez-Rubí and Daniel Ortuño-Sahagún

Introduction   263
Role of Nitric Oxide in Brain Physiology   263
Neuroinflammation and NO in Damaged Brain   265
NO Donors as Neuroprotective Agents in Brain Inflammation   275
Concluding Remarks   284
References   284

12. Synergistic Activities of Nitric Oxide and Various Drugs   **293**

Govindan Ravikumar and Harinath Chakrapani

Introduction   293
Delivery of NOS   296
NO and Anticancer Drug Cotreatment   297
NO Anticancer Drug Hybrids   299
NO-Based Antibacterial and Biofilm Dispersal Agents   303
Synergic Effect of NO with Antibiotics   304
Conclusion   307
References   307

13. Hydrogels for Topical Nitric Oxide Delivery   **313**

Mathilde Champeau, Amedea Barozzi Seabra and Marcelo Ganzarolli de Oliveira

Hydrogels with Dissolved Molecular NO Donors   314
Hydrogels with Covalently Attached NO Donors   323
Gels Based on Extemporaneous NO Release   325
Supramolecular NO-Releasing Hydrogels   326
Conclusion   328
References   328

Index   *331*

LIST OF CONTRIBUTORS

José L.B. Alves
Health Sciences Center, Federal University of Paraiba, João Pessoa, Paraíba, Brazil

Camille M. Balarini
Health Sciences Center, Federal University of Paraiba, João Pessoa, Paraíba, Brazil;
Biotechnology Center, Federal University of Paraíba, João Pessoa, Paraíba, Brazil

Divya Bhatnagar
New Jersey Center for Biomaterials and Rutgers, The State University of New Jersey,
Piscataway, NJ, United States

Valdir A. Braga
Biotechnology Center, Federal University of Paraíba, João Pessoa, Paraíba, Brazil

Harinath Chakrapani
Department of Chemistry, Indian Institute of Science Education and Research Pune,
Maharashtra, India

Mathilde Champeau
Institute of Chemistry, University of Campinas, UNICAMP, São Paulo, Brazil

Allen L. Chen
Tulane University School of Medicine, New Orleans, LA, USA

Josiane C. Cruz
Biotechnology Center, Federal University of Paraíba, João Pessoa, Paraíba, Brazil

Roberto S. da Silva
School of Pharmaceutical Sciences of Ribeirão Preto, University of São Paulo, Ribeirão Preto,
São Paulo, Brazil

Vinod B. Damodaran
New Jersey Center for Biomaterials and Rutgers, The State University of New Jersey,
Piscataway, NJ, United States

Marcelo Ganzarolli de Oliveira
Institute of Chemistry, University of Campinas, UNICAMP, São Paulo, Brazil

Nelson Durán
Institute of Chemistry, Biological Chemistry Laboratory, University of Campinas, Campinas,
São Paulo, Brazil; Brazilian Nanotechnology National Laboratory (LNNano-CNPEM),
Campinas, Brazil

Pascale L. Duvalsaint
Department of Physiology, Pharmacology, and Neuroscience, Sophie Davis School of
Biomedical Education, City University of New York School of Medicine, New York, NY,
United States

Kathleen Fontana
Department of Chemistry and Biochemistry, University of Windsor, Windsor, ON, Canada

Maria S. França-Silva
Biotechnology Center, Federal University of Paraíba, João Pessoa, Paraíba, Brazil

Megan C. Frost
Department of Biomedical Engineering, Michigan Technological University, Houghton, MI, United States

Marisol Godínez-Rubí
Departamento de Microbiología y Patología, Universidad de Guadalajara, Jalisco, México

Serap Gur
Department of Pharmacology, School of Pharmacy, Ankara University, Ankara, Turkey; Department of Pharmacology, Tulane University Health Sciences Center, New Orleans, LA, United States

Tania Henríquez
Department of Microbiology and Mycology, Institute of Biomedical Sciences, University of Chile, Santiago, Chile

Kitty Ka Kit Ho
School of Chemistry, University of New South Wales, Sydney, NSW, Australia

Philip J. Kadowitz
Department of Pharmacology, Tulane University Health Sciences Center, New Orleans, LA, United States

Khosrow Kashfi
Department of Physiology, Pharmacology, and Neuroscience, Sophie Davis School of Biomedical Education, City University of New York School of Medicine, New York, NY, United States

Naresh Kumar
School of Chemistry, University of New South Wales, Sydney, NSW, Australia

Samuel K. Kutty
School of Chemistry, University of New South Wales, Sydney, NSW, Australia

Tassia Joi Martins
Faculty of Philosophy, Sciences and Literature of Ribeirão Preto, University of São Paulo, Ribeirão Preto, São Paulo, Brazil

Massimiliano Marvasi
Middlesex University London, The Burroughs, London, United Kingdom

Bulent Mutus
Department of Chemistry and Biochemistry, University of Windsor, Windsor, ON, Canada

Laísa Bonafim Negri
School of Pharmaceutical Sciences of Ribeirão Preto, University of São Paulo, Ribeirão Preto, São Paulo, Brazil

Daniel Ortuño-Sahagún
Instituto de Investigación en Ciencias Biomédicas, Departamento de Biología Molecular y Genómica. Universidad de Guadalajara, Jalisco, México

Loyanne C.B. Ramos
School of Pharmaceutical Sciences of Ribeirão Preto, University of São Paulo, Ribeirão Preto, São Paulo, Brazil

Govindan Ravikumar
Department of Chemistry, Indian Institute of Science Education and Research Pune, Maharashtra, India

Melissa M. Reynolds
Department of Chemistry and School of Biomedical Engineering, Colorado State University, Fort Collins, CO, United States

Heather Rubin
Department of Chemistry and School of Biomedical Engineering, Colorado State University, Fort Collins, CO, United States

Amedea Barozzi Seabra
Center for Natural and Human Sciences, Federal University of ABC (UFABC), Santo André, São Paulo, Brazil; Institute of Chemistry, Biological Chemistry Laboratory, University of Campinas, Campinas, São Paulo, Brazil

CHAPTER 1

Nitric Oxide Derivative Ruthenium Compounds as NO-Based Chemotherapeutic and Phototherapeutic Agents

Laísa Bonafim Negri[1], Tassia Joi Martins[2], Loyanne C.B. Ramos[1] and Roberto S. da Silva[1]

[1]School of Pharmaceutical Sciences of Ribeirão Preto, University of São Paulo, Ribeirão Preto, São Paulo, Brazil
[2]Faculty of Philosophy, Sciences and Literature of Ribeirão Preto, University of São Paulo, Ribeirão Preto, São Paulo, Brazil

INTRODUCTION

In the context of chemistry, the term "complex" means a central atom surrounded by a set of linkers. The central atom is a metal center that performs the function of a Lewis acid capable of receiving electrons; this central atom combines with ligands that act as Lewis base which are capable of donating electrons. Among all the atoms, metallic elements are the most abundant and include species of blocks s, d, and f as well as some elements of block p (aluminum, gallium, indium, thallium, tin, lead, and bismuth). One of the most important characteristics of metals is their tendency to display stable oxidation states within each block, which allows extraction of these metals from their ores and facilitates their handling in the laboratory. Elements belonging to block d probably constitute one of the most versatile classes of compounds. Their diverse chemical properties have given rise to a wide range of compounds with application in several fields including medicinal chemistry (Atkins et al., 2009).

Medicinal Inorganic Chemistry originated in 1908 when Paul Ehrlich introduced the first studies on the structure–activity relationship of arsenic compounds (Atkins et al., 2009; Beraldo et al., 2008). Since then, several studies involving coordination compounds have been carried out aiming at their clinical applications. Essentially, the development of metal complexes for chemotherapeutic purposes involves hydrolysis, protein binding, membrane transportation, and interaction with the molecular target (Schwietert and MCccue, 1999). Many commercially available metal-based drugs have been successfully employed in the clinical setting. Among these drugs, platinum compounds such as cisplatin and its analogs (carboplatin and oxaliplatin) stand out in the treatment of ovarian cancer, head and neck cancer, bladder cancer, cervical cancer, and lymphomas (van Rijt and Sadler, 2009). The cytotoxicity of these compounds

Nitric Oxide Donors.

1

is related to the binding of platinum to a specific site in the DNA of the tumor cell, which triggers a cascade of reactions that culminate in cell death by apoptosis. Despite the wide use of these compounds in cancer therapy, severe side effects such as nausea, bone marrow suppression, and toxicity to the kidneys have emerged in patients (van Rijt and Sadler, 2009). These difficulties have motivated scientists to find different ways to employ metal-based compounds as drugs. One strategy involves applying metal ions that carry ligands with biological activity. Nitric oxide (NO) delivery agents are probably one of the best-known representatives of this class of complexes. Nitrosyl (NO^+) coordination compounds have found therapeutic application and have been the object of many studies since the 19th century (Szczepura and Takeuchi, 1990). Sodium nitroprusside ($Na_2[Fe(CN)_5(NO)]$) is an example of a metal complex that bears a nitrosyl ligand in the metal coordination sphere and can therefore release NO, being a useful compound in blood pressure control (Moncada et al., 1991; Stochel et al., 1998). However, this compound is only applied in medical emergency because it can also release cyanide (CN^-) in a reaction secondary to the reaction of pharmacological interest (Thomas et al., 2009). Researchers have therefore sought to control NO release from coordination compounds and make these complexes clinically feasible. In this sense, Flitney et al. (1996) conducted photochemical experiments with clusters such as $[Fe_4S_4(NO)_4]$ and $[Fe_4S_3(NO)_7]^-$. Photolysis at certain wavelengths promoted NO release. In addition, light induction and electrochemical reduction of coordinated nitrosyl could be interesting approaches that take the low affinity of the NO^0 ligand for some metal ions into account. For this reason, the search for nitrosyl ruthenium complexes capable of releasing NO in the organism via external stimulus has been quite intense. Togniolo et al. (2001) described spectroscopic and photochemical studies of the complex cis-$[RuCl(bpy)_2(NO)](PF_6)_2$ and demonstrated NO release in aqueous medium upon irradiation with laser at 355 nm ($\phi_{NO} = 0.98\,mol\,Einstein^{-1}$). Bonaventura et al. (2004) reported the physicochemical and photochemical properties of the complex trans-$[RuCl([15]aneN_4)(NO)]Cl_2$ in physiological pH. This ruthenium compound produced NO under reduction or light irradiation at 355 nm. Based on these results, our group has published numerous papers on nitrosyl ruthenium complexes. This chapter will address these publications and shall contribute to the description of the physicochemical and photochemical properties of nitrosyl ruthenium complexes for controlled NO release and their use in biological and physiological processes.

NITROSYL RUTHENIUM COMPLEXES AS NO DELIVERY AGENTS

Basic physical and chemical properties of some NO derivative species

NO is a paramagnetic molecule with an unpaired electron located in an antibonding π-orbital, as shown in Fig. 1.1.

Molecular orbital	NO^0	NO^+	NO^- (triplet)	NO^- (singlet)
$\sigma^*(2p)$				
$\pi^*(2p)$	↑		↑ ↑	↑ ↓
$\sigma(2p)$	↑↓	↑↓	↑↓	↑↓
$\pi(2p)$	↑↓ ↑↓	↑↓ ↑↓	↑↓ ↑↓	↑↓ ↑↓
$\sigma^*(2s)$	↑↓	↑↓	↑↓	↑↓
$\sigma(2s)$	↑↓	↑↓	↑↓	↑↓
$\sigma^*(1s)$	↑↓	↑↓	↑↓	↑↓
$\sigma(1s)$	↑↓	↑↓	↑↓	↑↓

Figure 1.1 Electron configuration of some nitric oxide derivative species.

$$2NO \longrightarrow N_2O_2$$

Scheme 1.1 Dimerization reaction of NO.

In aqueous solution, NO has solubility and diffusion rate of $1.9 \times 10^{-3}\,mol\,L^{-1}$ and about $50\,\mu M\,s^{-1}$, respectively (Ignarro, 2000; Wink et al., 1996). This colorless gas is stable in its pure form and, despite its radical nature, the formation of a dimer is thermodynamically unfavorable. Indeed, the energy involved (ΔH) in such dimerization (Scheme 1.1) corresponds to about $-2.6\,kcal\,mol^{-1}$, and the term $-T\Delta S$ is $+4.3\,kcal\,mol^{-1}$ at $1\,atm$ and $300\,K$, which gives a positive Gibbs free energy ($\Delta G = \Delta H - T\Delta S$) and shows that the reaction does not occur spontaneously under these conditions (Beckman et al., 1996). Low temperatures decrease the term $-T\Delta S$, the free energy becomes negative, and dimerization at lower temperatures is thus spontaneous.

NO tends to react quickly with some transition metals to establish a σ-bond between the metal and either nitrogen or oxygen (Richter-addo and Legzdins, 1992). In some cases and depending on the metal ion, in addition to forming the σ-bond, the d-orbital in the metal may interact with the empty $\pi^\star$ orbital in the NO derivative ligand. This interaction elicits back-bonding as in the case of the nitrosyl ligand (NO^+) (Fig. 1.2).

NO has rich biochemical diversity as a result of three known forms, namely NO^+, NO^0, and NO^- (nitroxyl), all of which present different physicochemical properties (Table 1.1).

In biological medium, nitrogen oxide derivative species react with oxygen to generate reactive oxygen and nitrogen species (RONS), as described for NO (Scheme 1.2).

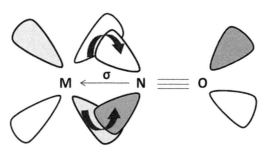

Figure 1.2 The back-bonding representation between metal ion and nitrosyl ligand.

Table 1.1 Bond length, vibrational energy, and reduction potential of free NO^+, NO^0, and NO^- species

	NO^+	NO^0	NO^-
Bond distance N(O) (Å)	1.06	1.15	1.26
$\nu_{(NO)}$ (cm^{-1})	2377	1875	1470
E vs ENH (V)	-1.20^a	–	-0.39^b

Source: Hughes (1999).
[a] $NO^{+/0}$.
[b] $NO^{0/-}$.

$$NO^0 \;+\; O_2 \qquad\qquad \rightarrow \; ONOO^\bullet \qquad\qquad (1)$$

$$NO^0 \;+\; ONOO^\bullet \qquad \rightarrow \; ONOONO \rightarrow 2NO_2 \qquad (2)$$

$$2NO_2 \;+\; 2NO^0 \qquad \rightarrow \; 2N_2O_3 \qquad\qquad (3)$$

$$\underline{2N_2O_3 \;+\; H_2O \qquad\qquad \rightarrow \; 4NO_2^- \;+\; 4H^+} \qquad (4)$$

$$4NO^0 \;+\; O_2 \;+\; 2H_2O \rightarrow \; 4NO_2^- \;+\; 4H^+ \qquad (5)$$

Scheme 1.2 NO oxidation reaction steps originating RONS (Ignarro, 2000).

Although NO^+, NO^0, and NO^- all have therapeutic implications (Shoman and Aly, 2016; Cheung et al., 2000), this chapter will focus on the biological mechanism related to NO delivery agents like $[Ru^{II}N_4L'NO^+]^{n+}$, paying special attention to the effect of L on the Ru-(N). This chapter will also consider biological properties of these complexes such as vasodilation, trypanocidal action, and cytotoxicity activities.

RUTHENIUM-NITROGEN OXIDE DERIVATIVES AS NO SOURCES

Several ruthenium-nitrogen oxide derivative compounds have been described as NO delivery agents (de Lima et al., 2014). Table 1.2 contains compounds that represent these agents and which will be the basis of this chapter. The physicochemical and photochemical characteristics of these compounds depend mainly on the interaction between Ru(II) and the ligands "L" in $[RuN_4L'NO_x]^{n+}$ as well as on the strength of the bond established between the metal ion and the nitrogen oxide derivative ligand. In general, all the ruthenium compounds discussed here are stable both in the solid state and in aqueous solution.

NITROSYL RUTHENIUM COMPLEXES AS NO DELIVERY AGENT BY REDUCTION PROCESS

Nitrosyl ruthenium complexes of the $[Ru^{II}N_4L'NO^+]^{n+}$ type are stable as complex salt but undergo pH-dependent electrophilic attack by hydroxide ion in aqueous solution to generate nitroruthenium(II) species (Sauaia and da Silva, 2003a) (Scheme 1.3).

Systematic tuning of the electrophilic character of coordinated NO^+ through modulation of the ancillary ligand "L" in $[RuN_4L'NO^+]^{n+}$ is possible (Cândido et al., 2015; Sauaia and da Silva, 2003a). The equilibrium constant for the conversion of nitrosyl species into nitro species depends on the π-acceptor character of ligand "L," and this rate constant decreases with increasing electron density in the nitrosyl ligand. On the steady state pK_{NO} affords as a result of this electrophilic attack. Fig. 1.3 shows the effect of pH on UV–visible spectrum of cis-$[Ru(bpy)_2(py)(NO)]^{3+}$ as representative of nitrosyl ruthenium(II) compounds. The pK_{NO} may help to describe the molecular formula of ruthenium-NO derivative species as a function of pH ($\{Ru^{II}\text{-}NO^+\}^{3+}$ or $\{Ru^{II}\text{-}NO_2^-\}^+$), as observed in Fig. 1.4 and Table 1.3 for some $[Ru^{II}N_4L'NO^+]^{n+}$ compounds.

The electrochemical reduction of $\{Ru^{II}\text{-}NO^+\}$ compounds is fascinating. Fig. 1.5 brings representative cyclic voltammograms (CV) of $[Ru^{II}N_4L'NO^+]^{n+}$ species. The CV of cis-$[Ru(bpy)_2(pz)(NO)]^{3+}$ in nonaqueous or aqueous medium showed two cathodic and anodic peaks namely "1" and "2" attributed to $NO^{+/0}$ and $NO^{0/-}$ process, respectively.

In aqueous solution, the reduction process is followed by chemical reaction centered on the NO derivative species. Fig. 1.6 depicts the controlled potential reduction of nitrosyl ruthenium compounds monitored by spectroelectrochemistry; spectral changes are evident.

The first electrolytic potential reduction conducted in the presence of the NO-sensor raises the peak current, which is consistent with increasing NO concentration (Fig. 1.6). These studies aid description of the electrochemical mechanism related to the $[Ru^{II}N_4L'NO^+]^{n+}$ species (Scheme 1.4). This mechanism leads to the assumption that nitrosyl ruthenium complexes release NO after the first reduction step.

Table 1.2 Chemical structure of some ruthenium-nitrogen oxide derivative species

Chemical structure	Chemical formula
	cis-[RuCl(H-dcbpy)$_2$(NO)]
	[Ru(bqdi)(terpy)(NO)]$^{3+}$
	trans-[RuCl([15]-ane$_4$)(NO)]$^{2+}$
	cis-[Ru(NO$_2$)L(bpy)$_2$]$^+$ **L$_1$** = pyridine **L$_2$** = 4-picoline **L$_3$** = 4-acetylpyridine

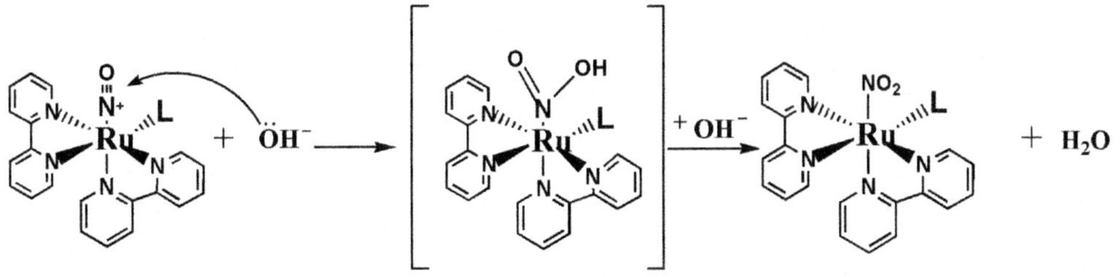

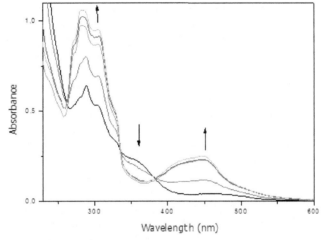

Scheme 1.3 Representative reaction of nitrosyl ruthenium complex with hydroxide ion.

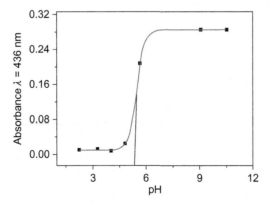

Figure 1.3 Spectral profile change with pH for the conversion of [Ru(bpy)(terpy)(NO)]$^{3+}$ in [Ru(NO$_2$)(bpy)(terpy)]$^+$. pH = 2.48, 4.30, 5.72, 7.16, and 10.29.

Figure 1.4 Effect of pH on the metal ligand charge transfer band of nitro ruthenium(II) species [Ru(bpy)$_2$(2-tFmepy)(NO)]$^+$.

Table 1.3 pK_{NO} values for some nitrosyl ruthenium compounds

Compounds	pK_{NO}
$[Ru(bpy)_2(4\text{-pic})(NO)]^{3+}$	3.30
$[Ru(bpy)_2(py)(NO)]^{3+}$	3.40
$[Ru(bpy)_2(4\text{-acpy})(NO)]^{3+}$	3.80
$Ru(bdqi\text{-COOH})(terpy)(NO)]^{3+}$	4.30
$[Ru)(bpy)_2(2\text{-pic})(NO)]^{3+}$	5.47
$[Ru(bpy)_2(2\text{-tFmepy})(NO)]^{3+}$	5.75
$[Ru(bpy)_2(2\text{-me-4-ampy})(NO)]^{3+}$	6.03

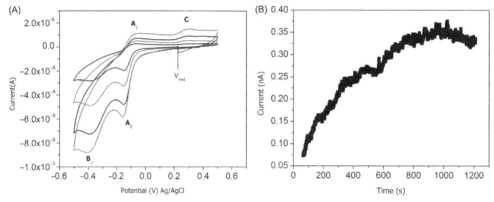

Figure 1.5 Cyclic voltammogram (A) and chronoamperogram (B) of $[Ru(terpy)(py\text{-SH})_2(NO)]^{3+}$ complex in buffer solution (pH = 2.03) with KCl (0.10 mol cm^{-3}) added as support electrolyte. $\nu = 200$, 100, 50, and 20 mV s^{-1}.

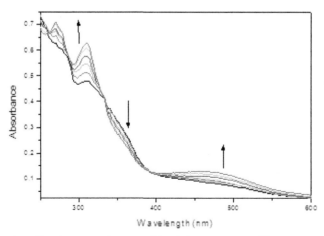

Figure 1.6 Spectral profile change with time of $[Ru(bdqi)(terpy)(NO)]^{3+}$ complex in buffer solution (pH = 4.50) after electrolysis at −0.30 V versus Ag/AgCl.

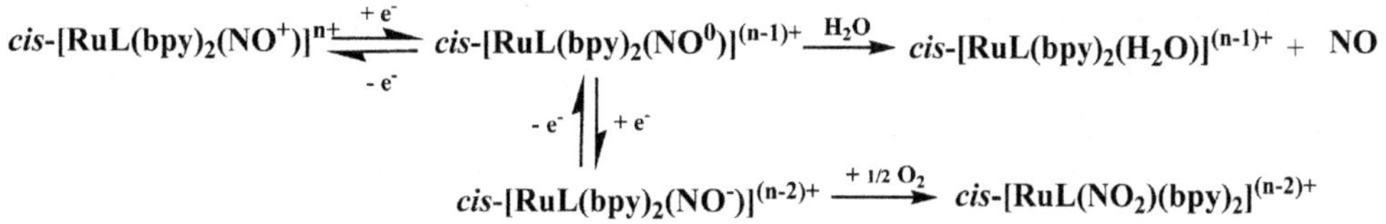

$$cis\text{-}[\text{RuL(bpy)}_2(\text{NO}^+)]^{n} \underset{-e^-}{\overset{+e^-}{\rightleftharpoons}} cis\text{-}[\text{RuL(bpy)}_2(\text{NO}^0)]^{(n-1)+} \xrightarrow{\text{H}_2\text{O}} cis\text{-}[\text{RuL(bpy)}_2(\text{H}_2\text{O})]^{(n-1)+} + \text{NO}$$

$$\big\updownarrow {\scriptstyle -e^-} {\scriptstyle +e^-}$$

$$cis\text{-}[\text{RuL(bpy)}_2(\text{NO}^-)]^{(n-2)+} \xrightarrow{+\,1/2\,\text{O}_2} cis\text{-}[\text{RuL(NO}_2)(\text{bpy})_2]^{(n-2)+}$$

Scheme 1.4 The electrochemical reduction mechanism of $cis\text{-}[\text{RuL(bpy)}_2(\text{NO})]^{n+}$ complex with nitric oxide release.

Synergistic effect between ruthenium complex and nitric oxide as a tool to increase cytotoxicity

We have investigated the cytotoxicity activity of several nitrosyl ruthenium species (Maranho et al., 2009). One of the example is $[\text{Ru(bdqi)(terpy)(NO)}]\text{Cl}_3$ as NO delivery agent and the effect of this complex on the melanoma cancer B16F10 cell line in aqueous solution as well as in solid lipid nanoparticles (SLN). The high affinity of ruthenium for NO is a marked feature of ruthenium chemistry. To understand the effect of NO as anticancer agent better, we have also evaluated the effect of $[\text{Ru(bdqi)(terpy)(H}_2\text{O})]^{2+}$ complex from a cytotoxic point of view. Fig. 1.7 illustrates the cell viability results obtained for these compounds in B16F10 cell line.

Only the complex $[\text{Ru(bdqi)(terpy)(H}_2\text{O})]^{2+}$ at very high concentration (250 μM) and after 48 hours of incubation exerts significant cytotoxicity. These results point to the dual action of $[\text{Ru(bdqi)(terpy)(NO)}]^{3+}$—NO release and formation of aquaruthenium(II) species. Intracellular reduction of $[\text{Ru(bdqi)(terpy)(NO)}]^{3+}$ should underlie NO release as judged by cytosolic NO concentration measurements (Fig. 1.8).

The intracellular NO release from $[\text{Ru(bdqi)(terpy)(NO)}]^{3+}$ gives rise to the formation of $[\text{Ru(bdqi)(terpy)(H}_2\text{O})]^{2+}$. Considering the described cytotoxicity of NO (Heinrich, 2013), decreased cell viability for $[\text{Ru(bdqi)(terpy)(NO)}]^{3+}$ should be expected. Based on the data from Fig. 1.7, it was not observed. It may be due to cellular uptake of nitrosyl ruthenium complex once encapsulation of $[\text{Ru(bdqi)(terpy)(NO)}]\text{Cl}_3$ into SLN increases cytotoxicity considerably (Fig. 1.9; Table 1.3).

In vitro cellular death with different concentrations of the nitrosyl ruthenium complex in B16F10 cancer cell line can be observed in Table 1.4. Clearly the SLN improve cellular uptake of ruthenium complex, which is followed by change on the biochemical parameters proportioned by intracellular chemical reactions with $[\text{Ru(bdqi)(terpy)(NO)}]\text{Cl}_3$. The biochemical pathway is under investigation but apparently one of the imminent process involves the DNA interaction with $[\text{Ru(bdqi)(terpy)(H}_2\text{O})]^{2+}$. It may guide discovering of new cancer drugs based on synergistic effect of NO and ruthenium compounds.

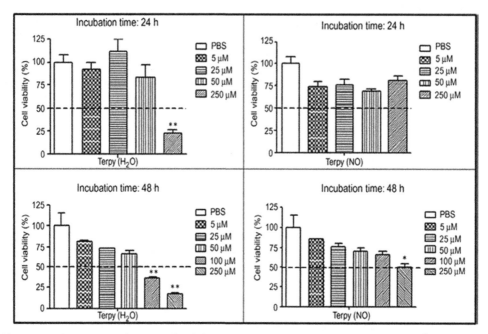

Figure 1.7 Comparation between [Ru(bdqi)(terpy)(NO)]Cl$_3$ and [Ru(H$_2$O)(bdqi)(terpy)]$^{2+}$ complexes in murine melanoma cells in different concentrations and incubation time. $*p < 0.05$ and $**p < 0.01$, different from control (Heinrich, 2013).

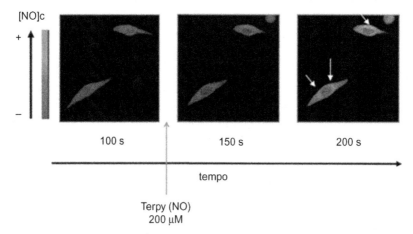

Figure 1.8 Cytosolic concentration of nitric oxide in murine melanoma cells (B16F10), using a fluorescent probe, DAF-2 DA (5 μM), after NO release from [Ru(bdqi)(terpy)(NO)]$^{3+}$ complex (Heinrich, 2013).

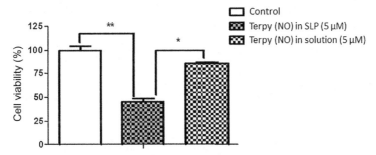

Figure 1.9 Cell viability in B16F10 cell of [Ru(bdqi)(terpy)(NO)]Cl$_3$ complex in aquo solution and [Ru(bdqi)(terpy)(NO)]Cl$_3$ encapsulated in solid lipid nanoparticles. *$p < 0.05$ and **$p < 0.01$, different from control (Heinrich, 2013).

Table 1.4 Cell viability (%) in murine melanoma cells (B16F10) of [Ru(bdqi)(terpy)(NO)]Cl$_3$ and [Ru(bdqi)(terpy)(H$_2$O)]$^{2+}$ complexes in solution and encapsulated in SLP in different concentration and incubation time (Heinrich, 2013)

% Cell viability in B16F10 cell line

[Ru^{II}L$_4$L′NO$^+$]$^{n+}$	[Ru(bdqi)(terpy)(H$_2$O)]$^{2+}$			[Ru(bdqi)(terpy)(NO)]$^{3+}$			[Ru(bdqi)(terpy)(NO)]$^{3+}$ encapsulated in SLN
Concentration (μM)	4 hours	24 hours	48 hours	4 hours	24 hours	48 hours	4 hours
5	100	100	80	100	75	86	45
25	110	110	75	110	75	75	–
50	100	100	70	100	75	70	–
100	–	–	30	–	–	65	–
250	85	25	20	100	75	50	–

PHOTO-NITRIC OXIDE RELEASE PROMOTED BY RUTHENIUM COMPLEXES

NO plays a critical role in biological medium: it can act for example as a vasorelaxant agent in blood vessels (Marchesi et al., 2012). The concentration–relaxation dependent (pD$_2$) and the maximum relaxant effect (E_{max}) depend on the molecular structure of the vasorelaxant agent. The NO derivative ligand bound to ruthenium also induces this process in isolated vessels of normotensive and hypertensive rats by producing NO after stimulation (Table 1.5).

Table 1.5 The vasorelaxation induced by nitric oxide from nitrosyl ruthenium complexes donors (Lunardi et al., 2009)

Complex $[Ru^{II}L_4L'NO^+]^{n+}$	pD_2	E_{max} (%)
trans-$[RuCl([15]\text{-}ane_4)(NO)]^{2+}$	5.03 ± 0.2	98.35 ± 1.22
$[Ru(bdqi)(terpy)(NO)]^{3+}$	6.47 ± 0.13	102.38 ± 0.38
cis-$[RuL(bpy)_2(NO)]^{n+}$	6.62 ± 0.12	101.2 ± 3.7

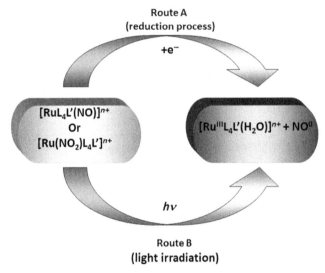

Figure 1.10 Ruthenium complexes as NO deliver agents.

Two mechanisms explain NO release from these ruthenium species (Fig. 1.10).

In route "a," the nitrosyl ruthenium complex releases NO by reduction in complexes of the $[RuL_4L'NO^+]^{n+}$ type or by oxygen transfer reaction in complexes of the $[Ru(NO_2)L_4L']^{n+}$ type. Fig. 1.11 summarizes the extensively explored pharmacologic mechanism of NO during vasorelaxation (de Lima et al., 2014).

Essentially, NO activates guanylate cyclase, which mediates vasorelaxation. The literature presents discussions involving vasodilation by reduction of nitrosyl ruthenium complexes (da Silva et al., 2015). Photo-vasodilation employing those species is less common and will be the focus of this discussion. As described in Fig. 1.10, route "b," aqueous nitrogen oxide derivative bound to ruthenium produces NO under light irradiation. All the studied nitrosyl ruthenium species release NO by excitation of the metal ligand charge transfer band (MLCT) centered on $d\pi(Ru^{II})$–$\pi^\star(NO)$, which generally emerges in the ultraviolet region (de Lima et al., 2006). The photochemical

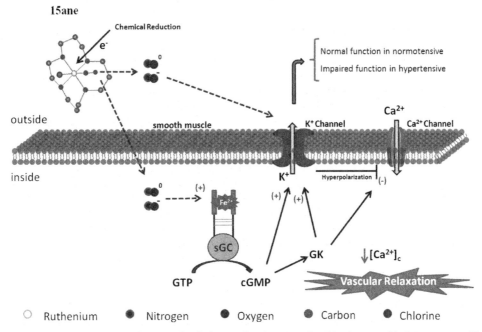

○ Ruthenium   ● Nitrogen   ● Oxygen   ● Carbon   ● Chlorine

Figure 1.11 Proposed NO release and cellular mechanisms involved in the vasodilation promoted by *trans*-[RuCl(15-aneN₄)(NO)]⁺ (15ane). The colored circles represent atoms in the chemical structure. The hydrogen atom has been omitted in the structure. sGC, soluble guanylyl cyclase; GK, G Kinase Protein; Ca²⁺, calcium; K⁺, potassium; [Ca²⁺]c, cytosolic calcium concentration. *Reproduced with permission from de Lima, R.G., Silva, B.R., da Silva, R.S., Bendhack, L.M., 2014. Ruthenium complexes as NO donors for vascular relaxation induction. Molecules 19, 9628–9654, MDPI Publishing Group.*

$$[Ru^{II}L_4L'(NO^+)]^{n+} \xrightarrow{\ h\nu\ } [Ru^{III}L_4L'(NO^0)]^{n+}$$

$$[Ru^{III}L_4L'(NO^0)]^{n+} + H_2O \longrightarrow [Ru^{III}L_4L'(H_2O)]^{n+} + NO$$

Scheme 1.5 The photochemical pathway of NO release from ruthenium complexes by UV light irradiation.

pathway described for these systems includes the formation of $[Ru^{III}L_4L'(H_2O)]^{n+}$, as shown in Scheme 1.5 (Sauaia et al., 2003c).

Nitrosyl ruthenium complexes rarely display bands in the visible region, which would result from contribution of the molecular orbital of NO⁺. Photostimulation of those compounds has been providing be useful vasodilator agents for in vitro analysis, although there is very little possibility to apply this knowledge for in vivo assays due dangerous use of UV light. In this way, we have succeeded to produce ruthenium complex, which presents bands with strong absorption on visible region. The complex [Ru(phthalocyanine)(NO₂)(NO)] is a representative species of such systems.

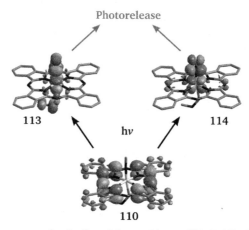

Figure 1.12 Optimized geometry for the B and Q transitions of [Ru(Pc)(ONO)(NO)] calculated by DFT and ZINDO/S. *Reproduced with permission from Carneiro, Z.A., de Moraes, J.C.B., Rodrigues, F.P., de Lima, R.G., Curti, C., da Rocha, Z.N., et al., 2011. Photocytotoxic activity of a nitrosyl phthalocyanine ruthenium complex—a system capable of producing nitric oxide and singlet oxygen. J. Inorg. Biochem. 105, 1035–1043, Elsevier Publishing Group.*

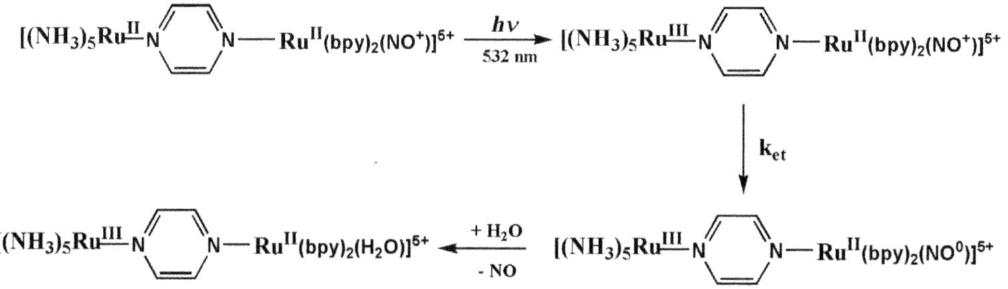

$$[(NH_3)_5Ru^{II}-N \diagdown N-Ru^{II}(bpy)_2(NO^+)]^{5+} \xrightarrow[\text{532 nm}]{h\nu} [(NH_3)_5Ru^{III}-N \diagdown N-Ru^{II}(bpy)_2(NO^+)]^{5+}$$

$$\downarrow k_{et}$$

$$[(NH_3)_5Ru^{III}-N \diagdown N-Ru^{II}(bpy)_2(H_2O)]^{5+} \xleftarrow[\text{- NO}]{\text{+ }H_2O} [(NH_3)_5Ru^{III}-N \diagdown N-Ru^{II}(bpy)_2(NO^0)]^{5+}$$

Scheme 1.6 Photo-induced electron transfer in ruthenium complexes to induce NO release (Sauaia et al., 2003b).

In aqueous solution, photolysis of this complex upon irradiation with light at 660 nm culminates in NO release. According to molecular orbital calculation (Fig. 1.12), the $\pi^\star$ orbital contributes to the transition centered in this region.

A similar mechanism occurs for some ruthenium compounds in which an antenna is responsible for the light absorption process. Photo-induced electron transfer is claimed to induce NO release (Scheme 1.6).

The binuclear ruthenium system illustrates how this system works for nitrosyl ruthenium compounds (Fig. 1.13) (Sauaia et al., 2003b).

The phthalocyanine–ruthenium compound as well as the binuclear ruthenium species have been proved to be useful tools as vasorelaxant agents by visible light irradiation (Carneiro et al., 2011). More recently, we have found be unnecessary have a covalent bond for photo-induced electron transfer between two ruthenium moieties.

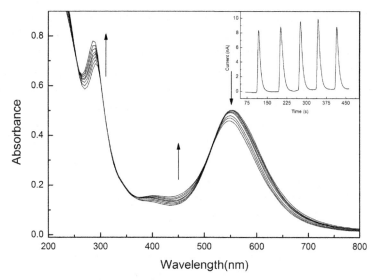

Figure 1.13 UV–visible spectra of the $[Ru^{II}(py)(NH_3)_4(pz)Ru^{II}(bpy)_2(NO)]^{5+}$ complex in acetate buffer solution (pH = 4.5) during flash-photolysis at λ_{irr} = 532 nm. Inset: Chronoamperograms of NO release as measured by the NO meter.

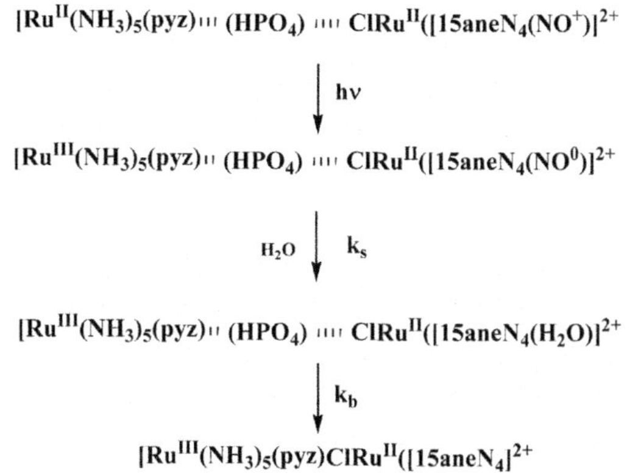

$$[Ru^{II}(NH_3)_5(pyz)\cdots(HPO_4)\cdots ClRu^{II}([15aneN_4(NO^+)]^{2+}$$

$$\downarrow h\nu$$

$$[Ru^{III}(NH_3)_5(pyz)\cdots(HPO_4)\cdots ClRu^{II}([15aneN_4(NO^0)]^{2+}$$

$$H_2O \downarrow k_s$$

$$[Ru^{III}(NH_3)_5(pyz)\cdots(HPO_4)\cdots ClRu^{II}([15aneN_4(H_2O)]^{2+}$$

$$\downarrow k_b$$

$$[Ru^{III}(NH_3)_5(pyz)ClRu^{II}([15aneN_4]^{2+}$$

Scheme 1.7 Supramolecular formation of ruthenium complexes mediated by phosphate bridge (da Silva et al., 2007).

Supramolecular systems for example constitute a strategy to construct electron donor–acceptor complexes. These sophisticated systems involve the design of two cationic ruthenium complexes, $[Ru(NH_3)_5(pyrazine)]^{2+}$ (antenna) and $[RuCl([15]aneN_4)(NO)]^{2+}$ (acceptor), linked via a phosphate bridge. These designs create ensembles that promote photo-induced electron transfer (Scheme 1.7).

Fig. 1.14 presents vasodilation experiments performed with this assembly.

New approaches based on a system that releases NO under visible light irradiation may have several applications including vasodilation control. The synthesized binuclear species incorporated into a sol–gel membrane gives a colorful material (Fig. 1.15A) that releases NO upon irradiation with light at 530 nm, as seen in Fig. 1.15B. It may constitute an interesting procedure to treat several body anomalies sensitive to NO, once ruthenium complexes on light-exposed skin may be photolyzed to free NO radicals.

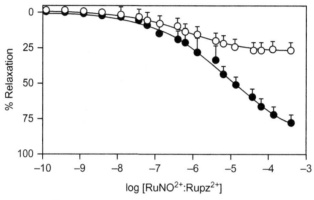

Figure 1.14 Effects of $RuNO^{2+}$:$Rupz^{2+}$ on rat thoracic aorta precontracted with phenylephrine without ($\bigcirc$) and with ($\bullet$) photolysis. The solution of $[RuCl([15]aneN_4)(NO)]^{2+}$/$[Ru(NH_3)_5(pyrazine)]^{2+}$ in phosphate buffer was cumulatively added (0.1 nM to 400 μM). *Reproduced with permission from da Silva et al., 2007, ACS Publishing Group.*

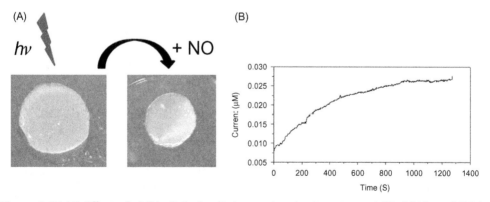

Figure 1.15 (A) Effect of visible light irradiation under the incorporated {$[Ru(NH_3)_5pz–(HPO_4)–Ru(bdq)(terpy)(NO)]]^{3+}$ sol–gel membrane. (B) NO-sensor measurement of "a."

IMAGE: A POWERFUL TOOL TO ELUCIDATE CYTOTOXICITY MECHANISMS OF NO DONOR AGENTS BASED ON RUTHENIUM COMPLEXES

Cellular imaging is a powerful tool to elucidate subcellular compartmentalization of novel drug complexes. Most importantly, image analysis can identify relationships between the molecular structure of the drug and its incorporation, distribution, and subcellular localization, all of which contribute to the distinction between the biochemical mechanisms involved in drug cytotoxicity. Nitrosyl ruthenium complexes, such as cis-[Ru(bpy)$_2$(4-pic)(NO)](PF$_6$)$_3$, release NO and have potential use as anticancer agents (Rodrigues et al., 2016). A recent study demonstrated the efficacy of a nitrosyl ruthenium complex as a prodrug that releases localized high flux NO upon activation, consequently inducing apoptosis in HepG2 hepatocarcinoma cells. 4-Amino-5-methylamino-2',7'-difluorofluorescein (DAF-FM) and H$_2$DCF-DA oxidation probes enable detection of NO and ROS by fluorescence (Fig. 1.16). The MTT assay aids evaluation of the viability of HepG2 cells, for which reduction in survival depends on drug dose and treatment time.

These results, subcellular fragmentation, and western blotting analyses further support that localized NO release mediates tumor cell death, and that cytotoxicity depends upon the intracellular concentration and subcellular localization of the ruthenium complex. While the biological mechanism of NO in HepG2 cells still remains to be completely unveiled, this compound induces apoptosis in HepG2 cells via cytochrome c release from mitochondria and activation of caspases 9/3. Recently we have hypnotized the possibility of synergistic effect involving NO and singlet oxygen as anticancer therapy modality (Maranho et al., 2009). It could be an interesting strategy to enhance photodynamic therapy (PDT).

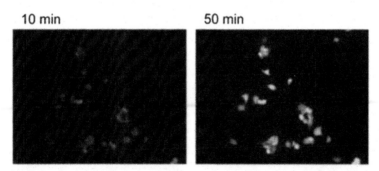

Figure 1.16 NO detection and ROS/RNS generation. Detection of NO released from 4pic/lip (100 μM) on HepG2 cells using DAF-FM fluorescent probe at 10 and 50 minutes. *Reproduced with permission from Rodrigues et al., 2016, Elsevier Publishing Group.*

PDT studies have shown that the efficiency of photosensitization processes depends on photophysical parameters such as the quantum yield of the triplet state as well as on the decay kinetics of the structural and functional parameters that influence drug incorporation, localization, and interaction with the cellular target. Several studies have reported that the efficacy and mechanism of action of photosensitizer drugs (PS) are directly related to the distance between the PS and the cellular target. This largely relies on the reactive species resulting from photostimulation, which should have a short lifetime and a small tendency to migrate and spread (Peng et al., 1991; Hilf, 2007; Mroz et al., 2011). The preferred location of the PS determines the efficiency of the photodynamic action and, most importantly, the type of cell death (Hsi et al., 1999; Murakami et al., 2015). Therefore, drug compartmentalization is the crucial parameter when elucidating the cytotoxic mechanisms of PDT.

A recent study evaluated the cytotoxicity of cis-[Ru(NO$_2$)(bodipy)(dcbpy)$_2$]$^{2+}$ and derivatives considering the possible application in PDT (dos Santos et al., 2016). Comparative analysis of the PS fluorescence emission pattern with the characteristics of standard probes helped to evaluate the intracellular localization of cis-[Ru(NO$_2$)(bodipy)(dcbpy)$_2$]$^{2+}$ (Fig. 1.17).

Cellular imaging revealed similarity between the fluorescence pattern of cis-[Ru(NO$_2$)(bodipy)(dcbpy-aminopropyl-β-lactose)$_2$]$^{2+}$ (Fig. 1.17), and the mitochondrion probes, which suggested cis-[Ru(NO$_2$)(bodipy)(dcbpy-aminopropyl-β-lactose)$_2$]$^{2+}$ localizes near mitochondria. In contrast, the fluorescence pattern of cis-[Ru(NO$_2$)(bodipy)(dcbpy-aminopropyl-β-lactose)$_2$]$^{2+}$ and the nuclear probe DAPI differed markedly, indicating nuclear cis-[Ru(NO$_2$)(bodipy)(dcbpy-aminopropyl-β-lactose)$_2$]$^{2+}$ localization is unlikely.

Mitochondria are important subcellular targets for many PS used in PDT, including the first-generation drugs. Mitochondrial damage induced by PS illumination leads to apoptotic cell death (Castano et al., 2004), which emphasizes the importance of PS subcellular location/target in PDT. In PDT, cytotoxic effects clearly depend

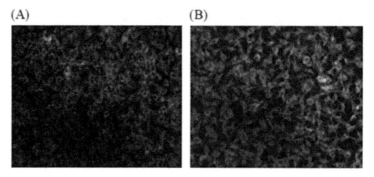

(A)   (B)

Figure 1.17 Fluorescence microscopy images of B16F10 cells after incubation with (A) cis-[Ru(NO$_2$)(bodipy)(dcbpy-aminopropyl-β-lactose)$_2$]$^{2+}$ (20 μM) for 1 hour and (B) 2 hours.

on light dose irradiation and intracellular concentration of the PS accumulating in the cancer cells. Despite the development of new PS with location-specific properties (Rosenkranz et al., 2000; Gomer, 1991; Kiesslich et al., 2013), most PS drugs do not display preferential affinity for tumor tissue. Accordingly, recent studies on PDT have dealt with the development of new PS agents with improved characteristics. The synergy of combined therapies and reactive species has been explored (Lunardi and Tedesco, 2005; Maranho et al., 2009) to target action on tumor cells while preserving healthy cells and minimizing undesirable effects (Reddi, 1997; Machado, 2000). Structural changes to PS and several drug delivery systems can increase PS selectivity and accumulation in target cells and tissues (Maranho et al., 2009; Carneiro et al., 2011).

Liposome delivery

Recently, liposomes have been shown to enhance the delivery of various drugs (Torchilin, 2005; Gregoriadis, 2007) and therefore improve therapeutic efficacy in many diseases (Sapra et al., 2005; Kapse-Mistry et al., 2014). Therefore, liposomes containing incorporated PS have been used in clinically approved treatments for many life-threatening diseases (Rodrigues et al., 2016). Liposomal delivery of NO donors has demonstrated that the nitrosyl complexes may act as a prodrug, delivering NO and contributing to tumor cell death (Eroy-reveles and Mascharak, 2009; Ostrowski and Ford, 2009; Maranho et al., 2009). Liposomal encapsulation of [Ru(phthalocyanine)(ONO)(NO)] increases tumor cell death as compared to the same complex in phosphate-buffered saline (Figs. 1.18 and 1.19). This might be due to enhanced

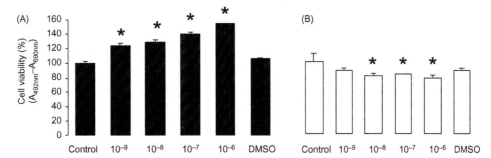

Figure 1.18 Effects of [Ru(pc)(ONO)(NO)] on B16F10 cell viability assessed by the MTT assay. (A) Cells were plated 2×10^4 cells/well in 96-well plates and treated for 3 hours with [Ru(pc)(ONO)(NO)] at 10^{-9}, 10^{-8}, 10^{-7}, and 10^{-6} M, diluted in PBS solution, pH 7.4. (B) Cells were rinsed with PBS followed by RPMI medium addition and irradiated ($4.0 \, \mathrm{J \, cm^{-2}}$). Irradiated cells remained incubated for 24 hours until MTT analyses. Black bars represent percentage of nonirradiated viable cells and white bars, irradiated viable cells. *$p < 0.05$ different from control. *Reproduced with permission from Carneiro, Z.A., de Moraes, J.C.B., Rodrigues, F.P., de Lima, R.G., Curti, C., da Rocha, Z.N., et al., 2011. Photocytotoxic activity of a nitrosyl phthalocyanine ruthenium complex—a system capable of producing nitric oxide and singlet oxygen. J. Inorg. Biochem. 105, 1035–1043, Elsevier Publishing Group.*

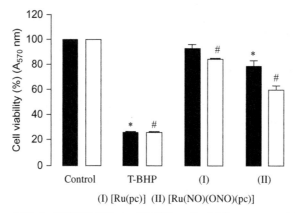

Figure 1.19 Effects of [Ru(pc)(ONO)(NO)] on B16F10 cell viability assessed by the MTT assay. Cells were treated with 1.0×10^{-6} M [Ru(pc)(ONO)(NO)] in liposomal vesicles, for 3 hours. After this time, cells were rinsed with PBS and a new RPMI medium was added, followed by irradiation with $4.0 \, J \, cm^{-2}$. Cell viability was assessed 3 hours after irradiation. Black bars represent percent viability of control cells that were not irradiated while white bars represent percent viability of irradiated cells. $p < 0.05$ * different from control without irradiation and # different from control with irradiation. *Reproduced with permission from Carneiro, Z.A., de Moraes, J.C.B., Rodrigues, F.P., de Lima, R.G., Curti, C., da Rocha, Z.N., et al., 2011. Photocytotoxic activity of a nitrosyl phthalocyanine ruthenium complex—a system capable of producing nitric oxide and singlet oxygen. J. Inorg. Biochem. 105, 1035–1043, Elsevier Publishing Group.*

intracellular accumulation of [Ru(phthalocyanine)(ONO)(NO)] prompted by the liposomal delivery system as well as possible protection of the complex against reducing biomolecules that could promote NO release after crossing of the membrane barrier (Carneiro et al., 2011).

Similarly, drug combinations that produce ROS and RNOS upon activation by photosensitization of the nitrosyl ruthenium complex (zinc phthalocyanine/[Ru(NH. NHq)(tpy)(NO)]$^{3+}$ associated with liposomal delivery seem to be promising new antitumor strategies (Fig. 1.20) (Maranho et al., 2009).

Recently, we have found that fluorescent nanoparticles could carry nitrosyl ruthenium compound and simultaneously show fluorescence activity (Franco et al., 2014). The use of quantum dots (QDs) is this kind of emerging strategy for drug imaging and photo-activated drug delivery. These semiconductor nanoparticles have unique size-dependent optical properties that include high absorbance coefficient and photoluminescence (PL). Customization of the surface of QDs either in core or core/shell structures by ligand exchange allows one to modulate their solubility, stability, and conjugation to biomolecules (DNA, proteins, etc.). In biomedical research, these characteristics not only highlight QDs as passive fluorescent bioprobes or labels for biological imaging but also as a nanoscaffold catering for therapeutic and diagnostic (theranostic)

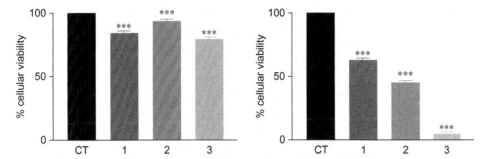

Figure 1.20 In vitro dark toxicity (A) and phototoxicity (B) for cells B16F10. CT, control; 1, zinc phytalocianine (1 μmol L^{-1}) liposome; 2, zinc phytalocianine/[Ru(NH.NHq)(tpy)(NO)]$^{3+}$ in the stealth liposome (1 mol L^{-1}/0.05 mmol L^{-1}); 3, zinc phytalocianine/[Ru(NH.NHq)(tpy)(NO)]$^{3+}$ (5 μmol L^{-1}/0.05 mmol L^{-1}) in the stealth liposome. $p < 0.05$ *** different from control. *Reproduced with permission from Maranho, D., de Lima, R.G., Primo, F.L., da Silva, R.S., Tedesco, A.C., 2009. Photoinduced nitric oxide and singlet oxygen release from ZnPc liposome vehicle associated with nitrosyl ruthenium complex. Synergistic effects in photodynamic therapy application. Photochem. Photobiol. 85, 705–713, Wiley Publishing Group.*

modalities (Lifeng and Xiaohu, 2008; Medintz et al., 2008). QDs have application in light harvesting systems. Researchers have proposed the use of QD as sensitizers for biological applications in PDT (Yaghini et al., 2009) and photo-activated drug delivery. It is noteworthy that light can control the photochemical reaction as well as the drug dosage (Neuman et al., 2008; Burks and Ford, 2012). Ford's group has reported NO release from nitro-chromium complex associated with QDs (CdTe/ZnS) (Burks and Ford, 2012). The photosensitization mechanism proposed these authors involved energy transfer that required absorption and PL spectral overlap between the acceptor and the donor. Another report of NO photodelivery system described charge transfer as the main process resulting from the photoactive conjugate formed between CdTe-MPA (MPA is mercaptopropionic acid) and the ruthenium nitrosyl complex *cis*-[Ru(4-ampy)(bpy)$_2$(NO)](PF$_6$)$_3$ (Ru−NO; where bpy is 2,2'-bipyridine and 4-ampy is 4-aminopyridine) (Franco et al., 2014). QDs alone are not toxic to the B16F10 tumor cell line (murine melanoma) even after preincubation for 24 hours. Confocal images aided analysis of the internalization and subcellular location of the CdTe-MPA nanoparticles in the cytoplasm; the nuclear probe Hoechst was used as reference (Fig. 1.21).

This CdTe QD···Ru−NO conjugate is a promising visible light-activated prodrug for targeted NO release.

CONCLUSION

In summary, these results illustrate the potential of QDs/coordination compounds conjugates as photochemical precursors for the delivery of bioactive substances to

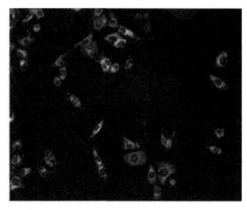

Figure 1.21 Overlap of fluorescence microscopy on B16F10 cells image incubated with CdTe-MPA at 10^{-9} M for 3 hours and Hoechst (Franco, 2014).

physiological targets. In addition, QD luminescence may allow tracking and localization of the NO delivery conjugate in cellular systems.

REFERENCES

Atkins, P., Overton, T., Rourke, J., Weller, M., Armstrong, F., 2009. Shriver and Atkins' Inorganic Chemistry, 5th ed Paperback. Oxford University Press.

Beckman, J.S., Koppenol, W.H., 1996. Nitric oxide, superoxide, and peroxynitrite: the good, the bad, the ugly. Am. Phys. Soc. 271, C1424–C1437.

Beraldo, H., Merce, A.L.R., Felcman, J., Recio, M.A.L., 2008. Aluminium and medicine. Molecular and Supramolecular Bioinorganic Chemistry: Applications in Medical Sciences. New Science Publishers, New York.

Bonaventura, D., Oliveira, F.D., Togniolo, V., Tedesco, A.C., da Silva, R.S., Bendhack, L.M., 2004. A macrocyclic nitrosyl ruthenium complex is a NO donor that induces rat aorta relaxation. Nitric Oxide 10, 83–91.

Burks, P.T., Ford, P.C., 2012. Quantum dot photosensitizers. Interactions with transition metal centers. Dalton Trans. 41 (42), 13030–13042.

Cândido, M.C.L., Oliveira, A.M., Silva, F.O.N., Holanda, A.K.M., Pereira, W.G., Sousa, E.H.S., et al., 2015. Photochemical and electrochemical study of the release of nitric oxide from [Ru(bpy)$_2$L(NO)](PF6)$_n$ complexes (L=imidazole, 1-methylimidazole, sulfite and thiourea), toward the development of therapeutic photodynamic agents. J. Braz. Chem. Soc. 26 (9), 1824–1830.

Carneiro, Z.A., de Moraes, J.C.B., Rodrigues, F.P., de Lima, R.G., Curti, C., da Rocha, Z.N., et al., 2011. Photocytotoxic activity of a nitrosyl phthalocyanine ruthenium complex—a system capable of producing nitric oxide and singlet oxygen. J. Inorg. Biochem. 105, 1035–1043.

Castano, A.P., Deminova, T.N., Hamblin, M.R., 2004. Mechanisms in photodynamic therapy: part one—photosensitizers, photochemistry and cellular localization. Photodiagn. Photodyn. Ther. 1, 279–293.

Cheung, W.S., Bhan, I., Lipton, S.A., 2000. Nitric oxide (NO.) stabilizes whereas nitrosonium (NO$^+$) enhances filopodial outgrowth by rat retinal ganglion cells in vitro. Brain Res. 868 (1), 1–13.

da Silva, R.S., Marchesi, M.S.P., Khin, C., Lunardi, C.N., Bendhack, L.M., Ford, P.C., 2007. The photoinduced electron transfer between the cationic complexes Ru(NH$_3$)$_5$pz^{2+} and trans-[Ru(NO$_2$)Cl([15] aneN$_4$]$^+$ mediated by phosphate ion. Visible light generation of nitric oxide for biological targets. J. Phys. Chem. B 111 (24), 6962–6968.

da Silva, R.S., de Lima, R.G., Machado, S.P., 2015. Design, reactivity, and biological activity of ruthenium nitrosyl complexes In: van Eldik, R. Olabe, J.A. (Eds.), Advances in Inorganic Chemistry, vol 67. Elsevier, Amsterdam, pp. 225–229.

de Lima, R.G., Sauaia, M.G., Bonaventura, D., Tedesco, A.C., Bendhack, L.M., da Silva, R.S., 2006. Influence of ancillary ligand L in the nitric oxide photorelease by the $[Ru(L)(tpy)(NO)]^{3+}$ complex and its vasodilator activity based on visible light irradiation. Inorg. Chim. Acta 359 (8), 2543–2549.

de Lima, R.G., Silva, B.R., da Silva, R.S., Bendhack, L.M., 2014. Ruthenium complexes as NO donors for vascular relaxation induction. Molecules 19, 9628–9654.

dos Santos J.S., Ramos L.C., Ferreira L.P., Campos V., de Rezende L.C.D., Emery F.S., et al., submitted for publication, 2016. Synthesis, photophysical characterization, cytotoxicity evaluation, and theranostic properties of cis-$[Ru(bodipy)(dcbpy-aminopropyl-\beta-lactose)_2(NO_2)]^{2+}$.

Eroy-reveles, A.A., Mascharak, P.K., 2009. Nitric oxide-donating materials and their potential in pharmacological applications for site-specific nitric oxide delivery. Future Med. Chem 1, 1497–1507.

Flitney, F.W., Megson, I.L., Thomson, J.L.M., Kennovin, G.D., Butler, A.R., 1996. Vasodilator responses of rat isolated tail artery enhanced by oxygen-dependent, photochemical release of nitric oxide from iron-sulphur-nitrosyls. Br. J. Pharmacol. 117, 1549–1557.

Franco L.P., 2014. Acomplamento quantum dot/complexos nitrosilos de rutênio em transferência eletrônica vetorial e em análise de imagem. Aspectos químicose biológicos relacionadosà produção de óxido nítrico. Thesis, Faculdade de Ciências Farmacêuticas de Ribeirão Preto, Ribeirão Preto, SP.

Franco, L.P., Cicillini, S.A., Biazzotto, J.C., Schiavon, M.A., Mikhailovsky, A.A., Burks, P., et al., 2014. Photoreactivity of a quantum dot–ruthenium nitrosyl conjugate. J. Phys. Chem. A 118 (51), 12184–12191.

Gomer, C., 1991. Preclinical examination of first and second generation photosensitizers used in photodynamic therapy. Photochem. Photobiol. 54 (6), 1093–1107.

Gregoriadis G., 2007. Liposome Technology, 3rd ed. Informa Healthcare, New York.

Heinrich T.A., 2013. Aspectos químicos fotoquímicos fotobiológicos de complexo rutênio-nitrosilo como precursor de óxido nítrico. Princípios de aplicação como agente citotóxico em linhagens de células tumorais. Thesis, Faculdade de Ciências Farmacêuticas de Ribeirão Preto, Ribeirão Preto, SP.

Hilf, R., 2007. Mitochondria are targets of photodynamic therapy, J. Bioenerg. Biomem 39 (1), 85–89.

Hsi, R.A., Rosenthal, D.I., Glatstein, E., 1999. Photodynamic therapy in the treatment of cancer: current state of the art. Drugs 57, 725–734.

Hughes, M.N., 1999. Relationships between nitric oxide, nitroxyl ion, nitrosonium cation and peroxynitrite. Biochim. Biophys. Acta (BBA) 1411 (2–3), 263–272.

Ignarro, J.L., 2000. Nitric Oxide: Biology and Pathobiology, ed, 1. Academic Press, San Diego, CA.

Kapse-Mistry, S., Govender, T., Srivastava, R., Yergeri, M., 2014. Nanodrug delivery in reversing multidrug resistance in cancer cells. Front. Pharmacol 5, 159.

Kiesslich, T., Gollmer, A., Maisch, T., Berneburg, M., Plaetzer, K., 2013. A comprehensive tutorial on in vitro characterization of new photosensitizers for photodynamic antitumor therapy and photodynamic inactivation of microorganisms. BioMed Res Int. 840417.

Lifeng, Q., Xiaohu, G., 2008. Emerging application of quantum dots for drug delivery and therapy. Expert Opin. Drug Deliv 5 (3), 263–267.

Lunardi, C.N., Tedesco, A.C., 2005. Synergic photosensitizer: a new trend in photodynamic therapy. Current Org. Chem 9, 813–821.

Lunardi, C.N., da Silva, R.S., Bendhack, L.M., 2009. New nitric oxide donors based on ruthenium complexes. Braz. J. Med. Biol. Res 42, 87–93.

Machado, A.E.H., 2000. Terapia fotodinâmica: princípios, potencial de aplicação e perspectivas. Química Nova 23 (2), 238–243.

Maranho, D., de Lima, R.G., Primo, F.L., da Silva, R.S., Tedesco, A.C., 2009. Photoinduced nitric oxide and singlet oxygen release from ZnPc liposome vehicle associated with nitrosyl ruthenium complex. Synergistic effects in photodynamic therapy application. Photochem. Photobiol. 85, 705–713.

Marchesi, M.S.P., Cicillini, S.A., Prazias, A.C.L., Bendhack, L.M., Batista, A.A., da Silva, R.S., 2012. Chemical mechanism of controlled nitric oxide release from trans-$[RuCl([15]aneN_4)NO](PF_6)_2$ as a vasorelaxant agent. Transit. Metal Chem. 37, 475–479.

Medintz, I.L., Mattoussi, H., Clapp, A.R., 2008. Potential clinical applications of quantum dots. Int. J. Nanomed. 3 (2), 151–167.

Moncada, S., Palmer, R.M.J., Higgs, E.A., 1991. Nitric oxide: physiology, pathophysiology, and pharmacology. Pharmacol. Rev. 43 (2), 109–142.

Mroz, P., Yaroslavsky, A., Kharkwal, G.B., Hamblin, M.R., 2011. Cell death pathways in photodynamic therapy of cancer, Cancers (Basel) 3 (2), 2516–2539.

Murakami, L.S., Ferreira, L.P., Santos, J.S., da Silva, R.S., Nomizo, A., Kuz'min, V.A., et al., 2015. Photocytotoxicity of a cyanine dye with two chromophores toward melanoma and normal cells. Biochim. Biophys. Acta. G, General Subjects 1850, 1150–1157.

Neuman, D., Ostrowski, A.D., Mikhailovsky, A.A., Absalonson, R.O., Strouse, G.F., Ford, P.C., 2008. Quantum dot fluorescence quenching pathways with Cr(III) complexes photosensitized NO production from trans-Cr(cyclam)(ONO)$^{2+}$. J. Am. Chem. Soc. 130, 168–175.

Ostrowski, A.D., Ford, P.C., 2009. Metal complexes as photochemical nitric oxide precursors: potential applications in the treatment of tumors. Dalton Trans. 48, 10660–10669.

Peng, Q., Moan, J., Farrants, G., Danielsen, H.E., Rimington, C., 1991. Localization of potent photosensitizers in human tumor LOX by means of laser scanning microscopy. Cancer Lett. 58 (1–2), 17–27.

Reddi, A.H., 1997. Bone morphogenetic proteins: an unconventional approach to isolation of first mammalian morphogens. Cytokine Growth Factor Rev. 8, 11–20.

Richter-addo, G.B., Legzdins, P., 1992. Metal Nitrosyls. Oxford University Press, New York.

Rodrigues, F.P., Carneiro, Z.A., Mascharak, P., Curti, C., da Silva, R.S., 2016. Incorporation of a ruthenium nitrosyl complex into liposomes, the nitric oxide released from these liposomes and HepG2 cell death mechanism. Coord. Chem. Rev. 306, 701–707.

Rosenkranz, A.A., Jans, D.A., Sobolev, A.S., 2000. Targeted intracellular delivery of photosensitizers to enhance photodynamic efficiency. Immunol. Cell Biol. 78, 452–464.

Sapra, P., Tyagi, P., Allen, T.M., 2005. Ligand-targeted liposomes for cancer treatment. Current Drug Deliv. 2 (4), 369–381.

Sauaia, M.G., da Silva, R.S., 2003. The reactivity of nitrosyl ruthenium complexes containing polypyridyl ligands. Transit. Metal Chem. 28, 254–259.

Sauaia, M.G., de Lima, R.G., Tedesco, A.C., da Silva, R.S., 2003a. Photoinduced NO release by visible light irradiation from pyrazine-bridged nitrosyl ruthenium complexes. J. Am. Chem. Soc. Commun. 125, 14718–14719.

Sauaia, M.G., Oliveira, F.S., Tedesco, A.C., da Silva, R.S., 2003b. Control of NO release by light irradiation from nitrosyl–ruthenium complexes containing polypyridyl ligands. Inorg. Chim. Acta 355, 191–196.

Schwietert, C.W., MCccue, J.P., 1999. Coordination compounds in medicinal chemistry. Coord. Chem. Rev. 184, 67–89.

Shoman, M.E., Aly, O.M., 2016. Nitroxyl (HNO): a reduced form of nitric oxide with distinct chemical, pharmacological, and therapeutic properties. Oxid. Med. Cell Longev. 1–15.

Stochel, G., Wanat, A., Kulis, E., Stasicka, Z., 1998. Light and metal complexes in medicine. Coord. Chem. Rev. 171, 203–220.

Szczepura, L.F., Takeuchi, K., 1990. Synthesis and characterization of novel (ciclopentadienyl)nitroruthenium complexes. Inorg. Chem. 29, 1772–1777.

Thomas, C., Svehla, L., Moffett, B.S., 2009. Sodium nitroprusside induced cyanide toxicity in pediatric patients. Expert Opin. Drug Saf. 8 (5), 599–602.

Togniolo, V.P., da Silva, R.S., Tedesco, A.C., 2001. Photo-induced nitric oxide release from chlorobis(2,2'-bipyridine) nitrosylruthenium(II) in aqueous solution. Inorg. Chim. Acta 316, 7–12.

Torchilin, V.P., 2005. Recent advances with liposomes as pharmaceutical carriers. Nat. Rev. Drug Discov. 4 (2), 145–160.

van Rijt, S.H., Sadler, P.J., 2009. Current applications and future potential for bioinorganic chemistry in the development of anticancer drugs. Drug Discov. Today 14, 1089–1097.

Wink, D.A., Grisham, M.B., Mitchell, J.B., Ford, P.C., 1996. Direct and indirect effects of nitric oxide in chemical reactions relevant to biology. Methods Enzymol. 268, 12–31.

Yaghini, E., Seifalian, A.M., Macrobert, A.J., 2009. Quantum dots and their potential biomedical applications in photosensitization for photodynamic therapy. Nanomedicine 4 (3), 353–363.

CHAPTER 2

Nitric Oxide Donors for Treating Neglected Diseases

Amedea Barozzi Seabra[1,2] and Nelson Durán[2,3]
[1]Center for Natural and Human Sciences, Federal University of ABC (UFABC), Santo André, São Paulo, Brazil
[2]Institute of Chemistry, Biological Chemistry Laboratory, University of Campinas, Campinas, São Paulo, Brazil
[3]Brazilian Nanotechnology National Laboratory (LNNano-CNPEM), Campinas, Brazil

NEGLECTED DISEASES

The diseases can be classified into global, neglected, and extremely neglected diseases. Cardiovascular, mental diseases, cancer, among others, are classified as global diseases, and are the main targets of the Research and Development efforts of pharmaceutical companies/research institutes (Chung et al., 2008). In contrast, neglected diseases afflict millions of people worldwide, and current drug therapy has limited efficacy and it is inappropriate, in most of the cases. Extremely neglected diseases threat poor people, in particular in developing countries and least developed countries, in which the population has restricted access to medical care and basic sanitary conditions. Unfortunately, most of the neglected and extremely neglected diseases are not included in the goals of Research and Development programs in pharmaceutical industries, and they not receive much attention from private and public sectors. Consequently, approximately 14 million people die annually from infectious diseases, mainly in developing and least developed countries (Chung et al., 2008). Neglected diseases and extremely neglected diseases are generally classified as neglected tropical diseases and cause stigma, discrimination, morbidity, and mortality (World Health Organization, 2010). In this chapter, the expression "neglected diseases" will be used to designate both neglected and extremely neglected diseases.

The majority of the neglected diseases includes parasitic infections. Some examples are malaria, leishmaniasis, trypanosomiasis, schistosomiasis, tuberculosis, dengue fever (DF), *Blastocystis hominis* infection, lymphatic filariasis, strongyloidiasis, Japanese encephalitis, helminth diseases, toxoplasmosis, onchocerciasis, ectoparasitic skin infections, leprosy, etc. Among them, tuberculosis and malaria are considered as major public health threat in endemic areas (Durán et al., 2009).

Although the Special Program for Research and Training in Tropical Diseases (TDR) contributed to the development of drugs against neglected diseases, most of the chemotherapeutics clinically employed nowadays were developed 25–40 years ago,

---

Nitric Oxide Donors.

which reflects the lack of interest on the development of new drugs/strategies to efficiently combat neglected diseases (Olliaro et al., 2015). In this context, new strategies/drugs to combat neglected diseases are urgently required.

WHY USING NO DONORS TO TREAT NEGLECTED DISEASES?

In recent years, the free radical NO has emerged as a key cytotoxic species involved in the pathogenesis of several diseases including parasitic neglected diseases (Anez et al., 2007; Schairer et al., 2012). NO is synthesized by the action of three enzymes: endothelial nitric oxide synthase (eNOS), neuronal nitric oxide synthase (nNOS), and inducible nitric oxide synthase (iNOS), by oxidation of L-arginine to L-citrulline (Bang et al., 2014; Brisboi et al., 2014). NO generated from eNOS or nNOS has physiological roles such as the control of vasodilation and cell communication. These enzymes are calcium-depended and produce relatively low concentrations of NO for short periods. In contrast, iNOS is calcium-independent isoform that produces relatively high amounts of NO (in the micro molar range) for long period of time (Bang et al., 2014; Brisboi et al., 2014). The NO produced by iNOS is the essential part of host immunological system against invading pathogens that cause neglected diseases. Indeed, upon activation, macrophages induce NO production by iNOS leading to cytotoxic effects against pathogens (Chouake et al., 2012; Förstermann and Sessa, 2012). NO regulates several immunological functions, including T- and B-cell proliferation, cytokine production, and leukocytes rolling (Bogdan, 2001). Therefore, NO produced by iNOS or exogenous NO possesses cytotoxic activity on several pathogens such as Metazoa and Protozoa (Ascenzi et al., 2003).

Upon infection, high concentrations of NO are synthesized by iNOS. NO and its oxidized species (NOx) display toxic effects toward invading pathogens. The harmful reactive oxygen species (ROS) generated by NO includes hydroxyl radicals (OH$^{\bullet}$), superoxide ($O_2^{\bullet-}$), and hydrogen peroxide (H_2O_2) (Brisboi et al., 2014; Chaturvedi and Nagar, 2009). Moreover, peroxynitrite ($OONO^-$) and dinitrogen trioxide (N_2O_3) are examples of harmful reactive nitrogen species (RNS) generated by NO in biological systems.

NO plays a key role in host defense against pathogenic microorganisms. However, upon infection with a neglected disease, the invading pathogen is able to compromise host macrophages impairing iNOS activity, and thus significantly decreasing NO production. With the decrease of NO production by iNOS, the invading pathogen enhances its infection in the host. In this context, the administration of exogenous NO donors is aimed to compensate the lack of endogenous NO production by compromised macrophages. The uses of exogenous NO donors as potent chemotherapeutic agents have been successfully applied in the treatment of several neglected diseases. Therefore, the next sections highlight the role of NO and NO donors in the pathogenesis of several important life-threatening neglected diseases.

MALARIA

The life-threatening parasitic disease Malaria is caused by the following *Plasmodium* (*P.*) protozoa species: *P. falciparum*, *P. vivax*, *P. ovale*, *P. knowlesi*, and *P. malariae*. Among them, *P. falciparum* is the agent of severe malaria (Bertinaria et al., 2011). This world-spread disease is transmitted by the female anopheles mosquitoes (Bertinaria et al., 2015). Malaria causes the death of approximately 2 million people annually, and it is estimated that 3 billion people are under the threat of this disease (Chung et al., 2008). The current treatment against malaria is based on artemisinin-combined therapies, which is reported to reduce 40% patient death (Siu and Ploss, 2015). Moreover, several drugs such as primaquine, chloroquine, atovaquone, halofantrine, amodiaquine, and mefloquine have been used (Chung et al., 2008). Among them, amodiaquine was reintroduced in the World Health Organization (WHO) model list of essential medicines (Rosenthal, 2001). In 2015, the Nobel Prize in Physiology or Medicine was awarded for the development of artemisinin and ivermectin as chemotherapeutics against malaria (Tambo et al., 2015). Consequently, the combat of neglected diseases gained significant attention from the Scientific Community. However, as in the case of other neglected diseases, the development of resistance to these chemotherapies is still a challenge to be overcome. At the moment, the only malaria vaccine is under phase III trial with low efficacy (average of 30%) (Siu and Ploss, 2015). In this context, malaria is still a worldwide epidemic disease and there is an increasing necessity in the design of new and effective therapies to efficiently combat the disease.

Severe malaria is the most aggressive form of the infection caused by *P. falciparum* increasing the incidence of mortality (Bertinaria et al., 2015; Cabrales et al., 2011). Cerebral malaria (CM) is a complication of severe malaria that causes the death of approximately 20% of infected people, in particular children (Bertinaria et al., 2015). Indeed, in the year of 2013, 95% of the 584,000 malaria deaths were due to CM (Eisenhut, 2015). In patients with CM, *P. falciparum* infects the red blood cells (RBCs) impairing the cerebral microcirculation causing hypoxia, ischemia, disruption of the blood–brain barrier, impairment of perfusion, edema, and coma (Mishra and Newton, 2009; Beare et al., 2009; Dondorp et al., 2008). Consequently, neuronal damages, brain swelling, and death are observed in patients with CM (Eisenhut, 2015). Therefore, new drugs to combat CM are urgently required.

NO and Malaria

NO plays a key role in the pathogenesis of malaria, in particularly in the case of CM caused by *P. falciparum* (Bertinaria et al., 2011). NO regulates the host immune defense against the parasite having an important antiplasmodial effect (Bertinaria et al., 2011). Indeed, low bioavailability of NO increases the pathogenesis of CM, while administration of exogenous NO donors or infusion of NO precursor L-arginine are reported to display toxic effects against *P. falciparum* (Anstey et al., 1996; Bertinaria et al., 2011, 2015; Lopansri et al., 2003; Yeo et al., 2007). Reduced NO bioavailability in patients with CM is associated with ischemia, endothelial dysfunction, hypoarginemia, cytoadherence of RBCs infected with *P. falciparum* to the endothelium, vasospasm, and elevated levels of cell-free hemoglobin, which is a NO scavenging in biological medium (Bertinaria et al., 2011; Cabrales et al., 2011; Eisenhut, 2015).

One of the sites of NO production is the endothelium, in which NO is synthesized by the action of the enzyme eNOS. As lipophilic, NO diffuses from the endothelium to smooth muscle cells where it binds to the heme moiety of guanosine triphosphate (GTP) leading to the formation of cyclic guanosine monophosphate (cGMP), producing the dilation of the blood vessel (Eisenhut, 2015). Therefore, regular NO production from eNOS is crucial to avoid complications caused by *P. plasmodium* in CM. However, patients with CM demonstrate significant decrease in NO levels, which contributes to the evolution of the disease. In this scenario, administration of NO donors represents a promising strategy to combat CM, reducing cerebral vasospasm.

NO Donors and Malaria

As low NO bioavailability is crucial to the pathogenesis of malaria, in particular to CM, important papers describe the administration of NO donors in the treatment of malaria. The anti *P. falciparum* effects of the NO donor–amodiaquine conjugates (NO–AQ) containing nitroxyl ($-ONO_2$) or furoxan (1,2,5-oxadiazole 2-oxide) substrates were evaluated (Bertinaria et al., 2011). The NO-based compounds were assayed in vitro toward both chloroquine sensitive, D10, and the chloroquine resistant, W-2 strains of *P. falciparum*. In addition, in vivo tests were performed using the well-established murine model of CM caused by *P. berghei* ANKA (PbA). The NO-conjugated compounds were found to increase mice survival from late-stage CM. Delayed mortality and improved animal survival were observed for infected mice with late-stage CM treated 1.4 mg per mouse per day of the NO-based conjugated during 5 days, in comparison with animals treated with only 1 mg per mouse of the antimalarial drug amodiaquine. Treatment with NO donor conjugates led to a reduction of 76% of parasitemia in 24 hours. Moreover, the administration of NO donor conjugates dilated precontracted rat aorta strips, indicating the ability of these NO donors to restore the vascular homeostasis, comprised in patients with CM (Bertinaria et al., 2011).

In a similar approach, hybrid compounds composed by dihydroartemisinin scaffold in combination with NO donors furoxan and diazeniumdiolate (NONOate) were prepared and evaluated in vitro and in vivo against CM (Bertinaria et al., 2015). The NO donor hybrid compounds dilated precontracted rat aorta strips, due to NO release, and demonstrated antiplasmodial effect against *P. berghei* ANKA, in a similar manner to classical antimalarial drugs (artesunate and artemether). In late-stage CM,

NO-based hybrid material increased mice survival from 27.5% to 51.6%, in comparison with artemether (Bertinaria et al., 2015). The artemisinin NO donor hybrid drug demonstrated promising effect in the treatment of CM, although more studies are necessary for the optimization of doses, pharmacokinetics, and further evaluation of the cerebral vascular responses to the treatment.

In order to compensate the low bioavailability of NO in CM, the NO donor dipropylenetriamine NONOate (DPTA/NO) was administrated in PbA-infected mice (Cabrales et al., 2011). In vivo study demonstrated that the exogenous administration of the NO donor enhanced hemodynamic performance, increased cerebral vascularization, reduced vasoconstriction and leukocyte accumulation, increased RBC velocities, attenuated the decrease in pial blood flow, decreased the risk of brain hemorrhages. Indeed, the NO donor avoided brain hemorrhages (1.4 vs 24.5 hemorrhagic foci per section) and inflammation (2.5 vs 10.9 adherent leukocytes per 100 μm vessel length), in comparison with control group (Cabrales et al., 2011).

One of the promising strategy to combat parasitic diseases including malaria is the inactivation of vital parasite proteases. In this sense, Sharma et al. (2004) evaluated the NO inhibition on the activity of the pepsin-like aspartic protease plasmepsin in *P. vivax*, which may be involved in hemoglobin degradation in *P. falciparum* infection. The authors demonstrated that different class of NO donors including (±) (*E*)-4-ethyl-2-[(*E*)-hydroxyimino]-5-nitro-3-hexenamide (NOR-3), sodium nitroprusside (SNP), and *S*-nitrosoglutathione (GSNO) inhibited plasmepsin activity in a dose-dependent manner. Nitrosylation of cysteine residues in the catalytic site of the enzyme, by the NO donors, is assumed to cause loss of enzyme activity. The NO donor concentration ($40 \mu mol L^{-1}$), employed to inactive plasmepsin activity, was found to be consistent with the concentration reported to kill malaria parasites (Sharma et al., 2004). Therefore, NO donors are able to nitrosylate free thiol groups present on the active site of parasite cysteine protease, leading to enzyme inactivation, which contributes to the death of the parasite.

Inactivation of parasite cysteine proteases is not the only antiparasitic mechanism of NO donors. Jeney et al. (2014) addressed the possible mechanisms involved in the NO activity in CM. The authors proposed that NO combats CM through: (1) induction of heme oxygenase-1 (HO-1), (2) the transcription factor nuclear factor erythroid 2-related factor 2 (NRF-2), and (3) CO production via hemecatabolism by HO-1. In their study, the application of the NO donor DPTA-NO in CM infected mice suppressed the pathogenesis of CM. However, NO bioavailability in CM is not enough to combat the disease, since cell-free hemoglobin readily scavenges NO and the expression of iNOS is reduced by cellular iron overload. The low bioavailability of NO can be compensated by the administration of exogenous NO donors, preventing the pathogenesis of CM (Jeney et al., 2014). Furthermore, NO donors were found to increase levels of reactive nitrogen intermediates, which are highly toxic against malaria parasite (Dascombe and Nahrevanian, 2003).

An interesting paper describes that the administration of high concentration (1 mg per mouse) of the NO donor DPTA-NO in PbA-infected mice partially prevented CM, however, a drop in blood pressure was observed (Martins et al., 2012). To avoid side effects related to the administration of high doses of the NO donor, sildenafil (0.1 mg per mouse) was added in combination with a lower dose of DPTA-NO (0.1 mg per mouse). The combination of sildenafil and DPTA-NO decreased the animal mortality (38 ± 10.6% mortality in treated group vs 82 ± 7.4% mortality in the control group). Moreover, lower doses of DPTA-NO alone were not efficient to prevent experimental CM, however, significant protection was achieved upon the combination of the NO donor with sildenafil, and no side effects were observed. Sidenafil, a PDE-5 inhibitor, increases brain levels of cGMP optimizing the L-arginine–NOS–NO pathway (Martins et al., 2012).

Wang et al. (2009) characterized the early stage of malaria infection by using two strains of mice infected with *P. yoelii*, a strain 17XL. The authors observed a significant production of the cytokine interferon gamma (INF-γ) in infected DBA/2 mice, in comparison with infected BALB/c mice. This difference was related with lower parasitemia, higher ratio of infected reticulocytes, and higher survival in DBA/2 mice. NO generation was in accordance to endogenous INF-γ driving Th1 immunity. Moreover, NO donors delayed the infection in a dose-dependent manner in BALB/c mice (Wang et al., 2009).

In C57Bl/6 mice infected (a common inbred strain of laboratory mouse) with PbA, the administration of exogenous NO donor decreased venular resistance and brain vascular inflammation (Zanini et al., 2011). The animals were treated twice a day with DPTA-NO or with saline solution. DPTA-NO treatment decreased the expression of endothelial cell adhesion molecules in the brain, the number of adherent platelets and leukocytes in pial vessels, and the inflammatory vascular resistance while prevented venular and arteriolar albumin leakages, in comparison with control groups. Fig. 2.1 shows the reduced number of leukocytes (A) and platelets (B) in pial vessels due to DPTA-NO administration in comparison with saline treatment.

However, higher dose of DPTA-NO (1 mg per mouse twice a day) caused marked hypotension (Zanini et al., 2011). In another study, in other to avoid drop in the blood pressure, the endogenous found NO donor, GSNO, was administered in PbA-infected mice in three different doses: high dose (3.5 mg), intermediate dose (0.35 mg), and low dose (0.035 mg) (Zanini et al., 2012). All GSNO doses decreased incidence of experimental CM in PbA-infected mice, decreased leukocyte accumulation, decreased edema formation, and avoid the incidence of brain hemorrhage. The high GSNO dose inhibited parasite growth but promoted hypotension. In contrast, intermediate and low doses of the NO donor showed mild effects on parasitemia, heart rate, and blood pressure, in comparison with control groups (saline-treated animals). Interestingly, lower dose of GSNO prevented CM with no changing in blood pressure nor in inhibition of parasite growth (Zanini et al., 2012). This study highlights the importance of the characterization

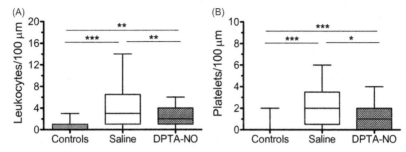

Figure 2.1 Mean number of adherent leukocytes (A) and platelet (B) in pial venules during *P. berghei* ANKA-infected mice on day 6 after infection and in uninfected control mice ($n = 5$). Saline-treated mice increased the number of adherent leukocytes and platelets, while DPTA-NO administration significantly reduced adherence. **$p < 0.01$; ***$p < 0.001$. *Reproduced with modification from Zanini, G.M., Cabrales, P., Barkho, W., Frangos, J.A., Carvalho, L.J.M., 2011. Exogenous nitric oxide decreases brain vascular inflammation, leakage and venular resistance during Plasmodium berghei ANKA infection in mice. J. Neuroinflammation 8, 66. http://www.jneuroinflammation.com/content/8/1/66, with permission from BioMed Central, under the Creative Commons Attribution License of an Open Access Article.*

of the optimization of NO donor dose to achieve the desire effect with minimum side effects such as hypotension. In this direction, further studies are still required.

In another study, a short-time NO donor (3-(2-hydroxy-1-methylethyl-2-nitroso-hydrazino) 1-propanamine, NOC5) and L-arginine (endogenous precursor of NO) were used in a model of BALB/c mice infected with the rodent malaria *P. yoelii*, a strain 17XL (Zheng et al., 2013). The preadministration of L-arginine decreased the levels of mRNA of the apical membrane antigen 1 (AMA 1) gene that encodes a protein responsible for host invasion. For animals treated with the NO donor, NOC5 was traperitoneally injected into Py17XL-infected mice after 5 days of infection, or incubated in vitro with purified PyXL schizonts. Either in vivo or in vitro treatments with the NO donor downregulated the transcript and protein levels of molecules related of the parasite invasion into host RBCs (merozoites surfaces protein 1 (MSP 1), AMA 1, and Py235). It should be noted that invasion of RBCs with malaria parasites is mandatory to the infection. MSP 1 and AMA 1 are essential ligands on the merozoite during parasite invasion process. In addition to MSP1 and AMA1, the 235-kDa rhoptry protein (Py235) of *P. yoelii* is also a blood stage antigen, which mediates immune evasion. Administration of NO donor downregulated the expression of parasite invasion-related molecules (MSP1 and AMA 1) (Zheng et al., 2013). The RBCs invasion by malaria parasites is an essential step to the development of the disease, and hence NO is a key factor to avoid this process. Taken together, administration of NO donors and/or NO donor combined with other drugs represents an interesting strategy to fight against malaria parasite, in particular to combat CM. However, more studies are necessary to better characterize the optimized doses to avoid possible side effects such as drop in the blood pressure.

LEISHMANIASIS

The vector-borne zoonotic disease Leishmaniasis is caused by various species of the protozoa of the genus *Leishmania* (*L.*). It is a macrophage-infected parasitic disease transmitted by sandflies (e.g., phebotomine). The following parasite species cause leishmaniasis: in the Old World, *L. major*, *L. tropica*, *L. aethiopica*, and *L. infantum*, and in the New World, *L. (Viannia) braziliensis*, *L. amazonensis*, *L. venezuelensis*, *L. mexicana*, *L. (V.) guyanensis*, *L. (V.) panamensis*, *L. pifanoi*, *L. (V.) peruviana*, *L. (V.) lainsoni*, and *L. (V.) shawi*. Different species of subgenus *L. amazonensis* and *L. Viannia* are found from Argentina to Mexico, mainly in the Amazon region of Brazil (Lindoso et al., 2012).

Clinically, there are three forms of leishmaniasis infection: visceral leishmaniasis (VL), which can be fatal if not treated, mucosal leishmaniasis (ML), and cutaneous leishmaniasis (CL). Leishmaniasis affects approximately 12 million people in 98 countries, and there are 70,000 deaths per year and 1.5–2 million new cases annually reported (Chung et al., 2008; den Boer et al., 2011). Chemotherapy for leishmaniasis is limited to pentavalent antimonials, amphotericin B, fluconazole, pentamidine, paramomycin, and milfetosine (Chung et al., 2008; den Boer et al., 2011; Marinho et al., 2015). The current therapy is scarce, causes side effects and resistance. Therefore, there is a necessity to develop new drugs/strategies to efficiently treat leishmaniasis with no side effects.

NO and Leishmaniasis

NO is directly involved in the pathogenesis of leishmaniasis. High iNOS expression was proved to reduce the number of parasites and is associated with parasite elimination (Palmeiro et al., 2012). Moreover, NO plays a vital role in maintaining the survivability of circulating RBCs (Chowdhury et al., 2010). As the Leishmania parasite compromises host macrophages impairing iNOS activity, and thus decreasing host NO production, the administration of exogenous NO donors is a promising chemotherapy approach to combat leishmaniasis (Sarkar et al., 2011). In this sense, exogenous NO donors might be able to compensate the decreased levels of endogenous NO production by infected macrophages.

NO Donors and Leishmaniasis

The leishmanicidal activity of the NO donors GSNO and *S*-nitroso-*N*-acetyl-L-cysteine (SNAC) was demonstrated in vitro toward *L. amazonensis* and *L. major* promastigotes (de Souza et al., 2006). Both NO donors showed a dose-dependent cytotoxicity against Leishmania protozoa. Indeed, in the case of GSNO incubation for 24 hours, the 50% inhibitory concentrations (IC_{50}) were found to be 68.8 ± 22.86 and $68.9 \pm 7.9\,\mu mol\,L^{-1}$ for *L. major* and *L. amanzonesis*, respectively. For SNAC treatment, the IC_{50} values of 54.6 ± 8.3 and $181.6 \pm 12.5\,\mu mol\,L^{-1}$ for *L. major* and *L. amanzonesis*, respectively, were obtained. The possible mechanism of leishmanicidal activity of the NO donors was proposed to involve the *S*-nitrosylation of cysteine moieties in the active site of parasite cysteine proteinases (de Souza et al., 2006). In a further study, the antileishmanial activity of GSNO was evaluated against intracellular amastigotes and GSNO was daily topically applied on the skin ulcers of *L. major* or *L. braziliensis* infected BALB/c mice (Costa et al., 2013). In vitro study showed a decrease in the number of intracellular *L. major* amastigotes upon GSNO treatment, and the detection of *S*-nitrosylated proteins indicated that the possible cytotoxic mechanism involves the *S*-nitrosylation of parasite cysteine proteases. In vivo experiments revealed that GSNO treatment suppressed lesion grown, promoted wound healing, and decreased parasite loading in a similar manner in comparison with treatment with amphotericin B, the traditional drug used in the treatment of leishmaniasis (Costa et al., 2013).

López-Jaramillo et al. (1998) firstly reported the successful administration of the NO donor *S*-nitroso-*N*-acetylpenicillamine (SNAP) (final concentration $200\,\mu mol\,L^{-1}$) based cream on lesions of patients with cutaneous leishmaniosis caused by *L. braziliensis*. The formulation was topically applied 4 times a day. Presence of granulation and an enhancement of ulcer were observed after 5 days of treatment in all patients. After 10 days, the ulcers were involved into granulation tissue, and after 30 days of treatment, the lesions were healed, due to the fact that NO donors, such as SNAP, accelerate and promote wound repair, in addition to their leishmanicidal effects. No side effects were observed upon SNAP topical treatment (López-Jaramillo et al., 1998). This work highlights the efficacy of topical applications of NO donors for cytotoxic effects and at the same time to enhance/accelerate wound healing process. Similarly, topical formulation, composed by acidified nitrite and ascorbic acid, able to generate NO was designed and applied on lesions caused by *L. tropica* on BALB/c mice (Davidson et al., 2000). However, a cure of only 12% of the patients and a healing of 28% of ulcer lesions were observed (Davidson et al., 2000). Both studies highlight the promising uses of topical administration of NO-releasing/generating formulations in the treatment of CL and indicate that more studies are necessary to optimize the direct release of therapeutic amounts of NO in an efficient treatment.

A promising tool to optimize the sustained and controlled NO release in the treatment of cutaneous leishmaniosis is the combination of NO donors with nanomaterials (Gutiérrez et al., 2016). Although the efficacy of NO donors against leishmania parasite has been demonstrated in several studies (de Souza et al., 2006; Costa et al., 2013), and despite the advances in the design of NO-releasing nanomaterials for diverse biomedical applications (Seabra and Durán, 2010, 2012; Seabra et al., 2015a,b), the combination of NO donors and nanomaterials is still poorly explored in the treatment of neglected diseases. Particularly, NO-releasing nanomaterials may find important applications in the treatment of CL. In this direction, López-Jaramillo et al. (2010) developed a nanofiber patch to topically release NO in the treatment of cutaneous leishmaniosis caused by *L. (V) panamensis*. By using electrospinning technique, the authors produced a multilayer transdermal patch capable to generate a sustained NO flux of $3.5\,\mu mol$ NO per cm^2 per day for 20 days. The NO-releasing nanomaterial is composed by ascorbic acid and acidified nitrite encapsulated in polymeric nanofibers. After 90 days follow-up, 40% of the patients treated with the NO-releasing nanomaterial was cured (López-Jaramillo et al., 2010). In a similar strategy, the NO donor SNP was encapsulated into biodegradable and biocompatible polymeric nanoparticles composed by poly(lactide-co-glycolide) (PLGA) along with doxorubicin (DOX) for the treatment of visceral Leishmania. A sustained NO release for over 72 hours from the polymeric nanoparticle was observed (Pandya et al., 2011).

The leishmanicidal effects of the NO donor isosorbide dinitrate (ISD) and the standard antileishmanial sodium stibogluconate (SAG) were evaluated in infected hamsters with VL, which decreases RBCs survival (Chowdhury et al., 2010). During VL, decreased NO bioavailability is observed due to low expression of iNOS. Upon treatment of the animals with ISD and SAG, NO level was replenished, which is important for host immunity against the parasite infection. Moreover, the combined treatment (ISD and SAG) deactivated caspase 3 via S-nitrosylation decreasing RBCs apoptosis in infected animals. The combined treatment also avoided infection-mediated ATP depletion and changed calcium homeostasis in RBCs (Chowdhury et al., 2010).

Recently, the NO donor (cis-[Ru(bpy)$_2$imN(NO)(PF$_6$)$_3$[Ru–NO]]) (Ru–NO complex) was injected intraperitoneally along with oral administration of *Brazilian propolis* during 30 days in BALB/c mice (Miranda et al., 2015). The animals were preinfected with *L. amazonensis* in their hind paw followed by treatment with the NO donor and *Brazilian propolis*. Ru–NO injection increased the NO levels in the infected hind paw (Fig. 2.2A), with no lesion exacerbation, indicating that NO released from the NO donor reached the center of the lesion. Analyses of the chronic inflammation revealed that the combination of Ru–NO and *Brazilian propolis* increased the efficacy of macrophages, decreased the number of parasitized cells, decreased the expression of tissue damage and the expression of proinflammatory markers. Moreover, the combined treatment promoted wound repair, increased the amount of fibroblasts (Fig. 2.2B), pro-healing cytokines and promoted collagen synthesis at the bed of the lesion (Miranda et al., 2015). As expected, NO donors not only display leishmanicidal acidity but also promote tissue repair. Interestingly, as shown in Fig. 2.2B, the Ru–NO complex plus *Brazilian propolis* treatment had a superior wound healing activity, as assayed for the number of fibroblast cells, in comparison with glucantime (N-methyl glucamine antimoniate), which is one of the antiparasitic agent for treating leishmaniasis in Brazil since 1940s.

Nahrevanian et al. (2007) demonstrated that the use of the NO donor GSNO on BALB/c mice, infected with leishmania, generated reactive nitrogen intermediates. These reactive nitrogen intermediates decrease the survival of leishmanial parasite

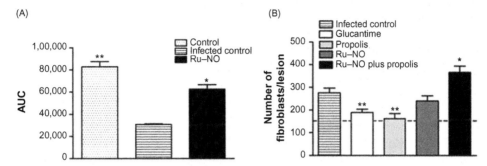

Figure 2.2 Administration of the Ru–NO complex increased the NO/peroxynitrite level, as evaluated by chemiluminescence technique in supernatant of macerated paw. The result was expressed as the mean ± SEM of five animals per group. Bars are represented by the medians of each group (AUC: area under the curve). (A) Administration of Ru–NO complex plus *Brazilian propolis* enhanced the wound healing process, as assayed for the number of fibroblasts analyzed in paw sections. The dotted line (---) represents the uninfected control (B). Significant difference relative to the infected control. *$p < 0.05$ and **$p < 0.01$, unpaired *t*-test. *Reproduced with modification from Miranda, M.M., Panis, C., Cataneo, A.H.D., da Silva, S.S., Kawakami, N.Y., Lopes, L.G.F., et al., 2015. Nitric oxide and brazilian propolis combined accelerates tissue repair by modulating cell migration, cytokine production and collagen deposition in experimental leishmaniasis. PLoS ONE 14 (10), e0125101. http://dx.doi.org/10.1371/journal.pone.0125101, with permission from PLoS, under the Creative Commons Attribution License of an Open Access Article.*

inside the macrophages. The same group demonstrated the antileishmanial activity of the NO donor trinitroglycerin (TNG) in BALB/c mice infected with *L. major* in cutaneous leishmaniasis model (Nahrevanian et al., 2009). TNG therapy increased NO levels, which reduced leishmaniasis and the pathology in infected mice, decreasing amastigote proliferation inside the host macrophages. The antileishmanial effect of TNG treatment reduced smears form lesions, spleens, livers, and lymph nodes, indicating that TNG may inhibit visceralization of *L. major* target organs (Nahrevanian et al., 2009). This study clearly demonstrated multiple roles of NO in the treatment of leishmaniasis.

Taken together, these studies demonstrate that NO donors have potent cytotoxic effects against cutaneous and visceral leishmaniosis, and in the case of CL, NO donors also have an important role in the promotion and acceleration of wound healing process.

TRYPANOSOMIASIS

The parasite *Trypanosoma brucei gambiense* or *T. b. rhodesiense* causes the neglected disease Human African trypanosomiasis, also known as sleeping sickness, which is transmitted to humans via the tsetse fly. It is estimated that 70 million people are at the infection risk, and 20,000 people are infected in Africa (Sutherland et al., 2015; Nagle et al.,

2014). American trypanosomiasis, or Chagas' disease, is caused by the protozoan parasite *Trypanosoma cruzi* (*T. cruzi*). Carlos Chagas discovered the protozoa in 1909 (Chung et al., 2008). Chagas' disease is still endemic in 21 countries in Latin America, 25 million people are living under the risk of infection in endemic countries and approximately 10–15 million people are contaminated, the disease remains essentially incurable (Rodríguez-Morales et al., 2015; Carvalho et al., 2012). The parasite life cycle is complex involving two multiplicative forms (epimastigote and amastigote) and one infective form (trypomastigote) (Coura and Castro, 2002). The parasite life cycle involves vertebrate and invertebrate hosts. Trypomastigotes infect the host macrophage. The human infection occurs through the metacyclic trypomastigote contaminated excreta of infected hematophages reduviid bug of the subfamily Triatominae (Seabra et al., 2015c).

The disease symptoms include fever, enlargement of the lymph nodes, liver and spleen, presence of subcutaneous edema, and signs of portal of entry of the protozoa via ocular membranes or skin (Rassi et al., 2010). In chronic stage of the disease, cardiac complications can be observed leading to thromboembolism, cardiac failure, and death (Rassi et al., 2010). Unfortunately, the most affected organ in trypanosomiasis is the heart (Rodríguez-Morales et al., 2015).

Treatment of Human African trypanosomiasis is based on two drugs (nifurtimox and SCYX-7158) that are under clinical trials (Nagle et al., 2014). Benznidazole and nifurtimox are the drugs used in the treatment of Chagas' disease (Pereira and Navarro, 2013). Both compounds induce significant side effects, and many *T. cruzi* strains are resistant to these drugs (Bern, 2015; Silva et al., 2009; Chung et al., 2008; Guedes et al., 2010; Machado et al., 2010). Unfortunately, although certain progress in the developing of new chemotherapies against Chagas' disease, the treatment has remained the same for more than 40 years, with no effective response for the chronic phase of the disease (Pereira and Navarro, 2013).

NO and Trypanosomiasis

NO synthesized via iNOS is reported to play a key role in the control of *T. cruzi* infection (Carvalho et al., 2012). Indeed, interleukin-12 (Aliberti et al., 1999) and tumor necrosis factor-alpha (TNF-α) (Cardillo et al., 1996) are produced during *T. cruzi* infection, yielding iNOS activation and gamma-interferon (INF-γ) synthesis (Aliberti et al., 1999). NO synthesized by iNOS displays trypanocidal activity (Waghabi et al., 2005; Silva et al., 2009). However, the parasite triggers the production of interleukin-10 and transforming growth factor beta, which downregulate NO synthesis through inhibition of INF-γ produced macrophage (Machado et al., 2000). Endogenous production of NO by host immune defense through iNOS activation is crucial to kill the parasite. However, the parasite downregulates NO production to promote the infection. Therefore, the administration of exogenous NO may have important role in the combat of *T. cruzi*.

NO Donors and Trypanosomiasis

Several reports describe the successful use of ruthenium (Ru)-based NO donors against trypanosomiasis. The NO donors trans-$[Ru(NO)(NH_3)_4L]^{n+}$, where L is N-heterocyclic H_2O, SO_3^{2-}, or triethyl phosphite were demonstrated to lyse *T. cruzi* in vitro and in vivo (Silva et al., 2009). The efficacy of Ru-based NO donors was assayed in a mouse model for acute Chagas' disease, via oral administration and intraperitoneal route. An optimized dose of 400 nmol per kg of body weight was obtained from dose–response curves. In addition, Ru–NO donors eliminated amastigote nests in myocardium tissue, with no side effects to the animals, indicating the efficient and safe uses of these NO donors against trypanosomatids (Silva et al., 2009).

The trypanosomiasis activity of the NO donor trans-$[RuCl([15]aneN_4)NO]^{2+}$ was evaluated in partially drug-resistant *T. cruzi* strain including the epimastigote (antiproliferative), trypomastigote, and amastigote forms (Guedes et al., 2010). For in vivo study, infected mice were treated during acute phase of the parasite infection. The Ru–NO donor was found to be 10- and 100-fold more efficient in comparison with traditional benznidazole treatment, against amastigotes and trypomastigotes, respectively. In addition, the NO donor led to a suppression of parasitemia and no animal death was observed during the NO treatment in the acute Chagas' disease. Upon the administration of the drugs alone (Ru–NO donor and benznidazole), parasitological cures were reported in only 30% of the animals treated with the NO donor (3.33 $\mu mol\,kg^{-1}\,day^{-1}$), and 40% of the animals treated with benznidazole (385 $\mu mol\,kg^{-1}\,day^{-1}$). However, the simultaneous administration of the NO donor and benznidazole led to 80% of parasitological cure with absence of myocarditis (Guedes et al., 2010). These results clearly indicate that NO donors can be administrated simultaneous with traditional chemotherapeutic drugs, such as benznidazole, to promote a synergist effect in the treatment of acute Chagas' disease.

In this direction, more recently, the trypanosomiasis activity of a ruthenium complex with benznidazole and NO ($RuBzNO_2$) was demonstrated in vitro and in vivo (Sesti-Costa et al., 2014). As expected, $RuBzNO_2$ was more effective in comparison with benznidazole (at same concentrations) in eliminating extracellular amastigote of *T. cruzi*, with no cytotoxicity toward mouse cells. Moreover, in vivo experiments revealed that $RuBzNO_2$ enhanced the survival of infected mice, decreasing the risk of heart damage in a superior manner in comparison with benznidazole treatment alone. Treatment with $RuBzNO_2$ also significantly reduced parasitism and tissue inflammation of infected animals, compared with benznidazole treatment. Indeed, the administration of $RuBzNO_2$ at 4 $\mu mol\,kg^{-1}$ enhanced the survival of infected mice in a superior manner compared with treatment with benznidazole alone, at the same concentration. Fig. 2.3A shows that NO is released from $RuBzNO_2$ inside the treated Vero cell line (cells derived from the kidney of the African green (vervet) monkey) (green dots).

(A)

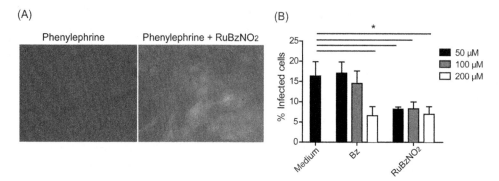

(B)

Figure 2.3 RuBzNO$_2$ is able to release NO inside the cell and kill the amastigotes (Vero cells), as observed from the conversion of the no fluorescent dye DAF-2DA to the fluorescent product DAF-2T. Vero cells were incubated with DAF-2 DA and the reducing agent phenylephrine in the presence or absence of RuBzNO$_2$. The formation of the fluorescent compound (DAF-2T), which occurs in the presence of NO in the cytoplasm, was assessed by fluorescence microscopy (A). RuBzNO$_2$ was able to inhibit the replication or survival of the amastigotes more efficiently than benznidazole because RuBzNO$_2$ at 50 µmol L^{-1} decreased the percentage of infected cells to the same extent that benznidazole did at 200 µmol L^{-1}. Cells were incubated with trypomastigote forms of *T. cruzi* for 24 hours at 37°C. The remaining parasites were washed from the culture, and the cells were incubated at 37°C for an additional 24 hours. Cells were stained with Giemsa dye, and the percentage of infected cells was determined by optical microscopy. The bars indicate means + SEM of triplicates and are representative of two independent experiments. * represents *p* < 0.05 (B). *Reproduced from Sesti-Costa, R., Carneiro, Z.A., Silva, M.C., Santos, M., Silva, G.K., Milanezi, C., et al., 2014. Ruthenium complex with benznidazole and nitric oxide as a new candidate for the treatment of Chagas Disease. PLoS Negl. Trop. Dis. 8 (10), e3207. http://dx.doi.org/10.1371/journal.pntd.0003207, with permission from PLoS, under the Creative Commons Attribution License of an Open Access Article.*

The released NO causes the amastigote death. Fig. 2.3B shows that RuBzNO$_2$ inhibited the survival or replication of the amastigotes, with superior efficient in comparison with benznidazole. Indeed, 50 µmol L^{-1} of RuBzNO$_2$ decreased the percentage of infected cells in a same extent that of 200 µmol L^{-1} of benznidazole administrate alone.

The observed trypanocidal activity of NO is assumed to involve oxidative stress mechanism of activated macrophages and infected cardiac myocytes (Sesti-Costa et al., 2014).

Similarly, the ability of the NO donor *cis*-[Ru(NO)(bpy)$_2$L]X$_n$ to lyse *T. cruzi* was demonstrated in vitro and in vivo (Silva et al., 2010). In fact, the Ru–NO donor complex showed inhibitory effects on *T. cruzi* glyceraldehyde 3-phosphate dehydrogenase (GAPDH) (IC$_{50}$ ranging from 89 to 153 µmol L^{-1}). Moreover, the inhibitory mechanism is assumed to involve the *S*-nitrosylation of cysteine moiety in the active site of the enzyme (Cys 166), as evidenced by enzyme crystal structure. A dose of 385 nmol kg^{-1} of the Ru–NO donor led to survival rates of 80% and 60% in infected mice with amastigote eradication in animal's myocardial tissue. The superior trypanocidal activity of Ru–NO complex was demonstrated by using 1000-fold lower doses of the NO donor, compared to clinically used doses of benznidazole, with potent in vitro and in vivo activity against *T. cruzi* (Silva et al., 2010).

Besides Ru–NO complex, *S*-nitrosothiols (RSNOs) were shown to have potent trypanocidal activity. Recently, Seabra et al. (2015c) successfully demonstrated the toxic effects of NO-releasing polymeric nanoparticles composed by chitosan/sodium tripolyphosphate (TPP). Chitosan/TPP nanoparticles with average hydrodynamic size of 300 nm efficiently encapsulated the thiol-containing low molecular weight molecule mercaptosuccinic acid (MSA) (encapsulation efficiency of 99%). Thiol groups of MSA-containing chitosan/TPP nanoparticles were nitrosated by the addition of sodium nitrite in acidified solution leading to the formation of *S*-nitroso-MSA-chitosan/TPP nanoparticles. The obtained nanoparticles were able to release free NO in a sustained manner. In vitro studies demonstrated that NO-releasing chitosan/TPP nanoparticles $(200\text{--}600\,\mu g\,mL^{-1})$ were toxic to replicative and noninfective epimastigote, and non-replicative and infective trypomastigote forms of *T. cruzi*. Furthermore, upon treatment with NO-releasing nanoparticles a significant decrease in the number of infected (with amastigotes) macrophage was reported (Seabra et al., 2015c). Further in vivo studies are necessary to confirm the potent effect of NO-releasing chitosan-based nanoparticles with minimum side effects.

The use of RSNOs against Chagas' disease was also reported by using topical application of SNAC to the antennae of instar nymphs of the blood-sucking bug *Rhodnius prolixus*, which is the main vector of Chagas' disease in Venezuela, Colombia, and Central America (Sfara et al., 2011). Upon treatment with the NO donor SNAC, the insects showed a living pigeon as the food source (Sfara et al., 2011).

The mechanisms of trypanocidal effect of NO donors were evaluated for GSNO, 3-morpholino-sydnonimine (SIN-1), and SNAP on the activity of cysteine proteinase of *T. brucei* (Steverding et al., 2009). The catalytic activity of purified *T. brucei* cysteine proteinase was inhibited by 50–90% upon incubation with $1.0\,mmol\,L^{-1}$ of the NO donors. However, the NO donors did not significantly inhibit the cysteine proteinase activity within the parasites (Steverding et al., 2009).

Taken together, these studies demonstrated the potent trypanocidal activity of different classes of NO donors, including Ru–NO complexes and RSNOs. The NO donors showed superior efficacy compared with traditional trypanocidal chemotherapy. Moreover, the combination of NO donors and benznidazole enhanced the toxic effects to *T. cruzi* parasite (in vitro and in vivo) with no observed side effects. The mechanisms of NO donor activity against *T. cruzi* are still not completely understood and more studies are necessary to elucidate this issue. Detailed studies are required to translate the NO donors from bench to clinical trials in the treatment of Chagas' disease.

SCHISTOSOMIASIS

Schistosomiasis, also known as Bilharzia, is caused by the parasitic flatworms Trematode of genus *Schistosoma* with five major species that infect humans: *S. mansoni*, *S. japonicum*, *S. haematobium*, *S. intercalatun*, and *S. Mekong* (Chung et al., 2008; Neves et al., 2015). WHO estimates that more than 250 million people are infected, 600 million people are under the risk of infection, and there are approximately 200,000 deaths per year (Chung et al., 2008; Neves et al., 2015; Guglielmo et al., 2014). The most affected areas in the world are South America, the Caribbean, the Middle East, and sub-Saharan Africa (Neves et al., 2015). The life cycle of the schistosoma parasites is complex having humans as definitive hosts and snail as intermediate host (Sayed et al., 2008). The human contamination occurs through direct skin contact with water containing parasite-contaminated snail (Chung et al., 2008). Poor sanitary conditions, lack of health polices, and poor understanding of the disease corroborate to the dissemination of schistosomiasis in countries under developing. Chemotherapy against schistosomiasis is based on the praziquantel (PZQ), metrifonate, and oxamniquine. Among them, PZQ is the drug of choice in most of the cases. However, as in the cases of other neglected disease treatments, side effects and tolerance are observed, there is no vaccine, hence the development of new drugs is urgently necessary (Neves et al., 2015).

NO Donors and Schistosomiasis

Hybrid compounds based on NO donor-PZQ and furoxan moieties were prepared and assayed against *S. mansoni* worms. The results showed the death of adult *S. mansoni* worms cultured upon treatment with NO donor hybrid PZQ drug (Guglielmo et al., 2014).

The NO donors (*trans*-[Ru(bpy)$_2$(NO)SO$_3$]CPF$_6$)-PF$_6$ and Na$_2$[Fe(CN)$_5$(NO)]-SNP, which are able to release NO upon activation in biological reducing medium, were used to treat BALB/c mice infected subcutaneously with Schistosoma BH strains (Pavanelli et al., 2014). The treatment with ruthenium-based NO donors increased animal survivals (Fig. 2.4), improving the mice resistance.

Histopathological analysis revealed a decrease in the influx of inflammatory cells in the hepatic tissues of treated animals. Overall, the use of NO donors decreased parasites proliferation, modulating the cytokine production and improving the host inflammatory response against to the parasite. Moreover, the authors reported an attenuation in the fibrosis of treated animals, and a reduction in the number of eggs and worms in host liver.

The authors suggest that NO plays a key role in the induction of host defense against the parasite infection, and this effect may be assigned to both generation of ROS, which are toxic to the parasite, and ability of NO to inactive vital parasite proteins through *S*-nitrosylation and/or nitration (Pavanelli et al., 2014).

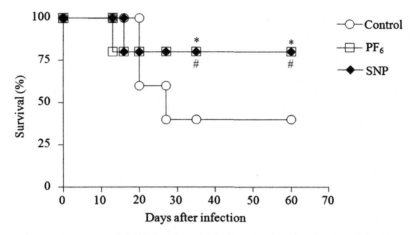

Figure 2.4 Survival curves of BALB/c mice treated with *trans*-[Ru(bpy)$_2$(NO)SO$_3$](PF$_6$)-PF$_6$ and Na$_2$[Fe(CN)$_5$(NO)]-SNP or control. BALB/c mice were infected by percutaneous exposure of tail skin with 70 cercariae of BH strain and treated with Ru–NO donors or control 20 days. The mortality of these mice was also evaluated. Data are representative of two independent experiments with 5 mice per group. The data shown represent the mean ± SEM of the results obtained. *$p < 0.05$ for PF$_6$ and #$p < 0.05$ for SNP-treated mice versus control mice. *Reproduced from Pavanelli, W.R., Silva, J.J.N., Panis, C., Cunha, T.M., Oliveira, F.J.A., Menezes, M.C.N.R., et al., 2014. Nitric oxide donors with therapeutic strategic in experimental Schistossomiasis mansoni. Am. J. Immunol. 10 (4), 225–239. http:// dx.doi.org/10.3844/ajisp.2014.225.239, with permission from Science Publications, under the Creative Commons Attribution License of an Open Access Article.*

TUBERCULOSIS

WHO estimates that 9 million new cases of tuberculosis are registered per year, and approximately half of millions of them is affected by multidrug resistant (MDR) *Mycobacterium tuberculosis* (*M. tuberculosis*) strain (World Health Organization, 2014). Tuberculosis is still a great challenge since it is the most serious infectious lung neglected disease in the world (Chung et al., 2008). The disease causes the deaths of approximately 1.3 million people per year (including human immunodeficiency virus (HIV)-positive people) (Glaziou et al., 2015). Although the disease is disseminated worldwide, it mainly affects Africa. Unfortunately, the combination of HIV and tuberculosis increases the death of individuals (Chung et al., 2008). MDR-tuberculosis is defined as resistance to isoniazid and rifampicin (the common drugs used in the treatment against tuberculosis). Extensively drug-resistant tuberculosis is resistant to fluoroquinolone or injectable second-line antituberculosis drug, in addition to isoniazid and rifampicin (D'Ambrosio et al., 2015; Yuen et al., 2015). New parches are necessary in the treatment of this neglected disease. In this scenario, NO donors may find important applications, since important reports demonstrate that NO controls the immunopathology of tuberculosis (Mishra et al., 2012).

NO Donors and Tuberculosis

The NO donor diethylenetriamine NONOate NO adduct (DETA/NO) was entrapped in biocompatible and biodegradable microparticles composed by PLGA, prepared by spray drying process (Verma et al., 2013). Inhalable PLGA microparticles containing 10% (w/w) of the NO donor were obtained and evaluated as intracellular NO delivery platform. The inhalable NO-releasing microparticles induced phagosome maturation and killed *M. tuberculosis*. Indeed, inhalation of NO-releasing PLGA microparticles in association with isoniazid and rifabutin significantly decreased the colony forming units (cfu) in sleeps and lungs by 4 log. Moreover, daily inhalation of the NO-releasing microparticles alone decreased $\log_{10}$ cfu in the lungs from 6.1 to 4.4 in 1 month treatment (Verma et al., 2013).

In a similar approach, inhalable PLGA microparticles contained NO donors (isosorbide mononitrate (ISMN), SNP, and DETA/NO) with high encapsulation efficiency (>90%) were able to intracellularly deliver NO to macrophages infected in vitro with *M. tuberculosis* (Verma et al., 2012). Cultured human monocytic cell line (THP-1) derived macrophages was able to phagocyte NO-releasing microparticles. A sustained NO secretion was detected in culture supernatant for up to 72 hours. Inhalable NO-releasing microparticles are able to target NO to macrophage, and locally NO released was observed to decrease *M. tuberculosis* cfu by 3-log in 24 hours, at low NO concentrations (Verma et al., 2012). The NO donors DETA/NO and spermine diazeniumdiolate (spermine-NONOate) were demonstrated to have bacteriostatic effects toward *M. tuberculosis* (Voskuil et al., 2011).

In the case of tuberculosis, the combination of microparticles and NO donors has demonstrated promising effects to treat the disease. The use of biodegradable and biocompatible polymeric particles able to carry and deliver therapeutic amounts of NO direct to the target site, where NO can have its therapeutic effects, represents a promising strategy that should be further explored. In this sense, novel nanomaterials composed by polymers and/or metals can be designed as versatile scaffolds in NO therapy to treat tuberculosis.

DENGUE FEVER

DF is an emerging arboviral disease caused by a group of related mosquito-borne flaviviruses (dengue virus) (Charnsilpa et al., 2005). The disease affects public health in developing countries, in particular Latin American, where it is endemic (Anez et al., 2007). Clinically, the disease ranges from asymptomatic for a certain period of time to life-threatening hemorrhagic dengue fever (Charnsilpa et al., 2005). Dengue is transmitted to human by the bite of infected mosquitoes of the genus *Aedes*, which inoculate the virus upon blood feeding on humans (Diallo et al., 2005; Ramos-Castañeda et al., 2008). Caution should be considered in the diagnosis of dengue virus infection, since other diseases, such as leptospirosis, viral syndrome, typhoid, measles, and influenza, may have similar symptoms (Mungrue, 2014). The disease affects approximately 2.5 billion people mainly in subtropical and tropical regions of the globe (Beatty et al., 2010). Antidengue virus activity was reported for flavonoids such as quercetin and fisetin (Zandi et al., 2012).

NO is reported to be an effective immune defender to inhibit viral dissemination. Immunological system through macrophages and cells in the macrophage lineage produces NO. NO synthesized by iNOS displays potent antiviral effects against both RNA and DNA virus, via multiple mechanisms including, S-nitrosylation of vital viral proteins, inhibition of viral genome synthesis, and apoptotic effects (Charnsilpa et al., 2005). Therefore, during human dengue virus infection, the host immunological system naturally produces NO via iNOS. Valero et al. (2002) firstly reported the presence of NO in human dengue infection. Increased levels of NO in patients contaminated with DF may arise from a NO protector effect during the infection. On the other side, some reports showed low levels of NO detected in hemorrhagic dengue fever, which may contribute to patient death (Anez et al., 2007). Therefore, the administration of exogenous NO donors may have a positive impact in the treatment against dengue virus.

NO Donors and DF

The impact of the NO donor SNAP on dengue virus replication in cultures of neuroblastoma cells has been demonstrated (Charnsilpa et al., 2005). The SNAP treatment suppressed viral RNA synthesis in infected cells, decreasing viral proteins and viral progeny production (Charnsilpa et al., 2005). Similar effects were reported for *Aedes albopictus* monolayer cell culture (C6/36 cells) infected with dengue virus and incubated with the NO donor SNP (Neves-Souza et al., 2005). Therefore, the administration of exogenous NO donors may lead to beneficial effects during dengue infection due to NO antivirus and proapoptotic effects (Anez et al., 2007). High levels of NO may be beneficial during dengue infection by its antiviral and apoptotic effects.

The consequences of administration of NO donors on infectious dengue virus serotype 2 production and RNA replication were analyzed in vitro (Charnsilpa et al., 2005). The NO donor SNAP (at concentrations 50, 75, and $100\,\mu mol\,L^{-1}$) was incubated with primary isolates of dengue virus (serotype 2), obtained from dengue infected patients, replicated in mouse neuroblastoma cells. SNAP treatment inhibited viral replication, in a dose-dependent manner. The lower doses of the NO donor (50 and $75\,\mu mol\,L^{-1}$) delayed and suppressed the virus replication, while the higher NO donor concentration tested ($100\,\mu mol\,L^{-1}$) fully inhibited the production of infectious agents up to 36 hours of monitoring (Charnsilpa et al., 2005). Hence, the administration of exogenous NO donor successfully downregulated virus replication, indicating that NO plays an important role in the pathogenesis of dengue, and it may decrease virus load in patients.

The importance of NO during dengue infection was demonstrated by using NO donors and NO inhibitors on dengue virus replication in *Aedes aegypti* and *Anopheles albimanus* (Ramos-Castañeda et al., 2008). Normally, dengue virus replicates at 3×10^6 genome copies/day/mosquito in *Aedes*. However, upon addition of $2\,mmol\,L^{-1}$ of the NO donor SNP to the infected blood meal, no evidence of virus genome accumulation was observed (Ramos-Castañeda et al., 2008). In a similar manner, the NO donor SNAP inhibited RNA accumulation, suggesting that this inhibitory effect of NO on RNA polymerase activity downregulates viral RNA synthesis (Takhampunya et al., 2006).

Although several evidences suggest an important defense role of NO against dengue virus, the research in this domain is still in the beginning and more studies are necessary to combat this neglected tropical viral disease.

BLASTOCYSTIS HOMINIS INFECTION

One of the most common human intestine parasitic disease is blastocystis, which is caused by a noninvasive, extracellular, luminal protozoan with controversial pathogenesis. The intestinal parasitic infection can be asymptomatic or leads to diarrhea, vomiting, and abdominal pain (Mirza et al., 2011). NO has been demonstrated to have toxic effects toward Blastocystis, however, the protozoan may have developed unique strategies to evade NO host defense by decreasing iNOS expression. Therefore, the administration of exogenous NO donor may have positive effects in the treatment of Blastocystis.

NO Donors and *Blastocystis homini* Infection

Exogenous NO generated in vitro from sodium nitrite ($NaNO_2$), at concentrations of 0.6, 0.8, and $1\,mmol\,L^{-1}$, were incubated with *B. hominis* (Eida et al., 2008). $NaNO_2$ inhibited the growth and lowered the viability of *B. hominis*. Apoptotic-like features were observed by transmission electron microscopy of culture of *B. hominis* after $NaNO_2$ treatment (Eida et al., 2008).

To demonstrate that NO production by intestinal cells plays an important role against Blastocystis infection, Mirza et al. (2011) revealed that two clinical isolates of Blastocystis (ST-7 (B) and ST-4 (WR-1) were susceptible to a range of NO donors. Indeed, the metronidazole-resistant isolate (ST-7(B)) was observed to be significantly sensitive to nitrosative stress. By using heterogeneous human epithelial colorectal adenocarcinoma cell (Caco-2) model of human intestinal epithelium, Blastocystis ST-7 (B) showed downregulation of iNOS and a dose-dependent inhibition of NO in the cells (Mirza et al., 2011). These results indicate an important effect of NO in the parasitic infection, and more studies are necessary to optimize the NO dose to be locally released with no side effects and to better understand the NO mechanism of action during the infection.

NO DONORS AND LYMPHATIC FILARIASIS

The neglected tropical disease Lymphatic filariasis, also known as elephantiasis tropica, is caused by the parasite *Brugia malayi* (*B. malayi*), which is transmitted to humans by mosquitoes, upon taking up the blood-borne microfilarial stage of the parasite (Schroeder et al., 2012). Lymphatic filariasis, in which the adult worms are found in the lymphatic system, is the most important filarial disease that affects humans (Chandy et al., 2011; Yadav and Srivastava, 2014). Chemotherapy is based on the administration of mebendazole, suramin, ivermectin, diethylcarbmazine, and flubendazole (Turner et al., 2010).

NO has been demonstrated to have important defense against several pathogens, including filaria. In fact, NO production by host defense system was demonstrated to be induced by *Th1* cytokine IFN-γ. Thus microfilariae of filarial nematodes might be a targeted NO release from NO donors (Pfaff et al., 2000). Indeed, microfilariae of the filarial species *Litomosoides sigmodontis* were found to be susceptible to NO generation from the NO donor SNAP, as well as NO produced from host-activated macrophages (Pfaff et al., 2000). Furthermore, in vivo studies with NO donor sodium 1-(*N,N*-diethylamino)diazen-1-ium-1,2-diolate (DEA/NO) led to a significant inhibition of the development of adult worms and caused alterations on worms motility (Rajan et al., 1996), suggesting a beneficial effect of administration of NO donors. Once again, further studies are necessary.

NO DONORS AND OTHER NEGLECTED DISEASES

The parasitic disease human strongyloidiasis is concentrated in tropical and subtropical regions of the Globe. More than 100 million people are contaminated with *Stronglyoides stercoralis* in several countries (Ruano et al., 2015). An impairment of host immune defense system increases the number of larvae invading virtually all organs and tissues, beside the intestine, leading to hyperinfection syndrome, which can promote secondary bacterial infection leading to death (Ruano et al., 2015). As in the case of other parasitic diseases, NO has important function in immunity to nematode infections due to its ability to regulate the host immunological system and/or to have a direct toxic effect to the parasite. In this context, the involvement of NO in naïve and immunosuppressed BALB/c mice infected with *S. venezuelensis* was characterized (Ruano et al., 2015). The NO donor LA419 [5-(6-nitro-oxi-hexahydrofuro[3,2,b]furan-3-1-il)thioacetate] was administrated to the infected animals in drinking water at $5\,mg\,kg^{-1}$ per day. In addition to the use of NO donor, iNOS inhibitor was administrated and induced a more severe disease with high parasite recovery. In contrast, infected animals treated with the NO donor showed significant fewer eggs in their feces, as well as decreased number of parasites in their guts and lungs and counts of eosinophil comparable to infected control animals (Ruano et al., 2015). Therefore, NO plays an important role in the host defense against Strongeoidiasis, since administration of NO donor decreased the intensity of *S. venezuelensis* infection.

Japanese encephalitis virus (JEV) is a member of the family Flaviviridae and leads to acute encephalitis with considerable high mortality rate in humans (Lin et al., 1997). In vitro study with the NO donor SNAP showed antiviral effects in JEV-infected mouse neuroblastoma cell line (N18), human neuronal NT-2 cell line, and baby hamster kidney (BHK-21) cells, as well as in persistently JEV-infected mouse lung epithelial progenitor cells (C2-2cells) (Lin et al., 1997).

The activity of NO donors in the defense against helminth diseases was demonstrated (Muro and Pérez-Arellano, 2010). The NO donor SNAP led to cell death in *Echinococcus granulosus* (Steers et al., 2001). SNP also damage the parasite *Toxiplasma gondii* tachyzoites causing apoptosis in toxoplasma gondiita chyzoites (Peng et al., 2003).

MECHANISMS OF ACTION OF NO AGAINST PATHOGENS THAT CAUSE NEGLECTED DISEASES

Although important progress in the study of antimicrobial actions of NO against several pathogens, the exact mechanism of NO toxicity is still not completely understood. Indeed, NO is assumed to have multiple pathways against invading microorganisms.

Several evidences indicate that one cytotoxic mechanism attributed to NO is rather due to the production of reactive oxygen/nitrogen species, including the powerful oxidant peroxynitrite ($ONOO^-$), dinitrogen trioxide (N_2O_3), and nitrogen dioxide (NO_2). Peroxynitrite is produced from the diffusion-controlled reaction between NO and another free radical, the superoxide ($O_2^{-\bullet}$), according to Eq. (2.1) (Anez et al., 2007; Heinrich et al., 2013; Pacher et al., 2007; Prolo et al., 2015).

$$^\bullet NO + O_2^{-\bullet} \rightarrow OONO^- \qquad (2.1)$$

These reactive oxygen/nitrogen species lead to direct oxidative reactions to the pathogen, including DNA and lipids damage, increasing the oxidative stress and induction of apoptosis (Gutierrez et al., 2009).

It should be noted that the best reported parasitic macromolecular targets for NO donors are the cysteine proteases (Ascenzi et al., 2003). Cysteine proteases are vital enzymes to the survival of the parasite and relevant in several aspects of parasite life cycle including host infection. *S*-nitrosylation of cysteine residues in the active site of cysteine proteinases by NO donor causes the inactivation of the enzyme. In antiparasitic chemotherapy, the development of cysteine proteinase targets is an efficient strategy, and NO donors are promising candidates. Fig. 2.5 shows the schematic representation of the possible toxic mechanisms of action of NO, synthesized by iNOS or generated by the administration of exogenous NO donors, against invading pathogens.

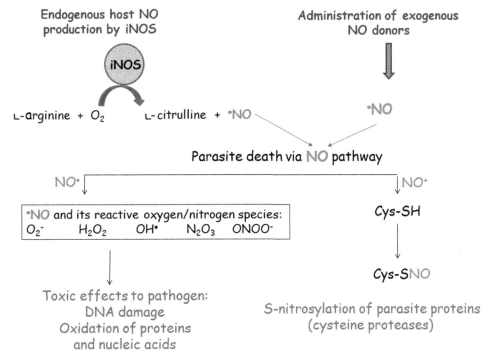

Figure 2.5 Schematic representation of the NO synthesized by inducible nitric oxide synthase (iNOS) in host cell system, or NO generated by the administration of NO donors on the formation of reactive oxygen species (ROS) and reactive nitrogen species (RNS), leading to antimicrobial activity against invading pathogens, and S-nitrosylation pathway, in which NO inactivates cysteine residues in the active site of cysteine proteinases, vital enzymes for pathogen survival.

FINAL REMARKS: CHALLENGES AND PERSPECTIVES

Neglected diseases still haunt humanity, mainly in developing countries. Most of drugs employed were developed 25–40 years ago, and the current bottleneck for the treatment of neglected diseases with the available chemotherapeutics is the increasing drug resistance and side effects. Therefore, it is necessary to design new strategies to overcome these issues.

As parasites are highly susceptible to oxidative stress, an emerging strategy to combat neglected diseases is the use of NO. Indeed, from the revised literature, NO synthesized by iNOS is the important mediator of host defense against several pathogenic microorganisms. However, the pathogens are able to downregulate the expression of iNOS, decreasing NO bioavailability, and thus enhancing the infection. Therefore, the administration of exogenous NO donors aims to compensate the deficiencies in the NO production by iNOS. Although the exact mechanism of action is still not completely understood, several publications described the promising administration of different classes of NO donors against a wide range of neglected diseases, as discussed in this chapter.

Recently, nanomaterials have been successfully used in medicine as efficient drug carrier systems. Although the potent uses of NO donors as chemotherapeutics in neglected diseases, and in spite of the great potential of nanomaterials to combat parasitic diseases (Durán et al., 2009), the combination of NO donors and nanotechnology for neglected diseases is only partly explored (Gutiérrez et al., 2016). The use of nanomaterials in combination with NO donors for biomedical applications represents a new avenue, since this combination may overcome several limitations related to direct administration of free NO donors, such as to increase the thermal and photochemical stability of the NO donor, decrease doses, increase efficiency, and minimize side effects (Seabra and Durán, 2010, 2012; Seabra et al., 2015a,b,c).

It is a challenge to create interest in governments and industries for the development of new drugs for the treatment of neglected diseases. As the administrations of NO donors have been demonstrated to be a promising approach, this chapter aims to be a source of inspiration for research and development based on the uses of NO donor therapies in the fight against neglected diseases.

ACKNOWLEDGMENTS

Support from INOMAT (CNPq), Brazilian Network on Nanotechnology (MCTI/CNPq), NanoBioss (MCTI), and FAPESP is acknowledged.

REFERENCES

Aliberti, J.C., Machado, F.S., Souto, J.T., Campanelli, A.P., Teixeira, M.M., Gazzinnelli, R.T., et al., 1999. β-Chemokines enhance parasite uptake and promote nitric oxide-dependent microbiostatic activity in murine inflammatory macrophages infected with *Trypanosoma cruzi*. Infect. Immun. 67, 4819–4826.

Anez, G., Valero, N., Mosquera, J., 2007. Role of nitric oxide in the pathogenesis of dengue. Dengue Bull 31, 118–123.

Anstey, N.M., Weinberg, J.B., Hassanali, M.Y., Mwaikambo, E.D., Manyenga, D., Misukonis, M.A., et al., 1996. Nitric oxide in Tanzanian children with malaria: inverse relationship between malaria severity and nitric oxide production/nitric oxide synthase type 2 expression. J. Exp. Med. 184, 557–567.

Ascenzi, P., Bocedi, A., Gradoni, L., 2003. The anti-parasitic effects of nitric oxide. IUBMB Life 55, 573–578.

Bang, C.S., Kinnunen, A., Karlsson, M., Onnberg, A., Soderquist, B., Persson, K., 2014. The antibacterial effect of nitric oxide against ESBL-producing uropathogenic *E. coli* is improved by combination with miconazole and polymyxin B nonapeptide. BMC Microbiol. 14, 65. http://www.biomedcentral.com/1471-2180/14/65.

Beare, N.A., Harding, S.P., Taylor, T.E., Lewallen, S., Molyneux, M.E., 2009. Perfusion abnormalities in children with cerebral malaria and malarial retinopathy. J. Infect. Dis 199, 263–271.

Beatty, M.E., Stone, A., Fitzsimons, D.W., Hanna, J.N., Lam, S.K., Vong, S., et al., 2010. Best practices in dengue surveillance: a report from the Asia-Pacific and Americas dengue prevention boards. PLoS Negl. Trop. Dis 4 (11), e890. http://dx.doi.org/10.1371/journal.pntd.0000890.

Bern, G., 2015. http://www.uptodate.com/contents/chagas-disease-antitrypanosomal-drug-therapy (accessed 12.07.15).

Bertinaria, M., Guglielmo, S., Rolando, B., Giorgis, M., Aragno, C., Fruttero, R., et al., 2011. Amodiaquine analogues containing NO-donor substructures: synthesis and their preliminary evaluation as potential tools in the treatment of cerebral malaria. Eur. J. Med. Chem 46, 1757–1767.

Bertinaria, M., Orjuela-Sanchez, P., Marini, E., Guglielmo, S., Hofer, A., Martins, Y.C., et al., 2015. NO-donor dihydroartemisinin derivatives as multitarget agents for the treatment of cerebral malaria. J. Med. Chem 58, 7895–7899.

Bogdan, C., 2001. Nitric oxide and the immune response. Nat. Immunol. 2, 907–916.

Brisboi, E.J., Bayliss, J., Wu, J., Major, T.C., Xi, C., Wang, S.C., et al., 2014. Optimized polymeric film-based nitric oxide delivery inhibits bacterial growth in a mouse burn wound model. Acta Biomater. 10, 4136–4142.

Cabrales, P., Zanini, G.M., Meays, D., Frangos, J.A., Carvalho, L.J.M., 2011. Nitric oxide protection against murine cerebral malaria is associated with improved cerebral microcirculatory physiology. J. Infect. Dis. 203, 1454–1463.

Cardillo, F., Voltarelli, J.C., Reed, S.G., Silva, J.S., 1996. Regulation of *Trypanosoma cruzi* infection in mice by gamma interferon and interleukin 10: role of NK cells. Infect. Immun. 64, 128–134.

Carvalho, C.M.E., Silverio, J.C., Silva, A.A., Pereira, I.R., Coelho, J.M.C., Britto, C.C., et al., 2012. Inducible nitric oxide synthase in heart tissue and nitric oxide in serum of *Trypanosoma cruzi*-infected rhesus monkeys: association with heart injury. PLoS Negl. Trop. Dis. 6 (5), e1644. http://dx.doi.org/10.1371/journal.pntd.0001644.

Chandy, A., Thakur, A.S., Singh, M.P., Mnigauha, A., 2011. A review of neglected tropical diseases: filariasis. Asian Pac. J. Trop. Med. 4, 581–586. doi:10.1016/S1995-7645(11)60150-8.

Charnsilpa, W., Takhampunya, R., Endy, T.P., Mammen, M.P., Libraty, D.H., Ubol, S., 2005. Nitric oxide radical suppresses replication of wild-type dengue 2 viruses in vitro. J. Med. Virol. 77, 89–95.

Chaturvedi, U.C., Nagar, R., 2009. Nitric oxide in dengue and dengue haemorrhagic fever: necessity or nuisance? FEMS Immunol. Med. Microbiol. 56, 9–24.

Chouake, J., Schairer, D., Kutner, A., Sanchez, D.A., Makdisi, J., Blecher-Paz, K., et al., 2012. Nitrosoglutathione generating nitric oxide nanoparyicles as an improved strategy for combating *Pseudomonas aeruginosa*-infected wounds. J. Drugs Dermatol. 11, 1471–1477.

Chowdhury, K.D., Sem, G., Biswas, T., 2010. Regulatory role of nitric oxide in the reduced survival of erythrocytes in visceral leishmaniasis. Biochim. Biophys. Acta 1800, 964–976.

Chung, M.C., Ferreira, E.I., Santos, J.L., Jeanine, G., Rando, D.G., Almeida, A.E., et al., 2008. Prodrugs for the treatment of neglected diseases. Molecules 13, 616–677.

Costa, I.S.F., de Souza, G.F.P., de Oliveira, M.G., Abrahamsohn, I.A., 2013. *S*-nitrosoglutathione (GSNO) is cytotoxic to intracellular amastigotes and promotes healing of topically treated *Leishmania major* or *Leishmania braziliensis* skin lesions. J. Antimicrob. Chemother. 68, 2561–2568.

Coura, J.R., Castro, S.L., 2002. A critical review on Chagas' disease chemotherapy. Mem. Inst. Oswaldo Cruz 97, 3–24.

D'Ambrosio, L., Centis, R., Sotgiu, G., Pontali, E., Spanevello, A., Migliori, G.B., 2015. New anti-tuberculosis drugs and regimens: 2015 update. Eur. Respir. Soc. (ERJ) Open Res. 1, 00010–02015. http://dx.doi.org/10.1183/23120541.00010-2015.

Dascombe, M., Nahrevanian, H., 2003. Drug modulation of NO in murine malaria pharmacological assessment of the role of nitric oxide in mice infected with lethal and nonlethal species of malaria. Parasite Immunol. 25, 149–159.

Davidson, R.N., Yardley, V., Croft, S.L., Konecny, P., Benjamin, N., 2000. A topical nitric oxide-generating therapy for cutaneous leishmaniasis. Trans. R. Soc. Trop. Med. Hyg. 94, 319–322.

den Boer, M., Argaw, D., Jannin, J., Alvar, J., 2011. Leishmaniasis impact and treatment access. Clin. Microbiol. Infect. 17, 1471–1477.

de Souza, G.F.P., Yokoyama-Yasunaka, J.K.U., Seabra, A.B., Miguel, D.C., de Oliveira, M.G., Uliana, S.R.B., 2006. Leishmanicidal activity of primary *S*-nitrosothiols against *Leishmania major* and *Leishmania amazonensis*: implications for the treatment of cutaneous leishmaniasis. Nitric Oxide 15, 209–216.

Diallo, M., Sall, A.A., Moncayo, A.C., Ba, Y., Fernandez, Z., Ortiz, D., et al., 2005. Potential role of sylvatic and domestic African mosquito species in dengue emergence. Am. J. Trop. Med. Hyg. 73, 445–449.

Dondorp, A.M., Ince, C., Charunwatthana, P., Hanson, J., van Kuijen, A., Faiz, M.A., et al., 2008. Direct in vivo assessment of microcirculatory dysfunction in severe falciparum malaria. J. Infect. Dis. 197, 79–84.

Durán, N., Marcato, P.D., Teixeira, Z., Durán, M., Costa, F.T.M., Brocchi, M., 2009. State of the art of nano-biotechnology applications in neglected diseases. Curr. Nanosci. 5, 396–408.

Eida, O.M., Hussein, E.M., Eida, A.M., El-Moamly, A.A., Salem, A.M., 2008. Evaluation of the nitric oxide activity against *Blastocystis hominis* in vitro and in vivo. J. Egypt. Soc. Parasitol. 38, 521–536.

Eisenhut, M., 2015. The evidence for a role of vasospasm in the pathogenesis of cerebral malaria. Malar. J. 14, 405. http://dx.doi.org/10.1186/s12936-015-0928-4.

Förstermann, U., Sessa, W.C., 2012. Nitric oxide synthases: regulation and function. Eur. Heart J. 33, 829–837.

Glaziou, P., Sismanidis, C., Floyd, K., Raviglione, M., 2015. Global epidemiology of tuberculosis. Cold Spring Harb. Perspect. Med. 5, a017798. http://dx.doi.org/10.1101/cshperspect.a017798.

Guedes, P.M.M., Oliveira, F.S., Gutierrez, F.R.S., Silva, G.K., Rodrigues, G.J., Bendhack, L.M., 2010. Nitric oxide donor trans-[RuCl([15]aneN$_4$)NO]$^{2+}$ as a possible therapeutic approach for Chagas' disease. Brit. J. Pharmacol. 160, 270–282.

Guglielmo, S., Cortese, D., Vottero, F., Rolando, B., Kommer, V.P., Williams, D.L., 2014. New praziquantel derivatives containing NO-donor furoxans and related furazans as active agents against *Schistosoma mansoni*. Eur. J. Med. Chem. 84, 135–145.

Gutierrez, F.R.S., Mineo, T.W.P., Pavanelli, W.R., Guedes, P.M.M., Silva, J.S., 2009. The effects of nitric oxide on the immune system during *Trypanosoma cruzi* infection. Mem. Inst. Oswaldo Cruz 104, 236–245.

Gutiérrez, V., Seabra, A.B., Reguera, R.M., Khandared, J., Calderón, M., 2016. New approaches from nano-medicine for treating leishmaniasis. Chem. Soc. Rev. 45, 152–168.

Heinrich, T.A., da Silva, R.S., Miranda, K.M., Switzer, C.H., Wink, D.A., Fukuto, J.M., 2013. Biological nitric oxide signalling: chemistry and terminology. Brit. J. Pharmacol. 169, 1417–1429.

Jeney, V., Ramos, S., Bergman, M.L., Bechmann, I., Tischer, J., Ferreira, A., 2014. Control of disease tolerance to malaria by nitric oxide and carbon monoxide. Cell Rep. 8, 126–136.

Lin, Y.L., Huang, Y.L., Ma, S.H., Yeh, C.T., Chiou, S.Y., Chen, L.K., et al., 1997. Inhibition of Japanese encephalitis virus infection by nitric oxide: antiviral effect of nitric oxide on RNA virus replication. J. Virol. 71, 5227–5235.

Lindoso, J.A.L., Costa, J.M.L., Queiroz, I.T., Goto, H., 2012. Review of the current treatments for leishmaniases. Res. Reports Trop. Med. 3, 69–77.

Lopansri, B.K., Anstey, N.M., Weinberg, J.B., Stoddard, G.J., Hobbs, M.R., Levesque, M.C., et al., 2003. Low plasma arginine concentrations in children with cerebral malaria and decreased nitric oxide production. Lancet 361, 676–678.

López-Jaramillo, P., Ruano, C., Rivera, J., Terán, E., Salazar-Irigoyen, R., Esplugues, J.V., 1998. Treatment of cutaneous leishmaniasis with nitric-oxide donor. Lancet 351, 1176–1177.

López-Jaramillo, P., Rincón, M.Y., García, R.G., Silva, S.U., Smith, E., Kampeerapappun, P., et al., 2010. A controlled, randomized-blinded clinical trial to assess the efficacy of a nitric oxide releasing patch in the treatment of cutaneous leishmaniasis by *Leishmania (V.) panamensis*. Am. J. Trop. Med. Hyg. 83, 97–101.

Machado, F.S., Martins, G.A., Aliberti, J.C.S., Mestriner, F.L.A.C., Cunha, F.Q., et al., 2000. *Trypanosoma cruzi*-infected cardiomuocytes produce chemokines and cytokines that trigger potent nitric oxide-dependent trypanocidal activity. Circulation 102, 3003–3008.

Machado, F.S., Tanowitz, H.B., Teixeira, M.M., 2010. New drugs for neglected infectious diseases: Chagas' disease. Br. J. Pharmacol. 160, 258–259.

Marinho, D.S., Casas, C.N.P.R., Pereira, C.C.A., Leite, I.C., 2015. Health economic evaluations of visceral leishmaniasis treatments: a systematic review. PLoS Negl. Trop. Dis. 9, e0003527. http://dx.doi.org/10.1371/journal.pntd.0003527.

Martins, Y.C., Zanini, G.M., Frangos, J.A., Carvalho, L.J.M., 2012. Efficacy of different nitric oxide-based strategies in preventing experimental cerebral malaria by *Plasmodium berghei* ANKA. PLoS ONE 7, e32048. http://dx.doi.org/10.1371/journal.pone.0032048.

Miranda, M.M., Panis, C., Cataneo, A.H.D., da Silva, S.S., Kawakami, N.Y., Lopes, L.G.F., et al., 2015. Nitric oxide and brazilian propolis combined accelerates tissue repair by modulating cell migration, cytokine production and collagen deposition in experimental leishmaniasis. PLoS ONE 14 (10), e0125101. http://dx.doi.org/10.1371/journal.pone.0125101.

Mirza, H., Wu, Z., Kidwai, F., Tan, K.S.W., 2011. A metronidazole-resistant isolate of *Blastocystis* spp. i Susceptible to nitric oxide and downregulates intestinal epithelial inducible nitric oxide synthase by a novel parasite survival mechanism. Infect. Immun. 79, 5019–5026.

Mishra, S.K., Newton, C.R., 2009. Diagnosis and management of the neurological complications of falciparum malaria. Nat. Rev. Neurol. 5, 189–198.

Mishra, B.B., Rathinam, V.A.H., Martens, G.W., Martinot, A.J., Kornfeld, H., Fitzgerald, K.A., 2012. Nitric oxide controls the immunopathology of tuberculosis by inhibiting NLRP3 inflammasome-dependent processing of IL-1β. Nat. Immunol. 14, 1. http://dx.doi.org/10.1038/ni.2474.

Mungrue, K., 2014. The laboratory diagnosis of dengue virus infection, a review. Adv. Lab. Med. Int. 4, 1–8.

Muro, A., Pérez-Arellano, J.L., 2010. Nitric oxide and respiratory helminthic diseases. J. Biomed. Biotechnol. Article ID 958108. http://dx.doi.org/10.1155/2010/958108.

Nagle, A.S., Khare, S., Kumar, A.B., Supek, F., Buchynskyy, A., Mathison, C.J.N., et al., 2014. Recent developments in drug discovery for leishmaniasis and human African trypanosomiasis. Chem. Rev. 114, 11305–11347.

Nahrevanian, H., Farahmand, M., Aghighi, Z., Assmar, M., Amirkhani, A., 2007. Pharmacological evaluation of anti-leishmanial activity by in vivo nitric oxide modulation in Balb/c mice infected with *Leishmania major* MRHO/IR/75/ER: an Iranian strain of cutaneous leishmaniasis. Exp. Parasitol. 116, 233–240.

Nahrevanian, H., Najafzadeh, M., Hajihosseini, R., Nazem, H., Farahmand, M., Zamani, Z., 2009. Anti-leishmanial effects of trinitroglycerin in BALB/C mice infected with *Leishmania major* via nitric oxide pathway. Korean J. Parasitol 47, 109–115.

Neves, B.J., Andrade, C.H., Cravo, P.V.L., 2015. Natural products as leads in Schistosome drug discovery. Molecules 20, 1872–1903.

Neves-Souza, P.C., Azeredo, E.L., Zagne, S.M., Vallsde-Souza, R., Reis, S.R., Cerqueira, D.I., et al., 2005. Inducible nitric oxide synthase (iNOS) expression in monocytes during acute dengue fever in patients and during in vitro infection. BMC Infect. Dis. 5, 64. http://dx.doi.org/10.1186/1471-2334-5-64.

Olliaro, P.L., Kuesel, A.C., Reeder, J.C., 2015. A changing model for developing health products for poverty-related infectious diseases. PLoS Negl. Trop. Dis. 9, e3379. http://dx.doi.org/10.1371/journal.pntd.0003379.

Pacher, P., Beckman, J.S., Liaudet, L., 2007. Nitric oxide and peroxynitrite in health and disease. Physiol. Rev. 87, 315–424.

Palmeiro, M.R., Morgado, F.N., Valete-Rosalino, C.M., Martins, A.C., Moreira, J., Quintella, L.P., 2012. Comparative study of the in situ immune response in oral and nasal mucosal leishmaniasis. Parasite Immunol. 34, 23–31.

Pandya, S., Verma, R.K., Misra, A., 2011. Nanoparticles containing nitric oxide donor with antileishmanial agent for synergistic effect against visceral leishmaniasis. J. Biomed. Nanotechnol. 7, 213–215.

Pavanelli, W.R., Silva, J.J.N., Panis, C., Cunha, T.M., Oliveira, F.J.A., Menezes, M.C.N.R., et al., 2014. Nitric oxide donors with therapeutic strategic in experimental *Schistossomiasis mansoni*. Am. J. Immunol. 10 (4), 225–239. http://dx.doi.org/10.3844/ajisp.2014.225.239.

Peng, B.W., Lin, J., Jiang, M.S., Zhang, T., 2003. Exogenous nitric oxide induces apoptosis in *Toxoplasma gondii* tachyzoites via a calcium signal transduction pathway. Parasitology 126, 541–550.

Pereira, P.C.M., Navarro, E.C., 2013. Challenges and perspectives of Chagas disease: a review. J. Venomous Animals Toxin Include. Trop. Dis. 19, 34.

Pfaff, A.W., Schulz-Key, H., Soboslay, P.T., Geiger, S.M., Hoffmann, W.H., 2000. The role of nitric oxide in the innate resistance to microfilariae of *Litomosoides sigmodontis* in mice. Parasite Immunol. 22, 397–405.

Prolo, C., Álvarez, M.N., Ríos, N., Peluffo, G., Radi, R., Romero, N., 2015. Nitric oxide diffusion to red blood cells limits extracellular, but not intraphagosomal, peroxynitrite formation by macrophages. Free Radic. Biol. Med. 87, 346–355.

Rajan, T.V., Porte, P., Yates, J.A., Keefer, L., Shultz, D.L., 1996. Role of nitric oxide in host defense against an extracellular, metazoan parasite, Brugia malayi. Infect. Immun. 64, 3351–3353.

Ramos-Castañeda, J., González, C., Jiménez, M.A., Duran, J., Hernández-Martínez, S., Rodríguez, M.H., et al., 2008. Effect of nitric oxide on dengue virus replication in *Aedes aegypti* and *Anopheles albimanus*. Intervirology 51, 335–341.

Rassi, A.J., Rassi, A., Marin-Neto, J.A., 2010. Chagas disease. Lancet 375, 1388–1402.

Rodríguez-Morales, O., Monteón-Padilla, V., Carrillo-Sánchez, C.C., Rios-Castro, M., Martínez-Cruz, M., Carabarin-Lima, A., et al., 2015. Experimental vaccines against Chagas Disease: a journey through history. J. Immunol. Res. Article ID 489758.

Rosenthal, P.J. (Ed.), 2001. Antimalarial Chemotherapy. Humana Press, Totowa, NJ.

Ruano, A.L., López-Abána, J., Fernández-Soto, P., Melo, A.L., Muro, A., 2015. Treatment with nitric oxide donors diminishes hyperinfection by *Strongyloides venezuelensis* in mice treated with dexamethasone. Acta Trop. 152, 90–95.

Sarkar, A., Saha, P., Mandal, G., Mukhopadhyay, D., Roy, S., Singh, S.K., et al., 2011. Monitoring of intracellular nitric oxide in leishmaniasis: its applicability in patients with visceral leishmaniasis. Cytom. Part A 79A, 35–45.

Sayed, A.A., Simeonov, A., Thomas, C.J., Inglese, J., Austin, C.P., Williams, D.L., 2008. Identification of oxadiazoles as new drug leads for the control of Schistosomiasis. Nat. Med. 14, 407–412.

Schairer, D.O., Chouake, J.S., Nosanchuk, J.D., Friedman, A.J., 2012. The potential of nitric oxide releasing therapies as antimicrobial agents. Virulence 3, 271–279.

Schroeder, J.H., Simbi, B.H., Ford, L., Cole, S.R., Taylor, M.J., Lawson, C., 2012. Live *Brugia malayi* microfilariae inhibit transendothelial migration of neutrophils and monocytes. PLoS Negl. Trop. Dis. 6 (11), e1914. http://dx.doi.org/10.1371/journal.pntd.0001914.

Seabra, A.B., Durán, N., 2010. Nitric oxide-releasing vehicles for biomedical applications. J. Mater. Chem. 20, 1624–1637.

Seabra, A.B., Durán, N., 2012. Nanotechnology allied to nitric oxide release materials for dermatological applications. Curr. Nanosci. 8, 520–525.

Seabra, A.B., de Lima, R., Calderón, M., 2015b. Nitric oxide releasing nanomaterials for cancer treatment: current status and perspectives. Curr. Top. Med. Chem. 15, 298–308.

Seabra, A.B., Justo, G.Z., Haddad, P.S., 2015a. State of the art, challenges and perspectives in the design of nitric oxide-releasing polymeric nanomaterials for biomedical applications. Biotechnol. Adv. 33, 1370–1379.

Seabra, A.B., Kitice, N.A., Pelegrino, M.T., Lancheros, C.A.C., Yamauchi, L.M., Pinge-Filho, P., et al., 2015c. Nitric oxide-releasing polymeric nanoparticles against *Trypanosoma cruzi*. J. Phys. Conf. Ser. 617, 012020. http://dx.doi.org/10.1088/1742-6596/617/1/012020.

Sesti-Costa, R., Carneiro, Z.A., Silva, M.C., Santos, M., Silva, G.K., Milanezi, C., et al., 2014. Ruthenium complex with benznidazole and nitric oxide as a new candidate for the treatment of Chagas Disease. PLoS Negl. Trop. Dis. 8 (10), e3207. http://dx.doi.org/10.1371/journal.pntd.0003207.

Sfara, V., Zerba, E.N., Alzogaray, R.A., 2011. Deterrence of feeding in *Rhodnius prolixus* (Hemiptera: Reduviidae) after treatment of antennae with a nitric oxide donor. Eur. J. Entomol. 108, 701–704.

Sharma, A., Eapen, A., Subbarao, S.K., 2004. Parasite killing in *Plasmodium vivax* malaria by nitric oxide: implication of aspartic protease inhibition. J. Biochem. 136, 329–334.

Silva, J.J.N., Pavanelli, W.R., Pereira, J.C.M., Silva, J.S., Franco, D.W., 2009. Experimental chemotherapy against *Trypanosoma cruzi* infection using ruthenium nitric oxide donors. Antimicrob. Agents Chemother. 53, 4414–4421.

Silva, J.J.N., Guedes, P.M.M., Zottis, A., Balliano, T.L., Silva, F.O.N., Lopes, L.G.F., et al., 2010. Novel ruthenium complexes as potential drugs for Chagas's disease: enzyme inhibition and in vitro/in vivo trypanocidal activity. Brit. J. Pharmacol 160, 260–269.

Siu, E., Ploss, A., 2015. Modeling malaria in humanized mice. Ann. N.Y. Acad. Sci 1342, 29–36.

Steers, N.J.R., Rogan, M.T., Heath, S., 2001. *In-vitro* susceptibility of hydatid cysts of *Echinococcus granulosus* to nitric oxide and the effect of the laminated layer on nitric oxide production. Parasite Immunol. 23, 411–417.

Steverding, D., Wang, X., Sexton, D.W., 2009. The trypanocidal effect of NO-releasing agents is not due to inhibition of the major cysteine proteinase in *Trypanosoma brucei*. Parasitol. Res. 105, 1333–1338.

Sutherland, C.S., Yukich, J., Goeree, R., Tediosi, F., 2015. A literature review of economic evaluations for a neglected tropical disease: human African trypanosomiasis ("sleeping sickness"). PLoS Negl. Trop. Dis. 5 (9), e0003397. http://dx.doi.org/10.1371/journal.pntd.0003397.

Takhampunya, R., Padmanabhan, R., Ubol, S., 2006. Antiviral action of nitric oxide on dengue virus type 2 replication. J. Gen. Virol. 87, 3003–3011.

Tambo, E., Khater, E.I.M., Chen, J.H., Bergquist, R., Zhou, X.N., 2015. Nobel prize for the artemisinin and ivermectin discoveries: a great boost towards elimination of the global infectious diseases of poverty. Infect. Dis. Poverty 4, 58. http://dx.doi.org/10.1186/s40249-015-0091-8.

Turner, J.D., Tendongfor, N., Esum, M., Johnston, K.L., Langley, R.S., Ford, L., et al., 2010. Macrofilaricidal activity after doxycycline only treatment of *Onchocerca volvulus* in an area of Loa loa coendemicity: a randomized controlled trial. PLoS Negl. Trop. Dis. 4 (4), e660. http://dx.doi.org/10.1371/journal.pntd.0000660.

Valero, N., Espina, L.M., Añez, G., Torres, E., Mosquera, J.A., 2002. Short report: increased level of serum nitric oxide in patients with dengue. Am. J. Trop. Med. Hyg. 66, 762–764.

Verma, R.K., Singh, A.K., Mohan, M., Agrawal, A.K., Verma, P.R.P., Gupta, A., et al., 2012. Inhalable microparticles containing nitric oxide donors: saying NO to intracellular *Mycobacterium tuberculosis*. Mol. Pharm. 9, 3183–3189.

Verma, R.K., Agrawal, A.K., Singh, A.K., Mohan, M., Gupta, A., Gupta, P., et al., 2013. Inhalable microparticles of nitric oxide donors induce phagosome maturation and kill *Mycobacterium tuberculosis*. Tuberculosis 93, 412–417.

Voskuil, M.I., Bartek, I.L., Visconti, K., Schoolnik, G.K., 2011. The response of *Mycobacterium tuberculosis* to reactive oxygen and nitrogen species. Front. Microbiol. 2, 105. http://dx.doi.org/10.3389/fmicb.2011.00105.

Waghabi, M.C., Keramidas, M., Feige, J.J., Araujo-Jorge, T.C., Bailly, S., 2005. Activation of transforming growth factor beta by *Trypanosoma cruzi*. Cell. Microbiol. 7, 511–517.

Wang, Q.H., Liu, Y.J., Liu, J., Chen, G., Zheng, W., Wan, J.C., et al., 2009. *Plasmodium yoelii*: Assessment of production and role of nitric oxide during the early stages of infection in susceptible and resistant mice. Exp. Parasitol. 121, 268–273.

World Health Organization, 2010. Working to overcome the global impact of neglected tropical diseases: First WHO report on neglected tropical diseases. Geneva, Switzerland.

World Health Organization, 2014. Global tuberculosis report 2014.

Yadav, S.K., Srivastava, S., 2014. Lymphatic filariasis: an overview. Global J. Pharm. 8, 656–664.

Yeo, T.W., Lampah, D.A., Gitawati, R., Tjitra, E., Kenangalem, E., McNeil, Y.R., et al., 2007. Impaired nitric oxide bioavailability and L-arginine reversible endothelial dysfunction in adults with falciparum malaria. J. Exp. Med. 204, 2693–2704.

Yuen, C.M., Jenkins, H.E., Rodriguez, C.A., Keshavjee, S., Becerra, M.C., 2015. Global and regional burden of isoniazid-resistant tuberculosis. Pediatrics 136, 1. http://dx.doi.org/10.1542/peds.2015-0172.

Zandi, K., Teoh, B.T., Sam, S.S., Wong, P.F., Mustafa, M.R., AbuBakar, S., 2012. Novel antiviral activity of baicalein against dengue virus. BMC Complement. Altern. Med. 12, 214. http://dx.doi.org/10.1186/1472-6882-12-214.

Zanini, G.M., Cabrales, P., Barkho, W., Frangos, J.A., Carvalho, L.J.M., 2011. Exogenous nitric oxide decreases brain vascular inflammation, leakage and venular resistance during *Plasmodium berghei* ANKA infection in mice. J. Neuroinflammation 8, 66. http://www.jneuroinflammation.com/content/8/1/66.

Zanini, G.M., Martins, Y.C., Cabrales, P., Frangos, J.A., Carvalho, L.J.M., 2012. *S*-nitrosoglutathione prevents experimental cerebral malaria. J. Neuroimmune Pharmacol. 7, 477–487.

Zheng, L., Feng, H., Liu, D., Pan, Y.-Y., Cao, Y.-M., 2013. The expression of malarial invasion-related molecules is affected by two different nitric oxide-based treatments. Folia Parasitol. 60, 213–217.

CHAPTER 3

Nitric Oxide-Donating Devices for Topical Applications

Kathleen Fontana and Bulent Mutus
Department of Chemistry and Biochemistry, University of Windsor, Windsor, ON, Canada

INTRODUCTION

Nitric oxide (NO) is a gaseous radical, important for signaling in both plants and animals (Kaiser, 2006). The role NO plays in the everyday health of organisms, as well as the role it can play in critical health situations have been well investigated. Of particular interest is the role NO plays in wound healing and the benefits it can have on biomedical devices and as a topical applicator.

The history of NO begins with its discovery by Joseph Priestley in 1772; however, it was not until two centuries later when research interests took off. In 1998, the Nobel Prize in Medicine was awarded to Robert F. Furchgott, Louis J. Ignarro, and Ferid Murad for their work with NO. For example, in 1987, the research group led by Dr. Ignarro published two papers in succession, the first declaring the chemical similarities between the endothelium-derived relaxing factor (EDRF) and the nitric oxide radical (NO) (Ignarro et al., 1987b). After exploring the link to EDRF, the second paper identified NO as being identical to EDRF in every way that comparisons were possible (Ignarro et al., 1987a). After NO was named "Molecule of the Year" in 1992, the boom of research interest that followed brings us to where we are today.

Much is known of NOs' role in human health, this includes smooth muscle relaxation and vasodilation as the EDRF. NO also plays a role in the immune system and the nervous system. It can stimulate and repress cell proliferation, provide protection from apoptosis, show neuroprotective effects, dilate blood vessels, prevent against thrombosis, kill bacteria, and more.

Nitric Oxide (NO) Synthases

A group of enzymes called nitric oxide synthases (NOSs) are responsible for much of the NO produced by eukaryotic cells. While these enzymes are mostly found in animals and are partly conserved in bacteria and in some algae, other organisms such as land plants use alternative methods to generate NO, which includes nitrite reduction (Jeandroz et al., 2016). At least three NOS isoforms have been isolated. This includes

Nitric Oxide Donors.

a calcium-independent inducible isoform (iNOS), discovered in macrophages, as well as two constitutive isoforms (cNOS), one originally discovered in vascular endothelial cells (eNOS) and a neuronal isoform found in the brain (nNOS); further research located various isoforms in other tissues (Treuer and Gonzalex, 2014; Rassaf et al., 2013). While regulated by intracellular calcium and exogenous calmodulin, NOSs require many cofactors to function. These include tetrahydrobiopterin, flavine mononucleotide (FMN), flavine adenine dinucleotide (FAD), nicotinamide adenine dinucleotide phosphate (NADPH), and oxygen (O_2) in addition to the enzymes' substrate, L-arginine (Treuer and Gonzalex, 2014).

The day-to-day functions of NO are made possible by the constitutive isoforms (cNOS) of NOS. The cNOS isoforms are always active and produce only the minimal amounts of NO that is required for signaling purposes.

The inducible nitric oxide synthase (iNOS) which produces large amounts of NO compared to the constitutive isoforms is induced during infection, inflammation, and wound healing (Kolios et al., 2004; Witte and Barbul, 2002). Inducers of iNOS include endotoxins and cytokines, which can be released by burn injuries, arthritis, and inflammatory bowel diseases (IBSs) (Willmota and Batha, 2003; Witte and Barbul, 2002). Upon stimulation, there is a several hour delay of NO release from iNOS, however, the enzyme can go on to actively produce NO for several days (Kolios et al., 2004).

Nitrite Reductases

One of nature's alternative to NOSs includes nitrite reductases (NIRs), which can be found in plants as well as some microbial species that use denitrification to create make energy by anaerobic respiration (Kaiser, 2006). In the denitrification pathway, NIRs are involved in the overall pathway that converts nitrate to nitrogen gas. Nitrate is reduced to nitrite by nitrate reductase and is then further reduced by NIRs to produce NO. Following this is a series of more reductions to ultimately produce nitrogen gas. In the dissimilatory nitrate reduction pathway, a nitrate reductase reaction is the first step to produce NO, which is an intermediate in the overall reaction scheme that yields ammonia (Bykov and Neese, 2015).

Both copper and iron containing NIRs exist in biological and ecological systems (Wasser et al., 2002). While iron NIRs are more abundant, copper NIRs have a greater physiological diversity since they are found in a greater variety of systems (Wasser et al., 2002). NIRs catalyze the reduction of nitrite to NO, with copper NIRs participating in a one election reduction (Wasser et al., 2002).

With copper in consideration, the general mechanism for the enzymatic process involves nitrite binding to Cu(I) with a hapticity of η^1-N (Wasser et al., 2002). Alternatively, nitrite can bind to copper (II) as well, with copper then being reduced to copper (I) after the nitrite has been bound (Wasser et al., 2002). The next step is the protonation of nitrite to produce water and copper-nitrosyl (Cu(I)-NO$^+$) (Wasser

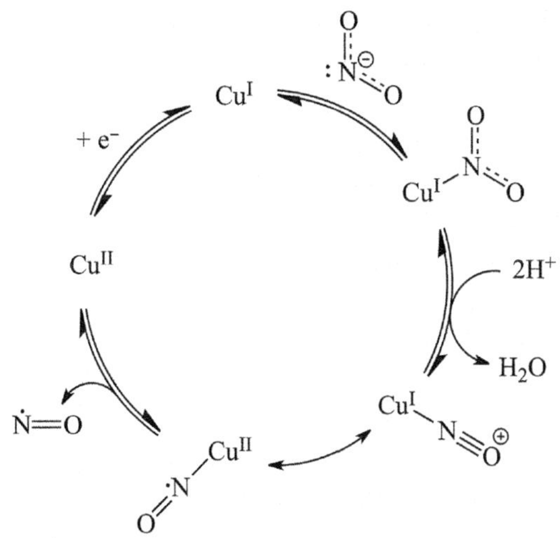

Figure 3.1 The catalytic cycle of copper reducing nitrite to nitric oxide. Beginning with copper (I), nitrite binds to the metal center, where it is subsequently protonated to make copper-nitrosyl, and is then reduced to nitric oxide. Nitric oxide is released and copper (II) can be reduced to copper (I) to complete the cycle.

et al., 2002; Halfen et al., 1998). While copper (I) oxidized to copper (II), NO is formed and readily released from the metal center (Wasser et al., 2002; Halfen et al., 1998). Copper (II) is then reduced to copper (I) which completes the catalytic cycle. The catalytic cycle is depicted in Fig. 3.1.

NO'S ROLE IN THE BODY

The NO produced by NOS activates soluble guanylate cyclase (sGC), an enzyme involved in vasodilation and is thought to be the only known receptor for NO (Willmota and Batha, 2003). Some of the cells that NO targets include vascular smooth muscle and platelets, where activating sGC in these cells means cyclic guanosine monophosphate (cGMP) levels are increased (Misko et al., 1993). The activation of cGMP controls vascular tone, participates in the control of blood pressure, and regulates the distribution of flow between vascular beds, including the brain, skeletal muscles, and penile erection (Willmota and Batha, 2003).

Another important role of NO is in neuronal transmissions. These include: (1) neurotransmitter release; (2) the increase in neuronal survival by reducing the intensity of calcium ion influx; (3) the promotion of capsular relaxation and inhibiting platelet aggregation; (4) the protection of endothelium against atherosclerosis; and (5) the participation in movements of the gut via control of smooth muscle contractions (Willmota and Batha, 2003; Kolios et al., 2004).

NO appears to either inhibit or activate particular enzymes, both to promote and prevent cellular inflammation and apoptosis and is both antiinflammatory and inflammatory through a few separate pathways (Willmota and Batha, 2003). For example, beyond sGC and cGMP, with the lone election of the radical, NO can go on to react with the superoxide anion radical (O_2^-) to result in peroxynitrite ($ONOO^-$) (Kolios et al., 2004; Willmota and Batha, 2003). Through oxidation, peroxynitrite can cause DNA base damage and sequentially apoptosis, lipid peroxidation which causes cell membrane damage, as well as altering protein conformations via nitration (Kolios et al., 2004; Willmota and Batha, 2003). NO can be a source of nitrite, however, the amount of nitrite made from NO is dwarfed by that source from the diet (Rassaf et al., 2013).

Depending on the source of NO, and the cellular environment, NO can exist either a free radical NO (NO^0), be reduced to nitroxyl (HNO/NO^-) or oxidized to the nitrosonium cation (NO^+) (Paulo et al., 2014). NO's reactive nature means it can also rapidly react with oxyhemoglobin to form methemoglobin and nitrate (Rassaf et al., 2013). NO can also interact with enzymes like ceruloplasmin, a multicopper oxidase, to produce NO^+, which is then hydrolyzed to nitrite (Rassaf et al., 2013).

NO has a central role in angiogenesis, the development of new blood vessels, a process that takes place during wound repair (Luo and Chen, 2005). In the vasculature, NO stimulates cell proliferation, provides protection from apoptosis, and is the mediator for the vascular endothelial growth factor (VEGF) (Luo and Chen, 2005). In addition, NO displays antiplatelet, antiatherosclerotic, hemodynamic, and neuroprotective properties (Willmota and Batha, 2003). In vitro studies have also shown NO to inhibit the Fenton reaction, which produces powerful oxidants, via its interactions with the superoxide anion (O_2^-) (Kolios et al., 2004).

During the wound healing process, L-arginine, the substrate of iNOS, is depleted not only by NOS but also by arginase, thus making arginine levels a limiting factor in NO release (Witte and Barbul, 2002). Arginase is an enzyme of the urea cycle that produces ornithine and urea from arginine and is found in large concentrations in wound fluid (Witte and Barbul, 2002). Studies indicate that NO is required in the early phases of healing and subsequently, NO plays less of a role.

The levels of NO produced from the iNOS isoform are higher than the levels from the cNOS isoforms. The cNOS produces short bursts of NO in the picomolar to nanomolar range, which satisfies the signaling requirements (Kolios et al., 2004). In contrast, the iNOS makes NO in micromolar quantities (Kolios et al., 2004). At these higher concentrations, NO displays toxicity effects that work to aid the immune system by killing microorganisms, viruses, and parasites (Willmota and Batha, 2003).

NO's Role in Wound Healing

NO has been found to be an important part of the wound healing process. By inhibiting NO at a wound site, it was found that there was decreased collagen being produced and a decreased formation of granulation tissue (Schäffer et al., 1996). This is due to the fact that NO is critical to collagen deposition, which is also related to the final mechanical strength of the healed wound (Witte and Barbul, 2002).

Macrophages are antimicrobial effector cells and immunoregulatory cells, which work to induce, suppress, and regulate adaptive immune responses (Rath et al., 2014). Macrophages can display two main phenotypes, an M1 phenotype for killing, fighting, proinflammation, and inhibition of proliferation, and an M2 phenotype for healing, fixing, antiinflammation, and promotion of proliferation (Rath et al., 2014). As an immune response, M1 macrophages release the higher levels of NO via arginine and iNOS for the purpose of fighting off infection (Rath et al., 2014). Existing in contrast is the M2 macrophage that expresses arginase, the enzyme that uses arginine to produce urea, and ornithine that can go on to make polyamines, proline, and more (Rath et al., 2014). The dichotomy between the two provides homeostasis for arginine metabolism in macrophages, one that controls iNOS NO output.

It has been hypothesized that eNOS NO favors the polarization of macrophages to the M2 phenotype, and thus favoring the healing and fixing of that type. Using cell culture studies, there has been evidence to support this hypothesis (Lee et al., 2015). Separate experiments showed that any increase levels of NO were found to attenuate M1 and increase M2 polarization, as well as NO from eNOS specifically (Lee et al., 2015).

Complications Resulting from Attenuation of NO Production

Certain impairments to human health have been known to cause decreases in NO production. Arginine is described as a semi-essential amino acid, meaning malnutrition would cause a lack of arginine available to be used as an NOS substrate (Witte and Barbul, 2002). Diabetes, radiation, and chronic steroid treatments have also been shown to affect NOS activity (Witte and Barbul, 2002). Specifically, diabetics are prone to chronic leg and foot ulcerations, which is caused by abnormalities in NO production (Martinez et al., 2009). This is because those with insulin-dependent and noninsulin-dependent diabetes mellitus have an impaired NO-mediated endothelium-dependent vasodilation (Williams et al., 1996). This leads to a decreased ability of the endothelium in making NO, which may contribute to the high prevalence of vascular disease in these patients (Williams et al., 1996). For metabolic disorders, including obesity and diabetes, the pathway from endothelial NO to vasodilator-stimulated phosphoprotein (VASP) signaling is compromised, thus making it a target for treatment (Lee et al., 2015).

It has also been shown that NO levels are elevated in individuals who are often exposed to hypoxic conditions such as the natives of Tibet who live at increased altitudes (Garrya et al., 2015). This supports the notion that NO treatments would be beneficial for those suffering from hypoxia. In addition, the antiinflammatory and antioxidant properties, as well as its ability to control blood flow can help after thrombosis, restenosis, ischemia, and for reperfusion treatments.

With the ischemia, hypoxia, and ATP depletion that is experienced poststroke, there is an upregulation of NOS which leads to increased NO levels observed in cerebrospinal fluid (Willmota and Batha, 2003). With high blood pressure present in 70% of stroke patients, NO treatments can be used as a preventative measure and can be used for poststroke to control blood pressure (Bath et al., 2015). Current research shows that NO is effective against lowering blood pressure, however, does not improve outcome after acute and subacute stroke via increasing functionality and decreasing recovery time (Bath et al., 2015, 2016). To find a correlation proving NO is effective for the recovery of a stroke, more research is necessary.

Regular cNOS activity contributes to a healthy digestive system. On the other side, the overexpression of iNOS and the chronic increased production of NO have been linked to IBS, and are considered detrimental (Kolios et al., 2004). In agreement with this sentiment is that the increase in endothelium-derived NOS (eNOS) experienced postbrain injury is neuroprotective, while iNOS can contribute to more damage (Garrya et al., 2015). This connection, found through animal models, has attributed the underlying differences between the isoforms to the timing, spatial location, and concentration of the NO released (Garrya et al., 2015). To reduce secondary brain damage in the animal models, NO donors have helped to boost the effects of eNOS and an iNOS inhibitor held promise for neurotherapeutic effects (Garrya et al., 2015).

NO treatments are used as complimentary therapy for many cardiopulmonary disorders, in addition to the newly approved treatment for newborns with pulmonary hypertension (Bhatraju et al., 2015). In particular, patients are benefiting from NO inhalation treatments for acute respiratory distress syndrome (ARDS) (Bhatraju et al., 2015). This technique is used in various stages of research and clinical trials to benefit patients of postcardiac arrest, extracorporeal membrane oxygenation (ECMO), cardiopulmonary bypass (CPB), sickle cell disease (SCD), acute chest syndrome (ACS), and lung and heart transplants (Bhatraju et al., 2015).

NO Releasing Biomedical Devices

There are many medical devices that are implanted post medical crisis for healing purposes. This includes stents, which are used after heart attacks to prevent shrinking of the arteries. Other examples include extracorporeal blood-loop circuits, electrical leads of pacemakers and defibrillators, intravascular oxygen sensors and catheters, as well as vascular grafts (Frost et al., 2005). These devices, which can be lifesaving and necessary, can also experience complications in vivo. Due to the risk of thrombus formation and thrombocytopenia, anticoagulants are prescribed to patients, some of which can be dangerous with long term use (Frost et al., 2005). Beyond blood clots and platelet aggregation, the required drugs can lead to hemorrhaging, and devices can lead to bleeding and endothelium damage (Frost et al., 2005).

In particular, stents are used during angioplasty surgery to treat and repair narrow, damaged, or blocked arteries. Most of the time, a metal mesh is used and this mesh can cause further damage to endothelium and smooth muscle cells, as well as acting as a site for protein adsorption (Frost et al., 2005). After protein absorbs, this can lead to platelet activation and adhesion, which then can cause thrombus formation (Frost et al., 2005). Coronary bare metal stents have a failure rate of 20–30%, with patients experiencing restenosis within a year after corrective surgery, because of this, stents that release drugs have been developed to make the implants more safe and biocompatible (Frost et al., 2005; McCarthy et al., 2016). Since NO is a potent antiplatelet agent and inhibitor of smooth muscle cell proliferation, a NO releasing stent would be beneficial (Frost et al., 2005). Having a stent that can release or generate NO locally may help medical implants fight against thrombosis and restenosis (Frost et al., 2005). One way to do this is to have polymers that have been doped or grafted with NO donor molecules, which release NO over extended periods of time, such as diazeniumdiolates and nitrosothiols, and then coat the stent in that polymer (Frost et al., 2005). Also possible is using *S*-nitrosothiols, which decompose to release NO under the influence of transition metal ions, including copper, nickel, and zinc, which can be incorporated into the fabrication of the stent (McCarthy et al., 2016).

Vascular grafts are used on damaged or diseased blood vessels, when surgeons need to redirect blood flow by replacing the blood vessel, oftentimes by using synthetic grafts (Keefer, 2003). These grafts can be prone to thrombus, especially since the damage to the endothelium results in impaired NO production from NOS (Keefer, 2003). One particular graft, which releases NO via functionalization with the enzyme galactosidase, was able to observe inhibited platelet adhesion and thrombus formation while the NO also saw improved tissue regeneration, remodeling, and physiological function (Wang et al., 2015).

Agents for Regulating NO

Delayed wound healing that is largely attributed to low bioavailability of NO could benefit from possible treatments including NO donors and NOS gene therapy (Luo and Chen, 2005). For example, lymphatic vessel dysfunction is an issue with patients dealing with diabetes, obesity, and high cholesterol levels and is caused by a low bioavailability of NO (Scallan et al., 2015). Low availability of bioactive NO could be caused by impaired production of NO or by an increase of NO inactivation by reactive oxygen species (Carpenter and Schoenfisch, 2012). As well, NO therapies could help with wound healing complications that arise with atopic dermatitis and peripheral vascular disease (Martinez et al., 2009). A deficiency in NO can be caused by injured or not properly functioning endothelium. This is the case for some cardiovascular issues, such as atherosclerosis, heart failure, hypertension, arterial thrombotic disorders, coronary heart disease, and stroke (Carpenter and Schoenfisch, 2012).

There are several therapeutic strategies to consider when trying to control NO levels. Depending on the aliment, increasing or decreasing NO levels would be beneficial. As discussed by Garrya et al., supplementing eNOS while inhibiting iNOS is a possible treatment path for controlling the release of NO in the best way possible to protect and prevent head injuries (Garrya et al., 2015). There are several strategies to decrease NO, if the situation warrants it, including selective and nonselective NOS inhibitors. Some nonselective inhibitors work by competing with arginine for the enzyme active site, these include NG-nitro-L-arginine (L-NNA), NG-monomethyl-L-arginine (L-NMMA), and NG-nitro-L-arginine methyl ester (L-NAME) (Willmota and Batha, 2003).

Another complex case for modulating NO is with cancer tumors. With large concentrations, in the micromolar region, NO contributes to reactive nitrogen species. This, together with reactive oxygen species wreak havoc inside the cell, including impairing cell functions and causing DNA base pair deamination which has been shown to be beneficial to tumor progression and survival (Carpenter and Schoenfisch, 2012). As well, elevated levels of NOS activity have been found in cancer cells where overexpression of NO has led to poor clinical outcomes (Carpenter and Schoenfisch, 2012). On the other hand, low concentrations, in the picomolar region, promote angiogenesis and are antiapoptotic, which also aid in tumor growth and nutrient delivery (Carpenter and Schoenfisch, 2012). Two options exist when looking at NO-based therapies for cancer. One of them is to increase concentrations of NO at the tumor site to initiate apoptosis, or necrosis, of cancer cells (Carpenter and Schoenfisch, 2012). Some NO donors have shown antitumor benefits, including diethylenetriamine NONOate (Fig. 3.2C), GTN, sodium nitroprusside, furoxan-based derivatives, and NO releasing aspirin (Carpenter and Schoenfisch, 2012). The other is to use NOS inhibitors via long-term systematic administration at the tumor that causes a decrease in tumor growth (Carpenter and Schoenfisch, 2012). This process has to be continued until the tumor is eradicated, if not, side effects of hypertension and regrowth of the tumor are the possibilities (Carpenter and Schoenfisch, 2012). As a cancer treatment, NO has an advantage of decreased toxicity toward healthy cells at the concentrations toxic to cancer cells (Carpenter and Schoenfisch, 2012).

When looking to increase NO production, there are several limiting factors that contribute to the production of NO. Some natural limiting factors for NO efficacy include the availability of NO synthase, as the response of iNOS is delayed by several hours after initiation. Whereas the other factors include the availability of arginine, which is used in other processes and is the substrate of NOS, and the stability of NO. With a short half-life and high reactivity, NO can degrade or be used up by reacting with the superoxide anion before reaching its target sites.

Using NO inhalation is especially useful for cardiac issues. By varying the concentration and partial pressures of NO and oxygen, desired treatment plans have the possibility to be highly specific and controlled (Bhatraju et al., 2015). Another option is to upregulate NOS enzymes, which can be done with statins (Willmota and Batha, 2003). Already used in vascular disease, statins are believed to increase the expression of endothelium NOS (eNOS) by posttranscriptional mechanisms (Willmota and Batha, 2003). For treating circulatory dysfunctions, inhalation of NO is not the best route since NO is quickly scavenged by hemoglobin, making it difficult to keep levels constant (Carpenter and Schoenfisch, 2012). Better therapies involve NO donors that provide an extended release of NO to maximize and prolong therapeutic levels of NO.

When dealing with malnutrition or lack of arginine and its role as the NOS substrate, increasing arginine levels through the diet or at a site of a wound would be beneficial. Some clinical trials showed an increase in endothelial function in patients with heart disease and high cholesterol, with the help of oral or intravenous L-arginine (Willmota and Batha, 2003). While other studies showed benefits from the antiplatelet properties of NO via oral or intravenous L-arginine (Willmota and Batha, 2003).

NO plays a role in wound healing and as such, it can be looked at as a means of aiding in the optimization of the process. The wound healing cascade begins immediately after injury and goes through the steps of blood coagulation, inflammation, cell proliferation, lesion contraction, and remodeling until the wound is fully healed, with NO playing a role through some wound healing processes as well as throughout everyday processes of tissue homeostasis (Carpenter and Schoenfisch, 2012). Providing NO at the site of injury aids the roles that NO plays, including helping angiogenesis, increasing collagen deposition, and cell proliferation. With high levels of NO, mirroring iNOS activity, NO can be antibacterial at a wound, while helping with immediate cell needs directly after injury. Particularly with wound dressings, which have traditionally been a passive dressings meant for protecting the wound from outside elements, these dressings have been made more active in recent years, playing a role in wound healing (Carpenter and Schoenfisch, 2012). Using NO releasing polymers and other donors, a wound dressing can do just that. For example, a hydrogel that releases NO can keep a moist environment while remaining permeable to oxygen, and the wound receives the benefit of supplemental NO. Some interesting results from a study using diabetic rats, simulating foot ulcers suffered by human diabetics, found that a NO releasing hydrogel improved the granulation and scar tissue thickness of the healed wound, with the only downside being a longer wound closer time (Carpenter and Schoenfisch, 2012).

With the short half-life and high reactivity of NO, the clever alternative is to use a low molecular weight NO donor to provide controlled release of NO in localized areas. Some organic nitrates are already widely used for medical proposes. For example, isosorbide mononitrate and glyceryl trinitrate are used to treat angina, anal fissures, heart failure, and pulmonary hypertension (Carpenter and Schoenfisch, 2012). However with these treatments, patients run the risk of building a tolerance and possible hypotension, as well as a headache as a possible side effect (Carpenter and Schoenfisch, 2012). Some issues with these low molecular weight donors can include them not releasing NO, or inactivation, before they reach their target site, not releasing NO in a sustained manner, and possible toxicity. To help balance these cons, higher molecular weight macromolecules can be used to provide a longer and more continuous delivery of NO to target sites, for therapeutic NO level (Carpenter and Schoenfisch, 2012). This includes drug carriers such as micelles, dendrimers, polymers, and nanocarriers.

NO Donors

Various NO donors exist for the purpose of releasing NO for biomedical purposes. This includes the NO donor molecule, with or without a backbone structure, whether that be a polymer or micelle to transport and control the release of NO. For example, nitroglycerin is an NO donor that has been widely established in vascular medicine (Willmota and Batha, 2003). There are several classes of donors, including S-nitrosothiols, which are a result of a reaction with acidified nitrite and thiols; a common one being S-nitrosoglutathione (GSNO) (Willmota and Batha, 2003). There are organic nitrites and nitrates that act as donors, inorganic nitroso compounds, sydnonimines, and hybrid NO donor drugs such as nitroaspirins (Willmota and Batha, 2003). A significant type of donor is diazeniumdiolates (NONOates), these molecules are the result of NO exposed to a nucleophile, with the end product being very flexible and predictable (Willmota and Batha, 2003). The properties of NONOates have led to their encouraging results in the fight to decrease infract size and prevent vasospasm due to stroke in animal studies (Willmota and Batha, 2003). Fig. 3.2 depicts some of these NO donors.

Some novel NO donors are metal nitrosyl complexes, since NO effectively binds to metal ions (Paulo et al., 2014). This includes some promising ruthenium complexes, since ruthenium has a high affinity for NO, that bind NO with a hapticity of η^1-N (Paulo et al., 2014). With a low cytotoxicity, they act as NO donors under external stimulation and can release NO at a specific biological target (Paulo et al., 2014). This is contrasted with NONOates, where free NO is released when dissolving on a first-order rate of spontaneous reversion to NO (Paulo et al., 2014).

One novel NO donor, a sydnonimine named SIN-1 (Fig. 3.2F), in conjunction with a novel oxygen scavenging hybrid compound, SA-2, has been used in animal models to lower elevated intraocular pressure that is associated with degeneration of the optic nerve and loss of retinal ganglion cells (Acharya et al., 2016). These particular compounds, by Acharya et al., found success in increasing the levels of the superoxide dismutase enzyme to protect photoreceptor cells from oxidative stress, as well as improving intraocular pressure (Acharya et al., 2016).

Some studies have used nitrite as an NO source, since heme iron, including hemoglobin and myoglobin heme groups, has the capabilities of converting nitrite to NO

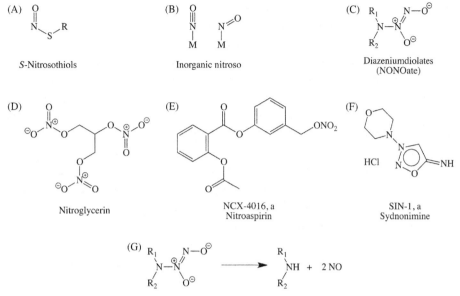

Figure 3.2 The general structures of various nitric oxide donors, including (A) *S*-nitrosothiols, (B) inorganic nitroso compounds, and (C) diazeniumdiolates (NONOates). The specific structures of (D) nitroglycerin, (E) NCX-4016, a nitroaspirin, and (F) SIN-1, a sydnonimine. (G) The reaction scheme of the acid-catalyzed dissociation of a NONOate to form two molecules of nitric oxide (Packer and Cadenas, 2005).

(Rassaf et al., 2013; Bice et al., 2016). Nitrate has been used as well, with some success (Bice et al., 2016). Most of the body's nitrite and nitrate is sourced from the diet and while nitrite can be converted to NO directly, nitrite must first be converted to nitrite. The body almost exclusively relies on consuming bacteria to do the job of converting nitrate (Rassaf et al., 2013).

Donor Carriers

There are many NO releasing macromolecules, designed for medical applications. Delivering NO releasing drugs, or any other drug, using macromolecules adds certain benefits to the delivery and utilization process. Drug carriers can add solubility to non-soluble or poorly soluble drugs, as well as in vitro and in vivo stability and biocompatibility (Reddy et al., 2015). The macromolecules help to decrease the body's tolerance and to help localize drug delivery, making the dosage more effective (Carpenter and Schoenfisch, 2012). This means that drug dosage can be decreased and a more controlled release of the drug can be achieved, depending on the breakdown of a particular carrier. Localizing delivery is generally done using active targeting techniques, with targeting moieties on the carrier that optimizes cellular uptake by target cells (Reddy et al., 2015). Some examples of these drug carrier macromolecules include polymers, micelles, liposomes, dendrimers, and nanoparticles.

Micelles and liposomes have very similar structures. As an example, soaps form micelles using organic chains, with a hydrophilic end and a hydrophobic chain. In aqueous solution, the hydrophobic end traps other hydrophobic substances in the center of a monolayered sphere. Liposomes can be made with a bilayer of phospholipids, with a hydrophilic head and a hydrophobic chain. The bilayer forms a sphere, trapping an aqueous environment in the core. With the core able to hold hydrophilic substances, the hydrophobic chains in the center of the bilayer are able to hold other hydrophobic substances. Micelles and liposomes can be made more complex for the purpose of drug delivery, by adding a protective layer or targeting moieties and peptides. They can also be made using a variety of synthetic materials and polymers. One major advantage of micelles is that they allow for a large amount of hydrophobic drugs to be carried, while keeping a small size and low toxicity (Reddy et al., 2015).

Dendrimers are highly branched nanoscale macromolecules and have been useful for the purpose of gene delivery and drug delivery (Gothwal et al., 2015). They are able to increase the physical stability of the drug and enhance the solubility of poorly soluble drugs (Gothwal et al., 2015). As man-made structures, their molecular structure, size, polydispersity, and available functional groups can be controlled to suite their particular purpose (Gothwal et al., 2015). Placing a drug, or an NO donor, in the hydrophobic center of a dendrimer means a delayed and sustained release, since it acts as a sink to hold the drug (Gothwal et al., 2015).

Polymers have been proven useful for generic drug carrying as well as for NO release on their own. An NO donor can be functionalized as part of a polymer, with donor groups covalently part of the polymers backbone, or added after the polymer, attached via covalent or noncovalent means (Frost et al., 2005). One example is the use of cysteine, which can be immobilized on polyethylene terephthalate (PET) and polyurethane surfaces, that can then go on to react with nitrosothiols in an NO-exchange reaction (Frost et al., 2005). Using polymer based materials for NO release add another level of versatility, since they could be used as coatings or films for biomedical devices, polymers have the flexibility to take on other shapes and properties can be easily manipulated if functional groups are available.

Nanocarriers of NO, and NO donors, can help to prolong the lifetime of treatment as well as offer a gradual release of NO. For example, NO gas trapped inside hydrogel–glass composite nanoparticles has been shown to release NO upon hydration (Carpenter and Schoenfisch, 2012). At pH 7.4, the release of NO for these particles has a half-life of 4 hours, while hamster models resulted in the particles staying in circulation for up to 6 hours before succumbing to the effects of macrophages (Carpenter and Schoenfisch, 2012). The extended release of NO means that less NO was needed, 30 times less, in comparison to just using NO donors; this study specifically compared it to the low molecular weight donors diethylenetriamine NONOate and dipropylenetriamine NONOate (Carpenter and Schoenfisch, 2012).

Some issues with artifical polymer and synthetic based drug carriers are that there may be a great difficulty in synthesizing the needed material in an industrial setting. Incorporation of the drug may be difficult in an industrial setting as there may be a high cost associated with the needed materials and processes, there may also be aggregation of macromolecules in a biological medium, and retention of the drug may be difficult (Reddy et al., 2015). The drug may diffuse out of or detach from the carrier, or become inactive before reaching the target site (Reddy et al., 2015). As well, being attached to a macromolecule may lead to liver toxicity, since it may be around for a longer time than the free drug would have been (Reddy et al., 2015).

TOPICAL APPLICATORS

Typical applicators include biologically active gels, pastes, creams, oils, liquid drops, liquid wash, patch, gauze, or bandages that are applied topically. There would be many opportunities to incorporate an NO donor into any of these. Topically applied NO has been known to increase dermal microcirculation, which could prove useful for those with impaired microcirculation, including those with diabetic foot syndrome and those with complications from skin grafting (Heuer et al., 2015).

Biofilms form from microorganisms accumulating on surfaces like catheters, implants, and on wounds (Wonoputri et al., 2015). They can lead to many infections that are difficult to treat due to antibiotic resistance and intrinsic defense mechanisms (Wonoputri et al., 2015). Disinfectants have been ineffective on controlling biofilm growth, so NO as an antibiofilm agent can be useful (Wonoputri et al., 2015). Low concentrations of NO have been shown to cause dispersal of biofilm and high concentrations can kill bacteria, without the risk of producing more resistant bacteria (Wonoputri et al., 2015). NO-based antimicrobials have become popular in their own right, due to its important role in the immune system and researchers' need for new antibiotics, since typical antibiotics develop resistance within 2 years and NO resistant bacteria have yet to be observed (Carpenter and Schoenfisch, 2012).

Fighting against biofilm formation is another benefit to use NO topically at a wound site, in addition to increased collagen deposition, antibacterial properties of high NO, and its influence on angiogenesis and cell proliferation.

Chitosan, a Versatile Biopolymer

Chitosan is a hydrophilic polysaccharide, made from D-glucosamine and N-acetyl-D-glucosamine monomers bound through $\beta(1-4)$ linkages (Islam et al., 2011). Chitosan is sourced from chitin, which is found in a variety of crustacean exoskeletons including crab, shrimp, and lobster (Islam et al., 2011). Chitin is found in many other organisms such as the exoskeletons of marine zooplankton like coral and jellyfish and in types of mollusks such as oysters and krill (Tharanathan and Kittur, 2003). It is present in insects like beetles, cockroaches, and butterflies, as well as in some fungi (Tharanathan and Kittur, 2003).

When starting with raw biomaterial, like shrimp shells for example, the raw material is stripped of minerals, proteins, and color to yield chitin that is low in impurities. Chitin is first purified of minerals using acid treatments (Percot et al., 2003). Next, proteins are removed using alkaline treatments and finally, color is removed from chitin by using either solvents or oxidants (Percot et al., 2003).

Purified chitin is the precursor in the preparation of chitosan. Chitin is made up of N-acetyl-D-glucosamine monomers and these monomers are deacetylated to D-glucosamine by using concentrated alkaline treatments (Islam et al., 2011). Converted chitin can be described as chitosan after 50% or more of the N-acetyl-D-glucosamine have been deacetylated, or until the material can be dissolved in 1% acetic acid (Islam et al., 2011). Chitosan used for academic purposes is typically 75%, or more, deacetylated. This conversion of chitin to chitosan can be seen in Fig. 3.3.

Chitosan is widely used for many things, including dietary supplements, water treatment, cosmetics, food processing, textiles, solid-state batteries, and agriculture (Dutta et al., 2004). It is used in biological applications as well such as in bandages, tissue engineering, wound dressing and healing, contact lenses, plastic surgery, drug delivery systems, and more (Dutta et al., 2004).

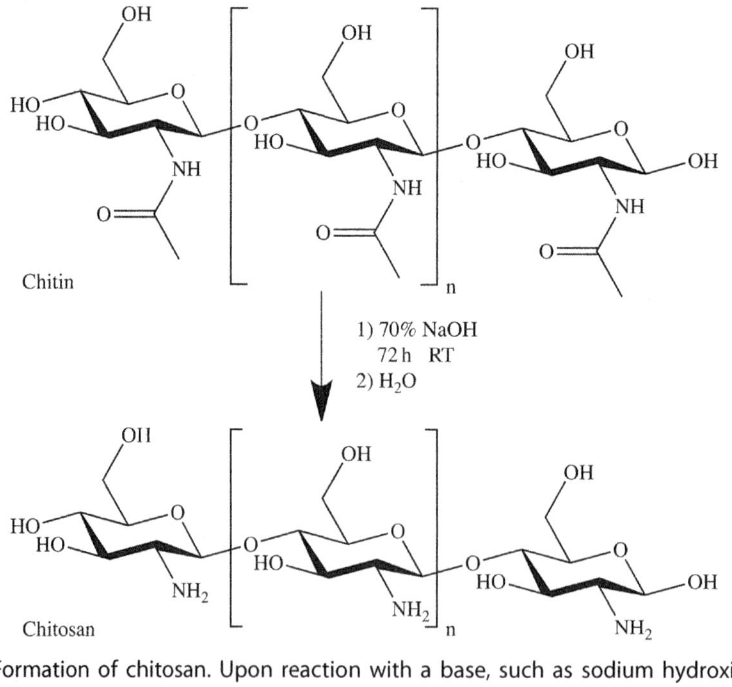

Figure 3.3 Formation of chitosan. Upon reaction with a base, such as sodium hydroxide, chitin is deacetylated and chitosan is formed.

Chitosan's versatility can be attributed to its properties. Chitosan is ideal for use in biological systems due to it being biocompatible, biodegradable, nontoxic, and antimicrobial (Dutta et al., 2004). One chitosan monomer has two reactive hydroxyl groups and an amine group that allows for mechanical and chemical properties to be easily manipulated (Dutta et al., 2004). Common in research is the addition of functional groups on the amine group of chitosan. This allows researchers to manipulate specific reactivity and properties of chitosan displays, catering it to fit a specific application (Rinaudo, 2006).

Chitosan has a semicrystalline structure, due to the large number of intramolecular and intermolecular hydrogen bonds that can be formed (Sogias et al., 2010). Due to chitosan's semicrystalline structure, its solubility is limited (Sogias et al., 2010). Chitin, on the other hand, is insoluble and crystalline, with monomers that have two hydroxyl groups, one amide and one acetyl group that can participate in hydrogen bonding. When chitosan is formed and the acetyl group is lost, the loss of hydrogen bonds weakens the structure, thus allowing chitosan to be soluble in aqueous solutions under pH 6.4. Since the pK_a of chitosan is 6.4, at pH levels below the pK_a, the amine is protonated, which further disturbs hydrogen bonding and allows for solubility (Schmuhl et al., 2001).

Chitosan also has the ability to chelate with various metal ions of copper, zinc, iron, mercury, manganese, and more. The ability of chitosan to chelate with metal ions increases with an increase in the degree of deacetylation (Rinaudo, 2006). The ability of chitosan to chelate is also influenced by the physical state of chitosan: powder, gel, film, or fiber (Rinaudo, 2006). Further, the mechanism of chitosan–metal complex formation depends on the pH of solution (Rinaudo, 2006). When comparing the differences of cross-linked and noncross-linked chitosans with copper absorption, Li et al. reports negligible differences in the infrared (FT-IR) spectra (Li and Bai, 2005). This suggests that the mechanism of adsorption for both cross-linked and noncross-linked chitosans is similar and FT-IR peaks show significant changes in the –N–H stretch and –C–N– stretch, from before and after copper absorption (Li and Bai, 2005). This suggests that the amine groups can be considered a main absorption site for metal ions.

Typically, cross-linking is done to increase the chemical stability of the polymer. Since chitosan is acid soluble, strengthening is needed for the chitosan product, whether that be a film or a particle, to withstand acidic conditions below a pH of 6.4. Chitosan chains can be cross-linked using many different reagents. Some common reagents include dialdehydes, such as glutaraldehyde, which react the amine group of chitosan (Ngah et al., 2005). Another type of cross-linking reagents is heterocyclic compounds, such as epichlorohydrin, which react with chitosan's hydroxyl and amino groups (Ngah et al., 2005). Other options include amides, such as N,N'-methylenebisacrylamide, which are used to cross-link chitosan as well as adding another element of support for the compound (Xu et al., 2015). Acids like sulfuric acid or tripolyphosphate are ionic cross-linking agents that react with amines (Xu et al., 2015). An additional group of cross-linking agents includes ethers such as ethylene glycol diglycidyl ether (EGDE) and crown ethers. EGDE has two epoxide rings, which are reactive due to their high amounts of steric strain. The epoxides of EGDE are capable of reacting with the amino, carboxyl, and hydroxyl functional groups under acidic or basic conditions (Xu et al., 2015; Schmuhl et al., 2001).

Copper Chitosan as an NO Donor

In mimicking NIRs, copper chitosan can act as an NO donor. It has now been established that chitosan absorbs copper atoms and that copper-based reductases catalyze the production of NO from nitrite. As well, increased NO production at a wound site has the potential to speed up the healing process. A study by Martinez et al. showed that NO releasing chitosan nanoparticles had beneficial effects on wound closure (Martinez et al., 2009). They attributed this to antimicrobial activity of chitosan and the promotion of collagen deposition from the NO (Martinez et al., 2009). Further applications for Martinez et al. NO releasing nanoparticles were for topical treatments for first aid purposes, ideal due to chitosan fighting infection and NO speeding up healing (Martinez et al., 2009).

Our group has examined the possibility of using copper chitosan particles, where copper is in its +1 state would reduce nitrite to NO while the reduction of Cu(II) back to Cu(I) could be accomplished by glucose, a reducing sugar. Since both glucose and nitrite are present in the blood, and would be present in wound fluid the Cu–chitosan particles should theoretically produce NO at the wound until the nitrite or glucose levels are depleted. Fig. 3.4 shows the reaction scheme for the regeneration of copper chitosan particles, producing NO and being reduced by glucose; while Fig. 3.1 shows a more in-depth catalytic cycle for copper-catalyzed NO formation.

Topical Copper Chitosan Particles

Chitosan can be dried out into films, in addition to being able to form particles. After being dissolved in acid, copper can be added to the solution to complex with chitosan's functional groups. The chitosan copper solution will form a spherical particle when dropped into a basic solution, where it is no longer soluble. The particles can then be cross-linked to increase their stability, then dried to form a truly solid particle. On a macro scale, millimeter size particles can be formed with reproducible sizes. For example, copper added to a 2% chitosan solution, dissolved in 1% acetic acid, pushed through a high gauged needle into 1 M NaOH will form uniform particles. When dried these particles have an average mass of 3.0 ± 0.3 mg for 10 particles.

On a smaller scale, microfluidic techniques can be used to make micro- and nano-sized particles. Using a flow focusing device, in a method adapted from Sugaya et al. (2013),

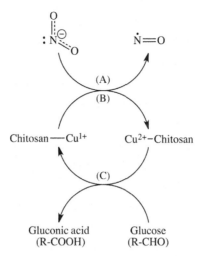

Figure 3.4 Catalytic cycle of copper chitosan: (A) nitric oxide reduced from nitrite; via (B) the oxidation copper chitosan particles; (C) copper being reduced by the oxidation of glucose.

consistent conditions can result in constantly sized microparticles, such as the particles in Fig. 3.5A and B that have an average size of around 3 microns.

We have demonstrated that the Cu–chitosan particles generated NO in the presence of nitrite and glucose. Authentic NO production was detected by using a Sievers Nitric Oxide Analyzer (NOA 280i). In preliminary experiments, mili-scale copper chitosan particles show promising results (Fig. 3.5). These experiments showed that chitosan particles lacking copper produced very little NO. Cu–chitosan produced ~140 nmol of NO per g of particle. The cross-linked particles produced much less NO (~25 nmol g^{-1} particle) (Fig. 3.5C). This can be explained by the fact that cross-linking ties up the chitosan free-amines that are chitosan-coordination sites for copper ions. While cross-linking is necessary for stability of the particles, the trend points to a decreased NO production. With copper-chitosan offering a wide application of topical possibilities it is a promising NO donor. It may be possible to fine-tune the release of NO by varying the degree of cross-linking to optimize particle stability whilst maintaining sufficient NO release to promote wound healing.

CONCLUSION

NO plays many roles in the human body, in many systems including the circulatory system, digestive system, immune system, and nervous system. It has influence in vasodilation, angiogenesis, cell proliferation, apoptosis, collagen deposition, and in prevention of bacterial invasion. A special look at copper-catalyzed NO production shows its

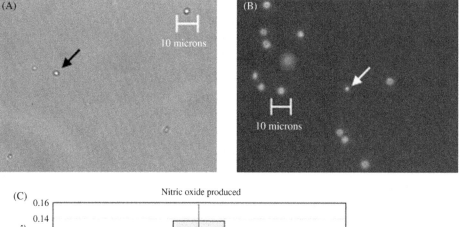

Figure 3.5 Copper chitosan microparticles and the detected NO produced from copper chitosan milliparticles. (A) Light microscope (DIC) images of copper microparticles; (B) fluorescent microscope image with dansyl chloride immobilized on the microparticles; (C) the NO detected by Sievers Nitric Oxide Analyzer (NOA 280i), from the gas produced after a 10-minute reaction period. The graph compares chitosan particles, copper noncross-linked particles, and copper cross-linked particles.

ability to produce NO while bound to a polymer NO donor carrier, chitosan. Copper chitosan particles have possible future applications as a topical applicator for wound healing; a film made of copper chitosan may have future applications in improving localized microcirculation.

REFERENCES

Acharya, S., Rogers, P., Krishnamoorthy, R.R., Stankowska, D.L., Dias, H.V., Yorio, T., 2016. Design and synthesis of novel hybrid sydnonimine and prodrug useful for glaucomatous optic neuropathy. Bioorg. Med. Chem. Lett. 26, 1490–1494.

Bath, P.M.W., Lisa Woodhouse, L., Scutt, P., Krishna, K., Wardlaw, J.M., Bereczki, D., et al., 2015. Efficacy of nitric oxide, with or without continuing antihypertensive treatment, for management of high blood pressure in acute stroke (ENOS): a partial–factorial randomised controlled trial. Lancet 385, 616–628.

Bath, P.M., Woodhouse, L., Krishnan, K., Anderson, C., Berge, E., Ford, G.A., et al., 2016. Effect of treatment delay, stroke type, and thrombolysis on the effect of glyceryl trinitrate, a nitric oxide donor, on outcome after acute stroke: a systematic review and meta-analysis of individual patient from randomised trials. Stroke Res. Treat., 2016, 1–12.

Bhatraju, P., Crawford, J., Hall, M., Lang, J.D.J., 2015. Inhaled nitric oxide: current clinical concepts. Nitric Oxide 50, 114–128.

Bice, J.S., Jones, B.R., Chamberlain, G.R., Baxter, G.F., 2016. Nitric oxide treatments as adjuncts to reperfusion in acute myocardial infarction: a systematic review of experimental and clinical studies. Basic Res. Cardiol. 111, 1–15.

Bykov, D., Neese, F., 2015. Six-electron reduction of nitrite to ammonia by cytochrome c nitrite reductase: insights from density functional theory studies. Inorg. Chem. 54, 9303–9316.

Carpenter, A.W., Schoenfisch, M.H., 2012. Nitric oxide release. Part II. Therapeutic applications. Chem. Soc. Rev. 41, 3742–3752.

Dutta, P.K., Dutta, J., Tripathi, V.S., 2004. Chitin and chitosan: chemistry, properties and applications. J. Sci. Ind. Res. 63, 20–31.

Frost, M.C., Reynolds, M.M., Meyerhoff, M.E., 2005. Polymers incorporating nitric oxide releasing/generating substances for improved biocompatibility of blood-contacting medical devices. Biomaterials 26, 1685–1693.

Garrya, P.S., Ezraa, M., Rowlanda, M.J., Westbrookb, J., Pattinson, K.T.S., 2015. The role of the nitric oxide pathway in brain injury and its treatment—from bench to bedside. Exp. Neurol. 263, 235–243.

Gothwal, A., Kesharwani, P., Gupta, U., Khan, I., Iqbal Mohd Amin, M.C., Banerjee, S., et al., 2015. Dendrimers as an effective nanocarrier in cardiovascular disease. Curr. Pharm. Des. 21, 4519–4526.

Halfen, J.A., Mahapatra, S., Wilkinson, E.C., Gengenbach, A.J., Young, V.G., Que, L., et al., 1998. Synthetic modeling of nitrite binding and activation by reduced copper proteins. characterization of copper(I)–nitrite complexes that evolve nitric oxide. J. Am. Chem. Soc. 118, 763–776.

Heuer, K., Hoffmanns, M.A., Demir, E., Baldus, S., Volkmar, C.M., RöHle, M., et al. 2015. The topical use of non-thermal dielectric barrier discharge (DBD): nitric oxide related effects on human skin, Nitric Oxide, 44, 52–60.

Ignarro, L.J., Burga, G.M., Wood, K.S., Byrns, R.E., Chaudhuri, G., 1987a. Endothelium-derived relaxing factor produced and released from artery and vein is nitric oxide. Proc. Natl. Acad. Sci. USA 84, 9265–9269.

Ignarro, L.J., Byrns, R.E., Burga, G.M., Wood, K.S., 1987b. Endothelium-derived relaxing factor from pulmonary artery and vein possesses pharmacologic and chemical properties identical to those of nitric oxide radical. Circ. Res. 61, 866–879.

Islam, M., Masum, S., Rahman, M., Molla, A.I., Shaikh, A.A., 2011. Preparation of chitosan from shrimp shell and investigation of its properties. Int. J. Basic Appl. Sci. 11, 77–80.

Jeandroz, S., Wipf, D., Stuehr, D.J., Lamattina, L., Melkonian, M., Tian, Z., et al., 2016. Occurrence, structure, and evolution of nitric oxide synthase-like proteins in the plant kingdom. Sci. Signal. 9, re2.

Kaiser, W.M., 2006. Nitric oxide production in plants. Plant Signal. Behav. 1, 46–51.

Keefer, L.K., 2003. Progress toward clinical application of the nitric oxide-releasing diazeniumdiolates. Annu. Rev. Pharmacol. Toxicol. 43, 585–607.

Kolios, G., Valatas, V., Ward, S.G., 2004. Nitric oxide in inflammatory bowel disease: a universal messenger in an unsolved puzzle. Immunology 113, 427–437.

Lee, W.J., Tateya, S., Cheng, A.M., Rizzo-Deleon, N., Wang, N.F., Handa, P., et al., 2015. M2 macrophage polarization mediates anti-inflammatory effects of endothelial nitric oxide signaling. Diabetes 64, 2836–2846.

Li, N., Bai, R., 2005. Copper adsorption on chitosan–cellulose hydrogel beads: behaviors and mechanisms. Separ. Purif. Technol. 42, 237–247.

Luo, J., Chen, A.F., 2005. Nitric oxide: a newly discovered function on wound healing. Acta Pharmacol. Sin. 26, 259–264.

Martinez, L.R., Han, G., Chacko, M., Mihu, M.R., Jacobson, M., Gialanella, P., et al., 2009. Antimicrobial and healing efficacy of sustained release nitric oxide nanoparticles against Staphylococcus aureus skin infection. J. Invest. Dermatol. 129, 2463–2469.

McCarthy, C.W., Guillory, R.J.I., Goldman, J., Frost, M.C., 2016. Transition metal mediated release of nitric oxide (NO) from S-nitroso-N-acetylpenicillamine (SNAP): potential applications for endogenous release of NO on the surface of stents via corrosion products. ACS Appl. Mater. Interf. 8, 10128–10135.

Misko, T.P., Schilling, R.J., Salvemini, D., Moore, W.M., Currie, M.G., 1993. A fluorometric assay for the measurement of nitrite in biological samples. Analyt. Biochem. 214, 11–16.

Ngah, W.S.W., Ab Ghani, S., Kamari, A., 2005. Adsorption behaviour of Fe(II) and Fe(III) ions in aqueous solution on chitosan and cross-linked chitosan beads. Biores. Technol. 96, 443–450.

Packer, L., Cadenas, E., 2005. Methods in enzymology. Nitric Oxide. In: Part, E. (Ed.), Illustrated. California Gulf Professional Publishing, San Diego, pp. 2005.

Paulo, M., Banin, T.M., de Andrade, F.A., Bendhack, L.M., 2014. Enhancing vascular relaxing effects of nitric oxide-donor ruthenium complexes. Future Med. Chem. 6, 825–838.

Percot, A., Viton, C., Domard, A., 2003. Optimization of chitin extraction from shrimp shells. Biomacromolecules 4, 12–18.

Rassaf, T., Ferdinandy, P., Schulz, R., 2013. Nitrite in organ protection. Br. J. Pharmacol. 171, 1–11.

Rath, M., Müller, I., Kropf, P., Closs, E.I., Munder, M., 2014. Metabolism via arginase or nitric oxide synthase: two competing arginine pathways in macrophages. Front. Immunol. 5, 532.

Reddy, B.P.K., Yadav, H.K.S., Nagesha, D.K., Raizaday, A., Karim, A., 2015. Polymeric micelles as novel carriers for poorly soluble drugs—a review. J. Nanosci. Nanotechnol. 16, 4009–4018.

Rinaudo, M., 2006. Chitin and chitosan: properties and applications. Progr. Polym. Sci. 31, 603–632.

Scallan, J.P., Hill, M.A., Davis, M.J., 2015. Lymphatic vascular integrity is disrupted in type 2 diabetes due to impaired nitric oxide signalling. Cardiovasc. Res. 107, 89–97.

Schäffer, M.R., Tantry, U., Gross, S.S., Wasserkrug, H.L., Barbul, A., 1996. Nitric oxide regulates wound healing. J. Surg. Res. 63, 237–240.

Schmuhl, R., Krieg, H.M., Keizer, K., 2001. Adsorption of Cu(II) and Cr(VI) ions by chitosan: kinetics and equilibrium studies. Water SA 27, 1–7.

Sogias, I.A., Khutoryanskiy, V.V., Williams, A.C., 2010. Exploring the factors affecting the solubility of chitosan in water. Macromol. Chem. Phys. 211, 426–433.

Sugaya, S., et al., 2013. Microfluidic production of single micrometer-sized hydrogel beads utilizing droplet dissolution in a polar solvent. Biomicrofluidics 7(5), 054120.

Tharanathan, R.N., Kittur, F.S., 2003. Chitin—the undisputed biomolecule of great potential. Crit. Rev. Food Sci. Nutr. 43, 61–87.

Treuer, A.V., Gonzalex, D.R., 2014. Nitric oxide synthases, S-nitrosylation and cardiovascular health: from molecular mechanisms to therapeutic opportunities (Review). Mol. Med. Rep. 11, 1555–1565.

Wang, Z., Lu, Y., Qin, K., Wu, Y., Tian, Y., Wang, J., et al., 2015. Enzyme-functionalized vascular grafts catalyze in situ release of nitric oxide from exogenous NO prodrug. J. Control. Release 210, 179–188.

Wasser, I.M., de Vries, S., Moënne-Loccoz, P., Schröder, I., Karlin, K.D., 2002. Nitric oxide in biological denitrification: Fe/Cu metalloenzyme and metal complex NOx redox chemistry. Chem. Rev. 102, 1201–1234.

Williams, S.B., Cusco, J.A., Roddy, M.-A., Johnstone, M.T., Creager, M.A., 1996. Impaired nitric oxide-mediated vasodilation in patients with non-insulin-dependent diabetes mellitus. Clin. Study 27, 567–574.

Willmota, M.R., Batha, P.M.W., 2003. The potential of nitric oxide therapeutics in stroke. Exp. Opin. Invest. Drugs 12, 455–470.

Witte, M.B., Barbul, A., 2002. Role of nitric oxide in wound repair. Am. J. Surg. 183, 406–412.

Wonoputri, V., Gunawan, C., Liu, S., Barraud, N., Yee, L.H., Lim, M., et al., 2015. Copper complex in poly(vinyl chloride) as a nitric oxide-generating catalyst for the control of nitrifying bacterial biofilms. ACS Appl. Mater. Interf. 7, 22148–22156.

Xu, L., Huang, Y.-A., Zhu, Q.-J., Ye, C., 2015. Chitosan in molecularly-imprinted polymers: current and future prospects. Int. J. Mol. Sci. 16, 18328–18347.

CHAPTER 4

Nitric Oxide Donors and Therapeutic Applications in Cancer

Khosrow Kashfi and Pascale L. Duvalsaint
Department of Physiology, Pharmacology, and Neuroscience, Sophie Davis School of Biomedical Education,
City University of New York School of Medicine, New York, NY, United States

INTRODUCTION

Nitric oxide (NO) is a free radical and one of the 10 smallest molecules found in nature. In 1992, the journal *Science* referred NO as the "Molecule of the Year" and in 1998 the Nobel Prize in Physiology and Medicine was awarded to Robert F. Furchgott, Louis J. Ignarro, and Ferid Murad for the major discoveries surrounding it. NO regulates a number of key biological functions, e.g., vascular relaxation (Ignarro et al., 1987), has antithrombolytic and antiinflammatory effects (Ignarro et al., 2002), is involved in neurotransmission, immune-response facilitation, and has a role in antipathogenic response (Geller and Billiar, 1998; Quinn et al., 1995; Bogdan, 2001). NO has a dichotomous role in cancer biology, with some reports suggesting that NO possesses antitumor properties, while others implicate NO in tumor promotion. For example, the production of endogenous NO is associated with apoptosis in cancer cells (Cui et al., 1994) characterized by upregulation of tumor suppressor p53, changes in expression of pro- and antiapoptotic Bcl-2 family members, activation of caspases, chromatin condensation, and DNA fragmentation (Brune et al., 1999). On the other hand, a large body of experimental and clinical data suggests a promoting role of NO in tumor progression and metastasis. Expression of iNOS has been reported in malignancies of the breast (Thomsen et al., 1995; Glynn et al., 2010; Basudhar et al., 2016), prostate (Klotz et al., 1998), lung (Masri et al., 2005), brain (Gallo et al., 1998), and colon (Ambs et al., 1998; Lagares-Garcia et al., 2001). This dual nature of NO has been attributed to low and high fluxes of NO (Wink et al., 1998; Ridnour et al., 2006; Xu et al., 2002; Miller and Megson, 2007; Choudhari et al., 2013; Vannini et al., 2015). The major sources of human exposure to NO are nitrovasodilators, which include nitroglycerine (glycerol trinitrate), amyl nitrite, isosorbide mono- and dinitrate, erythrityl tetranitrate, and sodium nitroprusside. These medications are taken sublingually, orally, or subcutaneously for the treatment of angina pectoris and other coronary artery diseases. Nitroglycerine has been used effectively for over 100 years, and the other organic medicinal nitrates have been available since the 1930s. In theory, these medications can inhibit or promote the development of cancer. For example, glyceryl trinitrate-induced apoptosis through activation caspases in three different colon cancer cell lines (Millet et al., 2002) and inhibited tumorigenesis in murine skin (Trikha et al., 2001). Further, isosorbite mononitrate and isosorbite dinitrate were shown to inhibit angiogenesis, tumor growth, and metastasis in mice (Pipili-Synetos et al., 1995). In contrast, feeding glyceryl trinitrate to F344 rats induced hepatocellular carcinomas that were characterized by K-ras point mutations (Tamano et al., 1996). Consistent with this finding is data showing elevated blood levels of nitrates and nitrites in patients with hepatocellular carcinoma (Notas et al., 2001).

Given the clinical importance and extensive use of NO-releasing vasodilators, the association between the commonly used NO-releasing nitrates and colorectal cancer risk was undertaken by using data from the Framingham Heart and Offspring studies (Muscat et al., 2005). The Framingham Heart study is a long-term multigenerational study, designed to identify genetic and environmental factors influencing the development of cardiovascular and other diseases (Dawber et al., 1963). Examination and testing of 5209 residents of Framingham, MA, was initiated in 1948 and with approximately 200 members of the original cohort currently alive and under follow-up. An offspring cohort and a third-generation cohort were subsequently added, followed more recently by two minority cohorts (Omni Group 1 and Omni Group 2) (Wei, 2015). Using these data, it was concluded that nitrovasodilators do not increase the risk of colorectal cancer (Muscat et al., 2005).

NO is synthesized and released intracellularly when L-arginine is oxidized to L-citruline by the enzyme nitric oxide synthase (NOS) (Moncada and Higgs, 1993; Ignarro, 1989) of which there are three isoforms (Fig. 4.1). Neuronal (nNOS also known as NOS1) and endothelial (eNOS also known as NOS3) are constitutive (cNOS) and are calcium-dependent forms of the enzyme that are regulated by negative feedback (Stuehr, 1997), phosphorylation (Dimmeler et al., 1999), and interaction with regulatory molecules (Sessa, 2004). cNOS releases low fluxes of NO over a short period of time, nanomolar concentrations for seconds or minutes, to regulate neural and vascular functions, respectively (Geller and Billiar, 1998; Alderton et al., 2001). The third isoform (iNOS also known as NOS2) is calcium independent, is regulated transcriptionally, and is induced by inflammatory cytokines, oxidative stress, hypoxia, and various endotoxins (Nathan and Xie, 1994; Kleinert et al., 2003). iNOS produces relatively large concentrations of NO, micromolar range and for longer intervals such as for hours or days (Michel and Feron, 1997; Kolios et al., 2004; Goligorsky et al., 2004) and is involved in immune surveillance.

CELLULAR ACTIONS OF NO

The chemical biology of NO is very diverse but can essentially be categorized as having two biological targets (Vasudevan and Thomas, 2014): (1) transition metals and (2) other free radicals (Fig. 4.2). These two reaction types tend to stabilize the

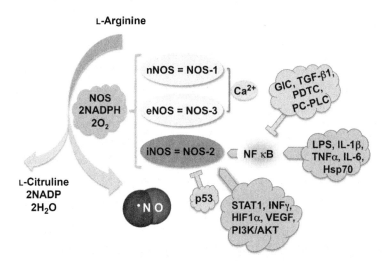

Figure 4.1 Biosynthesis of nitric oxide. NO is produced by three nitric oxide synthase (NOS) isoforms: neuronal, endothelial, and inducible (nNOS, eNOS, and iNOS) that catalyze the oxidation of L-arginine to L-citrulline. The transcription factor NF-κB is central to iNOS regulation. LPS, IL-1β, TNF-α, and IL-6 have been shown to induce iNOS, whereas glucocorticoids (GIC), transforming growth factor-β1 (TGFβ1), antioxidants (e.g., pyrrolidine dithiocarbamate, PDTC), and inhibitors of phosphatidylcho-line-specific phospholipase (PC-PLC) have been shown to inhibit iNOS expression by inhibiting NF-κB activation. iNOS expression is also directly induced by STAT1, INFγ, HIF1α, VEGF, PI3K/AKT, and inhibited by p53, reviewed in Vannini et al. (2015).

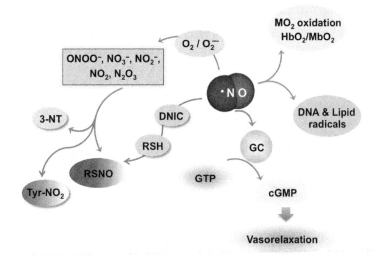

Figure 4.2 Proposed mechanism of action of NO. The biological effects of NO include direct effects where NO directly reacts with target molecules such as metal centers, DNA, and lipid radicals. Indirect effects are those where NO first reacts with other molecules such as O_2 or O_2^- to form nitrogen oxides that in turn go on to react with target molecules. Peroxynitrite ($ONOO^-$) can react with cysteine and/or tyrosine residues to form S-nitrosothiols (RSNO) and 3-nitrotyrosine (3-NT). RSNO can also be formed via N_2O_3 or through dinitrosyl iron complex (DNIC) which requires chelatable iron. NO reacts with the active site of soluble guanylate cyclase (sGC) and produces cyclic GMP leading to vasorelaxation.

unpaired electron on NO. The products of these two reaction types also fall into two categories: (*a*) indirect interactions of NO where other nitrogen oxides such as N_2O_3, NO_2, $ONOO^-$, NO_2^-, NO_3^- are formed first when NO reacts with molecular oxygen or superoxide anion (O_2^-) (Lancaster, 2003; Hill et al., 2010; Vasudevan and Thomas, 2014). These nitrogen oxides may in turn react with cellular targets; for example, peroxynitrite ($ONOO^-$) can react with cysteine and/or tyrosine residues to form *S*-nitrosothiols (RSNO) and 3-nitrotyrosine (3-NT), respectively (Beckman et al., 1990; Gow et al., 2004; Thomas et al., 2008). RSNO can also be formed via N_2O_3 (Broniowska and Hogg, 2012) or through dinitrosyl iron complex (DNIC) which requires chelatable iron (Bosworth et al., 2009). (*b*) Direct interactions where NO reacts with the active site of soluble guanylate cyclase (sGC) and produces cyclic GMP (cGMP). cGMP activates cGMP-dependent protein kinase G (PKG), which phosphorylates multiple substrates (Friebe and Koesling, 2003). The other two major downstream elements that may also be activated are cGMP-dependent gated ion channels and cGMP-dependent phosphodiesterase (Friebe and Koesling, 2003). In platelets, activation of the cGMP-dependent kinase phosphorylates a variety of substrates, and is involved in platelet adhesion and aggregation (Dangel et al., 2010). In general, an increase in cGMP leads to vasorelaxation of smooth muscles (Fig. 4.2) and decrease of platelet aggregation (Lincoln et al., 1996; Dangel et al., 2010).

Cancer as an Inflammatory Disease

The functional relationship between inflammation and cancer was first proposed by Rudolf Virchow in 1863, who noticed the presence of leukocytes in neoplastic tissues (Balkwill and Mantovani, 2001). Since then, a considerable body of evidence has supported the concept that tumors can originate at the sites of infection or chronic inflammation (Mueller and Fusenig, 2004). Acute inflammation is an adaptive host defense mechanism against infection or injury and is self-limiting, however, chronic inflammation may lead to various ailments including cancer (Schottenfeld and Beebe-Dimmer, 2006). For example, inflammatory diseases increase the risk of developing bladder, cervical, gastric, intestinal, esophageal, ovarian, prostate, and thyroid cancers (Mantovani et al., 2008). Hallmarks of cancer-related inflammation include the infiltration of white blood cells, tumor-associated macrophages, cytokines such as IL-1, IL-6, TNF-α, chemokines such as CCL2 and CXCL8, acceleration of cell cycle progression and cell proliferation, evasion from apoptosis, and stimulation of tumor angiogenesis (Colotta et al., 2009; Kundu and Surh, 2008). There are other links between inflammation and cancer such as tumors themselves secreting cytokines and chemokines (Ariztia et al., 2006) which lead to a positive feed forward loop between inflammation and cancer (Ben-Baruch, 2003); targeting the inflammatory mediators such as IL-1 (Voronov et al., 2003), TNF-α (Szlosarek and Balkwill, 2003), transcription factors such as NF-κB (Karin, 2006), and STAT3 (Yu et al., 2007) decreases the incidence and spread of cancer. Finally, regular use of nonsteroidal antiinflammatory drugs (NSAIDs) reduces the risk and mortality from a vast array of cancers (Kashfi, 2009).

NSAIDs PROTECT AGAINST CANCER, PROOF OF PRINCIPLE

Cancer prevention entered a new era upon observation that subjects using NSAIDs had lower incidence of colorectal cancer (Kune et al., 1988). Over 30 epidemiological studies, collectively describing results on greater than 1 million subjects, have established NSAIDs as the prototypical chemopreventive agents against many forms of cancer including colon (Thun et al., 1991; Benamouzig et al., 2003; Rothwell et al., 2012), breast (Takkouche et al., 2008; Yiannakopoulou, 2015; de Pedro et al., 2015), pancreas (Cui et al., 2014), prostate (Shebl et al., 2012), esophageal (Sun and Yu, 2011; Liao et al., 2012), head and neck (Becker et al., 2015), ovarian (Baandrup et al., 2015; Trabert et al., 2014), skin (Clouser et al., 2009; Elmets et al., 2010), and bladder (Daugherty et al., 2011; Nicastro et al., 2014). The most compelling case is against colorectal adenomas where three well-designed randomized, double-blind trials of aspirin established its chemopreventive effect (Baron et al., 2003; Benamouzig et al., 2003; Sandler et al., 2003). This chemoprevention at least conceptually was largely based on the effects of NSAIDs on the eicosanoid pathways, reviewed by Kashfi (2009) and Kashfi and Rigas (2005).

The general concept of cancer chemoprevention by NSAIDs cannot be overstated. However, all studies to date have failed to provide the details of the big picture: (1) which one of the nearly 30 NSAIDs is the most effective, (2) what is the optimal dose, and (3) what is the ideal schedule of administration? Even if such information existed, it would probably be of limited or no practical value. The reason is that, although we have unassailable proof of principle for chemoprevention, current NSAIDs cannot overcome two prohibitive limitations concerning their *safety* and *efficacy*. For example, for colon cancer, the one most thoroughly studied, NSAIDs can prevent at best 50% of the cases (Thun et al., 2002); all NSAIDs eventually cause some degree of gastrointestinal (GI) erosions which eventually may lead to ulcers, with most having cardiovascular (CV) and renal side effects. These considerations should be viewed against the fundamental distinction between chemotherapy and chemoprevention. In chemotherapy, it is accepted that substantial treatment-related toxicity exists to save the patient's life from a fully developed cancer. In contrast, chemopreventive agents will have to be prescribed to a healthy subject for a cancer that may never develop. Thus, for those subjects, safety and efficacy assumes a different value, and more demanding criteria than those applied to chemotherapy should be used.

SIDE EFFECTS ASSOCIATED WITH NSAID USE

Although the use of NSAIDs in general and aspirin in particular as a chemopreventive agent is highly convincing, what is absolutely clear is that aspirin may not be optimal for cancer prevention because of its shortcomings in safety and efficacy. The use of NSAIDs in general is limited by their significant toxicity, which includes (1) gastrointestinal, ranging from dyspepsia to gastrointestinal bleeding, obstruction, and perforation, (2) renal, and (3) cardiovascular, reviewed by Kashfi (2009). It is estimated that about 16,500 NSAID-related deaths occur among patients with rheumatoid arthritis or osteoarthritis every year in the United States. This figure is greater than the number of deaths from multiple myeloma, asthma, cervical cancer, or Hodgkin's disease. If deaths from gastrointestinal toxic effects from NSAIDs were tabulated separately in the National Vital Statistics reports, these effects would constitute the 15th most common cause of death in the United States (Wolfe et al., 1999). Yet these toxic effects remain mainly a "silent epidemic," with many physicians and most patients unaware of the magnitude of the problem.

NO-Based Compounds as Cancer Therapeutics

Given the overall roles of NO in various cancers, it is not surprising that NO has been exploited as an anticancer target. Different approaches include NO as a radiotherapy and chemotherapy sensitizer, NO-releasing drugs, and NOS inhibitors (Aranda et al., 2012). Tumors often have poorly developed vasculature, thus administered drugs may not reach their intended target. NO donors can enhance tumor blood flow and oxygen supply, thereby resensitizing it to chemotherapeutic agents or radiation therapy (Wink et al., 2008; Ning et al., 2012; Oronsky et al., 2016). NOS inhibitors, such as L–NAME, L-NIL, and 1400W, have been evaluated in several preclinical in vivo animal models for their antitumorigenic properties by blocking iNOS- and eNOS-derived NO production (Bian et al., 2012; Burke et al., 2013; Vasudevan and Thomas, 2014). There are also whole spectrums of synthetic "designer" NO-releasing compounds that are at various stages of development: some are preclinical, some are in different phases of clinical trials, and some have been abandoned for various reasons. While NOS inhibitors and NO donors have opposing effects, they can both serve as anticancer agents, as it has often been said, "the devil is in the details." Below, we have focused our review on some of the promising NO-releasing compounds that have utility against various cancers.

NITRIC OXIDE-RELEASING NSAIDs (NO-NSAIDs) AND THE RATIONALE FOR THEIR DEVELOPMENT

NO-NSAIDs also known as COX-inhibiting nitric oxide donators (CINODs) were developed to reduce the potential side effects associated with traditional NSAIDs (Wallace et al., 2002; Wallace and Soldato, 2003). The rationale for their development was largely based on the observations that in the GI system, NO can enhance the local mucosal defense mechanisms thus compensating for the decreases in prostaglandins

(PGs) that come about as a result of cyclooxygenase (COX) inhibition following an NSAID consumption. For example, NO increases blood flow, which reduces the effects of luminal irritants, and it also increases mucus and bicarbonate secretions (Wallace and Miller, 2000). Therefore, coupling an NO-releasing moiety to an NSAID might deliver NO to the site of NSAID-induced damage, thereby decreasing gastric toxicity. This strategy has been quite successful, with reports indicating that NO-NSAIDs are indeed safer than their corresponding parent NSAID to the GI mucosa of animals (Davies et al., 1997; Wallace et al., 1994a, 1994b; Elliott et al., 1995; Borhade et al., 2012; Gund et al., 2014; Nemmani et al., 2009; Pathan et al., 2010) and humans (Fiorucci et al., 2003a, 2003b). What is striking is that although NO-NSAIDs are GI safe, they still inhibit the formation of COX-1- and COX-2-derived PGs in vitro and in vivo (Cirino et al., 1996; Fiorucci et al., 1999, 2003a; Wallace et al., 1995).

STRUCTURAL FEATURES OF NITRATE NO-NSAIDs

NO-NSAIDs are traditional NSAIDs linked to an NO-releasing group via a chemical spacer. The three key structural components of this class of NO-NSAID are: the traditional NSAID moiety; the spacer, which can be either aliphatic or aromatic; and the NO-releasing group (a nitrate ester in this case) (Fig. 4.3).

All NO-NSAIDs tested were more potent in inhibiting the growth of various human cancer cell lines compared to their traditional counterparts (Kashfi et al., 2002). Aromatic spacers generated more potent NSAID derivatives than aliphatic ones (Kashfi et al., 2005; Nath et al., 2013), and positional isomerism greatly influenced all cell kinetic parameters that influence cellular mass. For example, the *ortho* and *para* positional isomers of NO-ASA were significantly more potent than the *meta* isomer in inhibiting the growth of HT-29 human colon cancer cells, inducing apoptosis, and inhibiting proliferation (Kashfi et al., 2005). A summary cartoon depicting the actions of NO-NSAIDs is shown in Fig. 4.4.

Second-Generation NO-NSAIDs

A second generation of NO-releasing aspirins soon followed the nitrate esters where use of furoxan derivatives was made as NO donors (Cena et al., 2003) (Fig. 4.5A). Unlike the nitrate esters which required enzymatic metabolism for NO release (Grosser and Schroder, 2000; Carini et al., 2002; Gao et al., 2005a), the furoxan–NSAID hybrids released NO in the presence of plasma, GSH, or albumin, i.e., through thiols mechanisms (Turnbull et al., 2006). These agents significantly inhibited COX-1 activity, reduced TNFα release from LPS-treated macrophages, and inhibited NF-κB activation (Turnbull et al., 2006, 2008). Another class of NO-releasing "aspirin-like" compounds has also been described in which the acetyl group on the aspirin has been replaced by acyl groups containing nitroxy NO-releasing moieties (Fig. 4.5B and C).

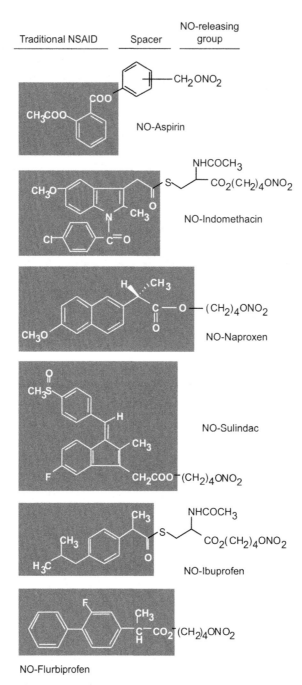

Figure 4.3 The chemical structures of NO-NSAIDs. The traditional NSAID is shown in the shaded box, the spacer molecule links the traditional NSAID to $-ONO_2$ which can release NO. *Reproduced with permission from Kashfi, K., Anti-inflammatory agents as cancer therapeutics. Adv. Pharmacol. 57: 31–89, 2009. Elsevier Books.*

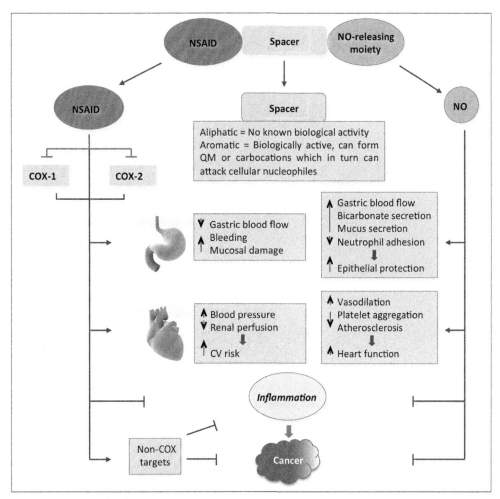

Figure 4.4 Mechanisms of action of NO-releasing NSAIDs. When hydrolyzed, NO-releasing NSAIDs produce the parent NSAID, the spacer, and the NO-releasing moiety from which NO is released. The NSAID component inhibits COX-1 and COX-2 resulting in compromised mucosal defense mechanisms, which may lead to ulcers. NSAIDs can reduce renal perfusion, which can lead to increases in blood pressure (BP) leading to cardiovascular (CV) damage. The released NO counteracts many of the detrimental effects of NSAIDs. These protective effects appear to be mediated through the cGMP pathway. NO enhances the mucosal defense mechanisms, causes vasodilation thus reducing BP leading to cardioprotective effects. Both the NSAID and NO have antiinflammatory effects, the former through inhibition of COX and latter through modulation of NF-κB, collectively these contribute to reduction in tumor mass.

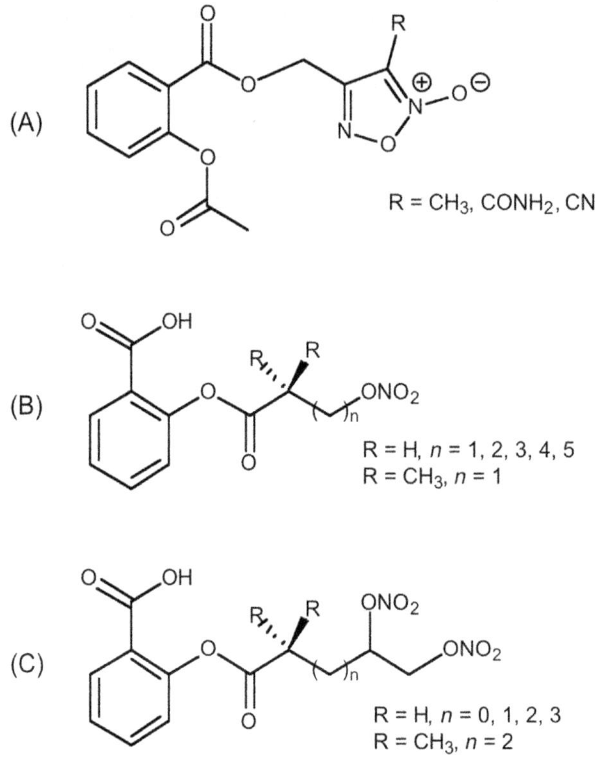

Figure 4.5 NO donor "Aspirin-like" compounds. A second generation of NO-releasing aspirin in which a furoxan derivative is the NO donor (A). In the "Aspirin-like" compounds, the acetyl group on the aspirin has been replaced by acyl groups containing nitroxy NO-releasing moieties (B and C). *Reproduced with permission from Kashfi, K., Anti-inflammatory agents as cancer therapeutics. Adv. Pharmacol. 57: 31–89, 2009. Elsevier Books.*

These compounds have reduced GI toxicity compared to aspirin and have exhibited strong antiinflammatory properties in the carrageenan-induced rat paw edema model (Lazzarato et al., 2008).

NO-RELEASING COXIBS

Selective cyclooxygenase-2 inhibitors (coxibs) such as celecoxib, rofecoxib, and valdecoxib were developed to overcome the GI safety concerns of traditional nonselective NSAIDs. In patients with a low risk for GI damage, coxibs have been successful in limiting the upper GI ulceration associated with traditional NSAIDs, reviewed by Solomon (2008). However, complete inhibition of COX-2 can result in downregulation of the beneficial vasodilator prostacyclin (PGI_2) with the potent effects of the platelet aggregator thromboxane A_2 unaffected. This eicosanoid imbalance can lead

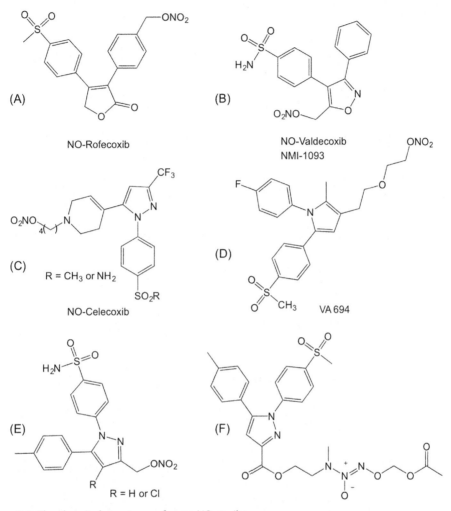

(A)

NO-Rofecoxib

(B)

NO-Valdecoxib
NMI-1093

(C)

R = CH₃ or NH₂

NO-Celecoxib

(D)

VA 694

(E)

R = H or Cl

(F)

Figure 4.6 The chemical structures of some NO-coxibs.

to an increased incidence of CV events (Funk and Fitzgerald, 2007). In fact, several large-scale clinical trials have shown that long-term use of coxibs and even traditional NSAIDs is associated with an increased risk of adverse myocardial events, reviewed by Kashfi and Rigas (2005) and Kashfi (2009). NO, like PGI_2, plays an important cyto-protective role in GI homeostasis by helping to enhance the local mucosal defense mechanisms. NO could also induce peripheral vasodilation to circumvent the eleva-tion in blood pressure exhibited by coxibs. As alluded to earlier, NO also inhibits both platelet aggregation and adhesion. Therefore, coxibs that release NO should exhibit a safer CV profile. Indeed recent data is inline with this thinking. Some examples of NO-releasing coxibs are shown in Fig. 4.6. These include NO-rofecoxib (Boschi et al.,

2010), NO-valdecoxib (Dhawan et al., 2005), NO-celecoxib (Chowdhury et al., 2010), VA 694 (Martelli et al., 2013) shown in Fig. 4.6A–D, respectively, and some others such as (pyrazoyl)benzenesulfonamides as derivatives of celecoxib (Bechmann et al., 2015) (Fig. 4.6E), and a diazen-1-ium-1,2-diolate (Abdellatif et al., 2008) (Fig. 4.6F, an example of a NONO-coxib). Recently this field has expanded rapidly as evidenced by the number of NO-coxibs reported (Abdellatif et al., 2010, 2015, 2016a, 2016b; Basudhar et al., 2015; Bakr et al., 2016; Biava et al., 2014; Gouda et al., 2014). Overall these agents are selective COX-2 inhibitors that release low to moderate levels of NO, have good antiinflammatory properties, suppressing hyperalgesia and edema. Some reports also address the CV beneficial effects of these agents. For example, in the vascular smooth muscle VA 694 exhibited NO-mediated relaxant effects and its chronic oral administration to young spontaneously hypertensive rats (SHRs) significantly slowed down the age-related development of hypertension which was associated with increased plasma levels of nitrates, stable end-metabolites of NO. Furthermore, it also caused a significant improvement of coronary flow and a reduction of endothelial dysfunction (Martelli et al., 2013). Reports are also emerging showing the anticancer properties of NO-coxibs. A series of novel pyrrolizine derivatives inhibited the growth of breast cancer MCF-7 and prostate cancer PC-3 cell lines, causing apoptosis as indicated by their ability to activate caspases-3/7 (Gouda et al., 2014).

CANCER PREVENTION WITH NO-NSAIDs

Preclinical studies have shown that NO-NSAIDs have chemopreventive properties. For example, NO-ASA, NO-sulindac, NO-ibuprofen, NO-indomethacin, NO-flurbiprofen, and NO-salicylic acid all displayed greater potency than their corresponding parent NSAID in inhibiting the growth of various cancer cells such as colon, prostate, lung, pancreas, skin, breast, multiple myeloma, and leukemia (Williams et al., 2001; Kashfi et al., 2002; Yeh et al., 2004; Huguenin et al., 2005; Tesei et al., 2005; Nath et al., 2005, 2009b, 2013; Khan et al., 2012; Chaudhary et al., 2013; Fionda et al., 2015). These studies underscored three general features of the NO-NSAID molecules: (1) NO enhanced the potency of each NSAID in inhibiting cancer cell growth; (2) this enhanced potency was seen in cancers of various tissue origin; and (3) NO-ASA was consistently the most potent NO-NSAID. What also became apparent was that the potency of an NSAID in inhibiting cell growth did not predict the potency of the corresponding NO-NSAID. For example, in comparing the growth inhibitory properties of ASA, sulindac, and indomethacin in a number of different cell lines, ASA was consistently the weakest of the three traditional NSAIDs, however, NO-ASA had the highest potency of all three NO derivatives, the magnitude of this exceeded those of the other NO-NSAIDs often by over 100-fold (Kashfi et al., 2002). The fact that nitration of different NSAIDs does not lead to the same cell growth inhibition, underscores the complexity of their effects on cell growth. The fact that the spacer molecule is not the same in all the NO-NSAIDs makes any efforts to define an underlying common mechanism difficult. The only shared structural feature amongst NO-NSAIDs is the NO-releasing moiety, in these cases that being a nitrate, the NSAID component is a general pharmacological function and not a structural feature. This property, however, cannot account for the enhanced potency of the NO-NSAIDs.

The growth inhibitory effects of NO-NSAIDs are due to enhanced cell death, reduced cell proliferation, and inhibition of cell cycle phase transitions. Of the NO-NSAIDs evaluated, NO-ASA was the only one that caused atypical cell death, e.g., necrosis (Kashfi et al., 2002, 2005) and this may also occur in vivo as well (Ouyang et al., 2006b).

The first study comparing the chemopreventive effects of ASA to that of *m*-NO-ASA (NCX-4016) was reported in 1998 (Bak et al., 1998). Using a rat model of colonic adenocarcinoma, the number of aberrant crypt foci (an early preneoplastic lesion) were counted after 4 weeks of treatment. Aspirin reduced the number of aberrant crypt foci by 64%, while *m*-NO-ASA produced an 85% reduction. The effect of *p*-NO-ASA (NCX 4040) was also evaluated in a *Min* (APC$^{Min/+}$) mouse model of intestinal cancer in which a truncating mutation in the Apc gene predisposes the mice to spontaneously develop intestinal tumors (Lipkin et al., 1999). In this study, tumor burden was used as an end point. Three weeks of treatment with *p*-NO-ASA had no effect on the animal weight or the GI integrity and there were no overt signs of toxicity. *p*-NO-ASA reduced tumor growth by 55% but it did not affect cell proliferation in small intestinal mucosa (Williams et al., 2004). Given that tumors had already formed when *p*-NO-ASA was administered, it may be concluded that this compound has chemotherapeutic as well as chemopreventive properties. In another *Min* mice study, *p*-NO-ASA was more efficacious than *m*-NO-ASA in reducing the number of intestinal tumors (Kashfi et al., 2005), a finding consistent with the potency ranking of these two positional isomers based on their effects on cultured cells (Williams et al., 2001; Kashfi et al., 2002, 2005; Yeh et al., 2004; Huguenin et al., 2005; Tesei et al., 2005; Nath et al., 2005, 2009b, 2013; Khan et al., 2012; Chaudhary et al., 2013; Fionda et al., 2015). In F344 rats treated with the carcinogen, azoxymethane to induce colon cancers, NO-indomethacin and *m*-NO-ASA significantly suppressed both tumor incidence and multiplicity, with NO-indomethacin being more effective (Rao et al., 2006). In a series of in vitro studies using various human colon cancer cell lines, 5-flurouracil (5-FU) or oxaliplatin, both of which are used for the clinical management of colorectal cancer (Meyerhardt and Mayer, 2005) had additive effects when combined with *p*-NO-ASA (Leonetti et al., 2006). Using colon cancer xenografts, simultaneous treatment with *p*-NO-ASA and oxaliplatin or oxaliplatin followed by *p*-NO-ASA showed the same additive effects. However, treatment with *p*-NO-ASA followed by oxaliplatin showed synergistic interactions possibly by sensitizing the cancer cells to the cytotoxic effects of the antitumor drug (Leonetti et al., 2006). Using Syrian golden hamsters and the carcinogen N-nitrosobis(2-oxopropyl)amine (BOP) to establish pancreatic cancer, compared to controls, m-NO-ASA reduced the incidence and multiplicity of pancreatic cancer by 88.9% and 94%, respectively, while ASA had no effect (Ouyang et al., 2006b). More recently, NO-naproxen at human equivalent doses (HED) was shown to be highly effective in reducing urinary bladder cancer using the hydroxybutyl(butyl) nitrosamine (OH-BBN) model, whereas NO-sulindac was inactive (Nicastro et al., 2014).

Molecular Targets of NO-ASA in Cancer

Apart from its effects on cell kinetics which determine cellular mass, NO-ASA has multiple pleiotropic effects that involve NF-κB, Wnt, NOS, mitogen-activated protein kinase (MAPK), COX-2, PPAR, drug metabolizing enzymes, reactive oxygen species, pro- and antiinflammatory cytokines, and S-nitrosylation, as briefly discussed below.

Cell Kinetics

NO-ASA inhibits cell proliferation (diminished expression of the proliferation marker PCNA) and enhances apoptosis, thus modulating cell renewal and cell death which are two determinants of cell growth. It also blocks transitions of the cells through the cell cycle, either between G_1 and S (Williams et al., 2001; Kashfi et al., 2002, 2005), G_2/M (Nath et al., 2009b), or G_0/G_1 (Nath et al., 2013).

NF-κB

p-NO-ASA profoundly affects the NF-κB–DNA interaction in cultured colon cancer cells decreasing it as early as an hour after treatment, similar results were obtained with pancreatic cancer cell lines (Williams et al., 2003, 2008). p-NO-ASA was also shown to suppress NF-κB activity in pre-B acute lymphoblastic leukemia (ALL) cell lines (Khan et al., 2012). Recently it was shown that activation of NF-κB was inhibited by both m-NO-ASA and p-NO-ASA isomers as demonstrated by decreases in NF-κB-DNA binding and luciferase activity at 24 hours. However, m-NO-ASA produced transient effects at 3 hours such as increased NF-κB–DNA-binding, increased levels of nuclear p50, even though both isomers inhibited IκB degradation. Increase in nuclear p50 by m-NO-ASA was associated with translocation of p50 into the nucleus as observed by immunofluorescence at 3 hours. In vivo, compared to controls, p-NO-ASA decreased NF-κB activation in xenografts of estrogen receptor negative, ER(−), breast cancer cells (Nath et al., 2015), and in intestinal epithelial cells of APC$^{min+/−}$ mice (Williams et al., 2008). p-NO-ASA can form a quinone methide intermediate (see Biological Actions of the Spacer section) and this has been shown to target the components of canonical NF-κB signaling, leading to the inhibition of NF-κB signaling (Pierce et al., 2016).

Wnt/β-Catenin Signaling

p-NO-ASA modulated the β-catenin signaling pathway in three different cancer cell lines of different origin (Gao et al., 2005b; Nath et al., 2003, 2005, 2009b). One of the significant downstream genes dependent on β-catenin/TCF-4 signaling is *Cyclin D1*, which has been implicated in carcinogenesis (Moon et al., 2004). The inhibition of this signaling pathway by low concentrations of *p*-NO-ASA, which was associated with reduced Cyclin D1 expression, suggests that it may represent an important disruptor in the carcinogenic process. Flurbiprofen benzyl nitrate (NBS-242, *p*-NO-flurbiprofen) cleaved β-catenin both in the cytoplasm and the nucleus of the human skin cancer cell line A-431, targeting cyclin D1 through caspase-3 activation (Nath et al., 2013). In an in vivo study, NO-indomethacin and *m*-NO-ASA inhibited β-catenin expression in the standard azoxymethane (AOM) rat model of colorectal cancer (Rao et al., 2006).

Nitric Oxide Synthase

iNOS may have an important role in tumor development since its increased expression has been reported in breast (Thomsen et al., 1995; Glynn et al., 2010; Basudhar et al., 2016), CNS (Cobbs et al., 1995), pancreas (Kasper et al., 2004), astrocytic gliomas (Hara and Okayasu, 2004), prostate (Aaltoma et al., 2001), acute myeloid leukemia (Brandao et al., 2001), lung (Masri et al., 2005), and colon (Ambs et al., 1998; Lagares-Garcia et al., 2001) tumors. iNOS is involved in the regulation of COX-2 (Landino et al., 1996; Perez-Sala and Lamas, 2001; Chaudhary et al., 2013) and its activity may be correlated with p53 mutations (Chiarugi et al., 1998). Importantly, the selective iNOS inhibitors *S,S'*-1,4-phenylene-bis(1,2-ethanediyl)bis-isothiourea (PBIT) (Rao et al., 1999) and ONO-1714 ((1S,5S,6R,7R)-7-chloro-3-imino-5-methyl-2-azabicyclo[4.1.0] heptane hydrochloride) (Takahashi et al., 2006) prevented colon cancer. Also, NOS inhibitors such as L-NAME, L-NIL, and 1400W have been evaluated in several preclinical in vivo animal models for their antitumorigenic properties by blocking iNOS- and eNOS-derived NO production (Bian et al., 2012; Burke et al., 2013; Vasudevan and Thomas, 2014). Various NO-NSAIDs inhibit the induction of iNOS in a macrophage cell line (Cirino et al., 1996). Studies with HT-29 colon cancer cells have shown that *p*-NO-ASA inhibits iNOS induction by cytokines (Williams et al., 2003). *p*-NO-ASA also potently inhibits the expression and enzymatic activity of iNOS in HT-29 cells (Spiegel et al., 2005). An NO chimera (GT-094), a nitrate containing an NSAID and disulfide pharmacophores, reduced iNOS levels in the AOM rat model of colorectal cancer (Hagos et al., 2007).

COX-2

The role of COX-2 and its inhibition in colon and other cancers has been the subject of considerable debate (Soh and Weinstein, 2003; Kashfi and Rigas, 2005; Kashfi, 2009). NO-NSAIDs in general and NO-ASA in particular were more potent than their corresponding traditional NSAIDs in inhibiting the growth of cultured HT-29 (expressing both COX-1 and COX-2) and HCT 15 (COX null) colon cancer cells (Williams et al., 2001; Yeh et al., 2004). Similar observations were made with the pancreatic cancer cell lines BxPC-3 (COX-1, COX-2 positive) and MIA PaCa-2 (COX null) (Kashfi et al., 2002). This raises a very important and interesting question about the role of COX in cancer, since NO-ASA was eqieffective and eqipotent in both COX positive and COX negative cell lines. In HT-29 cells, p-NO-ASA at concentrations around its IC_{50} value for growth inhibition increased COX-2 expression by nearly 9-fold. The induced enzyme was catalytically active (Williams et al., 2003). Similar findings were obtained with the DLD-1 colon cancer, BxPC-3 pancreatic cancer (Williams et al., 2003), and MCF-7 breast cancer cell lines (Nath et al., 2009). However, NO-indomethacin and m-NO-ASA also inhibited total COX including COX-2 activity and formation of PGE_2 in the AOM rat model of colon cancer (Rao et al., 2006). These results give caution in extrapolating cell culture data to in vivo. The mechanism(s) by which NO-ASA induces COX-2 is unclear, but may in part be PKC dependent (Nath et al., 2009).

Peroxisome Proliferator-Activated Receptor δ (PPARδ)

PPARδ mRNA was shown to be upregulated in colorectal carcinomas and endogenous PPARδ was transcriptionally responsive to PGI_2 (Gupta et al., 2004). This elevation of PPARδ expression in colorectal cancer cells was repressed by APC an effect mediated by β-catenin/Tcf-4-responsive elements in the PPARδ promotor (He et al., 1999). Sulindac blocked PPARδ from binding to its recognition sequences (He et al., 1999), and in SW480 colon cancer cells, sulindac sulfone decreased PPARδ expression more potently than the sulfide metabolite (Siezen et al., 2006). Taken together, these data suggest that NSAIDs may in part inhibit tumorigenesis through inhibition of PPARδ. In *Min* mice, m- and p-NO-ASAs inhibited the expression of PPARδ in both histologically normal and tumor tissues. m-NO-ASA suppressed PPARδ expression in normal mucosa by 23% and in neoplastic tissue by 41%; p-NO-ASA suppressed PPARδ expression in normal mucosa by 27% and in neoplastic tissue by 55% (Ouyang et al., 2006a).

Mitogen-Activated Protein Kinase (MAPK)

The MAPKs are a family of kinases that transduce signals from the cell membrane to the nucleus in response to a variety of stimuli modulating gene transcription and leading to biological response (Bode and Dong, 2004). MAPKs are required for controlling cell proliferation, differentiation, and death and their deregulation may lead to pathogenesis of many malignancies, including colon cancer. p-NO-aspirin treatment of colon cancer cells (HT-29 and SW480) activated JNK and p38 along with their respective downstream transcription factors, cJun and ATF-2. NO-ASA stimulation of p38 was biphasic, with an initial increase in phosphorylation within the first hour of treatment, and a second much stronger increase at 4 hours (Hundley and Rigas, 2006). Using UVB-induced skin tumors in SKH-1 hairless mice, NO–sulindac was shown to be chemopreventive, reducing levels of phosphorylated MAPKs Erk1/2, p38, and NJK1/2 (Chaudhary et al., 2013).

Xenobiotic Metabolizing Enzymes

Modulation of drug metabolizing enzymes, leading to elimination of endogenous and environmental carcinogens represents a successful strategy for cancer chemoprevention (Kwak et al., 2004). For example, dithiolethiones that induce phase II metabolizing enzymes inhibit tumorigenesis caused by environmental carcinogens in various animal models and in clinical trials, modulate the metabolism of the carcinogen, aflatoxin B1 (Kwak et al., 2004). In general, the induction of phase II enzymes is an adequate strategy for protecting mammals and their cells against carcinogens and other forms of electrophile and oxidants. Chemopreventive agents induce the expression of phase II genes through their effects on the Keap1–Nrf2 complex (Lee and Surh, 2005). In the nucleus, the transcription factor Nrf2, a member of the NF-E2 family, dimerizes with the small Maf protein and binds to the antioxidant response element (ARE), a *cis*-acting regulatory element in the promoter region of phase II enzymes. A cytoplasmic actin-binding protein, Keap1, is an inhibitor of Nrf2 that sequesters it in the cytoplasm. Inducers dissociate this complex, allowing Nrf2 to translocate to the nucleus. Studies evaluating the effects of NO-ASA on xenobiotic metabolizing enzymes have shown that *m*-NO-ASA induced the activity and expression of NAD(P)H:quinone oxireductase (NQO) and glutathione *S*-transferases (GSTs) in mouse hepatoma Hepa 1c1c7 and HT-29 human colon cancer cells (Gao et al., 2006). In *Min* mice, *m*-NO-ASA also induced the activities of NQO and GST in liver cytosolic and small intestine fractions but had no effect on the activity of these enzymes in the kidney, showing some degree of tissue of specificity (Gao et al., 2006). Expression of GST P1-1, GST A1-1, and NQO1 was induced in liver cytosols from *Min* mice, however, the expression of two phase I metabolizing enzymes, CYP450-1A1 and CYP450-2E1, was unaffected, suggesting that *m*-NO-ASA is a monofunctional inducer of phase II enzymes (Ouyang et al., 2006b). *m*-NO-ASA also induced the translocation of Nrf2 into the nucleus, an effect that paralleled the induction of NQO1 and GST P1-1 (Gao et al., 2006). Interestingly, *p*-NO-ASA can form a quinone methide intermediate (see Biological Actions of the Spacer section) and this has been shown to activate Nrf2 via covalent modification of Keap 1 (Dunlap et al., 2012) thus questioning the role of NO.

Oxidative Stress

Anticancer agents as part of their mechanism of action may induce reactive oxygen and nitrogen species (RONS). At low concentrations, RONS may protect the cell and at high concentrations can initiate biological damage, including cell death (Rigas and Sun, 2008). In evaluating the effects of p-NO-ASA in SW480 colon cancer cells, it was determined that the spacer in p-NO-ASA formed a conjugate with glutathione, depleting glutathione stores and thus induced a state of oxidative stress that led to apoptosis via activation of the intrinsic apoptosis pathway (Gao et al., 2005b). p-NO-ASA through induction of RONS also oxidized thioredoxin-1 (Trx-1) (Rigas and Sun, 2008), an oxidoreductase that is involved in redox regulation of cell signaling (Arner and Holmgren, 2006; Maulik and Das, 2008).

Modulation of Proinflammatory Cytokines

m-NO-ASA inhibits cytokine production from endotoxin-stimulated human monocytes and macrophages (Fiorucci et al., 2000a) and when administered to mice, decreased IL-1β, IL-8, IL-12, IL-18, TNF-α, and INFγ production and protected against concanavallin A-induced acute hepatitis (Fiorucci et al., 2000b). The effect exerted by NO-ASA on cytokine production was COX independent (Santucci et al., 1995). Locally generated NO contributes to limit inflammation by inhibiting generation of proinflammatory cytokines and/or by enhancing the production of antiinflammatory cytokines, IL-10 and TGFβ, resulting in downregulation of downstream mediators of inflammation such as COX and iNOS (Fiorucci et al., 2000b, 2001).

S-Nitrosylation

Posttranslational modifications of proteins are central molecular mechanisms that mediate signal transductions. One such modification is S-nitrosylation which involves the addition of NO to a cysteine thiol to form an S-nitroso-protein (SNO-protein) (Stamler et al., 1992). To put this in perspective, S-nitrosylation may be comparable to that of phosphorylation and ubiquitylation, which are critical cellular regulatory mechanisms (Lane et al., 2001; Lopez-Otin and Hunter, 2010). S-nitrosylation by nitrosative stress can regulate cellular homeostasis in order to maintain the balance between the induction (Son and Kim, 1995) and prevention of apoptosis (Li et al., 1997). NO can also modulate NF-κB activity at multiple steps in the signal transduction pathway via S-nitrosylation (Marshall et al., 2000) with several different NF-κB proteins including IκB, kinaseB, and p50 and p65 being regulated by this posttranslational modification (Reynaert et al., 2004). m-NO-ASA and NO-naproxen were shown to S-nitrosylate NF-κB both in vitro and in vivo, leading to a reduction in NF-κB protein levels (Chattopadhyay et al., 2010a). In HT-29 human cancer cells, m-NO-ASA and NO-naproxen S-nitrosylated NF-κB p65 in a time-dependent manner. This effect was NO specific as pretreatment of the cells with carboxy-PTIO, an NO scavenger, abrogated the increase in S-NO-NF-κB p65 protein levels (Chattopadhyay et al., 2010a; Kashfi, 2012).

BIOLOGICAL ACTIONS OF THE SPACER IN NO-ASA

The defining entity in all NO-NSAIDs has been the NO. Structure–activity studies with NO-ASA indicated that NO was pivotal for its anticancer effects (Kashfi and Rigas, 2005). However, careful re-examination regarding the contribution to the overall biological effect of each of the three structural components of NO-ASA in which the spacer joining the ASA to the NO-releasing moiety was aromatic, led to the surprising conclusion that the NO-releasing moiety was not required for the observed biological effects. Rather, the spacer was responsible for the biological actions of NO-ASA, with the NO-releasing moiety acting as a leaving group that facilitated the release and activation of the spacer to a quinone methide (QM) intermediate. Thus, o-NO-ASA and p-NO-ASA led to the formation of o-QM and p-QM intermediates, respectively. These QMs acted as powerful electrophiles such that the ASA component had little or no biological contribution (Dunlap et al., 2007; Hulsman et al., 2007; Kashfi and Rigas, 2007). On this basis, a series of o, p, and m ester-protected hydroxy benzyl phosphates (EHBPs) were synthesized in which the—ONO_2 leaving group was replaced by a substituted phosphate and the ASA was replaced by an acetate. Electron donating/withdrawing groups were also incorporated around the spacer to evaluate their effect on QM formation/stability and biological activity (Kodela et al., 2008a, 2011). EHBPs inhibit the growth of various human cancer cell lines, indicating an effect independent of tissue type exercising pleotropic effects involving cell death as well as cell cycle phase transitions and that a QM if formed is influenced by the nature of the substitutes about the benzyl spacer (Kodela et al., 2008a). Transient QMs and related electrophiles if formed during the metabolic activation of NO-ASA can lead to DNA alkylation and (reversible) adduct formation between the QM and deoxyadenosine, deoxyguanosine, deoxycytosine, or thymidine. Using dC which has the potential to form only one (reversible) adduct, the relative reversibility of QM reaction versus bioactivity was determined using a series of EHBPs (Kodela et al., 2008b) that indeed showed a reversible dC adduct was formed, and that electron donating/withdrawing groups significantly affected the rate of adduct formation/decomposition. As predicted, the m analog did not form a QM, a finding consistent with the proposed mechanism of action of NO-ASA (Kashfi and Rigas, 2007; Hulsman et al., 2007; Dunlap et al., 2007). As noted, m-NO-ASA cannot form a m-QM but rather it may form a putative zwitterion which in turn can react with biological nucleophiles. Also, there are a number of NO-releasing NSAIDs that have an aliphatic spacer, these compounds are biologically active albeit less so than their counterparts that have an aromatic spacer (Kashfi et al., 2002, 2005; Nath et al., 2013).

DIAZENIUMDIOLATE-BASED NO-RELEASING COMPOUNDS

Diazeniumdiolate prodrugs release NO upon hydrolysis or metabolic activation to form the parent anion, which further decomposes to release up to 2 moles of NO and the parent amine (Keefer, 2011; Nandurdikar et al., 2012) (Fig. 4.7A). The therapeutic applications of these prodrugs are diverse and largely depend on the O-2 protecting group ("R," Fig. 4.7A) and its mechanism of activation. For example, vinyl protected prodrug V-PYRRO/NO (Fig. 4.7B) is activated by cytochrome P450 to release NO and shows hepatoprotective properties against a variety of toxins (Liu et al., 2002). Glutathione (GSH)-activated arylated prodrug JS-K (Fig. 4.7C) has anticancer activity (Maciag et al., 2009). Recently, primary amine diazeniumdiolate prodrug AcOM-IPA/NO (Fig. 4.7D) (Andrei et al., 2010; Salmon et al., 2011) was reported to release nitroxyl (HNO), another potent bioeffector molecule with possible applications in treating heart failure, alcohol abuse, and cancer (Nagasawa et al., 1995; Paolocci et al., 2003; Basudhar et al., 2015). Secondary amine diazeniumdiolate ions are protonated at N-3 (see Fig. 4.7A for numbering) to release NO (Keefer, 2011), whereas primary amine diazeniumdiolates release nitroxyl (Andrei et al., 2010; Salmon et al., 2011) (HNO) on protonation at N-2. Below we have discussed the anticancer properties of these diazeniumdiolate prodrugs.

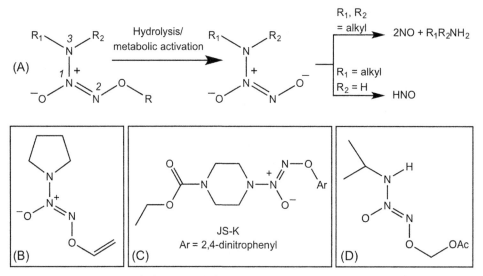

Figure 4.7 (A) Activation of diazeniumdiolate prodrugs to release NO or HNO. Structures of (B) V-PYRRO/NO, (C) JS-K, and (D) AcOM-IPA/NO. *Adapted and reproduced with permission from Nandurdikar, R.S., Maciag, A.E., Cao, Z., Keefer, L.K., Saavedra, J.E., 2012. Diazeniumdiolated carbamates: a novel class of nitric oxide donors. Bioorg. Med. Chem. 20, 2025–2029. Elsevier.*

NONO-NSAIDs

NONO-NSAIDs are based on linking an *N*-diazen-1-ium-1,2 diolate functional group to a classical NSAID (Fig. 4.8) yielding compounds that are not likely to lead to "nitrate tolerance" (Fung and Bauer, 1994; Csont and Ferdinandy, 2005; Hu et al., 2007) and also have the potential to generate 2 equivalents of NO (Velazquez et al., 2007). The production of NO from nitrate esters discussed earlier requires a three-electron reduction (Thatcher et al., 2004). However, NONO-NSAIDs do not require redox activation before the release of NO (Velazquez et al., 2007). Another attractive attribute of these classes of NO-releasing compounds is their rich derivatization chemistry that facilitates the targeting of NO to specific target organ and/or tissue site (Keefer, 2003).

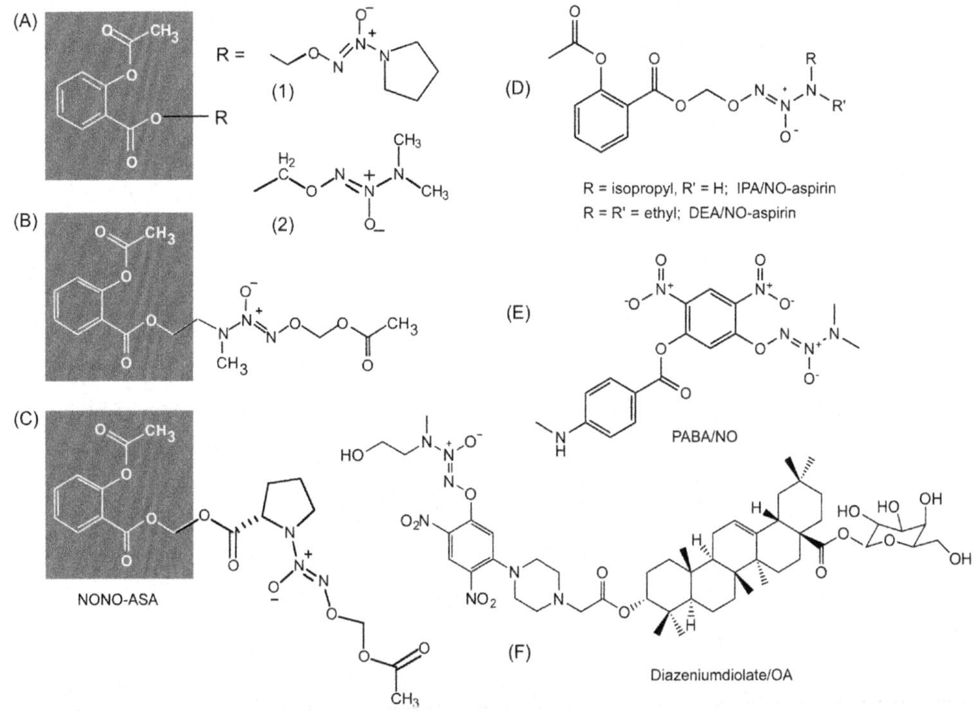

Figure 4.8 The chemical structures of NONO-aspirin, IPA/NO-aspirin, DEA/NO-aspirin, PABA/NO, and diazeniumdiolate/OA hybrid. Hybrid ester prodrugs possessing an 1-(pyrrolidin-1yl)diazen-1-ium-1,2-diolate, A(1), or 1-(*N,N*-dimethylamino)diazen-1-ium-1,2-diolate, A(2), moiety attached via a one-carbon methylene space to the carboxylic acid group of aspirin. In B, the NO-releasing moiety is O^2-acetoxymethy 1-[*N*-(2-hydroxyethyl)-*N*-methylamino]diazen-1-ium-1,2-diolate. (D), structural features of newer NONO-NSAIDs (IPA/NO-aspirin and DEA/NO-aspirin. (E), structural features of PABA/NO a diazeniumdiolate NO prodrug. (F), A oleanolic acid (OA) derivative of O^2-(2,4-dinitrophenyl) diazeniumdiolates. *Adapted with permission from Kashfi, K., 2009. Anti-inflammatory agents as cancer therapeutics. Adv. Pharmacol. 57: 31–89, 2009. Elsevier Books.*

The first agent reported in this compound class had a NONOate (O^2-unsubstituted N-diazen-1-ium-1,2-diolate) attached via a one-carbon methylene spacer to the carboxylic acid group of a traditional NSAID (aspirin, ibuprofen, indomethacin) (Velazquez et al., 2005) (Fig. 4.8A). In vitro, these agents did not inhibit the enzymatic catalytic activity of COX-1 or COX-2, however, they were equipotent to their traditional NSAID counterparts when evaluated in the carrageenan-induced rat paw antiinflammatory model. Also, unlike their traditional parent NSAID, these agents had no significant gastric toxicity when given orally (Velazquez et al., 2005). A series NONO-NSAIDs (aspirin, ibuprofen, indomethacin) was subsequently made that possessed O^2-acetoxymethyl-1-[N-(2-hydroxyethyl)-N-methylamino]diazen-1-ium-1,2-diolate moiety as the NO donor (2-HEMA/NO) (Velazquez et al., 2007) (Fig 4.8B). Here, the NO-donating moiety was attached via a two-carbon ethyl spacer to the carboxylic acid of the traditional NSAID, and because a secondary dialkylamine was used in their synthesis, the number of possible new NONO-NSAIDs was enormous. Like their predecessors, these agents were nonulcerogenic, and in vitro did not inhibit either COX-1 or COX-2 activity, but showed even better antiinflammatory properties, suggesting that they were acting as prodrugs, requiring metabolic activation by an esterase to release the parent NSAID. A potential limitation was that hydrolysis would also release one equiv. of the corresponding nitrosoamines that are biologically toxic. To overcome this concern, a second generation of O^2-acetoxymethyl-protected (PROLI/NO) releasing NONO-NSAIDs was developed where a diazeniumdiolate ion obtained from an amine such as L-proline was used, the N-nitroso derivative of which is nontoxic (Velazquez et al., 2008) (Fig. 4.8C). These agents were also nonulcerogenic, had better antiinflammatory properties, and effective analgesic activity. They also produced up to 1.9 mol of NO per mol of compound (Velazquez et al., 2008).

NONO-NSAIDs are an attractive class of compounds, however, there is no data on their chemopreventive potential. Based on the NO-NSAIDs, one might expect these compounds to have potent chemoprevention properties. The antiulcerogenic, antiinflammatory, analgesic, and antipyretic effects of an NONO-NSAID were compared directly to that of m-NO-ASA, together with effects on relevant biological markers such as gastric PGE_2 and lipid peroxidation levels, superoxide dismutase activity, and TNF-α levels. In all aspects, the two classes of compounds were similar suggesting that there may be a threshold for NO above which no further beneficial effects may be apparent by the higher levels of NO (Chattopadhyay et al., 2010b).

In a recent study, evaluating the potential use of NONO-NSAIDs as antimetastasis agents, the effects of NONO-aspirin and NONO-naproxen were compared to those exerted by their respective parent NSAIDs on avidities of human melanoma M624 cells (Cheng et al., 2012). Both NONO-NSAIDs, but not their corresponding parent NSAIDs, reduced M624 adhesion on vascular cellular adhesion molecule-1 (VCAM-1) and fibronectin under fluid flow conditions and static conditions, respectively.

The NONOate moiety on the NSAIDs was shown to be critical for their function in reducing the avidity of melanoma cells. These findings suggest that NONO-NSAIDs could potentially be a new class of antimetastasis drugs for cancer treatment.

HNO-NSAIDs

Decomposition of diazeniumdiolates (also known as NONOates) can lead to formation of nitroxyl (HNO) and/or NO (Keefer, 2011). HNO has emerged as an important pharmacological agent with beneficial effects in overcoming heart failure (Paolocci et al., 2003), preconditioning against myocardial infarction (Paolocci et al., 2001), and treating alcohol abuse (Nagasawa et al., 1995). Using Angeli's salt ($Na_2N_2O_3$) to generate HNO, the first anticancer activity of HNO was reported in 2008 (Norris et al., 2008) where it was shown that it suppressed the proliferation of both MCF-7 estrogen receptor (ER)-positive and MDA-MB-231 ER-negative human breast cancer cell lines under both hypoxic and normoxic conditions. In mice bearing ER-negative breast cancer xenografts, HNO treatment caused significant reductions in tumor growth which was accompanied by increases in apoptosis and suppression of angiogenesis (Norris et al., 2008). Recently, two new NONO-NSAIDs were prepared by derivatizing both a primary and secondary amine diazeniumdiolate with aspirin to produce O^2-(acetylsalicyloyloxymethyl)-1-(N-isopropylamino)-diazen-1-ium-1,2-diolate (IPA/NO-aspirin) (Fig. 4.8D) and O^2-(acetylsalicyloyloxymethyl)-1-(N,N-diethylamino)-diazen-1-ium-1,2-diolate (DEA/NO-aspirin) (Basudhar et al., 2013). Both of these agents had enhanced GI safety profiles, showed strong antiinflammatory properties, and exhibited significantly enhanced cytotoxicity compared to either aspirin or the parent diazeniumdiolate toward nonsmall cell lung carcinoma cells (A549), but they were not appreciably toxic toward endothelial cells (HUVECs) suggesting cancer-specific sensitivity. The HNO-NSAID prodrug also inhibited cylcooxgenase-2 and glyceraldehyde 3-phosphate dehydrogenase activity (GAPDH). Both IPA/NO-aspirin and DEA/NO-aspirin were also shown to reduce the growth of breast cancer cell lines MDA-MB-231, MDA-MB-468 (both ER-negative), and MCF-7 (ER-positive), more effectively than the parent compounds while not being appreciably cytotoxic in a related nontumorigenic MCF-10A cell line (Basudhar et al., 2015). Notably the cytotoxicity of these agents was about the same in ER-negative and ER-positive cell lines suggesting an effect independent of the estrogen status of the cells. In xenografts of GFP-transfected MDA-MB-231 cells, which allow noninvasive monitoring of tumor size, DEA/NO-aspirin caused a substantial reduction in fluorescence intensity, however, this was not statistically significant. On the other hand, treatment with IPA/NO-aspirin led to a significant decrease in both fluorescence intensity and tumor mass, suggesting that the HNO-donating derivative not only effectively inhibited tumor progression but was also tumoricidal (Basudhar et al., 2013). The high cytotoxicity of IPA/NO-aspirin was in part due to increased oxidant levels leading to DNA damage and to inhibition of GAPDH, which led to caspase-3-mediated induction of apoptosis.

JS-K and PABA/NO

O^2-(2,4-Dinitrophenyl)1-[(4-thoxycarbonyl)piperazin-1-yl]diazen-1-ium-1,2-diolate (JS-K, Fig. 4.7C) and O^2-[2,4-dinitro-5-(N-methyl-N-4-carboxyphenylamino) phenyl] 1-N,N-dimethylamino)diazen-1-ium-1,2-diolate (PABA/NO, Fig. 4.8E) are members of the diazeniumdiolate class of NO prodrugs that were designed to be activated for anticancer effects by glutathione S-transferase (GST)-induced release of NO (Keefer, 2010). The rationale for this was based on the observation that GST (specifically GST-π) a key phase II detoxification enzyme is frequently overexpressed in cancer tissue (Laschak et al., 2012; Bico et al., 1994). JS-K (Shami et al., 2003, 2009) and PABA/NO (Findlay et al., 2004; Saavedra et al., 2006) have shown promise as anticancer agents. PABA/NO showed strong antitumor activity comparable to that of cisplatin in subcutaneous A2780 human ovarian cancer xenografts in female immunodeficient mice (Findlay et al., 2004). Modulation of NO signaling may be a promising strategy in treating glioblastoma. U87 glioma cells exposed to PABA/NO showed a strong dose-dependent growth inhibitory effect in vitro, and a strong synergistic effect was observed after concomitant treatment with temozolomide (TMZ), but not with carboplatin (CPT) (Kogias et al., 2012). In an animal model where nude rats underwent stereotactic implantation of U87 glioma cells into the right striatum, PABA/NO administration by differing routes did not lead to an extension of survival time despite decreases in tumor proliferation and increases in apoptosis (Kogias et al., 2012).

JS-K has shown potent antileukemia activity in vitro using HL-60 and U937 cell lines and in vivo with HL-60 cells using a xenograft model (Shami et al., 2003). JS-K has also been shown to inhibit hepatoma Hep 3B cell proliferation (Ren et al., 2003), enhance arsenic and cisplatin-induced cytolethality in arsenic-transformed rat liver cells (Liu et al., 2004), and induce apoptosis in human multiple myeloma cell lines (Kiziltepe et al., 2007). In addition, JS-K was also shown to be effective against leukemia, renal, prostate, and brain cancer cells (Shami et al., 2006; Chakrapani et al., 2008), as well as both ER-positive and ER-negative breast cancer cell lines without affecting normal mammary epithelial cells (Mcmurtry et al., 2011). In further evaluating the metabolic actions of JS-K in order to decipher its mechanisms of cytotoxicity, it was determined that the activating step in the metabolism of JS-K in the cell was the dearylation of the diazeniumdiolate by GSH resulting in release of two equivalents of NO (Maciag et al., 2013). This caused rapid depletion of GSH, resulting in alterations in the redox potential of the cellular environment, which led to activation of the MAPK stress signaling pathways, and induction of apoptosis.

In the androgen receptor (AR)-positive prostate cancer (PCa) cell line 22Rv1, JS-K was able to reduce the intracellular concentration of functional AR (Laschak et al., 2012). This was most likely due to high intracellular levels of NO as demonstrated indirectly by high levels of nitrotyrosine in JS-K-treated cells. Moreover, JS-K diminished Wnt signaling in AR-positive 22Rv1 cells. In line with these observations, castration-resistant 22Rv1 cells were found to be more susceptible to the growth inhibitory effects of JS-K than the androgen-dependent LNCaP cells, which do not exhibit an active Wnt signaling pathway.

Malignant gliomas exhibit overexpression and genetic polymorphisms of the GST gene which influence the malignancy of the tumor and its response to chemo- or radiotherapy (Ezer et al., 2002; Juillerat-Jeanneret et al., 2008). Therefore, JS-K might be a suitable candidate for treating malignant gliomas. In human U87 glioma cells and primary glioblastoma cells, JS-K showed a dose-dependent cytotoxicity (Weyerbrock et al., 2012). Cell death was partially induced by caspase-dependent apoptosis, which could be blocked by the pancaspase inhibitors Z-VAD-FMK and Q-VD-OPH. Inhibition of GST by sulfasalazine, cGMP inhibition by ODQ (1H-[1,2,4]oxadiazolo[4,3-a]quinoxalin-1-one), and MEK1/2 inhibition by UO126 (bis[amino[(2-aminophenyl)thio]methylene]butanedinitrile) attenuated the antiproliferative effect of JS-K, suggesting the involvement of various intracellular death signaling pathways. JS-K also reduced the growth of U87 xenografts, showing reduced proliferation, increased apoptosis, and increased necrosis (Weyerbrock et al., 2012).

Using 18 different nonsmall cell lung cancer (NSCLC) cell lines, it was shown that JS-K was most effective against a subset of NSCLC cell lines where the endogenous levels of reactive oxygen and nitrogen species (ROS/RNS) were high (Maciag et al., 2011). Treatment of the lung adenocarcinoma cells with JS-K resulted in oxidative/nitrosative stress in cells with high basal levels of ROS/RNS, which, combined with the arylating properties of the compound, was reflected in glutathione depletion and alteration in cellular redox potential, mitochondrial membrane permeabilization, and cytochrome c release. JS-K treatment that was formulated in Pluronic P123 led to 75% reduction in the growth of H1703 lung adenocarcinoma cells in a xenograft model (Maciag et al., 2011). Of note, levels of peroxiredoxin 1 (PRX1) and 8-oxo-deoxy-guanosine glycosylase (OGG1) also correlated with JS-K sensitivity. Taken together, JS-K may have a role in personalized therapy for lung cancers characterized by high levels of ROS/RNS. PRX1 and OGG1 proteins, which can be easily measured, could function as biomarkers for identifying tumors sensitive to the therapy.

The protein β-catenin has a central role in the Wnt signaling pathway that regulates cell–cell adhesion and may promote leukemia cell proliferation (Chung et al., 2002). β-Catenin is highly expressed in Acute Lymphoblastic Leukemia (ALL), tumor lines of hematopoietic origin, and primary leukemia cells but is undetectable in normal peripheral blood T cells. Among the leukemia cell lines, β-catenin is expressed in high levels in Jurkat T cells (Chung et al., 2002). In this cell line, JS-K strongly inhibited its growth, reduced proliferation, caused G_2/M cell cycle arrest, which led to increased apoptosis (Nath et al., 2010). JS-K reduced the transcriptional activity of β-catenin and also reduced nuclear β-catenin protein levels but had no appreciable effect on levels in the cytosol. The underlying mechanism for these observations was shown to be through S-nitrosylation of nuclear β-catenin (Nath et al., 2010; Kashfi, 2012).

In order to improve the selectivity for cancer cells, and particularly to enhance the uncatalyzed stability of the GSTπ-activated O^2-(2,4-dinitrophenyl)diazeniumdiolates, novel hybrids comprising the latter and oleanolic acid (OA) have been prepared (Fu et al., 2013). The most potent of such hybrids is shown in Fig 4.8F. The rationale for these hybrids was based on the observations that (1) OA imparts additional hepatic selectivity and a synergetic biological profile to the GSTπ-activated moiety because of the liver-specific distribution and liver protective effects (Liu et al., 1998), (2) galactosyl moieties could be recognized by high levels of asialoglycoprotein receptors (ASGPR) expressed in human hepatocellular carcinoma (HCC) cells, further enhancing the selectivity and bioactivity (Stockert and Morell, 1983), (3) amino acids were employed to link the 2,4-dinitrophenyl ring with OA to avoid its hydrolysis, and the amine moiety may act as an electron donating group to increase the electron density of the 2,4-dinitrophenyl ring, thus slowing the uncatalyzed reaction with GSH. Indeed the hybrid shown in Fig 4.8F was shown to be very stable, it induced apoptosis in HepG2 cells by arresting the cell cycle at the G2/M phase, activating both the mitochondrion-mediated pathway and the MAPK pathway and enhancing the intracellular production of ROS (Fu et al., 2013).

NOSH-NSAIDs

Hydrogen sulfide (H_2S) like NO is a gasotransmitter of biological importance. It activates many intracellular pathways, it opens K_{ATP} channels and indirectly stimulates the cGMP pathway by inhibiting cGMP phosphodiesterases leading to vasodilation and angiogenesis (Papapetropoulos et al., 2009; Wang, 2012). Many H_2S-releasing NSAIDs have been developed and their utility as anticancer agents examined (Chattopadhyay et al., 2012a, 2012b, 2013; Kashfi and Olson, 2013; Elsheikh et al., 2014; Kodela et al., 2015b). However, similar to NO, H_2S has a dichotomous role in cancer, reviewed by Kashfi (2014) and Szabo (2016). Recently, a new class of antiinflammatory pharmaceuticals has been described where NO and H_2S-releasing moieties have been covalently attached to an NSAID backbone, thus releasing both NO and H_2S, these chimeras have been termed NOSH-NSAIDs (Kodela et al., 2012, 2013). The rationale for their development was based on the observations that H_2S much like NO has some of the same properties as PGs within the gastric mucosa (Fiorucci, 2009), thus it modulates some components of the mucosal defense systems leading to safer NSAIDs. Another consideration was that there could be synergy between NO and H₂S thus increasing potency and possibly efficacy. To date the synthesis of three such compounds, NOSH–aspirin (NBS-1120), NOSH–naproxen (AVT-219), and NOSH–sulindac (AVT-18A) (Fig. 4.9A–C) have been reported (Kodela et al., 2012, 2013).

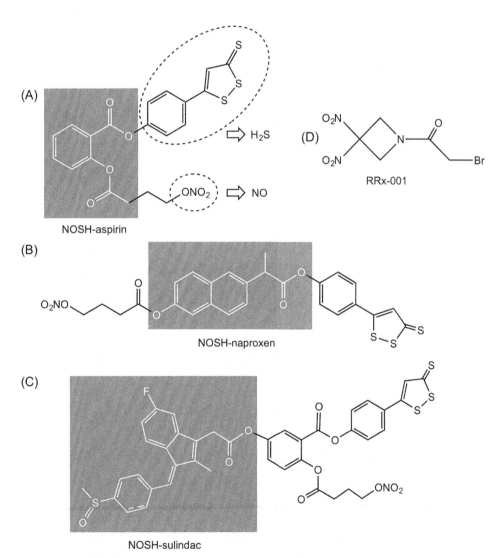

Figure 4.9 Structural components of (A) NOSH-aspirin, (B) NOSH-naproxen, (C) NOSH-sulindac, and (D) RRx-001. For the NOSH-NSAIDs, the parent compounds, aspirin, naproxen, and sulindac, are shown in the shaded boxes. ADT-OH (5-(4-hydroxyphenyl)-3H-1,2-dithiole-3-thione) releases H₂S and ONO₂ releases NO, both shown in dotted ellipses.

All three NOSH-NSAIDs were shown to have greater GI safety profiles compared to their parent counterparts and displayed strong antioxidant properties (Kashfi et al., 2015; Kodela et al., 2015a; Chattopadhyay et al., 2016). NOSH-aspirin (NBS-1120) was shown to have strong antiinflammatory (Kodela et al., 2012, 2015a), antipyretic, analgesic, and antiplatelet properties similar to its parent compound, aspirin (Kodela et al., 2015a). NOSH-aspirin inhibited the growth of 11 different human cancer cell lines of 6 different histological subtypes. The cell lines were that of colon (HT-29: COX-1 and COX-2 positive, HCT 15: COX null, and SW480: COX-1 positive, low levels of endogenous COX-2), breast (MCF-7: [ER(+)], MDA-MB-231 and SKBR3: [ER(-)]); T-cell leukemia (Jurkat), pancreatic (BxPC-3: both COX-1 and COX-2 positive, MIAPaCa-2: COX null), prostate (LNCaP), and lung (A549). The IC_{50}s for the growth inhibition was in the low to mid nanomolar ranges (Kodela et al., 2012). Using HT-29 colon cancer cells as a model, this growth inhibition was as a result of reductions in cell proliferation, G_0/G_1 cell cycle arrest, leading to increased apoptosis (Chattopadhyay et al., 2012c). The efficacy of NOSH-aspirin at different concentrations was also compared to that of aspirin using an in vivo xenograft mouse model of colon cancer chemoprevention (Kodela et al., 2015a). NOSH-aspirin dose-dependently inhibited tumor growth and tumor mass and was at least fivefold more potent than aspirin. Of note, NOSH-aspirin was also efficacious against established tumors in a xenograft model of colon cancer (Chattopadhyay et al., 2012c). Clearly, as a chemopreventive and chemotherapeutic agent, NOSH-aspirin is superior to aspirin both in terms of efficacy and safety.

Qualitatively similar results have been reported for both NOSH-naproxen (Chattopadhyay et al., 2016) and NOSH-sulindac (Kashfi et al., 2015). With regards to efficacy of NOSH-naproxen in a xenograft model of colon cancer, it is important to note that while treatment of the animals with NOSH-naproxen significantly reduced tumor growth and tumor mass with no overt sign of GI toxicity, naproxen-treated mice died due to GI bleeding (Chattopadhyay et al., 2016). A cartoon summarizing the classic pharmacological effects of NOSH-aspirin is depicted in Fig. 4.10.

RRx-001 AN AEROSPACE COMPOUND FOR CANCER TREATMENT

RRx-001 (Fig. 4.9D) also known as ABDNAZ (1-bromoacetyl-3,3-dinitroazetidine) is a novel aerospace-derived compound under active investigation as a chemo-, immuno- and radiosensitizer in phase II clinical trials (Zhao et al., 2015). This compound contains a unique high energetic organic nitro functional group, a gem dinitroazetidine that has not been used to date for medical and pharmaceutical applications. In an aerospace setting, compounds containing this energetic functionality, such as 1,3,3-trinitroazetidine, are designed to fragment explosively to propel rockets (Sikder et al., 2004). Modification of this structure by removing one of the nitro

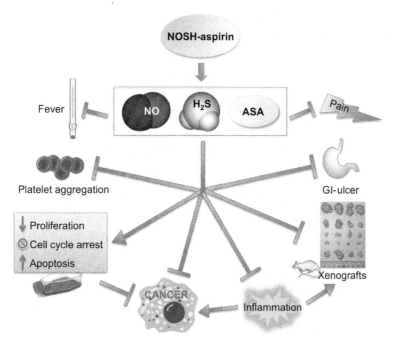

Figure 4.10 Pharmacological effects of NOSH-aspirin. NOSH-aspirin does not cause gastric damage, it is a potent antiinflammatory, and has antiplatelet, analgesic, and antipyretic activities. In vitro, NOSH-aspirin reduces cell proliferation, causes G_0/G_1 cell cycle arrest, leading to increased apoptosis; and in vivo it reduces tumor growth and tumor mass in mice.

groups and substituting it with a bromoacetate group resulted in RRx-001, a nonexplosive that may be used to treat cancer (Knox et al., 2009).

RRx-001 effectively inhibited the growth of 11 human cancer cell lines of varying tissue origins (Ning et al., 2012). The cell lines being oral squamous cell carcinoma (22B), pancreatic carcinoma (PANC-1), melanoma (M21), glioblastoma (U87 and SNB75), colon carcinoma (RKO), colorectal adenocarcinoma (HT-29), breast adenocarcinoma (MCF-7, ER-positive), renal cell carcinoma (A498), brain neuroblastoma (IMR32), nonsmall cell lung carcinoma (A549), and a murine squamous cell carcinoma (SCC VII). The cytotoxic activity of RRx-001 in these cell lines was enhanced under hypoxic conditions, which is an observable phenomena when a solid tumor exceeds a critical diameter of $1-2 \, mm^3$ (Oronsky et al., 2016). Using HL-60 cells (human promyelocytic leukemia cells) which have a doubling time of about 48 hours and HT-29 cells, RRx-001 induced intracellular ROS and RNS generation in a dose- and time-dependent manner (Ning et al., 2012). Pretreatment of the cells with the GSH precursor, N-acetylcysteine (NAC), inhibited RRx-001-induced apoptosis, whereas pretreatment with BSO (buthionine-(S,R)-sulfoxime, GSH synthesis inhibitor) increased RRx-001-induced apoptosis. It should be noted DNA damage (γH2AX expression) was increased in HT-29 and SCC cells as a result of RRx-001 exposure in a dose- and time-dependent manner. This implies that RRx-001 has utility as a chemotherapeutic agent but not as a chemopreventive one. The in vivo antitumor effect of RRx-001 was studied in mice bearing SCC VII tumors. This is an aggressive tumor with a volume doubling time of approximately 2 days. RRx-001 inhibited tumor growth and resulted in a significant increase in TGD (tumor growth delay) time (Ning et al., 2012). The therapeutic efficacy of combining RRx-001 with radiotherapy in tumor-bearing mice was also investigated. Treatment with either RRx-001, or local tumor radiation, or both showed that either radiation alone or RRx-001 alone inhibited tumor growth but that the combination significantly increased TGD. Treatment with RRx-001 either alone or in combination with radiation dramatically increased the rates of blood perfusion and blood volume of SCC VII tumors in a dose- and time-dependent manner which must be due to the generated NO (Ning et al., 2012).

RRx-001 differs from other NO-donating compounds in that the molecule induces local, endogenous, and biphasic production or release of NO, rather than fragmenting to release NO systemically. This activity is closely linked to the metabolism of RRx-001. Metabolic and disposition studies on RRx-001 have shown that, on infusion, the compound rapidly, irreversibly, and selectively binds to hemoglobin at a key NO binding site (Vitturi et al., 2013), and with glutathione (Scicinski et al., 2012, 2014) in directly increasing oxidative stress (Ning et al., 2015). While the RRx-001 glutathione adduct is rapidly excreted, RRx-001-bound hemoglobin remains in circulation for the duration of the lifetime of the red blood cell (Scicinski et al., 2015).

Results from a phase I clinical trial of RRX-001 in heavily pretreated patients with advanced/refractory solid tumors were notable for the lack of systemic toxicity at all doses tested and evidence of anticancer efficacy, with 15 of 25 patients enrolled in the study exhibiting partial response or stable disease (Reid et al., 2015). Furthermore, after progression on RRx-001 treatment, four patients were rechallenged with drugs to which they had previously acquired resistance, and all four exhibited resensitization (Reid et al., 2015).

Preliminary results from ongoing phase II clinical trials have recapitulated the favorable safety and tolerability profile from phase I and provided initial confirmation of antitumor priming to subsequent therapy post-RRx-001. Current reports include: RRx-001 pretreatment may sensitize or resensitize refractory small cell lung cancer patients to first-line chemotherapy (Carter et al., 2016a); low dose RRx-001 in combination with whole brain radiotherapy (WBRT) in the context of a dose escalation phase I/II trial acronymed BRAINSTORM (NCT 02215512) at the University of Michigan in two patients with melanoma brain metastases showed a decrease in tumor vascular permeability leading to increased oxygen delivery and increases in ROS, which led to radiation sensitization (Kim et al., 2016); RRx-001-induced intratumoral necrosis and immune infiltration in an EGFR-mutated NSCLC patient, with evidence of subsequent episensitization to platinum doublets in the form of a partial response (Carter et al., 2016b); a report describing a patient with metastatic epidermal growth factor receptor (EGFR) mutation-positive adenocarcinoma of the lung that continues to outlive stage IV diagnosis of NSCLC after treatment with RRx-001 (Brzezniak et al., 2016); a case report of a patient with chronic refractory diarrhea due to pulmonary neuroendocrine tumors (NET) treated with RRx-001 showed resolution of the diarrhea (Carter et al., 2015).

CONCLUDING REMARKS AND FUTURE DIRECTIONS

Given the enormous potential of NO in various aspects of medicine, it is quite surprising that not many NO-releasing agents have reached the marketplace. In the cancer arena, the reasons for this apparent lack of progress are multifactorial but primarily may be due to the focus that the scientific community has placed toward chemoprevention rather than chemotherapy. For example, phase I clinical trials with m-NO-aspirin (NCX-4016) developed by NicOx (Sophia Antipolis, France) confirmed its improved GI safety compared to aspirin and its efficacy to inhibit platelet aggregation (Fiorucci et al., 2003a, 2003b, 2004). However, its development as a chemopreventive drug in patients who are at high risk of developing colorectal cancer was terminated due to the putative genotoxicity of a potential metabolite (Nicox, 2007). Although the genotoxicity of this metabolite has not been confirmed by others or for that matter for the intact molecule, nevertheless the development of NCX-4016 as a chemopreventive agent was halted. Had the focus been on chemotreatment as a single agent alone or in combination with established chemotherapeutic agents, the outcome may have been quite different. Another NO-releasing NSAID, NO-naproxen (naproxcinod), which was being evaluated for treatment of knee and hip osteoarthritis (Schnitzer et al., 2005, 2010, 2011; Karlsson et al., 2009; Lohmander et al., 2005) has completed phase III clinical trials. However, concerns by the FDA about its GI and CV safety for long-term use did not lead to approval of the NDA application in 2010 without further studies (Nicox, 2010).

A major challenge for designing novel anticancer drugs is the generation of compounds with improved efficacy, lower side effects, and potential synergism with currently available chemotherapeutic drugs. All of the compounds reviewed in this chapter have the potential of being developed and formulated for chemotherapeutic applications against cancer. In particular, JS-K and RRx-001 are ahead of the curve in this regard. A relatively newcomer on the block are the NOSH-NSAIDs which are quite unique in utilizing both NO and H$_2$S. Research in this field is in its infancy; NO and H$_2$S share many similar actions, including modulation of leukocyte adherence to the vascular endothelium, both are "gasotransmitters," and both bind avidly to hemoglobin.

Do these similarities act to potentiate each other? NO and H$_2$S may interact together giving rise to a transient highly reactive species, which may ultimately be responsible for the actions of this class of compounds.

Certainly NO appears to have a role in modulating cancer therapy resistance, and this in itself holds a great deal of promise in improving patient outcomes, thus targeted cancer therapy with NO-based compounds is a feasible proposition. In the past 20 years or so, we have come a long way in our understanding of the biology of NO; the next decade should also prove to be quite fruitful in clinical applications of NO to cancer therapies such as chemotherapy, radiotherapy, and immunotherapy.

ACKNOWLEDGMENT

Supported in part by NIH grant R24 DA018055.

REFERENCES

Aaltoma, S.H., Lipponen, P.K., Kosma, V.M., 2001. Inducible nitric oxide synthase (iNOS) expression and its prognostic value in prostate cancer. Anticancer Res. 21, 3101–3106.

Abdellatif, K.R., Chowdhury, M.A., Dong, Y., Knaus, E.E., 2008. Diazen-1-ium-1,2-diolated nitric oxide donor ester prodrugs of 1-(4-methanesulfonylphenyl)-5-aryl-1H-pyrazol-3-carboxylic acids: synthesis, nitric oxide release studies and anti-inflammatory activities. Bioorg. Med. Chem. 16, 6528–6534.

Abdellatif, K.R., Chowdhury, M.A., Velazquez, C.A., Huang, Z., Dong, Y., Das, D., et al., 2010. Celecoxib prodrugs possessing a diazen-1-ium-1,2-diolate nitric oxide donor moiety: synthesis, biological evaluation and nitric oxide release studies. Bioorg. Med. Chem. Lett. 20, 4544–4549.

Abdellatif, K.R., Abdelgawad, M.A., Labib, M.B., Zidan, T.H., 2015. Synthesis, cyclooxygenase inhibition, anti-inflammatory evaluation and ulcerogenic liability of novel triarylpyrazoline derivatives as selective Cox-2 inhibitors. Bioorg. Med. Chem. Lett. 25, 5787–5791.

Abdellatif, K.R., Elsaady, M.T., Abdel-Aziz, S.A., Abusabaa, A.H., 2016a. Synthesis, cyclooxygenase inhibition and anti-inflammatory evaluation of new 1,3,5-triaryl-4,5-dihydro-1H-pyrazole derivatives possessing methanesulphonyl pharmacophore. J. Enzyme Inhib. Med. Chem. 1–11.

Abdellatif, K.R., Fadaly, W.A., Ali, W.A., Kamel, G.M., 2016b. Synthesis, cyclooxygenase inhibition, anti-inflammatory evaluation and ulcerogenic liability of new 1,5-diarylpyrazole derivatives. J Enzyme Inhib. Med. Chem. 31, 54–60.

Alderton, W.K., Cooper, C.E., Knowles, R.G., 2001. Nitric oxide synthases: structure, function and inhibition. Biochem. J. 357, 593–615.

Ambs, S., Merriam, W.G., Bennett, W.P., Felley-Bosco, E., Ogunfusika, M.O., Oser, S.M., et al., 1998. Frequent nitric oxide synthase-2 expression in human colon adenomas: implication for tumor angiogenesis and colon cancer progression. Cancer Res. 58, 334–341.

Andrei, D., Salmon, D.J., Donzelli, S., Wahab, A., Klose, J.R., Citro, M.L., et al., 2010. Dual mechanisms of HNO generation by a nitroxyl prodrug of the diazeniumdiolate (NONOate) class. J. Am. Chem. Soc. 132, 16526–16532.

Aranda, E., Lopez-Pedrera, C., De La Haba-Rodriguez, J.R., Rodriguez-Ariza, A., 2012. Nitric oxide and cancer: the emerging role of S-nitrosylation. Curr. Mol. Med. 12, 50–67.

Ariztia, E.V., Lee, C.J., Gogoi, R., Fishman, D.A., 2006. The tumor microenvironment: key to early detection. Crit. Rev. Clin. Lab. Sci. 43, 393–425.

Arner, E.S., Holmgren, A., 2006. The thioredoxin system in cancer. Semin. Cancer Biol. 16, 420–426.

Baandrup, L., Kjaer, S.K., Olsen, J.H., Dehlendorff, C., Friis, S., 2015. Low-dose aspirin use and the risk of ovarian cancer in Denmark. Ann. Oncol. 26, 787–792.

Bak, A.W., Mcknight, W., Li, P., Del Soldato, P., Calignano, A., Cirino, G., et al., 1998. Cyclooxygenase-independentv chemoprevention with an aspirin derivative in a rat model of colonic adenocarcinoma. Life Sci. 62 (L), 367–373.

Bakr, R.B., Azouz, A.A., Abdellatif, K.R., 2016. Synthesis, cyclooxygenase inhibition, anti-inflammatory evaluation and ulcerogenic liability of new 1-phenylpyrazolo[3,4-D]pyrimidine derivatives. J Enzyme Inhib. Med. Chem. 31, 6–12.

Balkwill, F., Mantovani, A., 2001. Inflammation and cancer: back to virchow? Lancet 357, 539–545.

Baron, J.A., Cole, B.F., Sandler, R.S., Haile, R.W., Ahnen, D., Bresalier, R., et al., 2003. A randomized trial of aspirin to prevent colorectal adenomas. N. Engl. J. Med. 348, 891–899.

Basudhar, D., Bharadwaj, G., Cheng, R.Y., Jain, S., Shi, S., Heinecke, J.L., et al., 2013. Synthesis and chemical and biological comparison of nitroxyl- and nitric oxide-releasing diazeniumdiolate-based aspirin derivatives. J. Med. Chem. 56, 7804–7820.

Basudhar, D., Cheng, R.C., Bharadwaj, G., Ridnour, L.A., Wink, D.A., Miranda, K.M., 2015. Chemotherapeutic potential of diazeniumdiolate-based aspirin prodrugs in breast cancer. Free Radic. Biol. Med. 83, 101–114.

Basudhar, D., Somasundaram, V., De Oliveira, G.A., Kesarwala, A., Heinecke, J.L., Cheng, R.Y., et al., 2016. Nos2-derived NO drives multiple pathways of breast cancer progression. Antioxid. Redox Signal.

Bechmann, N., Kniess, T., Kockerling, M., Pigorsch, A., Steinbach, J., Pietzsch, J., 2015. Novel (pyrazolyl) benzenesulfonamides with a nitric oxide-releasing moiety as selective cyclooxygenase-2 inhibitors. Bioorg. Med. Chem. Lett. 25, 3295–3300.

Becker, C., Wilson, J.C., Jick, S.S., Meier, C.R., 2015. Non-steroidal anti-inflammatory drugs and the risk of head and neck cancer: a case–control analysis. Int. J. Cancer 137, 2424–2431.

Beckman, J.S., Beckman, T.W., Chen, J., Marshall, P.A., Freeman, B.A., 1990. Apparent hydroxyl radical production by peroxynitrite: implications for endothelial injury from nitric oxide and superoxide. Proc. Natl. Acad. Sci. USA 87, 1620–1624.

Benamouzig, R., Deyra, J., Martin, A., Girard, B., Jullian, E., Piednoir, B., et al., 2003. Daily soluble aspirin and prevention of colorectal adenoma recurrence: one-year results of the APACC Trial. Gastroenterology 125, 328–336.

Ben-Baruch, A., 2003. Host microenvironment in breast cancer development: inflammatory cells, cytokines and chemokines in breast cancer progression: reciprocal tumor–microenvironment interactions. Breast Cancer Res. 5, 31–36.

Bian, K., Ghassemi, F., Sotolongo, A., Siu, A., Shauger, L., Kots, A., et al., 2012. Nos-2 signaling and cancer therapy. IUBMB Life 64, 676–683.

Biava, M., Battilocchio, C., Poce, G., Alfonso, S., Consalvi, S., Di Capua, A., et al., 2014. Enhancing the pharmacodynamic profile of a class of selective Cox-2 inhibiting nitric oxide donors. Bioorg. Med. Chem. 22, 772–786.

Bico, P., Chen, C.Y., Jones, M., Erhardt, J., Dirr, H., 1994. Class Pi glutathione S-transferase: Meisenheimer complex formation. Biochem. Mol. Biol. Int. 33, 887–892.

Bode, A.M., Dong, Z., 2004. Targeting signal transduction pathways by chemopreventive agents. Mutat. Res. 555, 33–51.

Bogdan, C., 2001. Nitric oxide and the immune response. Nat. Immunol. 2, 907–916.

Borhade, N., Pathan, A.R., Halder, S., Karwa, M., Dhiman, M., Pamidiboina, V., et al., 2012. NO-NSAIDs. Part 3: Nitric oxide-releasing prodrugs of non-steroidal anti-inflammatory drugs. Chem. Pharm. Bull. (Tokyo) 60, 465–481.

Boschi, D., Cena, C., Di Stilo, A., Rolando, B., Manzini, P., Fruttero, R., et al., 2010. Nitrooxymethyl-substituted analogues of rofecoxib: synthesis and pharmacological characterization. Chem. Biodivers. 7, 1173–1182.

Bosworth, C.A., Toledo Jr., J.C., Zmijewski, J.W., Li, Q., Lancaster Jr., J.R., 2009. Dinitrosyliron complexes and the mechanism(s) of cellular protein nitrosothiol formation from nitric oxide. Proc. Natl. Acad. Sci. USA 106, 4671–4676.

Brandao, M.M., Soares, E., Salles, T.S., Saad, S.T., 2001. Expression of inducible nitric oxide synthase is increased in acute myeloid leukaemia. Acta Haematol. 106, 95–99.

Broniowska, K.A., Hogg, N., 2012. The chemical biology of S-nitrosothiols. Antioxid. Redox Signal. 17, 969–980.

Brune, B., Von Knethen, A., Sandau, K.B., 1999. Nitric oxide (NO): an effector of apoptosis. Cell Death Differ. 6, 969–975.

Brzezniak, C., Schmitz, B.A., Peterson, P.G., Degesys, A., Oronsky, B.T., Scicinski, J.J., et al., 2016. Rrx-001-induced tumor necrosis and immune cell infiltration in an eGFR mutation-positive NSCLC with resistance to eGFR tyrosine kinase inhibitors: a case report. Case Rep. Oncol. 9, 45–50.

Burke, A.J., Sullivan, F.J., Giles, F.J., Glynn, S.A., 2013. The yin and yang of nitric oxide in cancer progression. Carcinogenesis 34, 503–512.

Carini, M., Aldini, G., Orioli, M., Maffei Facino, R., 2002. In vitro metabolism of a nitroderivative of acetylsalicylic acid (Ncx4016) by rat liver: LC and LC–MS studies. J. Pharm. Biomed. Anal. 29, 1061–1071.

Carter, C.A., Degesys, A., Oronsky, B., Scicinski, J., Caroen, S.Z., Oronsky, A.L., et al., 2015. Flushing out carcinoid syndrome: beneficial effect of the anticancer epigenetic agent RRx-001 in a patient with a treatment-refractory neuroendocrine tumor. Case Rep. Oncol. 8, 461–465.

Carter, C.A., Oronsky, B., Caroen, S., Scicinski, J., Degesys, A., Cabrales, P., et al., 2016a. Partial response in an RRx-001-primed patient with refractory small-cell lung cancer after a third introduction of platinum doublets. Case Rep. Oncol. 9, 285–289.

Carter, C.A., Oronsky, B.T., Caroen, S.Z., Scicinski, J.J., Cabrales, P., Reid, T., et al., 2016b. Partial response to platinum doublets in refractory eGFR-positive non-small cell lung cancer patients after RRx-001: evidence of episensitization. Case Rep. Oncol. 9, 62–67.

Cena, C., Lolli, M.L., Lazzarato, L., Guaita, E., Morini, G., Coruzzi, G., et al., 2003. Antiinflammatory, gastrosparing, and antiplatelet properties of new NO-donor esters of aspirin. J. Med. Chem. 46, 747–754.

Chakrapani, H., Goodblatt, M.M., Udupi, V., Malaviya, S., Shami, P.J., Keefer, L.K., et al., 2008. Synthesis and in vitro anti-leukemic activity of structural analogues of JS-K, an anti-cancer lead compound. Bioorg. Med. Chem. Lett. 18, 950–953.

Chattopadhyay, M., Goswami, S., Rodes, D.B., Kodela, R., Velazquez, C.A., Boring, D., et al., 2010a. NO-releasing NSAIDs suppress NF-kappaB signaling in vitro and in vivo through S-nitrosylation. Cancer Lett. 298, 204–211.

Chattopadhyay, M., Kodela, R., Nath, N., Barsegian, A., Boring, D., Kashfi, K., 2012a. Hydrogen sulfide-releasing aspirin suppresses NF-kappaB signaling in estrogen receptor negative breast cancer cells in vitro and in vivo. Biochem. Pharmacol. 83, 723–732.

Chattopadhyay, M., Kodela, R., Nath, N., Dastagirzada, Y.M., Velazquez-Martinez, C.A., Boring, D., et al., 2012b. Hydrogen sulfide-releasing NSAIDs inhibit the growth of human cancer cells: a general property and evidence of a tissue type-independent effect. Biochem. Pharmacol. 83, 715–722.

Chattopadhyay, M., Kodela, R., Olson, K.R., Kashfi, K., 2012c. NOSH-aspirin (NBS-1120), a novel nitric oxide- and hydrogen sulfide-releasing hybrid is a potent inhibitor of colon cancer cell growth in vitro and in a xenograft mouse model. Biochem. Biophys. Res. Commun. 419, 523–528.

Chattopadhyay, M., Nath, N., Kodela, R., Sobocki, T., Metkar, S., Gan, Z.Y., et al., 2013. Hydrogen sulfide-releasing aspirin inhibits the growth of leukemic Jurkat cells and modulates beta-catenin expression. Leuk. Res. 37, 1302–1308.

Chattopadhyay, M., Kodela, R., Duvalsaint, P.L., Kashfi, K., 2016. Gastrointestinal safety, chemotherapeutic potential, and classic pharmacological profile of NOSH-naproxen (AVT-219) a dual NO- and H_2S-releasing hybrid. Pharmacol. Res. Perspect. 4, E00224.

Chattopadhyay, M., Velazquez, C.A., Pruski, A., Nia, K.V., Abdellatif, K.R., Keefer, L.K., et al., 2010b. Comparison between 3-nitrooxyphenyl acetylsalicylate (NO-ASA) and O_2-(acetylsalicyloxymethyl)-1-(pyrrolidin-1-yl)diazen-1-ium-1,2-diolate (NONO-ASA) as safe anti-inflammatory, analgesic, antipyretic, antioxidant prodrugs. J. Pharmacol. Exp. Ther. 335, 443–450.

Chaudhary, S.C., Singh, T., Kapur, P., Weng, Z., Arumugam, A., Elmets, C.A., et al., 2013. Nitric oxide-releasing sulindac is a novel skin cancer chemopreventive agent for UVB-induced photocarcinogenesis. Toxicol. Appl. Pharmacol. 268, 249–255.

Cheng, H., Mollica, M.Y., Lee, S.H., Wang, L., Velazquez-Martinez, C.A., Wu, S., 2012. Effects of nitric oxide-releasing nonsteroidal anti-inflammatory drugs (NONO-NSAIDs) on melanoma cell adhesion. Toxicol. Appl. Pharmacol. 264, 161–166.

Chiarugi, V., Magnelli, L., Gallo, O., 1998. Cox-2, iNOS and P53 as play-makers of tumor angiogenesis (review). Int. J. Mol. Med. 2, 715–719.

Choudhari, S.K., Chaudhary, M., Bagde, S., Gadbail, A.R., Joshi, V., 2013. Nitric oxide and cancer: a review. World J. Surg. Oncol. 11, 118.

Chowdhury, M.A., Abdellatif, K.R., Dong, Y., Yu, G., Huang, Z., Rahman, M., et al., 2010. Celecoxib analogs possessing a N-(4-nitrooxybutyl)piperidin-4-yl or N-(4-nitrooxybutyl)-1,2,3,6-tetrahydropyridin-4-yl nitric oxide donor moiety: synthesis, biological evaluation and nitric oxide release studies. Bioorg. Med. Chem. Lett. 20, 1324–1329.

Chung, E.J., Hwang, S.G., Nguyen, P., Lee, S., Kim, J.S., Kim, J.W., et al., 2002. Regulation of leukemic cell adhesion, proliferation, and survival by beta-catenin. Blood 100, 982–990.

Cirino, G., Wheeler-Jones, C.P., Wallace, J.L., Del Soldato, P., Baydoun, A.R., 1996. Inhibition of inducible nitric oxide synthase expression by novel nonsteroidal anti-inflammatory derivatives with gastrointestinal-sparing properties. Br. J. Pharmacol. 117, 1421–1426.

Clouser, M.C., Roe, D.J., Foote, J.A., Harris, R.B., 2009. Effect of non-steroidal anti-inflammatory drugs on non-melanoma skin cancer incidence in the Skicap-Ak Trial. Pharmacoepidemiol. Drug Saf. 18, 276–283.

Cobbs, C.S., Brenman, J.E., Aldape, K.D., Bredt, D.S., Israel, M.A., 1995. Expression of nitric oxide synthase in human central nervous system tumors. Cancer Res. 55, 727–730.

Colotta, F., Allavena, P., Sica, A., Garlanda, C., Mantovani, A., 2009. Cancer-related inflammation, the seventh hallmark of cancer: links to genetic instability. Carcinogenesis 30, 1073–1081.

Csont, T., Ferdinandy, P., 2005. Cardioprotective effects of glyceryl trinitrate: beyond vascular nitrate tolerance. Pharmacol. Ther. 105, 57–68.

Cui, S., Reichner, J.S., Mateo, R.B., Albina, J.E., 1994. Activated murine macrophages induce apoptosis in tumor cells through nitric oxide-dependent or -independent mechanisms. Cancer Res. 54, 2462–2467.

Cui, X.J., He, Q., Zhang, J.M., Fan, H.J., Wen, Z.F., Qin, Y.R., 2014. High-dose aspirin consumption contributes to decreased risk for pancreatic cancer in a systematic review and meta-analysis. Pancreas 43, 135–140.

Dangel, O., Mergia, E., Karlisch, K., Groneberg, D., Koesling, D., Friebe, A., 2010. Nitric oxide-sensitive guanylyl cyclase is the only nitric oxide receptor mediating platelet inhibition. J. Thromb. Haemost. 8, 1343–1352.

Daugherty, S.E., Pfeiffer, R.M., Sigurdson, A.J., Hayes, R.B., Leitzmann, M., Schatzkin, A., et al., 2011. Nonsteroidal antiinflammatory drugs and bladder cancer: a pooled analysis. Am. J. Epidemiol. 173, 721–730.

Davies, N.M., Roseth, A.G., Appleyard, C.B., Mcknight, W., Del Soldato, P., Calignano, A., et al., 1997. No-naproxen vs. naproxen: ulcerogenic, analgesic and anti-inflammatory effects. Aliment. Pharmacol. Ther. 11, 69–79.

Dawber, T.R., Kannel, W.B., Lyell, L.P., 1963. An approach to longitudinal studies in a community: the Framingham study. Ann. NY. Acad. Sci. 107, 539–556.

De Pedro, M., Baeza, S., Escudero, M.T., Dierssen-Sotos, T., Gomez-Acebo, I., Pollan, M., et al., 2015. Effect of Cox-2 inhibitors and other non-steroidal inflammatory drugs on breast cancer risk: a meta-analysis. Breast Cancer Res. Treat. 149, 525–536.

Dhawan, V., Schwalb, D.J., Shumway, M.J., Warren, M.C., Wexler, R.S., Zemtseva, I.S., et al., 2005. Selective nitros(yl)ation induced in vivo by a nitric oxide-donating cyclooxygenase-2 inhibitor: a nobonomic analysis. Free Radic. Biol. Med. 39, 1191–1207.

Dimmeler, S., Fleming, I., Fisslthaler, B., Hermann, C., Busse, R., Zeiher, A.M., 1999. Activation of nitric oxide synthase in endothelial cells by Akt-dependent phosphorylation. Nature 399, 601–605.

Dunlap, T., Chandrasena, R.E., Wang, Z., Sinha, V., Wang, Z., Thatcher, G.R., 2007. Quinone formation as a chemoprevention strategy for hybrid drugs: balancing cytotoxicity and cytoprotection. Chem. Res. Toxicol. 20, 1903–1912.

Dunlap, T., Piyankarage, S.C., Wijewickrama, G.T., Abdul-Hay, S., Vanni, M., Litosh, V., et al., 2012. Quinone-induced activation of Keap1/Nrf2 signaling by aspirin prodrugs masquerading as nitric oxide. Chem. Res. Toxicol. 25, 2725–2736.

Elliott, S.N., Mcknight, W., Cirino, G., Wallace, J.L., 1995. A nitric oxide-releasing nonsteroidal anti-inflammatory drug accelerates gastric ulcer healing in rats. Gastroenterology 109, 524–530.

Elmets, C.A., Viner, J.L., Pentland, A.P., Cantrell, W., Lin, H.Y., Bailey, H., et al., 2010. Chemoprevention of nonmelanoma skin cancer with celecoxib: a randomized, double-blind, placebo-controlled trial. J. Natl. Cancer Inst. 102, 1835–1844.

Elsheikh, W., Blackler, R.W., Flannigan, K.L., Wallace, J.L., 2014. Enhanced chemopreventive effects of a hydrogen sulfide-releasing anti-inflammatory drug (Atb-346) in experimental colorectal cancer. Nitric Oxide 41, 131–137.

Ezer, R., Alonso, M., Pereira, E., Kim, M., Allen, J.C., Miller, D.C., et al., 2002. Identification Of glutathione S-transferase (GST) polymorphisms in brain tumors and association with susceptibility to pediatric astrocytomas. J. Neurooncol. 59, 123–134.

Findlay, V.J., Townsend, D.M., Saavedra, J.E., Buzard, G.S., Citro, M.L., Keefer, L.K., et al., 2004. Tumor cell responses to a novel glutathione S-transferase-activated nitric oxide-releasing prodrug. Mol. Pharmacol. 65, 1070–1079.

Fionda, C., Abruzzese, M.P., Zingoni, A., Soriani, A., Ricci, B., Molfetta, R., et al., 2015. Nitric oxide donors increase PVR/CD155 DNAM-1 ligand expression in multiple myeloma cells: role of DNA damage response activation. BMC Cancer 15, 17.

Fiorucci, S., 2009. Prevention of nonsteroidal anti-inflammatory drug-induced ulcer: looking to the future. Gastroenterol. Clin. North Am. 38, 315–332.

Fiorucci, S., Antonelli, E., Santucci, L., Morelli, O., Miglietti, M., Federici, B., et al., 1999. Gastrointestinal safety of nitric oxide-derived aspirin is related to inhibition of ice-like cysteine proteases in rats. Gastroenterology 116, 1089–1106.

Fiorucci, S., Santucci, L., Antonelli, E., Distrutti, E., Del Sero, G., Morelli, O., et al., 2000a. No-aspirin protects from T cell-mediated liver injury by inhibiting caspase-dependent processing of Th1-like cytokines. Gastroenterology 118, 404–421.

Fiorucci, S., Santucci, L., Cirino, G., Mencarelli, A., Familiari, L., Soldato, P.D., et al., 2000b. IL-1 beta converting enzyme is a target for nitric oxide-releasing aspirin: new insights in the antiinflammatory mechanism of nitric oxide-releasing nonsteroidal antiinflammatory drugs. J. Immunol. 165, 5245–5254.

Fiorucci, S., Distrutti, E., Ajuebor, M.N., Mencarelli, A., Mannucci, R., Palazzetti, B., et al., 2001. NO-Mesalamine protects colonic epithelial cells against apoptotic damage induced by proinflammatory cytokines. Am. J. Physiol. Gastrointest. Liver Physiol. 281, G654–G665.

Fiorucci, S., Santucci, L., Gresele, P., Faccino, R.M., Del Soldato, P., Morelli, A., 2003a. Gastrointestinal safety of NO-Aspirin (NCX-4016) in healthy human volunteers: a proof of concept endoscopic study. Gastroenterology 124, 600–607.

Fiorucci, S., Santucci, L., Wallace, J.L., Sardina, M., Romano, M., Del Soldato, P., et al., 2003b. Interaction of a selective cyclooxygenase-2 inhibitor with aspirin and NO-releasing aspirin in the human gastric mucosa. Proc. Natl. Acad. Sci. USA 100, 10937–10941.

Fiorucci, S., Mencarelli, A., Meneguzzi, A., Lechi, A., Renga, B., Del Soldato, P., et al., 2004. Co-administration of nitric oxide-aspirin (NCX-4016) and aspirin prevents platelet and monocyte activation and protects against gastric damage induced by aspirin in humans. J. Am. Coll. Cardiol. 44, 635–641.

Friebe, A., Koesling, D., 2003. Regulation of nitric oxide-sensitive guanylyl cyclase. Circ. Res. 93, 96–105.

Fu, J., Liu, L., Huang, Z., Lai, Y., Ji, H., Peng, S., et al., 2013. Hybrid molecule from O2-(2,4-dinitrophenyl) diazeniumdiolate and oleanolic acid: a glutathione S-transferase Pi-activated nitric oxide prodrug with selective anti-human hepatocellular carcinoma activity and improved stability. J. Med. Chem. 56, 4641–4655.

Fung, H.L., Bauer, J.A., 1994. Mechanisms of nitrate tolerance. Cardiovasc. Drugs Ther. 8, 489–499.

Funk, C.D., Fitzgerald, G.A., 2007. Cox-2 inhibitors and cardiovascular risk. J. Cardiovasc. Pharmacol. 50, 470–479.

Gallo, O., Masini, E., Morbidelli, L., Franchi, A., Fini-Storchi, I., Vergari, W.A., et al., 1998. Role of nitric oxide in angiogenesis and tumor progression in head and neck cancer. J. Natl. Cancer Inst. 90, 587–596.

Gao, J., Kashfi, K., Rigas, B., 2005a. In vitro metabolism of nitric oxide-donating aspirin: the effect of positional isomerism. J. Pharmacol. Exp. Ther. 312, 989–997.

Gao, J., Liu, X., Rigas, B., 2005b. Nitric oxide-donating aspirin induces apoptosis in human colon cancer cells through induction of oxidative stress. Proc. Natl. Acad. Sci. USA 102, 17207–17212.

Gao, J., Kashfi, K., Liu, X., Rigas, B., 2006. No-donating aspirin induces phase II enzymes in vitro and in vivo. Carcinogenesis 27, 803–810.

Geller, D.A., Billiar, T.R., 1998. Molecular biology of nitric oxide synthases. Cancer Metastasis Rev. 17, 7–23.

Glynn, S.A., Boersma, B.J., Dorsey, T.H., Yi, M., Yfantis, H.G., Ridnour, L.A., et al., 2010. Increased NOS2 predicts poor survival in estrogen receptor-negative breast cancer patients. J. Clin. Invest. 120, 3843–3854.

Goligorsky, M.S., Brodsky, S.V., Noiri, E., 2004. No bioavailability, endothelial dysfunction, and acute renal failure: new insights into pathophysiology. Semin. Nephrol. 24, 316–323.

Gouda, A.M., Abdelazeem, A.H., Arafa El, S.A., Abdellatif, K.R., 2014. Design, synthesis and pharmacological evaluation of novel pyrrolizine derivatives as potential anticancer agents. Bioorg. Chem. 53, 1–7.

Gow, A.J., Farkouh, C.R., Munson, D.A., Posencheg, M.A., Ischiropoulos, H., 2004. Biological significance of nitric oxide-mediated protein modifications. Am. J. Physiol. Lung Cell. Mol. Physiol. 287, L262–L268.

Grosser, N., Schroder, H., 2000. A common pathway for nitric oxide release from NO-aspirin and glyceryl trinitrate. Biochem. Biophys. Res. Commun. 274, 255–258.

Gund, M., Gaikwad, P., Borhade, N., Burhan, A., Desai, D.C., Sharma, A., et al., 2014. Gastric-sparing nitric oxide-releasable 'true' prodrugs of aspirin and naproxen. Bioorg. Med. Chem. Lett. 24, 5587–5592.

Gupta, R.A., Wang, D., Katkuri, S., Wang, H., Dey, S.K., Dubois, R.N., 2004. Activation of nuclear hormone receptor peroxisome proliferator-activated receptor-delta accelerates intestinal adenoma growth. Nat. Med. 10, 245–247.

Hagos, G.K., Carroll, R.E., Kouznetsova, T., Li, Q., Toader, V., Fernandez, P.A., et al., 2007. Colon cancer chemoprevention by a novel no chimera that shows anti-inflammatory and antiproliferative activity in vitro and in vivo. Mol. Cancer Ther. 6, 2230–2239.

Hara, A., Okayasu, I., 2004. Cyclooxygenase-2 and inducible nitric oxide synthase expression in human astrocytic gliomas: correlation with angiogenesis and prognostic significance. Acta Neuropathol. 108, 43–48.

He, T.C., Chan, T.A., Vogelstein, B., Kinzler, K.W., 1999. PPARdelta is an APC-regulated target of nonsteroidal anti-inflammatory drugs. Cell 99, 335–345.

Hill, B.G., Dranka, B.P., Bailey, S.M., Lancaster Jr., J.R., Darley-Usmar, V.M., 2010. What part of no don't you understand? some answers to the cardinal questions in nitric oxide biology. J. Biol. Chem. 285, 19699–19704.

Hu, R., Siu, C.W., Lau, E.O., Wang, W.Q., Lau, C.P., Tse, H.F., 2007. Impaired nitrate-mediated dilatation could reflect nitrate tolerance in patients with coronary artery disease. Int. J. Cardiol. 120, 351–356.

Huguenin, S., Vacherot, F., Fleury-Feith, J., Riffaud, J.P., Chopin, D.K., Bolla, M., et al., 2005. Evaluation of the antitumoral potential of different nitric oxide-donating non-steroidal anti-inflammatory drugs (NO-NSAIDs) on human urological tumor cell lines. Cancer Lett. 218, 163–170.

Hulsman, N., Medema, J.P., Bos, C., Jongejan, A., Leurs, R., Smit, M.J., et al., 2007. Chemical insights in the concept of hybrid drugs: the antitumor effect of nitric oxide-donating aspirin involves a quinone methide but not nitric oxide nor aspirin. J. Med. Chem. 50, 2424–2431.

Hundley, T.R., Rigas, B., 2006. Nitric oxide-donating aspirin inhibits colon cancer cell growth via mitogen-activated protein kinase activation. J. Pharmacol. Exp. Ther. 316, 25–34.

Ignarro, L.J., 1989. Endothelium-derived nitric oxide: actions and properties. FASEB J. 3, 31–36.

Ignarro, L.J., Buga, G.M., Wood, K.S., Byrns, R.E., Chaudhuri, G., 1987. Endothelium-derived relaxing factor produced and released from artery and vein is nitric oxide. Proc. Natl. Acad. Sci. USA 84, 9265–9269.

Ignarro, L.J., Napoli, C., Loscalzo, J., 2002. Nitric oxide donors and cardiovascular agents modulating the bioactivity of nitric oxide: an overview. Circ. Res. 90, 21–28.

Juillerat-Jeanneret, L., Bernasconi, C.C., Bricod, C., Gros, S., Trepey, S., Benhattar, J., et al., 2008. Heterogeneity of human glioblastoma: glutathione-S-transferase and methylguanine-methyltransferase. Cancer Invest. 26, 597–609.

Karin, M., 2006. NF-kappaB and cancer: mechanisms and targets. Mol. Carcinog. 45, 355–361.

Karlsson, J., Pivodic, A., Aguirre, D., Schnitzer, T.J., 2009. Efficacy, safety, and tolerability of the cyclooxygenase-inhibiting nitric oxide donor naproxcinod in treating osteoarthritis of the hip or knee. J. Rheumatol. 36, 1290–1297.

Kashfi, K., Olson, K.R., 2013. Biology and therapeutic potential of hydrogen sulfide and hydrogen sulfide-releasing chimeras. Biochem. Pharmacol. 85, 689–703.

Kashfi, K., Rigas, B., 2005. Non-Cox-2 targets and cancer: expanding the molecular target repertoire of chemoprevention. Biochem. Pharmacol. 70, 969–986.

Kashfi, K., Rigas, B., 2007. The mechanism of action of nitric oxide-donating aspirin. Biochem. Biophys. Res. Commun. 358, 1096–1101.

Kashfi, K., 2009. Anti-inflammatory agents as cancer therapeutics. Adv. Pharmacol. 57, 31–89.

Kashfi, K., 2012. Nitric oxide-releasing hybrid drugs target cellular processes through S-nitrosylation. For. Immunopathol. Dis. Therap. 3, 97–108.

Kashfi, K., 2014. Anti-cancer activity of new designer hydrogen sulfide-donating hybrids. Antioxid. Redox. Signal. 20, 831–846.

Kashfi, K., Borgo, S., Williams, J.L., Chen, J., Gao, J., Glekas, A., et al., 2005. Positional isomerism markedly affects the growth inhibition of colon cancer cells by nitric oxide-donating aspirin in vitro and in vivo. J. Pharmacol. Exp. Ther. 312, 978–988.

Kashfi, K., Chattopadhyay, M., Kodela, R., 2015. NOSH-sulindac (Avt-18a) is a novel nitric oxide- and hydrogen sulfide-releasing hybrid that is gastrointestinal safe and has potent anti-inflammatory, analgesic, antipyretic, anti-platelet, and anti-cancer properties. Redox Biol. 6, 287–296.

Kashfi, K., Ryan, Y., Qiao, L.L., Williams, J.L., Chen, J., Del Soldato, P., et al., 2002. Nitric oxide-donating nonsteroidal anti-inflammatory drugs inhibit the growth of various cultured human cancer cells: evidence of a tissue type-independent effect. J. Pharmacol. Exp. Ther. 303, 1273–1282.

Kasper, H.U., Wolf, H., Drebber, U., Wolf, H.K., Kern, M.A., 2004. Expression of inducible nitric oxide synthase and cyclooxygenase-2 in pancreatic adenocarcinoma: correlation with microvessel density. World J. Gastroenterol. 10, 1918–1922.

Keefer, L.K., 2003. Progress toward clinical application of the nitric oxide-releasing diazeniumdiolates. Annu. Rev. Pharmacol. Toxicol. 43, 585–607.

Keefer, L.K., 2010. Broad-spectrum anti-cancer activity of O-arylated diazeniumdiolates. For. Immunopathol. Dis. Therap. 1, 205–218.

Keefer, L.K., 2011. Fifty years of diazeniumdiolate research. From laboratory curiosity to broad-spectrum biomedical advances. ACS Chem. Biol. 6, 1147–1155.

Khan, N.I., Cisterne, A., Baraz, R., Bradstock, K.F., Bendall, L.J., 2012. Para-NO-aspirin inhibits NF-kappaB and induces apoptosis in B-cell progenitor acute lymphoblastic leukemia. Exp. Hematol. 40, 207–215. E1.

Kim, M.M., Parmar, H., Cao, Y., Knox, S.J., Oronsky, B., Scicinski, J., et al., 2016. Concurrent whole brain radiotherapy and RRx-001 for melanoma brain metastases. Neuro Oncol. 18, 455–456.

Kiziltepe, T., Hideshima, T., Ishitsuka, K., Ocio, E.M., Raje, N., Catley, L., et al., 2007. JS-K, a GST-activated nitric oxide generator, induces DNA double-strand breaks, activates DNA damage response pathways, and induces apoptosis in vitro and in vivo in human multiple myeloma cells. Blood 110, 709–718.

Kleinert, H., Schwarz, P.M., Forstermann, U., 2003. Regulation of the expression of inducible nitric oxide synthase. Biol. Chem. 384, 1343–1364.

Klotz, T., Bloch, W., Volberg, C., Engelmann, U., Addicks, K., 1998. Selective expression of inducible nitric oxide synthase in human prostate carcinoma. Cancer 82, 1897–1903.

Knox, S.J., Cannizzo, L., Warner, K., Wardle, R., Velarde, S. & Ning, S., 2009. Cyclic nitro compounds, pharmaceutical compositions thereof and uses thereof. United States Patent Application US7507842b2.

Kodela, R., Chattopadhyay, M., Kashfi, K., 2012. NOSH-aspirin: a novel nitric oxide-hydrogen sulfide-releasing hybrid: a new class of anti-inflammatory pharmaceuticals. ACS Med. Chem. Lett. 3, 257–262.

Kodela, R., Chattopadhyay, M., Kashfi, K., 2013. Synthesis and biological activity of NOSH-naproxen (AVT-219) and NOSH-sulindac (AVT-18A) as potent anti-inflammatory agents with chemotherapeutic potential. Medchemcomm, 4.

Kodela, R., Chattopadhyay, M., Nath, N., Cieciura, L.Z., Pospishil, L., Boring, D., et al., 2008a. Ester-protected hydroxybenzyl phosphates (EHBP) inhibit the growth of various cultured human cancer cells: evidence of a tissue type-independent effect. In: AACR: Seventh Annual International Conference on Frontiers in Cancer Prevention Research, Washington, DC.

Kodela, R., Chattopadhyay, M., Nath, N., Cieciura, L.Z., Pospishill, L., Boring, D., et al., 2011. Synthesis and biological activity of acetyl-protected hydroxybenzyl diethyl phosphates (EHBP) as potential chemotherapeutic agents. Bioorg. Med. Chem. Lett. 21, 7146–7150.

Kodela, R., Chattopadhyay, M., Velazquez-Martinez, C.A., Kashfi, K., 2015a. NOSH-aspirin (NBS-1120), a novel nitric oxide- and hydrogen sulfide-releasing hybrid has enhanced chemo-preventive properties compared to aspirin, is gastrointestinal safe with all the classic therapeutic indications. Biochem. Pharmacol. 98, 564–572.

Kodela, R., Nath, N., Chattopadhyay, M., Nesbitt, D.E., Velazquez-Martinez, C.A., Kashfi, K., 2015b. Hydrogen sulfide-releasing naproxen suppresses colon cancer cell growth and inhibits NF-kappaB signaling. Drug Des. Devel. Ther. 9, 4873–4882.

Kodela, R., Rokita, S.E., Boring, D., Crowell, J.A., Kashfi, K., 2008b. Bioactivated chemotherapeutic agents based on ester-protected hydroxybenzyl phosphates (EHBP) for reversible addition to cellular nucleophiles. In: AACR: Seventh Annual International Conference on Frontiers in Cancer Prevention Research, Washington, DC.

Kogias, E., Osterberg, N., Baumer, B., Psarras, N., Koentges, C., Papazoglou, A., et al., 2012. Growth-inhibitory and chemosensitizing effects of the glutathione-S-transferase-Pi-activated nitric oxide donor PABA/NO in malignant gliomas. Int. J. Cancer 130, 1184–1194.

Kolios, G., Valatas, V., Ward, S.G., 2004. Nitric oxide in inflammatory bowel disease: a universal messenger in an unsolved puzzle. Immunology 113, 427–437.

Kundu, J.K., Surh, Y.J., 2008. Inflammation: gearing the journey to cancer. Mutat. Res. 659, 15–30.

Kune, G., Kune, S., Watson, L., 1988. Colorectal cancer risk, chronic illnesses, operations, and medications: case–control results from the Melbourne Colorectal Cancer Study. Cancer Res. 48, 4399–4404.

Kwak, M.K., Wakabayashi, N., Kensler, T.W., 2004. Chemoprevention through the Keap1-Nrf2 signaling pathway by phase 2 enzyme inducers. Mutat. Res. 555, 133–148.

Lagares-Garcia, J.A., Moore, R.A., Collier, B., Heggere, M., Diaz, F., Qian, F., 2001. Nitric oxide synthase as a marker in colorectal carcinoma Am. Surg., 67709–713.

Lancaster Jr., J.R., 2003. Reactivity and diffusivity of nitrogen oxides in mammalian biology. In: Forman, H.J., Fukuto, J., Torres, M. (Eds.), Signal Transduction by Reactive Oxygen and Nitrogen Species: Pathways and Chemical Principles. Springer, Netherlands.

Landino, L.M., Crews, B.C., Timmons, M.D., Morrow, J.D., Marnett, L.J., 1996. Peroxynitrite, the coupling product of nitric oxide and superoxide, activates prostaglandin biosynthesis. Proc. Natl. Acad. Sci. USA 93, 15069–15074.

Lane, P., Hao, G., Gross, S.S., 2001. S-nitrosylation is emerging as a specific and fundamental posttranslational protein modification: head-to-head comparison with O-phosphorylation. Sci. STKE, 2001. Re1.

Laschak, M., Spindler, K.D., Schrader, A.J., Hessenauer, A., Streicher, W., Schrader, M., et al., 2012. Js-K, A glutathione/glutathione S-transferase-activated nitric oxide releasing prodrug inhibits androgen receptor and Wnt-signaling in prostate cancer cells. BMC Cancer 12, 130.

Lazzarato, L., Donnola, M., Rolando, B., Marini, E., Cena, C., Coruzzi, G., et al., 2008. Searching for new no-donor aspirin-like molecules: a new class of nitrooxy-acyl derivatives of salicylic acid. J. Med. Chem. 51, 1894–1903.

Lee, J.S., Surh, Y.J., 2005. Nrf2 as a novel molecular target for chemoprevention. Cancer Lett. 224, 171–184.

Leonetti, C., Scarsella, M., Zupi, G., Zoli, W., Amadori, D., Medri, L., et al., 2006. Efficacy of a nitric oxide-releasing nonsteroidal anti-inflammatory drug and cytotoxic drugs in human colon cancer cell lines in vitro and xenografts. Mol. Cancer Ther. 5, 919–926.

Li, J., Billiar, T.R., Talanian, R.V., Kim, Y.M., 1997. Nitric oxide reversibly inhibits seven members of the caspase family via S-nitrosylation. Biochem. Biophys. Res. Commun. 240, 419–424.

Liao, L.M., Vaughan, T.L., Corley, D.A., Cook, M.B., Casson, A.G., Kamangar, F., et al., 2012. Nonsteroidal anti-inflammatory drug use reduces risk of adenocarcinomas of the esophagus and esophagogastric junction in a pooled analysis. Gastroenterology 142, 442–452. E5; Quiz E22-3.

Lincoln, T., Cornwell, T., Komalavilas, P., Macmillan-Crow, L.A., Boerth, N.J., 1996. The nitric oxide-cyclic GMP signaling system. In: M, B. (Ed.), Biochemistry of Smooth Muscle Contraction. Academic Press, San Diego.

Lipkin, M., Yang, K., Edelmann, W., Xue, L., Fan, K., Risio, M., et al., 1999. Preclinical mouse models for cancer chemoprevention studies. Ann. NY. Acad. Sci. 889, 14–19.

Liu, J., Li, C., Qu, W., Leslie, E., Bonifant, C.L., Buzard, G.S., et al., 2004. Nitric oxide prodrugs and metal-lochemotherapeutics: JS-K and Cb-3-100 enhance arsenic and cisplatin cytolethality by increasing cellular accumulation. Mol. Cancer Ther. 3, 709–714.

Liu, J., Saavedra, J.E., Lu, T., Song, J.G., Clark, J., Waalkes, M.P., et al., 2002. O(2)-vinyl 1-(pyrrolidin-1-yl) diazen-1-ium-1,2-diolate protection against D-galactosamine/endotoxin-induced hepatotoxicity in mice: genomic analysis using microarrays. J. Pharmacol. Exp. Ther. 300, 18–25.

Liu, Y., Hartley, D.P., Liu, J., 1998. Protection against carbon tetrachloride hepatotoxicity by oleanolic acid is not mediated through metallothionein. Toxicol. Lett. 95, 77–85.

Lohmander, L.S., Mckeith, D., Svensson, O., Malmenas, M., Bolin, L., Kalla, A., et al., 2005. A randomised, placebo controlled, comparative trial of the gastrointestinal safety and efficacy of AZD3582 versus naproxen in osteoarthritis. Ann. Rheum. Dis. 64, 449–456.

Lopez-Otin, C., Hunter, T., 2010. The regulatory crosstalk between kinases and proteases in cancer. Nat. Rev. Cancer 10, 278–292.

Maciag, A.E., Chakrapani, H., Saavedra, J.E., Morris, N.L., Holland, R.J., Kosak, K.M., et al., 2011. The nitric oxide prodrug JS-K is effective against non-small-cell lung cancer cells in vitro and in vivo: involvement of reactive oxygen species. J. Pharmacol. Exp. Ther. 336, 313–320.

Maciag, A.E., Holland, R.J., Robert Cheng, Y.S., Rodriguez, L.G., Saavedra, J.E., Anderson, L.M., et al., 2013. Nitric oxide-releasing prodrug triggers cancer cell death through deregulation of cellular redox balance. Redox Biol. 1, 115–124.

Maciag, A.E., Saavedra, J.E., Chakrapani, H., 2009. The nitric oxide prodrug JS-K and its structural analogues as cancer therapeutic agents. Anticancer Agents Med. Chem. 9, 798–803.

Mantovani, A., Allavena, P., Sica, A., Balkwill, F., 2008. Cancer-related inflammation. Nature 454, 436–444.

Marshall, H.E., Merchant, K., Stamler, J.S., 2000. Nitrosation and oxidation in the regulation of gene expression. FASEB J. 14, 1889–1900.

Martelli, A., Testai, L., Anzini, M., Cappelli, A., Di Capua, A., Biava, M., et al., 2013. The novel anti-inflammatory agent Va694, endowed with both no-releasing and COX2-selective inhibiting properties, exhibits NO-mediated positive effects on blood pressure, coronary flow and endothelium in an experimental model of hypertension and endothelial dysfunction. Pharmacol. Res. 78, 1–9.

Masri, F.A., Comhair, S.A., Koeck, T., Xu, W., Janocha, A., Ghosh, S., et al., 2005. Abnormalities in nitric oxide and its derivatives in lung cancer. Am. J. Respir. Crit. Care Med. 172, 597–605.

Maulik, N., Das, D.K., 2008. Emerging potential of thioredoxin and thioredoxin interacting proteins in various disease conditions. Biochim. Biophys. Acta 1780, 1368–1382.

Mcmurtry, V., Saavedra, J.E., Nieves-Alicea, R., Simeone, A.M., Keefer, L.K., Tari, A.M., 2011. JS-K, a nitric oxide-releasing prodrug, induces breast cancer cell death while sparing normal mammary epithelial cells. Int. J. Oncol. 38, 963–971.

Meyerhardt, J.A., Mayer, R.J., 2005. Systemic therapy for colorectal cancer. N. Engl. J. Med. 352, 476–487.

Michel, T., Feron, O., 1997. Nitric oxide synthases: which, where, how, and why? J. Clin. Invest. 100, 2146–2152.

Miller, M.R., Megson, I.L., 2007. Recent developments in nitric oxide donor drugs. Br. J. Pharmacol. 151, 305–321.

Millet, A., Bettaieb, A., Renaud, F., Prevotat, L., Hammann, A., Solary, E., et al., 2002. Influence of the nitric oxide donor glyceryl trinitrate on apoptotic pathways in human colon cancer cells. Gastroenterology 123, 235–246.

Moncada, S., Higgs, A., 1993. The L-arginine-nitric oxide pathway. N. Engl. J. Med. 329, 2002–2012.

Moon, R.T., Kohn, A.D., De Ferrari, G.V., Kaykas, A., 2004. Wnt and beta-catenin signalling: diseases and therapies. Nat. Rev. Genet. 5, 691–701.

Mueller, M.M., Fusenig, N.E., 2004. Friends or foes—bipolar effects of the tumour stroma in cancer. Nat. Rev. Cancer 4, 839–849.

Muscat, J.E., Dyer, A.M., Rosenbaum, R.E., Rigas, B., 2005. Nitric oxide-releasing medications and colorectal cancer risk: the Framingham study. Anticancer Res. 25, 4471–4474.

Nagasawa, H.T., Kawle, S.P., Elberling, J.A., Demaster, E.G., Fukuto, J.M., 1995. Prodrugs of nitroxyl as potential aldehyde dehydrogenase inhibitors vis-a-vis vascular smooth muscle relaxants. J. Med. Chem. 38, 1865–1871.

Nandurdikar, R.S., Maciag, A.E., Cao, Z., Keefer, L.K., Saavedra, J.E., 2012. Diazeniumdiolated carbamates: a novel class of nitric oxide donors. Bioorg. Med. Chem. 20, 2025–2029.

Nath, N., Kashfi, K., Chen, J., Rigas, B., 2003. Nitric oxide-donating aspirin inhibits beta-catenin/T cell factor (TCF) signaling in SW480 colon cancer cells by disrupting the nuclear beta-catenin-TCF Association. Proc. Natl. Acad. Sci. USA 100, 12584–12589.

Nath, N., Labaze, G., Rigas, B., Kashfi, K., 2005. NO-donating aspirin inhibits the growth of leukemic Jurkat cells and modulates beta-catenin expression. Biochem. Biophys. Res. Commun. 326, 93–99.

Nath, N., Chattopadhyay, M., Pospishil, L., Cieciura, L.Z., Goswami, S., Kodela, R., et al., 2010. JS-K, a nitric oxide-releasing prodrug, modulates ß-catenin/TCF signaling in leukemic Jurkat cells: evidence of an S-nitrosylated mechanism. Biochem. Pharmacol. 80, 1641–1649.

Nath, N., Vassell, R., Chattopadhyay, M., Kogan, M., Kashfi, K., 2009. Nitro-aspirin inhibits MCF-7 breast cancer cell growth: effects on Cox-2 expression and Wnt/beta-catenin/TCF-4 signaling. Biochem. Pharmacol. 78, 1298–1304.

Nath, N., Liu, X., Jacobs, L., Kashfi, K., 2013. Flurbiprofen benzyl nitrate (NBS-242) inhibits the growth of A-431 human epidermoid carcinoma cells and targets beta-catenin. Drug Des. Devel. Ther. 7, 389–396.

Nath, N., Chattopadhyay, M., Rodes, D.B., Nazarenko, A., Kodela, R., Kashfi, K., 2015. Nitric oxide-releasing aspirin suppresses NF-kappaB signaling in estrogen receptor negative breast cancer cells in vitro and in vivo. Molecules 20, 12481–12499.

Nathan, C., Xie, Q.W., 1994. Nitric oxide synthases: roles, tolls, and controls. Cell 78, 915–918.

Nemmani, K.V., Mali, S.V., Borhade, N., Pathan, A.R., Karwa, M., Pamidiboina, V., et al., 2009. NO-NSAIDs: gastric-sparing nitric oxide-releasable prodrugs of non-steroidal anti-inflammatory drugs. Bioorg. Med. Chem. Lett. 19, 5297–5301.

Nicastro, H.L., Grubbs, C.J., Margaret Juliana, M., Bode, A.M., Kim, M.S., Lu, Y., et al., 2014. Preventive effects of NSAIDs, NO-NSAIDs, and NSAIDs plus difluoromethylornithine in a chemically induced urinary bladder cancer model. Cancer Prev. Res. (Phila) 7, 246–254.

Nicox, S.A., 2007. Nicox provides an update on NCX-4016. Sophia Antipolis, France [Online]. <http://www.nicox.com/news-media/press-releases/nicox-provides-an-update-on-ncx-4016/>. Updated June 18, 2007; Cited August 12, 2016.

Nicox, S.A. 2010. *FDA provides complete response letter to Nicox's new drug application for Naproxcinod.* [Online]. <http://www.marketwired.com/press-release/FDA-provides-Complete-Response-Letter-to-NicOxs-New-Drug-Application-for-naproxcinod-Paris-COX-1294017.htm>. Updated July 22, 2010; Cited August 12, 2016.

Ning, S., Bednarski, M., Oronsky, B., Scicinski, J., Saul, G., Knox, S.J., 2012. Dinitroazetidines are a novel class of anticancer agents and hypoxia-activated radiation sensitizers developed from highly energetic materials. Cancer Res. 72, 2600–2608.

Ning, S., Sekar, T.V., Scicinski, J., Oronsky, B., Peehl, D.M., Knox, S.J., et al., 2015. Nrf2 activity as a potential biomarker for the pan-epigenetic anticancer agent, RRx-001. Oncotarget 6, 21547–21556.

Norris, A.J., Sartippour, M.R., Lu, M., Park, T., Rao, J.Y., Jackson, M.I., et al., 2008. Nitroxyl inhibits breast tumor growth and angiogenesis. Int. J. Cancer 122, 1905–1910.

Notas, G., Xidakis, C., Valatas, V., Kouroumalis, A., Kouroumalis, E., 2001. Levels of circulating endothelin-1 and nitrates/nitrites in patients with virus-related hepatocellular carcinoma. J. Viral Hepat. 8, 63–69.

Oronsky, B., Scicinski, J., Ning, S., Peehl, D., Oronsky, A., Cabrales, P., et al., 2016. RRx-001, a novel dinitroazetidine radiosensitizer. Invest. New Drugs 34, 371–377.

Ouyang, N., Williams, J.L., Rigas, B., 2006a. No-donating aspirin isomers downregulate peroxisome proliferator-activated receptor (PPAR){Delta} expression in ApcMin/+ mice proportionally to their tumor inhibitory effect: implications for the role of PPARΔ in carcinogenesis. Carcinogenesis 27, 232–239.

Ouyang, N., Williams, J.L., Tsioulias, G.J., Gao, J., Iatropoulos, M.J., Kopelovich, L., et al., 2006b. Nitric oxide-donating aspirin prevents pancreatic cancer in a Hamster Tumor Model. Cancer Res. 66, 4503–4511.

Paolocci, N., Saavedra, W.F., Miranda, K.M., Martignani, C., Isoda, T., Hare, J.M., et al., 2001. Nitroxyl anion exerts redox sensitive positive cardiac inotropy in vivo by calcitonin gene-related peptide signaling. Proc. Natl. Acad. Sci. USA 98, 10463–10468.

Paolocci, N., Katori, T., Champion, H.C., St John, M.E., Miranda, K.M., Fukuto, J.M., et al., 2003. Positive inotropic and lusitropic effects of HNO/NO- in failing hearts: independence from beta-adrenergic signaling. Proc. Natl. Acad. Sci. USA 100, 5537–5542.

Papapetropoulos, A., Pyriochou, A., Altaany, Z., Yang, G., Marazioti, A., Zhou, Z., et al., 2009. Hydrogen sulfide is an endogenous stimulator of angiogenesis. Proc. Natl. Acad. Sci. USA 106, 21972–21977.

Pathan, A.R., Karwa, M., Pamidiboina, V., Deshattiwar, J.J., et al., 2010. Oral bioavailability, efficacy and gastric tolerability of P2026, a novel nitric oxide-releasing diclofenac in rat. Inflammopharmacology 18, 157–168.

Perez-Sala, D., Lamas, S., 2001. Regulation of cyclooxygenase-2 expression by nitric oxide in cells. Antioxid. Redox Signal. 3, 231–248.

Pierce, E.N., Piyankarage, S.C., Dunlap, T., Litosh, V., Siklos, M.I., Wang, Y.T., et al., 2016. Prodrugs bioactivated to quinones target NF-kappaB and multiple protein networks: identification of the quinonome. Chem. Res. Toxicol. 29, 1151–1159.

Pipili-Synetos, E., Papageorgiou, A., Sakkoula, E., Sotiropoulou, G., Fotsis, T., Karakiulakis, G., et al., 1995. Inhibition of angiogenesis, tumour growth and metastasis by the NO-releasing vasodilators, isosorbide mononitrate and dinitrate. Br. J. Pharmacol. 116, 1829–1834.

Quinn, A.C., Petros, A.J., Vallance, P., 1995. Nitric oxide: an endogenous gas. Br. J. Anaesth. 74, 443–451.

Rao, C.V., Kawamori, T., Hamid, R., Reddy, B.S., 1999. Chemoprevention of colonic aberrant crypt foci by an inducible nitric oxide synthase-selective inhibitor. Carcinogenesis 20, 641–644.

Rao, C.V., Reddy, B.S., Steele, V.E., Wang, C.X., Liu, X., Ouyang, N., et al., 2006. Nitric oxide-releasing aspirin and indomethacin are potent inhibitors against colon cancer in azoxymethane-treated rats: effects on molecular targets. Mol. Cancer Ther. 5, 1530–1538.

Reid, T., Oronsky, B., Scicinski, J., Scribner, C.L., Knox, S.J., Ning, S., et al., 2015. Safety and activity of RRx-001 in patients with advanced cancer: a first-in-human, open-label, dose-escalation phase 1 study. Lancet Oncol. 16, 1133–1142.

Ren, Z., Kar, S., Wang, Z., Wang, M., Saavedra, J.E., Carr, B.I., 2003. JS-K, a novel non-ionic diazeniumdiolate derivative, inhibits Hep 3b hepatoma cell growth and induces C-Jun phosphorylation via multiple map kinase pathways. J. Cell. Physiol. 197, 426–434.

Reynaert, N.L., Ckless, K., Korn, S.H., Vos, N., Guala, A.S., Wouters, E.F., et al., 2004. Nitric oxide represses inhibitory kappaB kinase through S-nitrosylation. Proc. Natl. Acad. Sci. USA 101, 8945–8950.

Ridnour, L.A., Thomas, D.D., Donzelli, S., Espey, M.G., Roberts, D.D., Wink, D.A., et al., 2006. The biphasic nature of nitric oxide responses in tumor biology. Antioxid. Redox. Signal. 8, 1329–1337.

Rigas, B., Sun, Y., 2008. Induction of oxidative stress as a mechanism of action of chemopreventive agents against cancer. Br. J. Cancer 98, 1157–1160.

Rothwell, P.M., Price, J.F., Fowkes, F.G., Zanchetti, A., Roncaglioni, M.C., Tognoni, G., et al., 2012. Short-term effects of daily aspirin on cancer incidence, mortality, and non-vascular death: analysis of the time course of risks and benefits in 51 randomised controlled trials. Lancet 379, 1602–1612.

Saavedra, J.E., Srinivasan, A., Buzard, G.S., Davies, K.M., Waterhouse, D.J., Inami, K., et al., 2006. PABA/NO as an anticancer lead: analogue synthesis, structure revision, solution chemistry, reactivity toward glutathione, and in vitro activity. J. Med. Chem. 49, 1157–1164.

Salmon, D.J., Torres De Holding, C.L., Thomas, L., Peterson, K.V., Goodman, G.P., Saavedra, J.E., et al., 2011. HNO and NO release from a primary amine-based diazeniumdiolate as a function of pH. Inorg. Chem. 50, 3262–3270.

Sandler, R.S., Halabi, S., Baron, J.A., Budinger, S., Paskett, E., Keresztes, R., et al., 2003. A Randomized trial of aspirin to prevent colorectal adenomas in patients with previous colorectal cancer. N. Engl. J. Med. 348, 883–890.

Santucci, L., Fiorucci, S., Di Matteo, F.M., Morelli, A., 1995. Role of tumor necrosis factor alpha release and leukocyte margination in indomethacin-induced gastric injury in rats. Gastroenterology 108, 393–401.

Schnitzer, T.J., Kivitz, A.J., Lipetz, R.S., Sanders, N., Hee, A., 2005. Comparison of the COX-inhibiting nitric oxide donator Azd3582 and rofecoxib in treating the signs and symptoms of osteoarthritis of the knee. Arthritis Rheum. 53, 827–837.

Schnitzer, T.J., Kivitz, A., Frayssinet, H., Duquesroix, B., 2010. Efficacy and safety of naproxcinod in the treatment of patients with osteoarthritis of the knee: a 13-week prospective, randomized, multicenter study. Osteoarthritis Cartilage 18, 629–639.

Schnitzer, T.J., Hochberg, M.C., Marrero, C.E., Duquesroix, B., Frayssinet, H., Beekman, M., 2011. Efficacy and safety of naproxcinod in patients with osteoarthritis of the knee: A 53-week prospective randomized multicenter study. Semin. Arthritis Rheum. 40, 285–297.

Schottenfeld, D., Beebe-Dimmer, J., 2006. Chronic inflammation: a common and important factor in the pathogenesis of neoplasia. CA Cancer J. Clin. 56, 69–83.

Scicinski, J., Oronsky, B., Cooper, V., Taylor, M., Alexander, M., Hadar, R., et al., 2014. Development of methods for the bioanalysis of RRx-001 and metabolites. Bioanalysis 6, 947–956.

Scicinski, J., Oronsky, B., Ning, S., Knox, S., Peehl, D., Kim, M.M., et al., 2015. NO to cancer: the complex and multifaceted role of nitric oxide and the epigenetic nitric oxide donor, RRx-001. Redox Biol. 6, 1–8.

Scicinski, J., Oronsky, B., Taylor, M., Luo, G., Musick, T., Marini, J., et al., 2012. Preclinical evaluation of the metabolism and disposition of RRx-001, a novel investigative anticancer agent. Drug Metab. Dispos. 40, 1810–1816.

Sessa, W.C., 2004. eNOS at a glance. J. Cell Sci. 117, 2427–2429.

Shami, P.J., Saavedra, J.E., Wang, L.Y., Bonifant, C.L., Diwan, B.A., Singh, S.V., et al., 2003. JS-K, a glutathione/glutathione S-transferase-activated nitric oxide donor of the diazeniumdiolate class with potent antineoplastic activity. Mol. Cancer Ther. 2, 409–417.

Shami, P.J., Saavedra, J.E., Bonifant, C.L., Chu, J., Udupi, V., Malaviya, S., et al., 2006. Antitumor activity of JS-K [O2-(2,4-dinitrophenyl) 1-[(4-ethoxycarbonyl)piperazin-1-yl]diazen-1-ium-1,2-diolate] and related O2-aryl diazeniumdiolates in vitro and in vivo. J. Med. Chem. 49, 4356–4366.

Shami, P.J., Maciag, A.E., Eddington, J.K., Udupi, V., Kosak, K.M., Saavedra, J.E., et al., 2009. JS-K, an arylating nitric oxide (NO) donor, has synergistic anti-leukemic activity with cytarabine (Ara-C). Leuk. Res. 33, 1525–1529.

Shebl, F.M., Sakoda, L.C., Black, A., Koshiol, J., Andriole, G.L., Grubb, R., et al., 2012. Aspirin but not ibuprofen use is associated with reduced risk of prostate cancer: a PLCO study. Br. J. Cancer 107, 207–214.

Siezen, C.L., Tijhuis, M.J., Kram, N.R., Van Soest, E.M., De Jong, D.J., Fodde, R., et al., 2006. Protective effect of nonsteroidal anti-inflammatory drugs on colorectal adenomas is modified by a polymorphism in peroxisome proliferator-activated receptor delta. Pharmacogenet. Genomics 16, 43–50.

Sikder, N., Sikder, A.K., Bulakh, N.R., Gandhe, B.R., 2004. 1,3,3-Trinitroazetidine (Tnaz), a melt-cast explosive: synthesis, characterization and thermal behaviour. J. Hazard. Mater. 113, 35–43.

Soh, J.W., Weinstein, I.B., 2003. Role of Cox-independent targets of NSAIDs and related compounds in cancer prevention and treatment. Prog. Exp. Tumor Res. 37, 261–285.

Solomon, D.H., 2008. Recommendations for use of selective and nonselective nonsteroidal antiinflammatory drugs: an American College of Rheumatology White Paper. Arthritis Rheum. 59, 1058–1073.

Son, K., Kim, Y.M., 1995. In vivo cisplatin-exposed macrophages increase immunostimulant-induced nitric oxide synthesis for tumor cell killing. Cancer Res. 55, 5524–5527.

Spiegel, A., Hundley, T.R., Chen, J., Gao, J., Ouyang, N., Liu, X., et al., 2005. No-donating aspirin inhibits both the expression and catalytic activity of inducible nitric oxide synthase in HT-29 human colon cancer cells. Biochem. Pharmacol. 70, 993–1000.

Stamler, J.S., Simon, D.I., Osborne, J.A., Mullins, M.E., Jaraki, O., Michel, T., et al., 1992. S-nitrosylation of proteins with nitric oxide: synthesis and characterization of biologically active compounds. Proc. Natl. Acad. Sci. USA 89, 444–448.

Stockert, R.J., Morell, A.G., 1983. Hepatic binding protein: the galactose-specific receptor of mammalian hepatocytes. Hepatology 3, 750–757.

Stuehr, D.J., 1997. Structure-function aspects in the nitric oxide synthases. Annu. Rev. Pharmacol. Toxicol. 37, 339–359.

Sun, L., Yu, S., 2011. Meta-analysis: non-steroidal anti-inflammatory drug use and the risk of esophageal squamous cell carcinoma. Dis. Esophagus 24, 544–549.

Szabo, C., 2016. Gasotransmitters in cancer: from pathophysiology to experimental therapy. Nat. Rev. Drug Discov. 15, 185–203.

Szlosarek, P.W., Balkwill, F.R., 2003. Tumour necrosis factor alpha: a potential target for the therapy of solid tumours. Lancet Oncol. 4, 565–573.

Takahashi, M., Mutoh, M., Shoji, Y., Sato, H., Kamanaka, Y., Naka, M., et al., 2006. Suppressive effect of an inducible nitric oxide inhibitor, Ono-1714, on aom-induced rat colon carcinogenesis. Nitric Oxide 14, 130–136.

Takkouche, B., Regueira-Mendez, C., Etminan, M., 2008. Breast cancer and use of nonsteroidal anti-inflammatory drugs: a meta-analysis. J. Natl. Cancer Inst. 100, 1439–1447.

Tamano, S., Ward, J.M., Diwan, B.A., Keefer, L.K., Weghorst, C.M., Calvert, R.J., et al., 1996. Histogenesis and the role of P53 and K-Ras mutations in hepatocarcinogenesis by glyceryl trinitrate (nitroglycerin) in male F344 rats. Carcinogenesis 17, 2477–2486.

Tesei, A., Ulivi, P., Fabbri, F., Rosetti, M., Leonetti, C., Scarsella, M., et al., 2005. In vitro and in vivo evaluation of Ncx 4040 cytotoxic activity in human colon cancer cell lines. J. Transl. Med. 3, 7.

Thatcher, G.R., Nicolescu, A.C., Bennett, B.M., Toader, V., 2004. Nitrates and no release: contemporary aspects in biological and medicinal chemistry. Free Radic. Biol. Med. 37, 1122–1143.

Thomas, D.D., Ridnour, L.A., Isenberg, J.S., Flores-Santana, W., Switzer, C.H., Donzelli, S., et al., 2008. The chemical biology of nitric oxide: implications in cellular signaling. Free Radic. Biol. Med. 45, 18–31.

Thomsen, L.L., Miles, D.W., Happerfield, L., Bobrow, L.G., Knowles, R.G., Moncada, S., 1995. Nitric oxide synthase activity in human breast cancer. Br. J. Cancer 72, 41–44.

Thun, M.J., Henley, S.J., Patrono, C., 2002. Nonsteroidal anti-inflammatory drugs as anticancer agents: mechanistic, pharmacologic, and clinical issues. J. Natl. Cancer Inst. 94, 252–266.

Thun, M.J., Namboodiri, M.M., Heath Jr., C., 1991. Aspirin use and reduced risk of fatal colon cancer. New Engl. J. Med. 325, 1593–1596.

Trabert, B., Ness, R.B., Lo-Ciganic, W.H., Murphy, M.A., Goode, E.L., Poole, E.M., et al., 2014. Aspirin, nonaspirin nonsteroidal anti-inflammatory drug, and acetaminophen use and risk of invasive epithelial ovarian cancer: a pooled analysis in the Ovarian Cancer Association Consortium. J. Natl. Cancer Inst., 106. Djt431.

Trikha, P., Sharma, N., Athar, M., 2001. Nitroglycerin: a NO donor inhibits Tpa-mediated tumor promotion in murine skin. Carcinogenesis 22, 1207–1211.

Turnbull, C.M., Cena, C., Fruttero, R., Gasco, A., Rossi, A.G., Megson, I.L., 2006. Mechanism of action of novel NO-releasing furoxan derivatives of aspirin in human platelets. Br. J. Pharmacol. 148, 517–526.

Turnbull, C.M., Marcarino, P., Sheldrake, T.A., Lazzarato, L., Cena, C., Fruttero, R., et al., 2008. A novel hybrid aspirin-NO-releasing compound inhibits TNFalpha release from LPS-activated human monocytes and macrophages. J. Inflamm. (Lond.) 5, 12.

Vannini, F., Kashfi, K., Nath, N., 2015. The dual role of iNOS in cancer. Redox. Biol. 6, 334–343.

Vasudevan, D., Thomas, D.D., 2014. Insights into the diverse effects of nitric oxide on tumor biology. Vitam. Horm. 96, 265–298.

Velazquez, C., Praveen Rao, P.N., Knaus, E.E., 2005. Novel nonsteroidal antiinflammatory drugs possessing a nitric oxide donor diazen-1-ium-1,2-diolate moiety: design, synthesis, biological evaluation, and nitric oxide release studies. J. Med. Chem. 48, 4061–4067.

Velazquez, C.A., Chen, Q.H., Citro, M.L., Keefer, L.K., Knaus, E.E., 2008. Second-generation aspirin and indomethacin prodrugs possessing an O(2)-(acetoxymethyl)-1-(2-carboxypyrrolidin-1-yl)diazenium-1,2-diolate nitric oxide donor moiety: design, synthesis, biological evaluation, and nitric oxide release studies. J. Med. Chem. 51, 1954–1961.

Velazquez, C.A., Praveen Rao, P.N., Citro, M.L., Keefer, L.K., Knaus, E.E., 2007. O2-acetoxymethyl-protected diazeniumdiolate-based NSAIDs (NONO-NSAIDs): synthesis, nitric oxide release, and biological evaluation studies. Bioorg. Med. Chem. 15, 4767–4774.

Vitturi, D.A., Sun, C.W., Harper, V.M., Thrash-Williams, B., Cantu-Medellin, N., Chacko, B.K., et al., 2013. Antioxidant functions for the hemoglobin Beta93 cysteine residue in erythrocytes and in the vascular compartment in vivo. Free Radic. Biol. Med. 55, 119–129.

Voronov, E., Shouval, D.S., Krelin, Y., Cagnano, E., Benharroch, D., Iwakura, Y., et al., 2003. IL-1 is required for tumor invasiveness and angiogenesis. Proc. Natl. Acad. Sci. USA 100, 2645–2650.

Wallace, J.L., Miller, M.J., 2000. Nitric oxide in mucosal defense: a little goes a long way Gastroenterology 119, 512–520.

Wallace, J.L., Soldato, P.D., 2003. The therapeutic potential of NO-NSAIDs. Fundam. Clin. Pharmacol. 17, 11–20.

Wallace, J.L., Reuter, B., Cicala, C., Mcknight, W., Grisham, M., Cirino, G., 1994a. A Diclofenac derivative without ulcerogenic properties. Eur. J. Pharmacol. 257, 249–255.

Wallace, J.L., Reuter, B., Cicala, C., Mcknight, W., Grisham, M.B., Cirino, G., 1994b. Novel nonsteroidal anti-inflammatory drug derivatives with markedly reduced ulcerogenic properties in the rat. Gastroenterology 107, 173–179.

Wallace, J.L., Mcknight, W., Del Soldato, P., Baydoun, A.R., Cirino, G., 1995. Anti-thrombotic effects of a nitric oxide-releasing, gastric-sparing aspirin derivative. J. Clin. Invest. 96, 2711–2718.

Wallace, J.L., Ignarro, L.J., Fiorucci, S., 2002. Potential cardioprotective actions of NO-releasing aspirin. Nat. Rev. Drug Discov. 1, 375–382.

Wang, R., 2012. Physiological implications of hydrogen sulfide: a Whiff exploration that blossomed. Physiol. Rev. 92, 791–896.

Wei, G. 2015. The Framingham Heart Study (Fhs). Project Period: 9/1/1948–5/1/2015 [Online]. Available from: <http://www.nhlbi.nih.gov/research/resources/obesity/population/framingham.htm> (accessed July 2016).

Weyerbrock, A., Osterberg, N., Psarras, N., Baumer, B., Kogias, E., Werres, A., et al., 2012. Js-K, a glutathione S-transferase-activated nitric oxide donor with antineoplastic activity in malignant gliomas. Neurosurgery 70, 497–510. Discussion 510.

Williams, J.L., Borgo, S., Hasan, I., Castillo, E., Traganos, F., Rigas, B., 2001. Nitric oxide-releasing non-steroidal anti-inflammatory drugs (NSAIDs) alter the kinetics of human colon cancer cell lines more effectively than traditional nsaids: implications for colon cancer chemoprevention. Cancer Res. 61, 3285–3289.

Williams, J.L., Nath, N., Chen, J., Hundley, T.R., Gao, J., Kopelovich, L., et al., 2003. Growth inhibition of human colon cancer cells by nitric oxide (NO)-donating aspirin is associated with cyclooxygenase-2 induction and beta-catenin/t-cell factor signaling, nuclear factor-kappaB, and NO synthase 2 inhibition: implications for chemoprevention. Cancer Res. 63, 7613–7618.

Williams, J.L., Kashfi, K., Ouyang, N., Del Soldato, P., Kopelovich, L., Rigas, B., 2004. NO-donating aspirin inhibits intestinal carcinogenesis in Min (Apc(Min/+)) mice. Biochem. Biophys. Res. Commun. 313, 784–788.

Williams, J.L., Ji, P., Ouyang, N., Liu, X., Rigas, B., 2008. NO-donating aspirin inhibits the activation of NF-kappaB in human cancer cell lines and Min mice. Carcinogenesis 29, 390–397.

Wink, D.A., Ridnour, L.A., Hussain, S.P., Harris, C.C., 2008. The reemergence of nitric oxide and cancer. Nitric Oxide 19, 65–67.

Wink, D.A., Vodovotz, Y., Laval, J., Laval, F., Dewhirst, M.W., Mitchell, J.B., 1998. The multifaceted roles of nitric oxide in cancer. Carcinogenesis 19, 711–721.

Wolfe, M.M., Lichtenstein, D.R., Singh, G., 1999. Gastrointestinal toxicity of nonsteroidal antiinflammatory drugs. N. Engl. J. Med. 340, 1888–1899.

Xu, W., Liu, L.Z., Loizidou, M., Ahmed, M., Charles, I.G., 2002. The role of nitric oxide in cancer. Cell Res. 12, 311–320.

Yeh, R.K., Chen, J., Williams, J.L., Baluch, M., Hundley, T.R., Rosenbaum, R.E., et al., 2004. NO-donating nonsteroidal antiinflammatory drugs (NSAIDs) inhibit colon cancer cell growth more potently than traditional NSAIDs: a general pharmacological property? Biochem. Pharmacol. 67, 2197–2205.

Yiannakopoulou, E., 2015. Aspirin and NSAIDs for breast cancer chemoprevention. Eur. J. Cancer Prev. 24, 416–421.

Yu, H., Kortylewski, M., Pardoll, D., 2007. Crosstalk between cancer and immune cells: role of Stat3 in the tumour microenvironment. Nat. Rev. Immunol. 7, 41–51.

Zhao, H., Ning, S., Scicinski, J., Oronsky, B., Knox, S.J., Peehl, D.M., 2015. Epigenetic Effects of RRx-001: a possible unifying mechanism of anticancer activity. Oncotarget 6, 43172–43181.

CHAPTER 5

Nitric Oxide Donors and Penile Erectile Function

Serap Gur[1,3], Allen L. Chen[2] and Philip J. Kadowitz[3]
[1]Department of Pharmacology, School of Pharmacy, Ankara University, Ankara, Turkey
[2]Tulane University School of Medicine, New Orleans, LA, USA
[3]Department of Pharmacology, Tulane University Health Sciences Center, New Orleans, LA, United States

INTRODUCTION

Since the discovery of nitric oxide (NO) as a critical role in 1987, research on this signaling molecule has expanded in many directions. NO is a short-lived, soluble, and freely diffusible gas and an important intercellular mediator, and maintains the normal function. NO potentially acts, depending on the concentration, and mediates physiological signaling (e.g., neurotransmission or vasodilatation), whereas higher concentrations mediate immune/inflammatory actions and are neurotoxic (Calabrese et al., 2007).

NO regulates the vascular permeability (Di Lorenzo et al., 2013) and maintains the vascular tone (Kelm and Schrader, 1990). It also prevents thrombosis (Cines et al., 1998), inflammation (Cirino et al., 2006), and inhibits the growth (Kapadia et al., 2008) and mitogenesis of smooth muscle cells (Sarkar et al., 1997) causing smooth muscle hyperplasia (Fig. 5.1). The mechanisms involved lie in the alteration of signaling pathway in the endothelial cells (Tabit et al., 2010), enhancement of oxidative stress (Tesfamariam and Cohen, 1992), protein kinase C activation (Naruse et al., 2006), and endothelial inflammatory activation (Piga et al., 2007). Soluble guanylyl cyclase (sGC) and potassium channels are the primary targets of the NO species released from the inorganic compounds.

The requirement for the amelioration of NO deficiency has generated demand for exogenous NO from chemical sources, called NO donors, for in vitro studies to discover into NO-mediated biological processes. NO donors are widely used in the clinical management of hypertension and cardiovascular diseases (Scatena et al., 2010). Also, NO donors were found to possess strong antioxidant properties and modulate the inflammatory response.

Present studies indicate the NO donors may serve as a promising class of compounds for the treatment of erectile dysfunction (ED), but concerns regarding neurotoxicity and a narrow therapeutic window remain. In this chapter, we will primarily

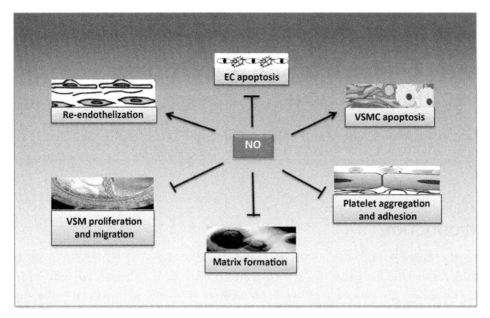

Figure 5.1 Beneficial properties of nitric oxide (NO). EC, endothelial cell; VSMC, vascular smooth muscle cell.

focus on the NO aimed at increasing its production, supply, or amplifying the down-stream signaling pathways of NO and provide an overview of NO donors on NO biology in erectile function.

NITRIC OXIDE SYNTHASE (NOS) ISOFORMS

NO is synthesized by one of the three different isoforms of NOS that differ in enzy-matic activity, in how they are regulated (transcriptional, translational, and posttrans-lational mechanisms), and in how they are expressed/compartmentalized in tissues as well as the subcellular level. Endothelial NOS (eNOS), neuronal NOS (nNOS), and inducible NOS (iNOS) are three isoforms of NOS.

eNOS was found mainly in the endothelium, promoting the pathways for NO formation. However, it is in the vascular smooth muscle that NO interacts with the ferrous state of the heme prosthetic center of the enzyme sGC. eNOS is essential in facilitating the achievement and maintenance of full penile erection (Hurt et al., 2002).

nNOS is expressed in nonadrenergic, noncholinergic (NANC) cells of nerve fibers of the corpora cavernosa (Burnett, 1995; Maggi et al., 2000). The release of NO from nNOS-containing nerves can mediate for the NO-induced nitrergic erectile response (Hurt et al., 2012; Burnett et al., 1996). The role of nNOS and NO release from the cavernous nerves has been studied in knockout mice and by the use of nNOS inhibitors (Hurt et al., 2012). Each of the separate neurons of the cavernosal nerve may contain more than one neurotransmitter. There are at least three components in the cavernosal nerves innervating the penis: (1) an NANC (nitrergic) component which is responsible for the release of NO, (2) a component that releases the sympathetic system neurotransmitter noradrenaline that plays a role in penile detumescence, and (3) a cholinergic component responsible for the release of acetylcholine (ACh) (Lue et al., 2000). ACh released from cholinergic nerve terminals can stimulate muscarinic receptors on the endothelium to increase intracellular calcium concentration combines with calmodulin to activate eNOS. The NO that is released from the endothelium diffuses into the corporal smooth muscle cells and activates sGC. The erectile response to intracavernosal injection of ACh has attenuated the response to intracavernosal injection of the NO donor. Sodium nitroprusside (SNP) was not altered at the time the response to cavernosal nerve stimulation was significantly attenuated (Lasker et al., 2013a). ACh released from cholinergic nerves in response to cavernosal nerve stimulation may mediate 30% of the erectile response in the rat (Lasker et al., 2013a; Senbel et al., 2012).

iNOS has also been detected in the corpora cavernosa though it is most likely localized in smooth muscle cells. It is associated with inflammatory effects and other pathologic changes of the penis (Vernet et al., 1995). Furthermore, iNOS is produced by inflammatory stimuli in leukocytes.

THE ROLE OF NO IN ERECTILE PHYSIOLOGY

The erectile response is triggered by the release of NO from the terminals of NANC or nitrergic nerve fibers and the endothelium of the corpora cavernosa and penile small arteries (Lue et al., 2000). NO release from the endothelium of the corpora cavernosa and penile small arteries is controlled by the activation of M3 muscarinic receptors and an increase in intracellular calcium in the endothelium, which activates eNOS. This increased the conversion of L-arginine to L-citrulline and NO (Fig. 5.2) (Traish et al., 1995).

NO then diffuses across the smooth muscle membrane and binds to the reduced heme iron (Fe^{2+}) on the β-subunit of sGC, which catalyzes the production of cGMP from GTP (Fig. 5.2) (Bellamy et al., 2002). The most prominent natural target of NO is sGC (Arnold et al., 1977), whose activation produces cyclic guanosine monophosphate (cGMP) when NO binds to a reduced heme group in the enzyme (Bredt and Snyder, 1992). Then, cGMP activates cGMP-dependent protein kinase (PKG), which may induce additional effects on the second messenger systems (Fig. 5.2). Metabolism of cGMP by phosphodiesterases suppresses or terminates NO/sCG signaling

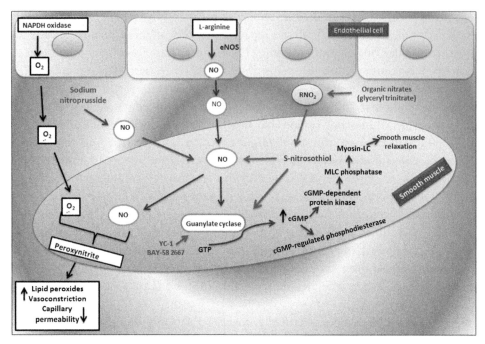

Figure 5.2 Transformation pathways of a nitric oxide (NO) in the penile tissue. After stimulation of endothelial cells, endothelial NO synthase (eNOS) is activated, which transforms L-arginine to NO. NO diffuses to smooth muscle cells of penile blood vessels. Inside smooth muscle, NO activates guanylate cyclase, transforming guanosine triphosphate (GTP) to cyclic guanosine monophosphate (cGMP). This cGMP, acting via myosin light chains (MLC) phosphatase leads to relaxation of smooth muscle cell. NO may be also delivered by exogenous NO donors, such as sodium nitrate, nitroprusside, or other organic nitrates and then it acts the same way, as endogenous NO. Thiols (R-SH), e.g., glutathione (GSH) cooperate with NO.

(Kleppisch, 2009). Dysregulation of NO levels plays a significant role in ED, with both low and excess NO levels inhibiting erectile function.

The cGMP that is formed decreases the intracellular calcium ion concentration in penile smooth muscle by activation of a protein kinase that phosphorylates specific proteins and potassium ion channels, resulting in hyperpolarization of the muscle cell membrane (Burnett, 1995). The sequestration of cytosolic calcium induces relaxation of the corporal smooth muscle and vasodilation of penile small arteries (Andersson and Wagner, 1995). Decreased intracellular calcium results in blood filling the sinusoids of the corpora cavernosa and increases intracavernosal pressure, which restricts venous outflow.

Erection is provided by compression of small veins of the tunica albuginea. The increase in blood flow to the penis increases shear stress, and further NO release is mediated by a flow-mediated mechanism, and a myogenic response in the veins

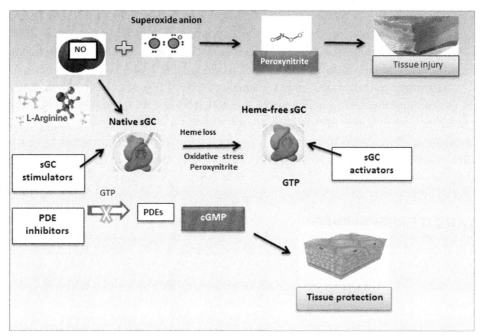

Figure 5.3 NO/cGMP signaling cascade. NO endogenously produced by NO synthases or released from exogenously applied NO donors activates NO-sensitive GC and leads to increased synthesis of cGMP. This intracellular messenger in turn modulates the activity of cGMP-dependent kinases, cGMP-gated ion channels, and cGMP-regulated phosphodiesterases. These effectors are involved in the regulation of several physiological functions in the cardiovascular and nervous systems. YC-1 and BAY 58-2667 represent new classes of activators of NO-sensitive GC.

draining the corpora cavernosa (Andersson and Wagner, 1995). This all contributes to the development of penile erection.

Numerous pathophysiologic processes and increased oxidative stress from the production of reactive oxygen species (ROS) can reduce NO formation and bioavailability (Evgenov et al., 2006). Superoxides can react with NO to form peroxynitrite ($ONOO^-$) (Fig. 5.3). Both peroxynitrite and ROS are potent oxidants that may cause protein nitration and oxidation, lipid peroxidation, enzyme inactivation, and DNA oxidative damage (Calcerrada et al., 2011). In particular, the oxidation of low-density lipoprotein (LDL) produces oxidized LDL, further reducing NO bioavailability by dephosphorylation of eNOS (Fleming et al., 2005; Zhou et al., 2013) and reduction of eNOS mRNA level (Searles, 2006).

ROS have been implicated in several pathophysiologic conditions such as atherosclerosis, diabetes, hypertension, coronary artery disease, and congestive heart failure of which many ED patients have comorbid conditions (Stasch et al., 2006). Oxidative stress can induce ED by reducing NO bioavailability (through the generation of superoxide anion or peroxynitrite) and by altering the redox state of sGC, by oxidizing its heme iron, rendering the enzyme unresponsive to NO (Evgenov et al., 2006). Oxidative stress generates the oxidant species such as superoxide anion and peroxynitrite, from NO oxidation, and is capable of oxidizing the typically reduced heme on sGC, attenuating the enzyme's responsiveness to NO (Evgenov et al., 2006; Stasch et al., 2006). Superoxide anion and peroxynitrite have been shown to impair endothelial function and to cause ED (Beckman and Koppenol, 1996; Laight et al., 2000). A decrease in oxidative stress has been demonstrated to increase erectile activity in many studies in animal models of ED (Keegan et al., 1999; Agarwal et al., 2006; Shukla et al., 2009).

ED AND ITS PREVALENCE

ED is a chronic condition that significantly reduces the quality of life in middle-aged and older men and represents a challenge for our healthcare system. The prevalence of chronic ED in the United States varies, mostly due to differences in methodology and populations in epidemiologic studies. ED adversely affects up to 20% of all men and is the most commonly treated sexual disorder. The overall prevalence of ED in men aged 20 years in the United States is estimated to be 18.4% to 22%, steadily increasing with age to 77.5% in those aged 75 years (Saigal et al., 2006; Laumann et al., 2007).

The International Men's Attitudes to Life Events and Sexuality (MALES) study in the United States found that the prevalence of ED in this population was 22% (Rosen et al., 2004). The Massachusetts Male Aging Study (MMAS) showed that the combined prevalence of minimal, moderate, and complete ED in men aged 40–70 years was 52%, with complete impotence increasing from 5% to 15% from 40 to 70 years (Feldman et al., 1994). A larger study in the United States found the age-standardized prevalence of ED, using conservative criteria of poor to very poor function, was 33%, ranging from about 25% among those younger than 59 years to 61% in individuals older than 70 years (Bacon et al., 2003).

NO DONORS

NO donors are a heterogeneous group of compounds which have the ability to release NO or an NO-related species, such as the nitrosonium ion (NO^+) or the nitroxyl anion (NO^-), in vitro or in vivo independently of its endogenous sources. A vast array of NO donors including nitrates, nitrites, N-nitroso, C-nitroso, S-nitroso, certain heterocyclics, metal/NO complexes, and diazeniumdiolates have been reported (Jalbuena and Tyrer, 1991; Wimalawansa, 2010). The primary determinant of these effects is the manner in which NO is released, the amount of NO produced, and the time during which it is released. Despite the fact that these compounds are considered as NO donors, they show different pharmacokinetic and dynamic properties that lead to the type and extent of their biological actions (Miller and Megson, 2007). Moreover, some NO donors generate alternative products that may arise during their metabolism. In this context, it is important to underline that co-generation of superoxide anion might represent a limit to the therapeutic potential of NO donors and their metabolites which can even be used as peroxynitrite generators (Menconi et al., 1998).

NO donors allow the introduction of defined amounts with defined rates of release of the test compound. Also, NO donors display pH dependence and sensitivity to oxidants and reductants, which influence what NO derivatives are produced, and in the most recent generation also offer organelle targeting for NO release.

The following are the NO donors most frequently employed in clinical and basic research: organic nitrates (e.g., nitroglycerin, isosorbide-5-mononitrate, nicorandil, pentaerythritol tetranitrate); sodium nitroprusside (SNP), S-nitrosothiols [e.g., S-nitroso-N-acetylpenicillamine (SNAP) and S-nitrosoglutathione]; sydnonimines (e.g., molsidomine, SIN-1); NONOates (JS-K, Spermine NONOate, and Proli-NONOate), and SNP (Scatena et al., 2010). The NO donors have been shown to relax several isolated strips (Serafim et al., 2012). In earlier studies, SNP, nitroglycerin, and SNAP, which are nitrovasodilators known to generate NO, also caused marked concentration-dependent relaxation of corpus cavernosum (CC) (Bush et al., 1992).

NO DONORS FOR ERECTILE FUNCTION

An overview of the preclinical literature regarding the effects of NO donors in ED is provided in Table 5.1.

SODIUM NITRITE (NaNO$_2$)

NaNO$_2$ by acting as a storage form of NO has beneficial pharmacologic actions. Lasker et al. (2010) investigated the effects of NaNO$_2$ on erectile function in the rat. NaNO$_2$ intracavernosally produced dose-related increases in intracavernosal pressure and decreased in systemic arterial pressure. NaNO$_2$ was 1000-fold less potent than SNP in increasing intracavernosal pressure (see Table 5.1) (Lasker et al., 2010). Relaxation induced by NaNO$_2$ and nifedipine was reduced in older animals (Gocmen et al., 1997). Transplant patients frequently present with ED. To test any interference effect of cyclosporine A, which is commonly used in the posttransplantation management, Ragazzi et al. investigated the in vitro functional responses of CC from

Table 5.1 Several nitric oxide donors and erectile function

Substances	Animal models	Effects	Results	References
Sodium nitrite	Rat	Increases in intracavernosal pressure	1000-fold less potent than sodium nitroprusside in increasing i.c. pressure	Lasker et al. (2010)
SNAP	Diabetic monkey	Relaxation	Enhanced the effect of sildenafil on CC	Taher et al. (2000)
SNP	Heart failure–prone rats	Involvement with potassium channels (Kv 7)	A role in erectile function and an early indicator of cardiovascular disease in ED	Jepps et al. (2016)
SIN–1	Human CC	Increases in whole-cell outward K(+) currents	Physiologic regulation of human corporal smooth muscle tone	Insuk et al. (2003)
Spermine NONOate	Mouse CC	Relaxation	Relatively slow to develop, and it was reversible and reproducible	Ertug et al. (2014)
BAY 41-2272	Obese mice	Improved CC relaxation	Reducing oxidative stress	Nunes et al. (2015)
Oximes (FAL and FAM)	Rat	Inducing erection by sGC	Resistant to oxidative stress	Pauwels et al. (2015)
FAL, FAM, CAOx	Mice	FAL–/FAM-induced		Pauwels et al. (2014)
BAY 60-2770	Rat (cavernosal nerve injury)		Potent erectile activity	Lasker et al. (2013b)

SNAP, S-nitroacetylpenicillamine; FAL, formaldoxime; FAM, formamidoxime; CAOx, cinnamaldoxime; CC, corpus cavernosum; SIN–1, 3–morpholinosydnonimine.

cyclosporine A-treated rabbits. The relaxation produced by $NaNO_2$ was diminished in cyclosporine A-treated rabbits (Ragazzi et al., 1996).

S-NITROACETYLPENICILLAMINE

SNAP is an *S*-nitrosothiol and is used as a model for the general class of *S*-nitrosothiols which have received much attention in biochemistry because NO and some organic nitroso derivatives serve as signaling molecules in living systems, especially related to vasodilatation (Zhang and Hogg, 2005). In previous study, NO donor SNAP injected intracavernosally on penile erection in cats increased the cavernosal pressure and penile length in a dose-dependent manner (Wang et al., 1994). The maximal effects on cavernosal pressure and penile length induced by *S*-nitrosocysteine (NO-CYS) and SNAP, respectively, were eightfold and fivefold increases in pressure, and 45% and 34% increases in length when compared with baseline values (Wang et al., 1994). Furthermore, SNAP enhanced the relaxation effect of sildenafil on CC of diabetic monkey (Ragazzi et al., 1996; Taher et al., 2000).

SODIUM NITROPRUSSIDE

SNP is a compound consisting of an iron core surrounded by five cyanide ion (CN^-) molecules and one molecule of the nitrosonium ion (NO^+). SNP does not liberate NO spontaneously in vitro but does require partial reduction (one-electron transfer) by a variety of reducing agents present in membrane cells. However, SNP can release, in aqueous solution, several oxidants and free radical species, such as iron, cyanide, superoxide, hydrogen peroxide (H_2O_2), and hydroxyl radical in direct proportion to its concentration (Dawson et al., 1991; Terwel et al., 2000). Because of the nitrosative and prooxidant potential inherent in the different NO donor, these have been widely used as models of neuronal damage (Miller and Megson, 2007; Scatena et al., 2010).

Human cavernosal smooth muscle cells incubated to SNP (0–0.8 mM) displayed a dose-dependent decline in DNA and ATP synthesis with an increase in the levels of 8-iso PGF2-α and about an eightfold increase in nitrite accumulation. These results showed that the NO released by SNP (>0.8 mM) caused a significant cytotoxicity to human cavernosal smooth muscle cells, related to increased oxidative stress to these cells (Rajasekaran et al., 2001). In a recent study, relaxation induced by SNP were reduced by blocking Kv 7 channels with linopirdine in penile arteries and CC (Jepps et al., 2016).

ORGANIC NITRATES

Organic nitrates, the most common NO donors utilized in coronary artery disease, require enzymatic bioactivation to deliver NO (Thatcher et al., 2004). Their effect is principally at the vascular level, by increasing venous capacitance and coronary vasodilation (Scatena et al., 2010). Clinically approved NO donors such as glyceryl trinitrate (GTN), isosorbide dinitrate (ISDN), isosorbide-5-mononitrate, amyl nitrite, and pentaerythrityl tetranitrate are in use despite associated nitrate tolerance (Cerqueira-Silva et al., 2008; Cappelleri et al., 2016).

In approximately 75% of precontracted corporal tissue strips derived from impotent patients, the nitroglycerine-induced response was biphasic, consisting of a rapid relaxation response that reached steady state before onset of a more slowly developing regaining of tension, termed the desensitization response (Christ et al., 1995). Prolonged exposure to a high concentration of isosorbide dinitrate resulted in tolerance being developed against its relaxant activity in strips of CC, precontracted by phenylephrine (Uma et al., 1998).

Prolonged in vitro exposure of CC strips obtained from control and diabetic groups to high concentrations of isosorbide dinitrate caused significant desensitization to the relaxant effect of the drug (Yildirim et al., 2000). So, prolonged exposure of corporal tissue to the agents like nitroglycerine, used for the treatment of impotence, may render ineffective therapy in diabetic erectile impotence.

The oral PDE-5 inhibitors have provided safe, effective treatment of ED for some men. However, a large proportion of men who have ED do not respond to PDE-5 inhibitors or become less responsive or less satisfied as the duration of therapy increases. In addition, men who are receiving organic nitrates and nitrates, such as amyl nitrate, cannot take PDE-5 inhibitors because of nitrate interactions. For example, the use of PDE-5 inhibitors in the presence of oral nitrates is strictly contraindicated in diabetic men, as in nondiabetic subjects (Tamas and Kempler, 2014).

SYDNONIMINES

Sydnonimines release NO spontaneously without enzymatic activity. 3-Morpholino-sydnonimine (SIN-1, linsidomine) stimulates guanylate cyclase and efficiently inhibits contractions in rabbit isolated CC (Sparwasser et al., 1994). Therefore, these compounds are considered peroxynitrite donors more than NO donor and are utilized as nitrosative stress inducers in experimental models. In an earlier study, the SIN-1 induced increases in whole-cell outward K(+) currents in the physiologic regulation of human corporal smooth muscle tone (Insuk et al., 2003). The combination of SIN-1 plus vasoactive intestinal peptide (VIP) intracavernous injected induced erectile responses with the increased mean maximal intracavernous pressure in an in vivo rabbit model (Sazova et al., 2002).

OXIMES

Oxidative stress occurs, impairing NO-mediated relaxations in many disease conditions. Hence, the influence of oxidative stress on oxime-induced effects is also of interest. Oximes induce erection which is mediated by sGC (Pauwels et al., 2015). Because of their NO-donating capacities, oxime derivatives have shown to offer some therapeutic perspectives for the treatment of ED as well as cardiovascular diseases (Pauwels et al., 2015). Oximes such as formaldoxime (FAL) and formamidoxime (FAM) maintained on normal intracavernosal pressure which is not influenced by oxidative stress (Pauwels et al., 2015). The oxime-induced relaxations are resistant to oxidative stress, which enhances their therapeutic potential for the treatment of ED (Pauwels et al., 2015). Furthermore, NO-donating oximes relaxed mice CC (Pauwels et al., 2014).

SPERMINE NONOate

Spermine NONOate is a diazenium diolate NO donor compound derived from the polyamine spermine. Spermine NONOate has a potent relaxant action in cavernous tissue, and this effect can be potentiated by oral sildenafil treatment (Ertug et al., 2014). The pharmacologic profile has investigated the effects of an NO donor spermine NONOate in the cavernous tissue from mice treated or untreated with sildenafil. Spermine NONOate relaxed mouse CC, which was relatively slow to develop, and it was reversible and reproducible.

S-NITROSOTHIOLS

S-nitrosothiols are a heterogeneous group characterized by a nitroso group attached by a single chemical bond to the sulfur atom of a thiol (Miller and Megson, 2007). *S*-nitrosoglutathione (GSNO) is an important intermediary in organic nitrate metabolism. The remaining nitrosothiols are synthetic. These compounds act as NO carriers, NO reservoirs, and intermediates in protein nitrosylation. They also have the ability to transfer the different NO species through chains of thiols, without releasing the NO molecule itself. This feature diminishes the possibility of NO reacting with O_2^- to form $ONOO^-$ or that of reacting with other molecules to nitrosylate these. GSNO dose-dependently increased EFS-induced amplitudes of relaxation of isolated human CC and the in vitro formation of cGMP (Seidler et al., 2002).

NITROSAMINES

Nitrosamine-based NO-releasing agents (NO donors) provide this unique and innovative alternative to possible treatments for long-term NO deficiency-related diseases.

N-nitrosamines have the advantages of improved compatibility in intra-arterial devices and can serve as valuable tools in biomedical research (Kozai et al., 2014; Kyriss and Schneider, 2013). On the other hand, it is likely to control the release status of NO by altering the spacer groups of these NO donors.

Endothelial dysfunction characterized by endogenous NO deficiency made 56% of patients affected by ED decline treatment with PDE-5 inhibitors. The study compared the effect of SNP and two substances of the nitrosyl-ruthenium complex, *cis*-[$Ru(bpy)_2(SO_3)(NO)$]PF-6-9 (FONO) and *trans*-[$Ru(NH_3)_4$(caffeine)(NO)]$C1_3$ (LLNO1) on relaxation of rabbit CC smooth muscle. In CC samples, FONO1 produced no significant effect, but LLNO1 and SNP-induced dose–dependent relaxation with comparable potency and maximum effect. The substances tested were shown to activate soluble guanylate cyclase and release intracellular NO (Cerqueira-Silva et al., 2008).

Veliev et al. (2009) investigated the effect of intracorporeal administration of new NO donor dinitrosyl iron complex in a bilateral cavernous nerve crush injury group compared to a sham-operated group and an untreated group nerve crush injury group. These authors observed a stabilization of fibroblast proliferation and improvement of smooth muscle mitotic activity in the group treated with intracavernosal injections of NO donor.

PDE-5 INHIBITORS

cGMP is hydrolyzed by cGMP phosphodiesterases, of which PDE-5 is most abundant (Manganiello, 2003). PDE-5 inhibitors are the first-line treatment for ED, and it has been reported that 50% of ED patients benefit from treatment with PDE-5 inhibitors (Berner et al., 2006). It was observed that corpora cavernosal relaxation responses to EFS were enhanced by zaprinast, a PDE-5 inhibitor (Rajfer et al., 1992). This was an important discovery, as it was observed that PDE-5 inhibitors have minimal effect on the systemic arterial pressure but induce penile erection in patients. This led to the development of the first PDE-5 inhibitor sildenafil (Viagra) in 1998 for the treatment of ED (Goldstein et al., 1998). However, some patients do not respond adequately to these oral agents (McMahon et al., 2006). To be effective, PDE-5 inhibitor therapy requires some basal level of endogenous NO formation in the penile tissues. The causes of severely impaired NO formation and failure with the PDE-5 inhibitors may include severity of ED, surgical procedures that cause nerve damage, hypogonadism, incorrect drug dosage, and psychosocial factors (McMahon et al., 2006).

Long-term tadalafil administration preserves relaxation responses probably by affecting through the NO/cGMP pathway in experimental spinal cord transection-applied rats (Toksoz et al., 2013). Our recent results from in vivo and in vitro studies in human and rat suggested that a new PDE-5 inhibitor avanafil promotes the CC relaxation and penile erection via NO-cGMP pathway (Gur et al., 2015).

sGC STIMULATORS AND ACTIVATORS

sGC stimulators are NO-independent, heme-dependent agents that directly activate sGC by binding to a different site than NO, increasing the catalytic activity of reduced sGC which increases cGMP formation and induces penile erection (Hsieh et al., 2003; Schmidt et al., 2003; Evgenov et al., 2006). Therefore, sGC stimulators are effective in conditions in which the NO-sGC-cGMP pathway is depleted and thus are theoretically advantageous over PDE-5 inhibitors. sGC stimulators are relatively free of off-target effects and do not lose their efficacy with prolonged use (Nossaman et al., 2012). In pathophysiological situations where NO formation or bioavailability is severely depressed, two developed classes called sGC stimulators and sGC activators can ameliorate erectile function (Gur et al., 2010).

sGC is a redox-sensitive enzyme. Excessive oxidative stress or nitrosylation can change the redox status of sGC from the typically reduced heme iron to an oxidized heme, rendering it less active and less responsive to NO. Because sGC stimulators require the presence of the heme group, which is often absent in oxidative disease states, drug development also focused on creating molecules that could activate heme-deficient sGC. Oxidized sGC subsequently loses its heme moiety, after which point it will eventually be degraded by the proteasome (Meurer et al., 2009).

YC-1 was the first sGC stimulator. This drug intracavernosally promoted erectile responses to cavernosal nerve stimulation and systemic apomorphine injections in rats (Mizusawa et al., 2002). BAY 41-2272 produced weak erectile effects when administered intravenously to conscious rabbits as well as orally by measuring the uncovered penile length. When BAY 41-2272 was applied and followed by an iv dose of SNP, erectile responses were potentiated (Bischoff et al., 2003). BAY 41-8543 can synergize with NO to increase the catalytic activity of the enzyme and has been shown to induce an erection in the rat (Lasker et al., 2013b). The erectile response to the intracavernosal injection of the sGC stimulator BAY 41-8543 is not affected by NOS inhibition with L-NAME induced hypertension (Burnett et al., 1992; Mizusawa et al., 2002) or by cavernosal nerve crush injury in rats (Lasker et al., 2013a). When NO bioavailability is severely depressed and erectile function is greatly diminished, the intracavernosal injection of BAY 41-8543 can restore normal erectile responses (Lasker et al., 2013a). In response to cavernosal nerve stimulation in the rat after iv the sGC stimulator BAY 60-4552 and the PDE-5 inhibitor vardenafil produced preerectile effects in chronic cavernosal nerve crush injury model (Oudot et al., 2011; Albersen et al., 2013).

In 2002, BAY 58-2667 (cinaciguat), an amino dicarboxylic acid, was identified and demonstrated to be the first NO- and heme-independent sGC activator (Stasch et al., 2002). Cinaciguat is a heme mimetic that can bind and replace the endogenous heme site in sGC (Schmidt et al., 2004). Hence, it is the heme-free sGC that is the target for sGC activators such as cinaciguat (Roy et al., 2008).

In severe ED when sGC is not responsive to NO or sGC stimulators, the heme-and NO-independent sGC activators can be useful (Schmidt et al., 2009). It was also demonstrated that BAY 60-2770 could produce a normal erectile response when the cavernosal nerves were injured (Lasker et al., 2013a). Therefore, in pathological conditions such as severe ED when NO formation or bioavailability is impaired, and sGC is unresponsive to NO stimulation, sGC stimulators, and sGC activators show a new strategy of therapy that can lead normal penile erectile function after patients do not respond to more traditional treatments.

The development of cGC stimulators and activators came with the recognition that inducing smooth muscle relaxation in conditions of decreased NO bioavailability would be of value in chronic ED.

NO-DONATING STATINS

Statins have also been seen to implement their nonlipid effects by promoting tetra-hydrobiopterin (BH4) synthesis with an increase of NO bioavailability. Furthermore, NO-donating statins in experimental studies have shown to produce better therapeutic effects than their parent's drugs. Also, it has been shown that NO-donating statin induced less myotoxicity, the most common side effect due to medication with statins. The earlier studies have displayed the superior therapeutic results of NO-donating statins while producing fewer adverse effects.

Another approach for providing additional NO consists of incorporating NO moieties into the structure of existing therapeutic agents, including statins. The promotion of BH4 synthesis and eNOS coupling in the vasculature may be important to manage ED by increasing NO bioavailability.

Both laboratory and clinical studies have attempted to demonstrate the therapeutic effects of statin through the promotion of NO formation in ED models. For example, pravastatin improves NO-mediated CC relaxations of aged rats probably by inhibiting NADPH oxidase/Rho kinase pathways, and this effect does not seem to be associated with lipid lowering effect of this drug (Dalaklioglu et al., 2014). Atorvastatin, but not vitamin E, is a promising drug for sildenafil nonresponders (El-Sisi et al., 2013). While ED is a common problem for elderly males, higher serum lipid levels may have important role in the pathogenesis. Therefore NO-donating statins may be significant strategy in urological and primary care practice by including questions on ED during routine consultations and relevant clinical protocols.

CONCLUSIONS AND PERSPECTIVES

As a gasotransmitter, NO plays an essential role in vascular function and redox biology. For decades, our understanding of NO biology has evolved through empirical medications, identification of sGC as a receptor that transduces NO signaling in various pathways, bioavailable metabolites, effects on protein posttranslational modifications, and novel therapeutic approaches. Without a doubt, the future NO research will yield exciting and important new discoveries in the years to come. Therefore, there exists a clear need for NO donors that can release NO in a slow, sustainable, and rate-tunable fashion. According to the reported properties, NO has a promising therapeutic potential and, currently, there is great interest in the design of NO-donating derivatives for the transport and release of NO within the body. For this reason, any drug or pro-drug that can release NO must be carefully tested to determine the release profile of NO under biological conditions.

ED is a common and debilitating disorder seen in over 50% of men older than 70 years. While the current medications have profoundly impacted the treatment of ED over the past 50 years, the newer atypical anti-ED has not fulfilled initial expectations, and enormous challenges remain in long-term treatment of this debilitating disease. Oral PDE-5 inhibitors are the first-line therapy for ED. The physiological pathways responsible for erections have been extensively studied, and much advancement has been made since the introduction of PDE-5 inhibitors.

The present analysis suggests that NO donors might be a promising class of compounds for the treatment of ED. It is clear that NO concentrations should exist physiologically which is essential for developing a quantitative understanding of NO signaling, for performing experiments with NO that emulate reality, and for knowing whether or not NO concentrations become abnormal in disease states. Several independent lines of evidence suggest that NO operates physiologically at levels that are orders of magnitude lower than the near-micromolar order once considered correct. However, it is important to underline that co-generation of superoxide anion might represent a limit to the therapeutic potential of NO donors and their metabolites which can even be used as peroxynitrite generators (Menconi et al., 1998).

NO has been incorporated into statins. NO-donating statins are shown to possess stronger antiinflammatory, antithrombotic, and anti-proliferative actions than their parent drugs. Also, these newly formed drugs offer better therapeutic effects in peripheral ischemia, provide improved cardioprotection, and reduce the risk of adverse reactions. Thus, new forms of a statin with higher therapeutic potential and lower adverse effects may be beneficial in ED patients with cardiovascular diseases.

$NaNO_2$ can play a role as an NO donor that increases erectile activity. Currently, no prophylactic treatment is applied at the time of cavernous nerve injury. The benefits of these therapeutic modalities by using NO donors have been demonstrated in preclinical as well as in clinical studies. Oximes are new agents of molecules with potential for the therapy of ED. Spermine NONOate may be considered as an attractive treatment for ED in pathologic disorders with a lack of endogenous NO production. S-nitrosothiols can show benefits in the pharmacotherapy of ED.

Underlying these activities is the relentless pursuit to find the desirable NO donors, which can ameliorate symptoms arising from pathological NO deficiencies. Hopefully, this review provides an excellent introduction to investigators new to the field of NO and NO donors that will help them deftly approach the study of these molecules in the erectile function.

REFERENCES

Agarwal, A., Nandipati, K.C., Sharma, R.K., Zippe, C.D., Raina, R., 2006. Role of oxidative stress in the pathophysiological mechanism of erectile dysfunction. J. Androl. 27, 335–347.

Albersen, M., Linsen, L., Tinel, H., Sandner, P., van Renterghem, K., 2013. Synergistic effects of BAY 60-4552 and vardenafil on relaxation of corpus cavernosum tissue of patients with erectile dysfunction and clinical phosphodiesterase type 5 inhibitor failure. J. Sex Med. 10, 1268–1277.

Andersson, K.E., Wagner, G., 1995. Physiology of penile erection. Physiol. Rev. 75, 191–236.

Arnold, W.P., Mittal, C.K., Katsuki, S., Murad, F., 1977. Nitric oxide activates guanylate cyclase and increases guanosine 3':5'-cyclic monophosphate levels in various tissue preparations. Proc. Natl. Acad. Sci. USA 74, 3203–3207.

Bacon, C.G., Mittleman, M.A., Kawachi, I., Giovannucci, E., Glasser, D.B., Rimm, E.B., 2003. Sexual function in men older than 50 years of age: results from the health professionals follow-up study. Ann. Intern. Med. 139, 161–168.

Beckman, J.S., Koppenol, W.H., 1996. Nitric oxide, superoxide, and peroxynitrite: the good, the bad, and ugly. Am. J. Physiol. 271, C1424–C1437.

Bellamy, T.C., Wood, J., Garthwaite, J., 2002. On the activation of soluble guanylyl cyclase by nitric oxide. Proc. Natl. Acad. Sci. USA 99, 507–510.

Berner, M.M., Kriston, L., Harms, A., 2006. Efficacy of PDE-5-inhibitors for erectile dysfunction. A comparative meta-analysis of fixed-dose regimen randomized controlled trials administering the International Index of Erectile Function in broad-spectrum populations. Int. J. Impot. Res. 18, 229–235.

Bischoff, E., Schramm, M., Straub, A., Feurer, A., Stasch, J.P., 2003. BAY 41-2272: a stimulator of soluble guanylyl cyclase induces nitric oxide-dependent penile erection in vivo. Urology 61, 464–467.

Bredt, D.S., Snyder, S.H., 1992. Nitric oxide, a novel neuronal messenger. Neuron 8, 3–11.

Burnett, A.L., 1995. Nitric oxide control of lower genitourinary tract functions: a review. Urology 45, 1071–1083.

Burnett, A.L., Donowitz, M., Marshall, F.F., 1996. Inhibition of transport processes of intestinal segments following augmentation enterocystoplasty in rats. J. Urol. 156, 1872–1875.

Burnett, A.L., Lowenstein, C.J., Bredt, D.S., Chang, T.S., Snyder, S.H., 1992. Nitric oxide: a physiologic mediator of penile erection. Science 257, 401–403.

Bush, P.A., Aronson, W.J., Buga, G.M., Rajfer, J., Ignarro, L.J., 1992. Nitric oxide is a potent relaxant of human and rabbit corpus cavernosum. J. Urol. 147, 1650–1655.

Calabrese, V., Mancuso, C., Calvani, M., Rizzarelli, E., Butterfield, D.A., Stella, A.M., 2007. Nitric oxide in the central nervous system: neuroprotection versus neurotoxicity. Nat. Rev. Neurosci. 8, 766–775.

Calcerrada, P., Peluffo, G., Radi, R., 2011. Nitric oxide-derived oxidants with a focus on peroxynitrite: molecular targets, cellular responses and therapeutic implications. Curr. Pharm. Des. 17, 3905–3932.

Cappelleri, J.C., Tseng, L.J., Luo, X., Stecher, V., Lue, T.F., 2016. Simplified interpretation of the erectile function domain of the international index of erectile function. J. Sex Med. 13, 690–696.

Cerqueira-Silva, C.B., Moreira, C.N., Figueira, A.R., Correa, R.X., Oliveira, A.C., 2008. Detection of a resistance gradient to Passion fruit woodiness virus and selection of "yellow" passion fruit plants under field conditions. Genet. Mol. Res. 7, 1209–1216.

Christ, G.J., Kim, D.C., Taub, H.C., Gondre, C.M., Melman, A., 1995. Characterization of nitroglycerine-induced relaxation in human corpus cavernosum smooth muscle: implications to erectile physiology and dysfunction. Can. J. Physiol. Pharmacol. 73, 1714–1726.

Cines, D.B., Pollak, E.S., Buck, C.A., Loscalzo, J., Zimmerman, G.A., Mcever, R.P., et al., 1998. Endothelial cells in physiology and in the pathophysiology of vascular disorders. Blood 91, 3527–3561.

Cirino, G., Distrutti, E., Wallace, J.L., 2006. Nitric oxide and inflammation. Inflamm. Allergy Drug Targets 5, 115–119.

Dalaklioglu, S., Sahin, P., Tasatargil, A., Celik-Ozenci, C., 2014. Pravastatin improves the impaired nitric oxide-mediated neurogenic and endothelium-dependent relaxation of corpus cavernosum in aged rats. Aging Male 17, 259–266.

Dawson, V.L., Dawson, T.M., London, E.D., Bredt, D.S., Snyder, S.H., 1991. Nitric oxide mediates glutamate neurotoxicity in primary cortical cultures. Proc. Natl Acad. Sci. USA 88, 6368–6371.

Di Lorenzo, A., Lin, M.I., Murata, T., Landskroner-Eiger, S., Schleicher, M., Kothiya, M., et al., 2013. eNOS-derived nitric oxide regulates endothelial barrier function through VE-cadherin and Rho GTPases. J. Cell Sci. 126, 5541–5552.

El-Sisi, A.A., Hegazy, S.K., Salem, K.A., Abdelkawy, K.S., 2013. Atorvastatin improves erectile dysfunction in patients initially irresponsive to Sildenafil by the activation of endothelial nitric oxide synthase. Int. J. Impot. Res. 25, 143–148.

Ertug, F.P., Singirik, E., Buyuknacar, H.S., Gocmen, C., Secilmis, M.A., 2014. Pharmacological profile of a nitric oxide donor spermine NONOate in the mouse corpus cavernosum. Turk. J. Med. Sci. 44, 569–575.

Evgenov, O.V., Pacher, P., Schmidt, P.M., Hasko, G., Schmidt, H.H., Stasch, J.P., 2006. NO-independent stimulators and activators of soluble guanylate cyclase: discovery and therapeutic potential. Nat. Rev. Drug Discov. 5, 755–768.

Feldman, H.A., Goldstein, I., Hatzichristou, D.G., Krane, R.J., Mckinlay, J.B., 1994. Construction of a surrogate variable for impotence in the Massachusetts Male Aging Study. J. Clin. Epidemiol. 47, 457–467.

Fleming, I., Mohamed, A., Galle, J., Turchanowa, L., Brandes, R.P., Fisslthaler, B., et al., 2005. Oxidized low-density lipoprotein increases superoxide production by endothelial nitric oxide synthase by inhibiting PKCalpha. Cardiovasc. Res. 65, 897–906.

Gocmen, C., Ucar, P., Singirik, E., Dikmen, A., Baysal, F., 1997. An in vitro study of nonadrenergic-noncholinergic activity on the cavernous tissue of mouse. Urol. Res. 25, 269–275.

Goldstein, I., Lue, T.F., Padma-Nathan, H., Rosen, R.C., Steers, W.D., Wicker, P.A., 1998. Oral sildenafil in the treatment of erectile dysfunction. Sildenafil Study Group. N. Engl. J. Med. 338, 1397–1404.

Gur, S., Kadowitz, P.J., Hellstrom, W.J., 2010. Exploring the potential of NO-independent stimulators and activators of soluble guanylate cyclase for the medical treatment of erectile dysfunction. Curr. Pharm. Des. 16, 1619–1633.

Gur, S., Sikka, S.C., Pankey, E.A., Lasker, G.F., Chandra, S., Kadowitz, P.J., et al., 2015. Effect of avanafil on rat and human corpus cavernosum. Andrologia 47, 897–903.

Hsieh, G.C., O'neill, A.B., Moreland, R.B., Sullivan, J.P., Brioni, J.D., 2003. YC-1 potentiates the nitric oxide/cyclic GMP pathway in corpus cavernosum and facilitates penile erection in rats. Eur. J. Pharmacol. 458, 183–189.

Hurt, K.J., Musicki, B., Palese, M.A., Crone, J.K., Becker, R.E., Moriarity, J.L., et al., 2002. Akt-dependent phosphorylation of endothelial nitric-oxide synthase mediates penile erection. Proc. Natl. Acad. Sci. USA 99, 4061–4066.

Hurt, K.J., Sezen, S.F., Lagoda, G.F., Musicki, B., Rameau, G.A., Snyder, S.H., et al., 2012. Cyclic AMP-dependent phosphorylation of neuronal nitric oxide synthase mediates penile erection. Proc. Natl. Acad. Sci. USA 109, 16624–16629.

Insuk, S.O., Chae, M.R., Choi, J.W., Yang, D.K., Sim, J.H., Lee, S.W., 2003. Molecular basis and characteristics of KATP channel in human corporal smooth muscle cells. Int. J. Impot. Res. 15, 258–266.

Jalbuena, R.C., Tyrer, H.W., 1991. Accuracy of electronic deposition of cells onto microscope slides using a cytometric positioning system. Biomed. Sci. Instrum. 27, 9–19.

Jepps, T.A., Olesen, S.P., Greenwood, I.A., Dalsgaard, T., 2016. Molecular and functional characterization of Kv 7 channels in penile arteries and corpus cavernosum of healthy and metabolic syndrome rats. Br. J. Pharmacol. 173, 1478–1490.

Kapadia, M.R., Chow, L.W., Tsihlis, N.D., Ahanchi, S.S., Eng, J.W., Murar, J., et al., 2008. Nitric oxide and nanotechnology: a novel approach to inhibit neointimal hyperplasia. J. Vasc. Surg. 47, 173–182.

Keegan, A., Cotter, M.A., Cameron, N.E., 1999. Effects of diabetes and treatment with the antioxidant alpha-lipoic acid on endothelial and neurogenic responses of corpus cavernosum in rats. Diabetologia 42, 343–350.

Kelm, M., Schrader, J., 1990. Control of coronary vascular tone by nitric oxide. Circ. Res. 66, 1561–1575.

Kleppisch, T., 2009. Phosphodiesterases in the central nervous system. Handb. Exp. Pharmacol (191), 71–92. http://dx.doi.org/10.1007/978-3-540-68964-5_5. Review. PMID: 19089326.

Kozai, D., Kabasawa, Y., Ebert, M., Kiyonaka, S., Firman, Otani, Y., Numata, T., et al., 2014. Transnitrosylation directs TRPA1 selectivity in N-nitrosamine activators. Mol. Pharmacol. 85, 175–185.

Kyriss, T., Schneider, N.K., 2013. The development of scientific consultants: how the tobacco industry creates controversy on the carcinogenicity of tobacco-specific nitrosamines. Tob. Control. 22, e3.

Laight, D.W., Carrier, M.J., Anggard, E.E., 2000. Antioxidants, diabetes and endothelial dysfunction. Cardiovasc. Res. 47, 457–464.

Lasker, G.F., Matt, C.J., Badejo Jr., A.M., Casey, D.B., Dhaliwal, J.S., Murthy, S.N., et al., 2010. Intracavernosal administration of sodium nitrite as an erectile pharmacotherapy. Can. J. Physiol. Pharmacol. 88, 770–776.

Lasker, G.F., Pankey, E.A., Allain, A.V., Dhaliwal, J.S., Stasch, J.P., Murthy, S.N., et al., 2013a. Analysis of erectile responses to BAY 41-8543 and muscarinic receptor stimulation in the rat. J. Sex Med. 10, 704–718.

Lasker, G.F., Pankey, E.A., Frink, T.J., Zeitzer, J.R., Walter, K.A., Kadowitz, P.J., 2013b. The sGC activator BAY 60-2770 has potent erectile activity in the rat. Am. J. Physiol. Heart Circ. Physiol. 304, H1670–H1679.

Laumann, E.O., West, S., Glasser, D., Carson, C., Rosen, R., Kang, J.H., 2007. Prevalence and correlates of erectile dysfunction by race and ethnicity among men aged 40 or older in the United States: from the male attitudes regarding sexual health survey. J. Sex Med. 4, 57–65.

Lue, T.F., Brant, W.O., Shindel, A., Bella, A.J., 2000. Sexual dysfunction in diabetes. In: De Groot, L.J., Beck-Peccoz, P., Chrousos, G., Dungan, K., Grossman, A., Hershman, J.M., Koch, C., Mclachlan, R., New, M., Rebar, R., Singer, F., Vinik, A., Weickert, M.O. (eds.), Endotext. South Dartmouth, MA.

Maggi, M., Filippi, S., Ledda, F., Magini, A., Forti, G., 2000. Erectile dysfunction: from biochemical pharmacology to advances in medical therapy. Eur. J. Endocrinol. 143, 143–154.

Manganiello, V., 2003. Cyclic nucleotide phosphodiesterase 5 and sildenafil: promises realized. Mol. Pharmacol 63 (6), 1209–1211. Review. No abstract available. PMID: 12761329.

McMahon, C.G.1, Carson, C.C., Fischer, C.J., Wang, W.C., Florio, V.A., Bradley, J.D., 2006. Tolerance to the therapeutic effect of tadalafil does not occur during 6 months of treatment: A randomized, double-blind, placebo-controlled study in men with erectile dysfunction. J. Sex Med 3 (3), 504–511. PMID: 16681476.

Menconi, M.J., Unno, N., Smith, M., Aguirre, D.E., Fink, M.P., 1998. Nitric oxide donor-induced hyperpermeability of cultured intestinal epithelial monolayers: role of superoxide radical, hydroxyl radical, and peroxynitrite. Biochim. Biophys. Acta 1425, 189–203.

Meurer, S., Pioch, S., Pabst, T., Opitz, N., Schmidt, P.M., Beckhaus, T., et al., 2009. Nitric oxide-independent vasodilator rescues heme-oxidized soluble guanylate cyclase from proteasomal degradation. Circ. Res. 105, 33–41.

Miller, M.R., Megson, I.L., 2007. Recent developments in nitric oxide donor drugs. Br. J. Pharmacol. 151, 305–321.

Mizusawa, H., Hedlund, P., Brioni, J.D., Sullivan, J.P., Andersson, K.E., 2002. Nitric oxide independent activation of guanylate cyclase by YC-1 causes erectile responses in the rat. J. Urol. 167, 2276–2281.

Naruse, K., Rask-Madsen, C., Takahara, N., Ha, S.W., Suzuma, K., Way, K.J., et al., 2006. Activation of vascular protein kinase C-beta inhibits Akt-dependent endothelial nitric oxide synthase function in obesity-associated insulin resistance. Diabetes 55, 691–698.

Nossaman, B., Pankey, E., Kadowitz, P., 2012. Stimulators and activators of soluble guanylate cyclase: review and potential therapeutic indications. Crit. Care Res. Pract. 2012, 290805. http://dx.doi.org/10.1155/2012/290805. PMID: 22482042.

Nunes, K.P., et al., 2015. Beneficial effect of the soluble guanylyl cyclase stimulator BAY 41-2272 on impaired penile erection in db/db−/− type II diabetic and obese mice. J. Pharmacol. Exp. Ther. 353 (2), 330–339.

Oudot, A., Behr-Roussel, D., Poirier, S., Sandner, P., Bernabe, J., Alexandre, L., et al., 2011. Combination of BAY 60-4552 and vardenafil exerts proerectile facilitator effects in rats with cavernous nerve injury: a proof of concept study for the treatment of phosphodiesterase type 5 inhibitor failure. Eur. Urol. 60, 1020–1026.

Pauwels, B., Boydens, C., Brouckaert, P., van de Voorde, J., 2015. Oximes induce erection and are resistant to oxidative stress. J. Sex Med. 12, 906–915.

Pauwels, B., Boydens, C., Decaluwe, K., Van De Voorde, J., 2014. NO-donating oximes relax corpora cavernosa through mechanisms other than those involved in arterial relaxation. J. Sex Med. 11, 1664–1674.

Piga, R., Naito, Y., Kokura, S., Handa, O., Yoshikawa, T., 2007. Short-term high glucose exposure induces monocyte-endothelial cells adhesion and transmigration by increasing VCAM-1 and MCP-1 expression in human aortic endothelial cells. Atherosclerosis 193, 328–334.

Ragazzi, E., Meggiato, C., Chinellato, A., Italiano, G., Pagano, F., Calabro, A., 1996. Chronic treatment with cyclosporine A in New Zealand rabbit: aortic and erectile tissue alterations. Urol. Res. 24, 323–328.

Rajasekaran, M., Hellstrom, W.J., Sikka, S.C., 2001. Nitric oxide induces oxidative stress and mediates cytotoxicity to human cavernosal cells in culture. J. Androl. 22, 34–39.

Rajfer, J., Aronson, W.J., Bush, P.A., Dorey, F.J., Ignarro, L.J., 1992. Nitric oxide as a mediator of relaxation of the corpus cavernosum in response to nonadrenergic, noncholinergic neurotransmission. N. Engl. J. Med 326 (2), 90–94. PMID: 1309211.

Rosen, R.C., Fisher, W.A., Eardley, I., Niederberger, C., Nadel, A., Sand, M., et al., 2004. The multinational Men's Attitudes to Life Events and Sexuality (MALES) study: I. Prevalence of erectile dysfunction and related health concerns in the general population. Curr. Med. Res. Opin. 20, 607–617.

Roy, B., Mo, E., Vernon, J., Garthwaite, J., 2008. Probing the presence of the ligand-binding haem in cellular nitric oxide receptors. Br. J. Pharmacol. 153, 1495–1504.

Saigal, C.S., Wessells, H., Pace, J., Schonlau, M., Wilt, T.J., Urologic Diseases in America Project, 2006. Predictors and prevalence of erectile dysfunction in a racially diverse population. Arch. Intern. Med. 166, 207–212.

Sarkar, R., Gordon, D., Stanley, J.C., Webb, R.C., 1997. Dual cell cycle-specific mechanisms mediate the antimitogenic effects of nitric oxide in vascular smooth muscle cells. J. Hypertens. 15, 275–283.

Sazova, O., Kadioglu, A., Gurkan, L., Kayaarasi, Z., Bross, S., Manning, M., et al., 2002. Intracavernous administration of SIN-1 + VIP in an in vivo rabbit model for erectile function. Int. J. Impot. Res. 14, 44–49.

Scatena, R., Bottoni, P., Pontoglio, A., Giardina, B., 2010. Pharmacological modulation of nitric oxide release: new pharmacological perspectives, potential benefits and risks. Curr. Med. Chem. 17, 61–73.

Schmidt, H.H., Schmidt, P.M., Stasch, J.P., 2009. NO- and haem-independent soluble guanylate cyclase activators. Handb. Exp. Pharmacol. 309–339.

Schmidt, P., Schramm, M., Schroder, H., Stasch, J.P., 2003. Mechanisms of nitric oxide independent activation of soluble guanylyl cyclase. Eur. J. Pharmacol. 468, 167–174.

Schmidt, P.M., Schramm, M., Schroder, H., Wunder, F., Stasch, J.P., 2004. Identification of residues crucially involved in the binding of the heme moiety of soluble guanylate cyclase. J. Biol. Chem. 279, 3025–3032.

Searles, C.D., 2006. Transcriptional and posttranscriptional regulation of endothelial nitric oxide synthase expression. Am. J. Physiol. Cell Physiol. 291, C803–C816.

Seidler, M., Uckert, S., Waldkirch, E., Stief, C.G., Oelke, M., Tsikas, D., et al., 2002. In vitro effects of a novel class of nitric oxide (NO) donating compounds on isolated human erectile tissue. Eur. Urol. 42, 523–528.

Senbel, A.M., Hashad, A., Sharabi, F.M., Daabees, T.T., 2012. Activation of muscarinic receptors inhibits neurogenic nitric oxide in the corpus cavernosum. Pharmacol. Res. 65, 303–311.

Serafim, R.A., Primi, M.C., Trossini, G.H., Ferreira, E.I., 2012. Nitric oxide: state of the art in drug design. Curr. Med. Chem. 19, 386–405.

Shukla, N., Hotston, M., Persad, R., Angelini, G.D., Jeremy, J.Y., 2009. The administration of folic acid improves erectile function and reduces intracavernosal oxidative stress in the diabetic rabbit. BJU Int. 103, 98–103.

Sparwasser, C., Drescher, P., Will, J.A., Madsen, P.O., 1994. Smooth muscle tone regulation in rabbit cavernosal and spongiosal tissue by cyclic AMP- and cyclic GMP-dependent mechanisms. J. Urol. 152, 2159–2163.

Stasch, J.P., Schmidt, P., Alonso-Alija, C., Apeler, H., Dembowsky, K., Haerter, M., et al., 2002. NO- and haem-independent activation of soluble guanylyl cyclase: molecular basis and cardiovascular implications of a new pharmacological principle. Br. J. Pharmacol. 136, 773–783.

Stasch, J.P., Schmidt, P.M., Nedvetsky, P.I., Nedvetskaya, T.Y., H, S.A., Meurer, S., et al., 2006. Targeting the heme-oxidized nitric oxide receptor for selective vasodilatation of diseased blood vessels. J. Clin. Invest. 116, 2552–2561.

Tabit, C.E., Chung, W.B., Hamburg, N.M., Vita, J.A., 2010. Endothelial dysfunction in diabetes mellitus: molecular mechanisms and clinical implications. Rev. Endocr. Metab. Disord. 11, 61–74.

Taher, A., Birowo, P., Kamil, S.T., Shahab, N., 2000. Relaxation effect of nitric oxide-donor on diabetic penile smooth muscle in vitro. Clin. Hemorheol. Microcirc. 23, 277–281.

Tamas, V., Kempler, P., 2014. Sexual dysfunction in diabetes. Handb. Clin. Neurol. 126, 223–232.

Terwel, D., Nieland, L.J., Schutte, B., Reutelingsperger, C.P., Ramaekers, F.C., Steinbusch, H.W., 2000. S-nitroso-N-acetylpenicillamine and nitroprusside induce apoptosis in a neuronal cell line by the production of different reactive molecules. Eur. J. Pharmacol. 400, 19–33.

Tesfamariam, B., Cohen, R.A., 1992. Free radicals mediate endothelial cell dysfunction caused by elevated glucose. Am. J. Physiol. 263, H321–H326.

Thatcher, G.R., Nicolescu, A.C., Bennett, B.M., Toader, V., 2004. Nitrates and NO release: contemporary aspects in biological and medicinal chemistry. Free Radic. Biol. Med. 37, 1122–1143.

Toksoz, S., Erdem, S.R., Peskircioglu, C.L., Keskin, U., 2013. The effect of long-term oral tadalafil treatment on corpus cavernosum function in an experimental spinal cord transection rat model. Spinal Cord 51, 663–667.

Traish, A.M., Palmer, M.S., Goldstein, I., Moreland, R.B., 1995. Expression of functional muscarinic acetylcholine receptor subtypes in human corpus cavernosum and in cultured smooth muscle cells. Receptor 5, 159–176.

Uma, S., Yildirim, S., Sarioglu, Y., Butuner, C., Yildirim, K., 1998. Tolerance to isosorbide dinitrate in isolated strips of rabbit corpus cavernosum. Int. J. Androl. 21, 364–369.

Veliev, E.I., Vanin, A.F., Kotov, S.V., Shiahlo, V.K., 2009. Modern aspects of pathophysiology and prevention of erectile dysfunction and cavernous fibrosis after radical prostatectomy. Urologiia, 46–51.

Vernet, D., Cai, L., Garban, H., Babbitt, M.L., Murray, F.T., Rajfer, J., et al., 1995. Reduction of penile nitric oxide synthase in diabetic BB/WORdp (type I) and BBZ/WORdp (type II) rats with erectile dysfunction. Endocrinology 136, 5709–5717.

Wang, R., Domer, F.R., Sikka, S.C., Kadowitz, P.J., Hellstrom, W.J., 1994. Nitric oxide mediates penile erection in cats. J. Urol. 151, 234–237.

Wimalawansa, S.J., 2010. Nitric oxide and bone. Ann. NY. NY Acad. Sci. 1192, 391–403.

Yildirim, S., Ayan, S., Sarioglu, Y., Gultekin, Y., Uma, S., 2000. Does diabetes mellitus affect the progress of tolerance to isosorbide dinitrate (ISDN) in corporal tissue? Nitric Oxide 4, 29–34.

Zhang, Y., Hogg, N., 2005. S-Nitrosothiols: cellular formation and transport. Free Radic. Biol. Med. 38, 831–838.

Zhou, J., Abid, M.D., Xiong, Y., Chen, Q., Chen, J., 2013. ox-LDL downregulates eNOS activity via LOX-1-mediated endoplasmic reticulum stress. Int. J. Mol. Med. 32, 1442–1450.

CHAPTER 6

Nitric Oxide Donors in Nerve Regeneration

Vinod B. Damodaran[1,†], Divya Bhatnagar[1], Heather Rubin[2] and Melissa M. Reynolds[2]

[1]New Jersey Center for Biomaterials and Rutgers, The State University of New Jersey, Piscataway, NJ, United States
[2]Department of Chemistry and School of Biomedical Engineering, Colorado State University, Fort Collins, CO, United States

INTRODUCTION

Nitric oxide (NO) is a key signaling molecule involved in various critical activities related to nervous system functioning such as intracellular and intercellular messaging. Experimentation that aids in understanding the particular role of NO in such pathways has become increasingly popular in recent years. While recovery of the nervous system after trauma is still not entirely understood, many believe NO may be an important factor for regulating fundamental processes associated with such traumas (Zochodne and Levy, 2005). However, various studies showed that NO signaling provides a switching mechanism between the degenerative and regenerative states of neuronal remodeling mediated by the available NO concentration in situ (Rabinovich et al., 2016; Awasaki and Ito, 2016). In brief, according to Rabinovich et al. high NO levels promote the neurodegeneration while low levels support the neuroregeneration. Given the dual nature of NO, a detailed description of the role and mechanism of NO in both neuroregeneration and neurodegeneration is crucial for the proper design of an NO-releasing prodrug or device for the desired application. A comprehensive narration of these aspects is provided in the first section of this chapter. Subsequently, a detailed and complete summary of various NO donors evaluated for nerve regenerations is presented.

NO IN NEUROREGENERATION

Neurotransmission

NO functions as a critical neurotransmitter in both central nervous system (CNS) and peripheral nervous system (PNS) by cyclic guanosine monophosphate (cGMP)-dependent mechanisms. Garthwaite et al. (1988) were the first to report the role of NO as a neurotransmitter. They observed activation of brain NMDA (*N*-methyl-D-aspartate) receptors by glutamate resulted in the release of NO, which had similar

†Deceased

Nitric Oxide Donors.

properties as endothelium-derived relaxing factor (EDRF), a local hormone that causes smooth muscle relaxation through a cGMP-dependent pathway. NO is a free radical gas which simply diffuses from nerve terminals into adjacent cells and forms covalent linkages to a soluble guanylyl cyclase (sGC) enzyme or other protein to stimulate their activity (Cary et al., 2006; Garthwaite, 2010). Fig. 6.1 presents a schematic illustration of the signaling mechanism of NO through cGMP pathway (Francis and Corbin, 2005). The covalent binding of NO to sGC causes 100- to 200-fold activation of the enzyme, which increases the conversion of GTP to cGMP, resulting in increased production of cGMP and initiation of a cGMP signaling pathway (Francis et al., 2010). Elevation in cGMP can further activate cGMP-dependent protein kinase (PKG), other phosphodiesterases, cGMP regulated ion channels, and transcription factors through which NO influences smooth muscle relaxation, synaptic plasticity, neurotransmission, and neurosecretion (Krumenacker et al., 2004; Arnold et al., 1977).

NO donors can further stimulate the release of classical neurotransmitters such as acetylcholine. In the nucleus accumbens (NAc), the glutamatergic neurons can release

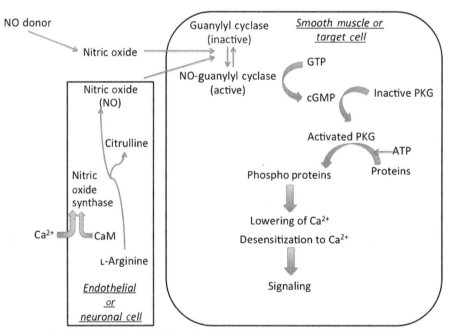

Figure 6.1 NO signaling through cGMP pathway. NO is synthesized from L-arginine by nitric oxide synthases (NOSs) through an increase in the cellular Ca^{2+} concentration or through the binding of Ca^{2+}/calmodulin (CaM) complex. NO then diffuses into the extracellular space and vascular smooth muscle cell and activates guanylyl cyclase to synthesize cGMP from GTP. The increase in the cGMP level enables the PKG phosphotransferase activity and leads to the signaling activity. *Adapted from Francis, S.H., Corbin, J.D., 2005. Phosphodiesterase-5 inhibition: the molecular biology of erectile function and dysfunction. Urol. Clin. North Am. 32, 419–429, Elsevier.*

glutamate where cGMP mediates the NO-induced release of acetylcholine by enhancing the glutamate outflow (Prast et al., 1998). Lorrain et al. showed that NO can increase the release of dopamine (DA) and serotonin in the rat medial preoptic area by an sGC-cGMP-dependent pathway (Lorrain and Hull, 1993). N-methyl-D-aspartate (NMDA) receptors and NO donors have also proved to stimulate the noradrenaline (NE) release in the hippocampus, both in vivo and in vitro, whereas NOS inhibitors have shown to decrease NE levels (Lonart et al., 1992).

NEURONAL GROWTH AND SYNAPTIC PLASTICITY

The increases in NOS expression during development play a role in synapse formation and maturation. Disruption of the NO signaling during the development phase can introduce changes in synapse morphology and interfere with neuronal growth (Gibbs et al., 2001; Godfrey and Schwarte, 2010; Seidel and Bicker, 2000). Long-term potentiation (LTP) and long-term depression (LTD) are the two mechanisms, which affect the efficacy of a synapse and synaptic plasticity is important for recognition memory. NO that is produced in the postsynaptic cell and acts on the presynaptic sites is recognized as a retrograde diffusible messenger. Tamagnini et al. (2013) showed that NO-dependent signaling is critical in perirhinal cortex-dependent visual recognition memory in the rats. Cooke et al. (2013) studied the role of endogenous NO signaling in neuronal growth and synaptic remodeling after nerve injury in CNS. They demonstrated that blocking sGC activity completely suppresses neurite extension and synaptic remodeling after nerve crush, showing the importance of cGMP in these processes. However, they concluded that the effects of cGMP-dependent protein kinase inhibition on neurite growth and synaptic remodeling were mediated by separate signaling pathways.

NEUROPROTECTANT

NO can have beneficial effects on recovery from nerve injury. Yuan et al. (2009) used NOS blocker, N-nitro-L-arginine methyl ester (L-NAME) after hypophysectomy in the adult Sprague–Dawley (SD) rats. They found that the L-NAME treatment effectively blocked the regeneration of magnocellular neurons of the rodent hypothalamus suggesting that the induced increase of nNOS expression enhances the regenerative ability of magnocellular neurons following hypophysectomy. NO also plays a role in axonal elongation (Rentería and Constantine-Paton, 1999; Van Wagenen and Rehder, 1999). NO-cGMP transduction pathway is known to play a vital role in regulating axonal growth and neuronal migration. Stern and Bicker (2008) showed that NO/cGMP promotes axonal regeneration in an insect embryo, whereas inhibiting NO or sGC delays regeneration. NO-induced cGMP immunostaining shown in Fig. 6.2 confirms the serotonergic neurons as direct targets of NO.

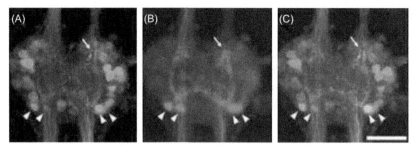

Figure 6.2 NO-induced cGMP immunoreactivity in serotonergic and other neurons. (A) Confocal image of an abdominal ganglion A6 of a 65% embryo stained for NO-induced cyclic GMP, 2-hour postcrush. Severed axons have already retreated into the ganglion (arrow). (B) Confocal image of serotonin immunoreactivity in the same preparation as in (A). Arrowheads point to the four serotonergic cells. (C) Superimposed images of (A) and (B), revealing colocalization of serotonin and NO-induced cGMP in cell bodies (arrowheads) and axon stumps (small arrow). Scale bar: 100 lm, anterior is to the top. *Reproduced with permission from Stern, M., Bicker, G., 2008a. Nitric oxide regulates axonal regeneration in an insect embryonic CNS. Dev. Neurobiol. 68, 295–308, John Wiley & Sons.*

In the PNS, the three isoforms of NOS are overexpressed after transection of the nerve. High levels of NO are maintained at the lesion site by different cellular sources that act on different targets to benefiting peripheral nerve regeneration (González-Hernández and Rustioni, 1999a). nNOS further is colocalized with growth associated protein-43 (GAP-43) in growth cones, thus indicating its involvement in nerve regeneration (González-Hernández and Rustioni, 1999b). It has been shown that the local release of NO following peripheral nerve injury is a crucial factor that can cause regenerative delay and delayed Wallerian degeneration in nNOS knockout mice after peripheral nerve injury (Keilhoff et al., 2002).

NO promotes cell survival and neuroprotection by different pathways as shown in Fig. 6.3 (Calabrese et al., 2007). NO activates cyclic AMP-responsive-element-binding protein (CREB) and Akt, two proteins that are mainly involved in neuroprotection through the stimulation of the soluble guanylate cyclase (sGC)–cyclic GMP (cGMP)–protein kinase G (PKG) pathway (Contestabile and Ciani, 2004; Riccio et al., 2006). It is known that prolonged stimulation of N-methyl-D-aspartate receptor (NMDAR) causes toxic cell death (Guix et al., 2005). By S-nitrosylation of thiols from cysteines such as caspase 3 and the NR1 and NR2 subunits of the NMDAR receptor, the intracellular Ca^{2+} influx responsible for cell death is inhibited thus leading to a decrease in cell death (Guix et al., 2005). Under pro-oxidative conditions, excess NO and reactive nitrogen species (RNS) can be formed. As a secondary mechanism, in the brain, NO induces the activity of hemeoxygenase 1 (HO-1) that is known to have a neuroprotective function. The upregulation of HO-1 is followed by an increase in biliverdin, the precursor of the powerful antioxidant and antinitrosative molecule bilirubin (Mancuso et al., 2003; Mancuso, 2004; Motterlini et al., 2002).

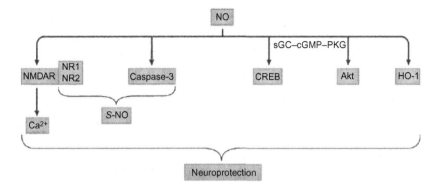

Nature Reviews | Neuroscience

Figure 6.3 Neuroprotective effects of nitric oxide. *nNOS, neuronal nitrogen oxide synthase; S-NO,
S-nitrosylation. Reproduced with permission from Calabrese, V., Mancuso, C., Calvani, M., Rizzarelli, E.,
Butterfield, D.A., Stella, A.M.G., 2007. Nitric oxide in the central nervous system: neuroprotection versus
neurotoxicity. Nat. Rev. Neurosci. 8, 766–775, Nature Publishing Group.*

NO IN NEURODEGENERATION

When produced in excess, NO can become toxic (Guix et al., 2005). In peripheral nerve regeneration, localized increase of NO may be toxic to the regenerating
axons (Zochodne et al., 1997). At a higher concentration, NO can lead to cell death
by increasing peroxynitrite ($ONOO^-$) concentrations. NO can react with oxygen in
an oxidative–reductive reaction to form toxic compounds that belong to RNS family. Peroxynitrite is formed by one such reaction between NO and superoxide anion
and has been known to cause "nitrosative stress" or cellular damage (Pacher et al., 2007;
Ridnour et al., 2004). These nitrosative stresses have been known to be involved in many
neurodegenerative disorders such as Parkinson's disease (PD), Huntington's disease (HD),
Alzheimer's disease (AD), amyotrophic lateral sclerosis (ALS), multiple sclerosis (MS), and
ischemia since all these diseases exhibit oxidative stress. Nitrotyrosination that alters the
normal activity of proteins by inducing conformational changes has also been indicated
in several neurodegenerative disorders (Guix et al., 2005). Fig. 6.4 illustrates the various
mechanisms demonstrating the neurotoxic effects of NO (Calabrese et al., 2007).

ALZHEIMER'S DISEASE

Proteins such as α-enolase have been shown to being specifically oxidized in the brains
of people with Alzheimer's disease (Castegna et al., 2002). α-Enolase together with
the increased nitration of triosephosphate isomerase causes altered glucose tolerance
and metabolism, a possible mechanism exhibited in patients with Alzheimer's disease.

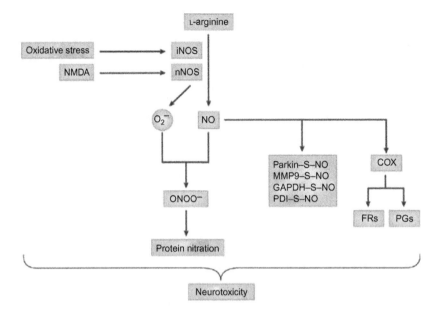

Nature Reviews | Neuroscience

Figure 6.4 Neurotoxic effects of nitric oxide (NO). If it is produced in excess, or if a cell is in a pro-oxidant state, NO has cytotoxic effects. It is well established that NO can react with superoxide anions (O_2^-; produced by inducible nitric oxide synthase (iNOS) under inflammatory conditions or neuronal nitric oxide synthase (nNOS), as in the case of excitotoxicity) to form peroxynitrite ($ONOO^-$), an anion with strong oxidant properties. As a consequence of the interaction between peroxynitrite and cellular components, protein nitration takes place, resulting in damage to cellular components. The NO-mediated S-nitrosylation (S-NO) of certain substrates, such as matrix metalloproteinase 9 (MMP9), parkin, GAPDH, and protein-disulfide isomerase (PDI), has been proposed to be a novel mechanism through which NO becomes neurotoxic. NO also activates the hemoprotein cyclooxygenase (COX). During its catalytic cycle, COX generates free radicals (FRs) and prostaglandins (PGs), both of which have strong proinflammatory features. *NMDA, N- methyl-D-aspartate. Reproduced with permission from Calabrese, V., Mancuso, C., Calvani, M., Rizzarelli, E., Butterfield, D.A., Stella, A.M.G., 2007. Nitric oxide in the central nervous system: neuroprotection versus neurotoxicity. Nat. Rev. Neurosci. 8, 766–775, Nature Publishing Group.*

Acetylcholinesterase inhibitors and NMDAR antagonists are the few available treatments to manage the cognitive deficit functions in AD. AD is a neurodegenerative disorder correlated with oxidative stress, and it has been shown that NO plays a role in the progression of the disease. AD patients have been found to have amyloid beta–peptide (Aβ) in their brains, which is characterized to form intracellular neurofibrillary tangles. Aβ fibrils on their reaction with NO can induce reactive oxidative species (ROS) formation that can have toxic effects due to the production of peroxynitrite (Miranda et al., 2000). Amyloid$^\beta$ (A$^\beta$) is a critical factor involved in the pathogenesis of Alzheimer's disease. Continuous infusion of intracerebroventricular A$^\beta$1–40 induces iNOS leading to peroxynitrite formation followed by tyrosine nitration of proteins in the hippocampus. This can result in an impairment of nicotine-evoked acetylcholine (ACh) release and memory deficits. To achieve a delay in AD progression, reducing Aβ production, cholinergic deficit and targeting NOS isoforms that damage the brain cells are important considerations.

PARKINSON'S DISEASE

PD is a neurodegenerative disorder associated with aging resulting from progressive loss of DA-producing brain cells within the substantia nigra. Several studies have indicated that oxidative damage in PD is mediated by the formation of peroxynitrite (Guix et al., 2005). DA release is modulated by NMDAR stimulation. Higher NO concentrations can decrease NMDA-induced DA levels (Kegeles et al., 2002). Tyrosine hydroxylase is an enzyme that promotes the conversion of tyramine to dopamine. This enzyme can be inhibited by peroxynitrite through nitrotyrosination (Kuhn and Geddes, 2002). Peroxynitrite can also contribute to the depletion of major cellular antioxidant defense and has also been implicated in the apoptosis of dopaminergic neurons in PD (Naoi and Maruyama, 2001). Another pathway of PD involves Parkin, an enzyme protein that plays an important role in the ubiquitin–proteasome system and acts as a regulator of protein breakdown. It has been shown that S-nitrosocysteine derived NO can nitrosylate parkin. Parkin mutation has been acknowledged as a common pathway of sporadic and hereditary Parkinson's disease (Miklya et al., 2014).

NO derived from the endogenous nitrosothiol S-nitrosocysteine can S-nitrosylate matrix metalloproteinase 9 (MMP9), causing neuronal apoptosis. MMP9s are implicated in the pathogenesis of neurodegenerative disorders (stroke, AD, multiple sclerosis) (Gu et al., 2002). A similar mechanism has been proposed for Parkinson's disease. The GAPDH–SIAH1 (E3 ubiquitin ligase) complex is formed when NO S-nitrosylates GAPDH, thereby inducing apoptosis (Hara et al., 2006). During the neurodegenerative diseases such as cerebral ischemia, denatured proteins generated by protein-disulfide isomerase (PDI) can accumulate and be toxic to neurons. In the brains of AD and PD patients, S-nitrosylation of cysteines and the inhibition of PDI enzymatic activity have been observed (Uehara et al., 2006).

HUNTINGTON'S DISEASE

Huntington's disease is another neurodegenerative disease that causes progressive breakdown of nerve cells in the brain, particularly in the striatum and cerebral cortex. It is a genetic disorder that can cause dementia. Researchers have reported that oxidative and nitrosative damage in the basal ganglia could be the trigger for HD (Browne and Beal, 2006). Development of HD is also affected by an excessive production of NO that can result in the oxidative fragmentation of DNA in striatal neurons (Butterfield et al., 2001). Other studies have also implicated the increased production of oxidative stress products in HD patients (Sorolla et al., 2008; Lin and Beal, 2006).

In brief, NO can have both neuroprotective and neurotoxic effects. Here we have reviewed the role of NO in neuroregeneration as well as in neurodegeneration especially under oxidation and nitrotyrosination of functional proteins. Peroxynitrite formation is the primary reason underlining all the neurodegenerative diseases. Therefore, prevention of peroxynitrite formation without modifying the activity of NOS is one possible way to counter neurodegenerative disorders.

EXAMPLES OF NO DONORS EVALUATED FOR NERVE REGENERATION

NO donors have demonstrated the ability to increase NO bioavailability, which may yield desirable effects with regard to angiogenesis, vasodilation, neurogenesis, and synaptic plasticity—all of which are central for neurorepair. Consequently, various NO donors (Fig. 6.5) have been explored along with other important NO regulators to probe the pathways and provide mechanistic insight surrounding NO influence on the recovery of the central, peripheral, and autonomic nervous systems. Herein, results and discussions of such findings are highlighted.

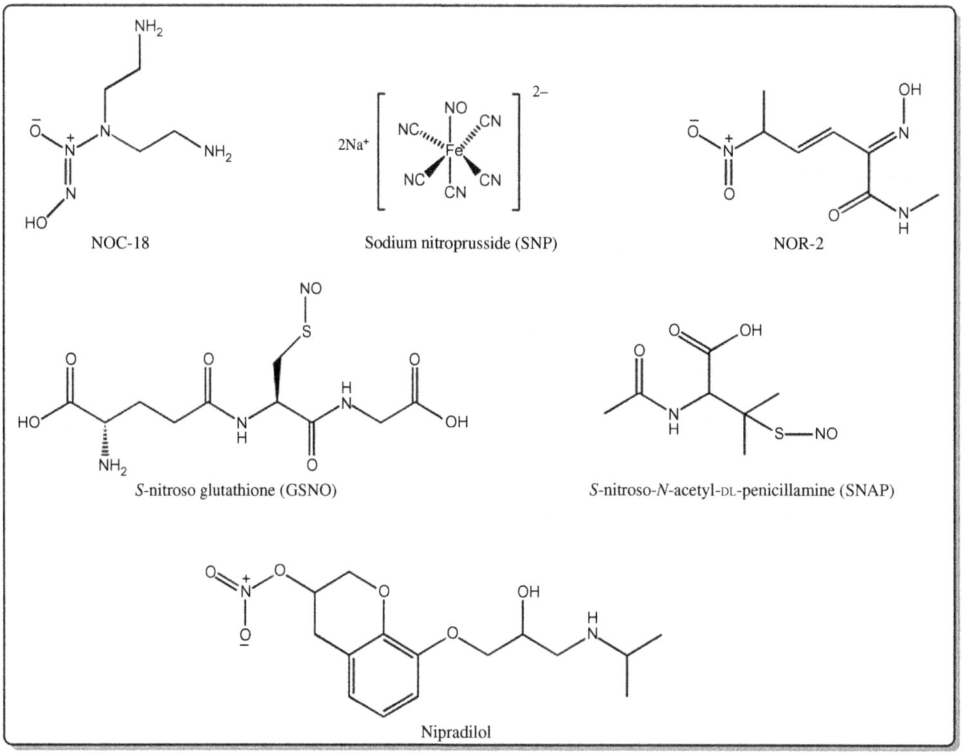

Figure 6.5 Various NO donors evaluated for nerve regeneration.

Hydra regeneration studies on the NO-cGMP pathway with NOC-18

NO has been recognized as vital to activating sGC (Beuve et al., 2016). The resultant increase in intracellular cGMP affects many biochemical processes including activation of protein kinase G (PKG), notorious for regulating cellular processes (Sellak et al., 2013). Notably, the NO–cGMP pathway may be initiated during the differentiation of embryonic stem cells. To probe the potential for the NO–cGMP pathway to influence proliferation and differentiation events during development and regeneration, the involvement of NO in the regeneration of *Hydra vulgaris* was evaluated (Colasanti et al., 2009). An archaic invertebrate with a nervous system, known as the *Hydra*, possesses an impressive capability to regenerate. A *Hydra* body is comparable to an early-stage embryo containing ectodermal, endodermal, and interstitial stem cells. Together, the three stem cell bodies and the ability to rapidly regenerate make *Hydra* an ideal system for studying morphogenetic processes including development and regeneration. An initial control study with regenerating *Hydra* found NO concentrations steadily increased during *Hydra* regeneration, peaking 36 hours after damage, as determined by the Griess Assay. The Griess Assay estimates NO concentrations based on nitrite ions in solution, however, this method has proven inaccurate for measuring NO (Hunter et al., 2013). Although the exact amount of NO may not be accurate, the outcomes still suggest NO release plays a vital role in regeneration. Expanding upon this study, the heads of the specimen, consisting of the hyposome and tentacles, were removed under a microscope and placed in culture plates for observation. Samples were dosed daily with either the NO donor 3,3-*bis*(aminoethil)-1-hydroxy-2-oxo-1-triazene (NOC-18); the NOS inhibitor L-nitroarginine methyl ester (L-NAME); the guanylate cyclase inhibitor 1H [1,2,4]oxadiazolo-[4,3-a]quinoxa-lin-1-one (ODQ); or the protein kinase G inhibitor KT-5823. Nitric oxide synthase (NOS) encompasses a set of enzymes that catalyze the release of NO from endogenous sources like L-arginine; therefore, areas with high concentrations of these enzymes are believed to be more active with regard to NO release. Under a microscope, tentacles per head were monitored for growth. The decapitated *Hydra* treated with NOC-18 showed significantly increased growth rates in comparison to the control and L-NAME treated samples (Fig. 6.6). After just 2 days, NOC-18 remarkably showed tentacle growth, whereas no tentacles were visible in the control and L-NAME treated samples. Similarly, the diminished growth rate was observed when samples were treated with ODQ or KT-5823, suggesting that sGC and PKG are both involved in the NO pathway of *Hydra* regeneration. The authors also demonstrated the impact of NO on cell proliferation and discovered that DNA synthesis was accelerated in the presence of NOC-18 and impeded when treated with L-NAME. These results demonstrate NO also plays an important role in cell proliferation of *Hydra* regeneration. NOC-18 is a slow releasing (20 hours half-life) NO donor where the rate of release is attributed to the structure. NOC-18 is an ideal NO donor

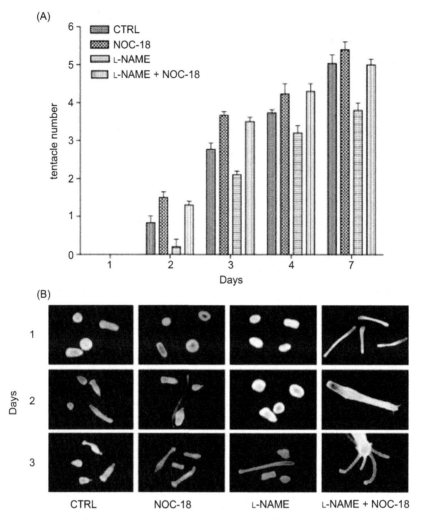

Figure 6.6 NO donor and NOS inhibitor affect *Hydra* regeneration. *Hydra* were treated with the NO donor NOC-18, the NOS inhibitor L-NAME, or both. NOC-18 treated samples enhanced regenerative rate, whereas L-NAME treatment inhibited the regeneration process. *Reproduced with permission from Colasanti, M., Mazzone, V., Mancinelli, L., Leone, S., Venturini, G., 2009. Involvement of nitric oxide in the head regeneration of Hydra vulgaris. Nitric Oxide 21, 164–170, Elsevier.*

since the amine byproduct formed has no known interferences with cellular activities; however, since NO release is spontaneous and the donor is reported to rapidly degrade it is uncertain as to how much NO was administered in the *Hydra* regeneration experiments. Even though effective NO concentrations are not conveyed in this study, it is clear that the NO-cGMP/PKG pathway is vital to cnidarian *Hydra* regeneration, in turn supporting the apparent involvement of this pathway in early embryogenesis, likely translatable to more complex animals.

LOCUST STUDIES ON NO-cGMP WITH NOC-18 AND SNP

Insects are acknowledged as invertebrates with complex nervous systems, many of which capable of regeneration in both the PNS and CNS. Since regeneration of the CNS in advanced vertebrates following axonal injury has been observed with minimal success, insects present a unique opportunity to study these processes. The locust has displayed axonal regeneration subsequent to nerve crush injury and the mechanism of such regeneration has been explored by Stern and coworkers in an attempt to evaluate how the NO-cGMP system is integrated into this physiological event (Patschke et al., 2004; Stern and Bicker, 2008). The foremost goal was to determine if the locust embryo exemplifies an appropriate model system for investigating factors that manipulate CNS regeneration. In one study, the connective tissues on the right side were carefully crushed with forceps without affecting the neural sheath of the locust embryos (mostly 65–70% developed). The researchers focused on serotonergic neurons of the *Locusta*, where typically neurons reform around 40% embryogenesis. The expression of SERT, serotonin reuptake transporter, as well as physiological serotonin synthesis allowed for developmental tracking of neurons via staining. In the crushed region, there were four serotonin-immunoreactive axons. Typically, after a few hours, neurons began to appear. In most cases, after 24 hours minor serotonin-immunoreactive processes were noticeable, and after 96 hours immunofiber regeneration was achieved, many of which reaching the adjacent ganglia. Regenerated fibers were visually obvious due to thinner appearance and abnormal growth directions and configuration (not as straight) compared to normal neurons. The number of regenerated serotonin-immunoreactive neurites in the connective tissue that grew beyond the crush site was counted and compared to those on the uncrushed control side. In culture postcrush, an average of 44.5% regeneration occurred in the first 48 hours. Meaning, 48-hour postcrush a great degree of nerve regeneration in the locust was achieved, rendering this system an ideal model on which to investigate the uncertainties surrounding how an increase in NOC-18 can influence such CNS regeneration events.

As described previously with the *Hydra* studies, the NO-cGMP pathway is involved in regeneration of the nervous system in cnidarian invertebrates. To assess if a similar mechanism was at play in the CNS regeneration event of locusts, NO and cGMP concentrations were methodically manipulated. When NOC-18 was employed as an NO donor, a significant increase in regeneration was observed. Concentrations of cGMP within the crushed samples were increased when incubated with membrane-permeable 8Br-cGMP. This ultimately led to a significant increase in the rate of regeneration. When the NO scavenger carboxy-PTIO was employed, significantly less axons displayed axonal regrowth over 48 hours. Furthermore, utilizing the inhibitor of guanylyl cyclase, ODQ, regeneration was significantly inhibited; however, the coincubation with ODQ and 8Br-cGMP resulted in a restoration of regrowth to near normal levels. This data indicates

NO–cGMP regulation is pertinent to axon regeneration in insect embryos. Using fluorescence labeling, the authors observed colocalization of NO initiated cGMP and serotonin in the cell bodies and neurites of S1 and S2 neurons, signifying NO likely directly targets serotonergic neurons. These results distinguish the NO donor NOC-18 as both a positive initiator of neurite outgrowth during development and beneficial in axonal regeneration after locust nerve crush, but once again the amount of NO required necessitates further investigation.

Sodium nitroprusside (SNP) is recognized as an NO donating drug often used as a vasodilator. Building upon the above study, SNP was used to determine if NOS is indeed localized to the aforementioned areas proposed for NO activity after nervous system injury. SNP also aided in investigating what markers are most helpful for identifying this activity. The mechanism of NO release, however, appears to vary widely based on other small molecules present and can lead to a variety of byproducts (Grossi and D'Angelo, 2005). Therefore, it is not surprising that unlike NOC-18, SNP had little to no effect on regeneration rates.

In another study, nervous systems of the *Locusta* were carefully dissected and subjected to abdominal ganglia nerve crush, without damaging the local sheath (Stern et al., 2010). NADPH disphosphorase (NADPHd) staining and immunostaining against universal nitric oxide synthase (uNOS), citrulline, and NO-induced cGMP were used to assess components of the NO–cGMP pathway. NADPHd staining has been used to indicate NOS presence, however, it does not necessarily indicate NOS activity, ergo a byproduct of NO release known as citrulline was monitored to better represent the physiological activity of the existing NOS. Since all citrulline-immunoreactive cells also stained with NADPHd, and immunostained for NOS, it was determined NADPHd staining is an adequate method to detect nitrergic cells. At 30% embryo development, NOS was detected with NADPHd, suggesting the NO–cGMP pathway may be engaged early on during the development and formation of the locust CNS. On the other hand, NO activity and appearance as determined by the citrulline and NOS expression did not appear until late stages of the neuropil regeneration. This suggests that NO is unlikely to contribute to the formation in the neuropil region but may, however, be active during synapse formation or plasticity in later stages of development. Additional testing with NO donors may once again aid in a better understanding of the role of NO involved versus the impact of the specified chosen donor.

GOLDFISH OPTIC NERVE REGENERATION STUDIED WITH NOR-2 AND SNAP

In the search to better understand the mechanism of NO in nerve regeneration, more complex vertebrate systems are being investigated. The NO–cGMP pathway impacts nerve regeneration following injury in both the *Hydra* and the *Locusta*. Primary studies in more advanced vertebrate systems would necessitate the ability to monitor the NO-related influence in an isolable system capable of regeneration. Unlike mammalian retina, fish retinal ganglion cells (RGCs) can survive and regrow axons postnerve damage. Axonal regrowth begins 5–6 days after damage, tectum reinnervation occurs within 5–6 weeks, and visual functions are restored within 5–6 months. Therefore, axonal elongation during nerve regeneration in goldfish both in vitro and in vivo warranted investigation. In this study, performed by Koriyama and coworkers, the authors attempted to elucidate the role of the NO-cGMP pathway in goldfish RGCs and to clarify the specific mode of action (Koriyama et al., 2009). Since NADPH activity is directly and proportionally associated with NO activity, NADPHd staining was performed and revealed intense staining in the photoreceptor cells and horizontal cells, with mild staining in the inner nuclear layer. Positive cells significantly increased within 5 days following axonomy and peaked after 20 days, while normal levels were restored after 40 days. Nitrite concentration was also assessed as an indicator of NO production using the Griess method. Nitrite production in the retina echoed the NADPH findings, showing a significant increase after 5 days, peaking around 20 days. The addition of an NOS inhibitor reduced nitrite concentration by nearly 24%, indicating that NO production was uniquely regulated during optic nerve regeneration. Following similar patterns as observed before, nNOS in the ganglion cell layer (GCL) also exhibited increased concentrations noticeable around 5 days, and peaking after 20 days following nerve lesion.

The regulatory effect of NO influence was manipulated using a series of NO donors and inhibitors (Fig. 6.7). As monitored via positive staining with anti-GAP-43 immunochemistry, retinal explant control cultures having spontaneous neurite outgrowth were quantified and designated as 100% regrowth. Employing the NO donor NOR-2 at doses between 50 and 100 μM, increased neurite outgrowth by 150% was achieved. Similar effects were observed when the NO donor S-nitroso-N-acetyl penicillamine (SNAP) was dosed at a slightly higher level of 500 μM. On the other hand, NOS inhibitors like L-NAME significantly reduced neurite outgrowth by 40–60% at doses of 200–400 μM. By adding an NO donor back to a culture containing L-NAME, outgrowth rates could be restored to nearly 80–90% outgrowth. The NO scavenger c-PTIO was added at varying time points following NOR-2 stimulation in order to highlight optimal NO bioavailability. When added at the same time as NOR-2, neurite growth was entirely blocked. After 3 hours, outgrowth was moderately blocked, and after 6 hours c-PITO had little to no effect on neurite outgrowth. Plainly, NO surges seem most crucial to neurite regeneration within the first few hours postnerve injury. This also suggests that NO bioavailability in the CNS can be increased to promote neurite outgrowth after damage using exogenous sources of NO from NOR-2 and SNAP, however, it is not clear, what the concentrations of NO available were with each donor, as NO donor concentration is not directly correlated with NO availability.

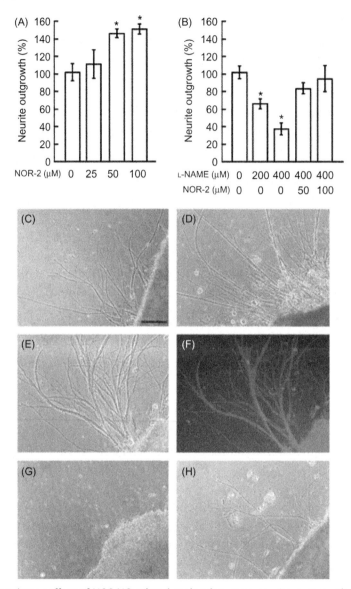

Figure 6.7 Regulatory effect of NOS-NO related molecules upon neurite outgrowth in the goldfish retinal explant culture. (A and B) Quantitative data for NOS-NO related drugs on neurite outgrowth were obtained from axotomized goldfish retinal explant culture relative to the control (no treatment, 100%). (A) NO generator, NOR-2 induced the neurite outgrowth. *$p < 0.05$ versus control ($n = 4$). (B) Universal NOS inhibitor, L-NAME inhibited the neurite outgrowth. *$p < 0.05$ versus control ($n = 4$). (C–H) Photomicrographs of neurite outgrowth under various conditions: (C) control, (D) NOR-2, (E) SNAP, (F) immunohistochemistry of anti-GAP-43 antibody in (E), (G) L-NAME, (H) L-NAME plus NOR-2. *Reproduced with permission from Koriyama, Y., Yasuda, R., Homma, K., Mawatari, K., Nagashima, M., Sugitani, K., et al., 2009. Nitric oxide-cGMP signaling regulates axonal elongation during optic nerve regeneration in the goldfish in vitro and in vivo. J. Neurochem. 110, 890–901, John Wiley & Sons.*

Additionally, the four analogs of NOR have varying effects on cell toxicity even at very low doses. Depending on the length of exposure to these donors, adverse effects can arise (Yamamoto et al., 2000). As a result, time scale studies shown may not be entirely accurate and could be strengthened by verification with another NO donor like SNAP.

In goldfish retina, nNOS is at least in part responsible for NO production following axonomy to promote neurite outgrowth. Concentrations of nNOS-specific inhibitor ETPI as low as 10 μM suppressed neurite outgrowth by 40–60%. Samples treated with small interfering RNS (siRNA), shown to specifically target region of goldfish nNOS mRNA, resulted in a 60% decrease in outgrowth. Whether NO acting via the NO-cGMP pathway was directly responsible for neurite outgrowth here remained unclear thus far, meriting further exploration. The selective sGC inhibitor ODQ was added and found to decrease neurite outgrowth. When, however, membrane-permeable dbcGMP was added, ODQ effects were reversed. Additionally, the PKG-specific inhibitor Rp-cGMPS dose-dependently inhibited neurite outgrowth. It was found that dbcGMP dosed samples showed significantly increased rates of regeneration (70%) after injury compared to controls (40%), suggesting that dbcGMP undoubtedly facilitates goldfish optic nerve regeneration in vivo. In all, these results strongly suggest that the NO-cGMP pathway plays a penultimate role in optic nerve regeneration in goldfish, and that an increase in NO or NO targeted receptors during the first few hours following nerve injury may lead to enhanced regeneration to the optic nerve.

REGENERATING RGCs OF CATS WITH NIPRADILOL

Nipradilol (8-[2-hydroxy-3-(isopropylamino)propoxy]-3-chromanol,3-nitrate) is an NO donor as well as a selective beta adrenergic receptor blocker, commonly administered to treat cardiovascular hypertension conditions (Imai et al., 1988; Toshio and Iguchi, 1998). It is used clinically in eastern Asia as an antiglaucoma drug. Following axotomization of the optic nerve (OpN) in adult cats, RGCs are capable of just 2–5% regeneration. These levels are too low to recover visual function, for this reason a method to enhance RGC regeneration after OpN transection could lead to a restoration of lost functionalities, including acute vision. Preliminary studies investigating neurotropic effects found nipradilol to be neuroprotective for axotomized RGCs in adult cats (Watanabe et al., 2006). Hence, neuroprotection for axotomized RGCs in adult cats was reaffirmed and the connection between the donor's potential to release NO and the blockage of adrenoreceptors with nipradilol was investigated by Watanabe et al. (2006). The drug was injected into the vitreous humor at concentrations of 10 μM, 0.1 μM, and 0.01 μM to determine the optimal dosage. Injection concentrations of 10 μM and 0.01 μM yielded a subtle increase in survival ratios compared to controls. Consequently, injections were carried out at 0.1 μM, with mean survival ratios after

7 days of 1.4-fold, and at 14 days 2.6-fold compared to controls. The increase in survival ratios with the nipradilol-treated samples exemplifies the drugs ability to overall enhance RGC regeneration.

RGCs in cats can be classified into one of three categories: alpha, beta, or NAB based on soma size and dendrite morphology. Having confirmed nipradilol's ability to induce RGC generation, it was next explored as to which specific RGCs were increasing. Cells with large somas and thick primary dendrites branching in wide fields were classified as alpha, while cells containing medium-sized somas and bushy dendrites in small fields were classified as beta. All others were termed NAB. Using systematic LY (Lucifer Yellow) injections into RGCs in three patches, the ratio of alpha, beta, and NAB cells was attained. While alpha cell counts matched controls in nipradilol-injected retinae, the drug appeared to promote the survival of both beta and NAB cells, as evidenced by a 5.1-fold increase and a 2.0-fold increase, respectively, observed after 14 days.

Next, whether nipradilol was capable of enhancing the regeneration of cat RGCs including the elongation of cat RGCs after transplanting a segment of PN into the stump of transected cat OpN was explored. The number of RGCs having regenerated axons 10 mm or longer as well as the ratio of RGCs with 20 mm or longer axons versus RGCs having axons 10–20 mm were calculated. Initial experimentation showed the mean number of regenerating RGCs after 6 weeks could be enhanced when 0.1 μM nipradilol was injected; however, no increase compared to controls was observed when 1 μM was used. Showing once again that concentration seems to be vital for regeneration capabilities with nipradilol. Timing was also deemed critical to regeneration, as injections immediately before transection did not influence regeneration, rather injections 30 minutes prior to transection were ideal. Increased numbers of RGCs were observed in the nipradilol-injected retina than the controls with generally smaller somal sizes, indicating the survival population consisted mostly of beta cells or NAB cells. Targeted LY injections found that after 4 weeks the number of alpha, beta, and NAB cells all increased 3.4-fold, 4.1-fold, and 6.8 fold, respectively, raising the possibility the nipradilol is capable of increasing all three cell types in cat RGCs. After 4 weeks, the 20R/10R ratio increased 2.2-fold compared to controls. By 6 weeks, ratios matched that of controls. Taken together, these results suggest nipradilol-injected retinae elongated faster within 4 weeks compared to controls and nipradilol increases the number of regenerating alpha, beta, and NAB RGCs.

It was evident that nipradilol effectively promoted RGC regeneration in adult cats, therefore an effort to elucidate the mechanism by which nipradilol promoted axonal regeneration in adult cat RGCs including the contribution from NO donation was investigated. Firstly, an additional NO donor, SNP (100 μM) was evaluated and found to increase regenerating RGCs 3.2-fold in 6 weeks, but the 20R/10R ratio was not enhanced as observed with nipradilol. This suggests that NO was involved in the increase in a number of RGC cells, but is not likely the only contributing factor for axonal elongation. Additionally, c-PTIO, an NO scavenger, was added followed by nipradilol prior to transplantation of a PN to the cut optic nerve to reduce NO concentrations potentially donated by nipradilol and an insignificant increase in RGCs was described. Lastly, denitro-nipradilol in place of nipradilol was used, effectively reducing the NO release ability that nipradilol exploits. These studies revealed a significant decrease in the efficacy of regeneration. In conclusion, nipradilol at specific doses benefits regenerating RGC rates in cat retina including alpha, beta, and NAB cells while also enhancing axonal elongation. NO release also proves quintessential to the overall mechanism in conjunction with other contributing factors of RGC regrowth in cat retina. NO release from nipradilol is not observed in the absence of cells and has been linked to dependency on glutathione S-transferase (Hayashi and Iguchi, 1998). Therefore, it seems that both NO release and nipradilol are important for regeneration in these studies.

RGCs can also be classified as ON-center or OFF-center based on the depth ratio. Following the results of the aforementioned study, nipradilol effects on survival and axonal regeneration of OFF-center RGCs in adult cats were also investigated (Yata et al., 2007). Neurolucida analysis, using three-dimensional imaging, was used to identify ON cells and OFF cells based on dendritic ramification. RGCs with dendrites spreading 0–26% were classified as OFF cells, while those spreading dendrites 30–78% in depth were classified as ON cells. OFF cells regenerate axons less efficiently than ON cells. When RGCs underwent intravitreal injection of brain-derived neurotropic factor (BDNF) + ciliary neurotropic factor (CNTF) + forskolin (BCF), BCF influenced the axonal regeneration of ON cells, but not OFF cells. Nipradilol promotes RGC regeneration fourfold, but retains ON/OFF ratios consistent with saline; 61% ON cells and 38% OFF cells. The ON cells increased 5.4-fold, while OFF cells increased 8.2-fold, suggesting that nipradilol is capable of promoting both ON and OFF cell regeneration. NO donor nipradilol enhances ON and OFF RGCs in both alpha and beta cell types, whereas BCF increased only ON RGCs in both cell types. Additionally, 14 days after OPN transection, fewer OFF RGCs were observed in the OpN-transected retina than the intact, suggesting that OFF RGCS are more vulnerable than ON to transection. This may result in lower survival, however, it was found that nipradilol promotes neuroprotection, as indicated by proportions consistent with controls after transection as well as axonal regeneration resulting in an overall increase in RGCs including alpha, beta, and NAB cells. These studies show nipradilol is a distinctive donor that may offer a potential for recovering the function of acute vision.

REINNERVATION AFTER PENILE NERVE CRUSH STUDIES WITH SNP IN RATS

There remains a poor understanding of how the parasympathetic nervous system (PSNS) responds physiologically to injury. Often neurons of the PSNS are isolated to target organs making selective damage experimentally challenging. To investigate penile nerve injury regeneration, penile nerves in rats were either isolated and manually crushed or cut completely on one side (Nangle and Keast, 2007). The ability of the erectile tissue to recover following nerve injury when SNP was added, found noradrenergic and nitrergic axons became widely prevalent in proximal regions to injury. As such, their results suggest that vasodilator therapies employed early after penile nerve injury may be quintessential to expediting functional recovery following trauma. A subsequent study examined the response to NO using SNP following penile nerve crush (Nangle et al., 2009). By assessing contractile responses of the tissues and analyzing nerve morphology, SNP sensitivity was reduced in samples containing metabolic syndrome. Metabolic syndrome is associated with aging conditions leading to erectile dysfunction. These studies suggest that NO sensitivity remains fairly consistent during nerve injury; however, rats that had a decreased sensitivity to NO were significantly less likely to recover erectile function. In conclusion, NO regulation arises as a strategic aspect in penile nerve recovery after trauma.

EXOGENOUS SNAC ON MOTOR FUNCTIONAL RECOVERY IN RATS

Nerve injuries that do not completely transect the nerve encompass the most commonly observed peripheral nerve injuries. Traditionally, steroids have been widely administered to treat nervous system injury; although, it has been hypothesized that supplementation of NO may enhance peripheral nerve regeneration (Park et al., 2005). Since ischemia/reperfusion (I/R) appears to downregulate eNOS and nNOS in the rat sciatic nerve, it seems likely that NO production is hindered. Thus an exogenous source of NO may influence motor functional recovery of the reperfused rat sciatic nerve. I/R injury was achieved invoking a 100-g load over 2 hours. The effects of the steroid methylprednisolone (MP) versus NO via S-nitroso-N-acetylcysteine (SNAC) were then evaluated in rats by means of a walking track test, muscle contractile test, muscle weight analysis, and histology. Each rat was continuously infused intravenously with MP, SNAC, or PBS. The material blood pressure only slightly decreased in all animals and the heart rate remained stable for both the PBS and SNAC-treated groups; although significant bradycardia was found in the MP-treated group. The walking track test in brief allowed the animals to walk down a corridor leaving behind blue footprints, where the distance between toes and overall print length could be attained and used to calculate an overall quantification of the sciatic functional index (SFI). A score of 0 represented normal nerve function, while -100 indicated complete dysfunction. SNAC-treated rats began to show signs of recovery at day 7 with an SFI of -84.2 ± 13.8, while the MP- and PBS-treated rats remained paralyzed. The muscle contractile test involved removal of the extensor digitorum longus (EDL) muscle, then attached by suture to an isometric force transducer on one end and attached to a glass hook at the bottom of a field-stimulating electrode on the distal end. Maximal twitch force was measured and the resultant contractile force was expressed as a percentage of the normal contralateral muscle contractile force. In SNAC-treated rats, maximal twitch force was 80% as early as day 21 and 94% by day 42, while the MP- and PBS-treated rats recovered twitch force at 76% and 60%, respectively. The SNAC-treated rats demonstrated a significantly greater isometric force produced at all frequencies attained. In each experimental limb muscle atrophy was apparent, hence the weight of the muscle was recorded and expressed as a percentage of the normal contralateral muscle. The wet weight for the SNAC-treated rats once again was continuously higher, while MP- and PBS-treated rats progressively exhibited degenerated muscle weight until day 21, and showed weight gains near 10% less than the SNAC-treated samples by day 42. In all, the results of the walking track test, muscle contractile test, and muscle weight analysis indicate that although MP marginally aids in reducing the damaging effects of I/R injury in rats, the NO donor SNAC shows greater potential for maximizing recovery.

Histologic examinations corroborated SNAC's potential to enhance recovery after I/R injury in rats. The blind histologic examinations indicated more Wallerian degeneration, myelin debris, and acute inflammatory responses were present in MP- and PBS-treated rats than in SNAC-treated specimen at day 5. Uniquely, the SNAC-treated rats also appeared to have near normal myelin sheath thickness and regenerated axons at day 42. Taken together, these results show improved early regeneration of axons in the reperfused peripheral nerve as well as improved axonal regeneration due to exogenous NO. In conclusion, NO is preferential in the early stages of recovery from peripheral nerve I/R injury. NO may assist by scavenging superoxide free radicals, which can be detrimental to early stages of reperfusion or by inducing necessary vasodilation. Unfortunately, the concentrations of NO release were once again overlooked. Stability among different S-nitrosothiols in solution must take the concentration effect carefully into account since half-lives of primary RSNOs found in vivo can vary widely by their intrinsic structural properties (de Oliveira et al., 2002). Moreover, peripheral nerve I/R injury may result in undesirable inhibition of NO production, vital to the recovery of the nervous system during this type of injury, and S-nitrosothiol NO donors like SNAC may aid in countering the resultant deleterious effects.

S-NITROSOGLUTATHIONE (GSNO) EFFECTS AFTER I/R INJURY IN RATS

Traumatic brain injury (TBI) often results in adverse neurobehavioral dysfunctions attributed to oxidative injury of neurogenic vasculature (Cornelius et al., 2013). Such dysfunctions lead to an undesirable progression of damage. Reactive oxygen and nitrogen species in brain and endothelial cells complicate studies of the mechanistic pathways invoking such damage. Peroxynitrite is a highly ROS formed rapidly from superoxide and NO. It is associated with a series of unwanted effects including oxidatively uncoupling endothelial NOS causing a depletion of L-arginine. Increasing peroxynitrite concentrations increase tetrahydrobiopterin upregulation of ICAM-1 compromising the BBB and increasing edema. Additionally, resultant nitration of tyrosine in proteins can be physiologically problematic. The formation of peroxynitrite during TBI significantly depletes bioavailability of NO vital to initiating angiogenic processes. Since the NO donor GSNO has shown potential to protect the CNS through decreasing reactive oxygen species, the ability for GSNO to influence both short and long-term neurorepairs following TBI in rats was explored (Khan et al., 2011).

The rat controlled cortical impact (CCI) model used mimics contusion type injury, representative of near 40% of all TBIs. It mirrors effects of human brain injuries such as inflammation, BBB deviations, neuron loss, motor control loss, and memory degradation. The direct interaction between GSNO and peroxynitrite was explored by monitoring 3-nitrotyrosine (3-NT) expression using dot blot analysis and immunohistochemistry (IHC), 4 hours after TBI in the injured penumbra region (Fig. 6.8). Moreover, brain tissues were stained for 4-hydroxynonenal (4-HNE)—a consequence of lipid peroxidation, as well as ferritin—capable of catalyzing the formation of a highly reactive hydroxyl radical with peroxynitrite. TBI-induced expression of 3-NT was significantly hindered in the GSNO-treated samples. After 2 weeks, 3-NT expression was markedly increased in the injured penumbra. The GSNO-treated samples, however, once again displayed a reduction in 3-NT expression. Samples treated with peroxynitrite forming 3-morpholino sydnonimine (SIN-1) on the other hand displayed significantly increased 3-NT expression compared to the TBI group. Plasma analyses echoed these findings. Concurrently, plasma NO levels were monitored indirectly by way of nitrate/nitrite concentration with a NO fluorometric assay kit from Biovison. As expected, peroxynitrite concentrations exhibited an inverse relationship with NO. A significant drop in NO concentration was observed with SIN-1 treated samples as compared to the TBI plasma. The lipid peroxidation product, 4-HNE, levels in plasma were significantly lower in the GSNO-treated samples, while SIN-1 treated samples displayed an increased expression of 4-HNE. Once again, traumatic penumbra resonated these results. It appears by increasing NO through implementing GSNO it is possible to decrease peroxynitrite concentrations after TBI in both brain tissue and plasma.

If peroxynitrite was responsible for damage after TBI, an increase in brain water content (edema) would be expected. BBB leakage evaluated by edema was determined 24 hours after CCI and extravasations of Evan's blue dye were highest in SIN-1 treated samples when compared to TBI and much lower still in GSNO treated. Indicating, peroxynitrite did cause neurovascular oxidative injury during TBI. In the TBI group, brain tissue water content increased. In the SIN-1 treated sample, water content was further increased. Conversely, GSNO-treated samples contained less edema than the

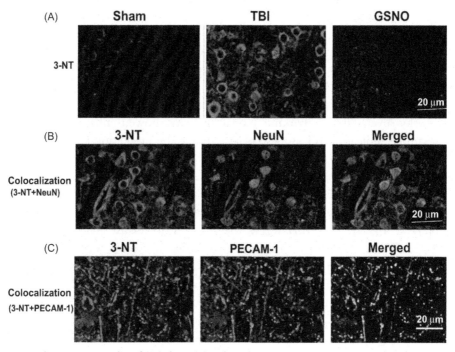

Figure 6.8 Photomicrographs of IHC show (A) reduced 3-NT (red) in the traumatic penumbra region in the presence of GSNO. Sham brain does not show 3-NT-positive cells. Colocalization (yellowish, merged) of 3-NT (red color) with either neuronal marker NeuN (green) (B) or endothelial cell marker PECAM-1 (green) (C) indicates that both neurons and endothelial cells have increased expression of 3-NT. *Reproduced from Khan, M., Sakakima, H., Dhammu, T.S., Shunmugavel, A., Im, W.-B., Gilg, A.G., et al., 2011. S-nitrosoglutathione reduces oxidative injury and promotes mechanisms of neurorepair following traumatic brain injury in rats. J. Neuroinflam. 8, 78; licensee BioMed Central Ltd.*

TBI samples. Therefore, while a boost in peroxynitrite increases edema, greater GSNO concentrations protect neurovascular tissues during TBI.

An increase in edema may concur with BBB damage. BBB leakage is influenced by a surge in mRNA levels of ICAM-1; therefore, one way to assess damage to the BBB and potential for GSNO to induce neurorepair is to monitor ICAM-1 levels in the traumatic penumbra area. GSNO-treating samples prevented the increase of ICAM-1 levels suggesting GSNO may stimulate neurorepair and reduce inflammation.

Glutathione (GSH) is a precursor to GSNO and an endogenous antioxidant to peroxynitrite involved in endothelial function. To better understand the relationship between peroxynitrite and GSH, GSH levels were monitored. The TBI brain had significantly depleted levels of GSH compared to the controls and GSNO samples. SIN-1 levels were significantly lower than the TBI control levels. Peroxynitrite reduces the concentration of GSH, which in turn may be partially responsible for its damaging effects.

Histopathology, demyelination, and axonal integrity were assessed using hematoxylin and eosin (H&E), luxol fast blue (LFB), and Bielschowsky silver impregnation, respectively, 14 days after the TBI. H&E studies revealed that GSNO prevented TBI-induced tissue deformation and immune cell infiltration. While controls showed a decline in myelin, the GSNO-treated samples did not have nearly the same degree of declination, which supports that GSNO may induce the neurorepair process. The Bielschowsky silver staining, marking axonal integrity, showed enhanced staining over the TBI sample, indicating GSNO also plays a role in the protection of axonal integrity following TBI.

BDNF, TrkB, and synaptophysin are important factors associated with neurorepair. Expression of BDNF and its receptor TrkB promotes neuronal survival and axonal growth, while synaptophysin facilitates synaptogenesis. A significant decrease in synaptophysin, BDNF, and TrkB was witnessed in TBI versus Sham control groups. GSNO-treated samples displayed a significant increase in expression of all three factors compared to the TBI samples. Hence, GSNO contributes to the neurorepair process at least in part by inducing these vital neurotropic factors following TBI. In all, treatment with GSNO after TBI depletes peroxynitrite levels, increases NO and GSH bioavailability, and safeguards the BBB, while simultaneously supporting neurorepair. Overall, this evidence suggests that GSNO can be effective in protecting the CNS by reducing reactive oxygen species and enhancing both short- and long-term neurorepairs.

GSNO INFLUENCES ANGIOGENESIS IN RATS

Stroke is amongst the top five leading causes of death both in the United States and globally. A stroke leads to neuronal damage that can ultimately cause irreversible neuronal death. As such, a better understanding of factors central to neurorepair associated with stroke is a necessary step toward enabling the therapeutic design to minimize disabilities and enhance recovery for stroke victims. It is generally accepted that NO induces vasodilation as well as regulates several angiogenic and potential neurorepair factors such as hypoxia-inducible factor-1 alpha (HIF-1α). Limited data identifies the specific mechanisms of such neurorepair processes. In response, Khan and coworkers investigated the mechanism by which the NO donor GSNO may influence neurorepair specifically through the HIF-1α/VEGF/angiogenesis pathway (Khan et al., 2015). HIF-1 is a regulatory nuclear transcription factor capable of invoking tissue survival by inducing GLUT, a cellular metabolism enzyme, as well as influencing angiogenic enzymes like VEGF, VEGFR1, and angiopoietin. Stabilizers of HIF-1α, like GSNO, may promote neuroprotection, angiogenesis, and neurotropins. In rats, cerebral ischemia and reperfusion (stroke) was induced by left middle cerebral artery occlusion (MCAO). Animals were sacrificed either 7 or 14 days later and brain tissue was prepared for histological and immunohistochemical analyses. Shortly after reperfusion, the

GSNO group was injected with GSNO, followed by feeding of GSNO daily until the time of sacrifice. Some animals were also treated with 2-methoxyestradiol (2-ME), an inhibitor of HIF-1α.

Analyses were performed to assess the weight, neurological state, and motor behavior of the rats. Animals showing paralysis 1 hour after MCAO were graded on a scale of 0–4, where a 0 was awarded for normal functionality and 4 awarded when leaning continuously to the contralateral side or when simply no spontaneous motor activity was observed. Modified neurological severity scores (mNSS) were determined to reflect asymmetries. A body swing test was also used where the animals were held vertical and a swing recorded whenever the rat moved its head out of vertical alignment. A normal healthy rat tends to have an equal number of swings on both sides. Furthermore, motor coordination and balance were evaluated on an accelerating rotarod task. Each rat was placed on the rotarod cylinder and speed was increased. The mean speed which the animal fell off out of the drum was noted. After 14 days of GSNO treatments, tolerated rotation speed, body swing behavior, mNSS, as well as body weights significantly improved compared to controls. As a result, the NO donor GSNO can be associated with improving motor function after stroke in rats.

The physiological impact of GSNO following induced stroke was also assessed. Immunohistochemical and histological studies entailed brain tissue stained for HIF-1α, VEGF, and PECAM-1; cell proliferation, examined using Ki67; cell counting, and immunostained sections. The degree of neuronal damage was determined by cresyl violet (NISSL) staining, which highlights key structural features of the neurons. Both immunohistochemical staining and Western blot analyses were performed to determine protein concentrations. 14 days following injury, angiogenic mediators HIF-1α, VEGF, and PECAM-1 increased in the injured areas. Advantageously to neurorepair, treatment with GSNO significantly increased the expression of these mediators and promoted new vessel formation, as evidenced by a substantial increase in expression of vessel density markers, including laminin and GSL-1. Additionally, Ki67 expression was also significantly enhanced in GSNO-treated samples compared to controls, suggesting a possible increase in angiogenic processes including cell proliferation was being induced.

When the inhibitor of HIF-1α, 2-ME, was administered, the previously mentioned beneficial effect on brain infarctions due to GSNO, was blocked, rotarod performances mimicked controls, and neurological scores were impeded. These results suggest that the interaction between GSNO and HIF-1α is vital for recovery following cerebral IR. Cell transfection with siRNAs, known to silence HIF-1α, reduced the expression of angiogenic factor VEGF, compared with untreated cells, and inhibited GSNOs capillary-like structure. These results support the hypothesis that HIF-1α proangiogenic activity is stimulated by GSNO. Consequently, GSNO-mediated stabilization of HIF-1α may be an effective approach to enhance neurorepair processes, ultimately reducing stroke-induced disabilities.

CONCLUSION

A review of the recent literature suggests that NO appears to be involved in the process of nerve regeneration. NOC-18 was found to directly target serotonergic neurons and initiate the cGMP/PKG pathway early on in *Hydra* regeneration. In the *Locusta*, both NOC-18 and SNP endorse axonal regrowth and neurite outgrowth, but perhaps in later stages of development after the initial reformation. SNP was used to discover that after penile nerve injury, recovery was hindered by insensitivity to NO, suggesting NO was in some way necessary for the recovery of penile function after injury. Nipradilol was determined as an effective drug only when nitrosated, indicating it was most likely the NO release that was responsible for nipradilol's ability to enhance RCG recovery in cats following optic nerve trauma. SNAC was capable of inducing motor functional recovery in rats following damage to the sciatic nerve. Following induced stroke, GSNO was determined as an appropriate source of additional NO to reduce the effects of superoxide peroxynitrite, minimizing damaging effects after TBI. It was also found that GSNO could stabilize HIF-1α, consequently enhancing neurorepair. These studies provide interesting precedence for further investigating NO influence on neuronal regeneration.

Zochodne et al. (1997) investigated the effect of NOS on the peripheral nerve regeneration of a transected sciatic nerve in mice. Based on aforementioned observations during CNS regeneration, researchers anticipated when dosed with L-NAME, an NOS inhibitor, antagonistic impression on nerve regeneration of myelinated fibers from the proximal stump would occur. The results, however, reflected the contrary. Both electrophysiological and morphometric evaluations illustrate PNS regeneration in mice benefits from inhibition of NOS. It is possible that local levels of NO release in these studies may have been excessively high, leading to toxic adverse effects. NO concentrations can be crucial to efficacy, therefore, in order to fully understand the implications of using each NO donor above, a direct method to determine NO concentration in these instances is of utmost importance. As seen in the GSNO experimentation, other reactive species like peroxynitrite or proteins can succumb to rapid reaction with NO, effectively depleting detectable NO. Additionally, each type of NO donor yields different byproducts, as seen with SNP and nipradilol that may be at least in part responsible for the present results, not necessarily solely caused by NO. NO drugs are constantly expanding, leading to a greater understanding of the difference in NO loading, release kinetics and mechanisms associated with each (Miller and Megson, 2007; Joslin et al., 2013; Damodaran et al., 2012). Although the studies mentioned reveal NO is a vital signaling molecule undoubtedly involved in nerve regeneration, thus far research has just begun to shed light on NO donor potential in this field. Each NO donor exhibits unique physical and chemical properties resulting in differentiated physiological consequences. Additional consideration of NO release rates and concentration-dependent pathways warrants further exploration and may lead to original specialized therapeutic methods better adapted to promote nerve regeneration.

IN MEMORIAM: VINOD B. DAMODARAN

Vinod was a great scientist, a fine chemist, a wonderful colleague, and an exemplar in his field. He was taken from this world too soon and will always be missed.

REFERENCES

Arnold, W.P., Mittal, C.K., Katsuki, S., Murad, F., 1977. Nitric oxide activates guanylate cyclase and increases guanosine 3′:5′-cyclic monophosphate levels in various tissue preparations. Proc. Natl. Acad. Sci. USA 74, 3203–3207.

Awasaki, T., Ito, K., 2016. Neurodevelopment: regeneration switch is a gas. Nature 531, 182–183.

Beuve, A., Wu, C.G., Cui, C.L., Liu, T., Jain, M.R., Huang, C., et al., 2016. Identification of novel S-nitrosation sites in soluble guanylyl cyclase, the nitric oxide receptor. J. Proteom. 138, 40–47.

Browne, S.E., Beal, M.F., 2006. Oxidative damage in Huntington's disease pathogenesis. Antioxid. Redox Signal. 8, 2061–2073.

Butterfield, D., Howard, B.J., Lafontaine, M.A., 2001. Brain oxidative stress in animal models of accelerated aging and the age-related neurodegenerative disorders, Alzheimer's disease and Huntington's disease. Curr. Med. Chem. 8, 815–828.

Calabrese, V., Mancuso, C., Calvani, M., Rizzarelli, E., Butterfield, D.A., Stella, A.M.G., 2007. Nitric oxide in the central nervous system: neuroprotection versus neurotoxicity. Nat. Rev. Neurosci. 8, 766–775.

Cary, S.P., Winger, J.A., Derbyshire, E.R., Marletta, M.A., 2006. Nitric oxide signaling: no longer simply on or off. Trends Biochem. Sci. 31, 231–239.

Castegna, A., Aksenov, M., Aksenova, M., Thongboonkerd, V., Klein, J.B., Pierce, W.M., et al., 2002. Proteomic identification of oxidatively modified proteins in Alzheimer's disease brain. Part I: creatine kinase BB, glutamine synthase, and ubiquitin carboxy-terminal hydrolase L-1. Free. Radic. Biol. Med. 33, 562–571.

Colasanti, M., Mazzone, V., Mancinelli, L., Leone, S., Venturini, G., 2009. Involvement of nitric oxide in the head regeneration of Hydra vulgaris. Nitric Oxide 21, 164–170.

Contestabile, A., Ciani, E., 2004. Role of nitric oxide in the regulation of neuronal proliferation, survival and differentiation. Neurochem. Int. 45, 903–914.

Cooke, R.M., Mistry, R., Challiss, R.J., Straub, V.A., 2013. Nitric oxide synthesis and cGMP production is important for neurite growth and synapse remodeling after axotomy. J. Neurosci. 33, 5626–5637.

Cornelius, C., Crupi, R., Calabrese, V., Graziano, A., Milone, P., Pennisi, G., et al., 2013. Traumatic brain injury: oxidative stress and neuroprotection. Antioxid. Redox Signal. 19, 836–853.

Damodaran, V.B., Place, L.W., Kipper, M.J., Reynolds, M.M., 2012. Enzymatically degradable nitric oxide releasing S-nitrosated dextran thiomers for biomedical applications. J. Mater. Chem. 22, 23038–23048.

de Oliveira, M.G., Shishido, S.M., Seabra, A.B., Morgon, N.H., 2002. Thermal stability of primary S-nitrosothiols: roles of autocatalysis and structural effects on the rate of nitric oxide release. J. Phys. Chem. A 106, 8963–8970.

Francis, S.H., Busch, J.L., Corbin, J.D., 2010. cGMP-dependent protein kinases and cGMP phosphodiesterases in nitric oxide and cGMP action. Pharmacol. Rev. 62, 525–563.

Francis, S.H., Corbin, J.D., 2005. Phosphodiesterase-5 inhibition: the molecular biology of erectile function and dysfunction. Urol. Clin. North Am. 32, 419–429.

Garthwaite, J., 2010. New insight into the functioning of nitric oxide-receptive guanylyl cyclase: physiological and pharmacological implications. Mol. Cell. Biochem. 334, 221–232.

Garthwaite, J., Charles, S.L., Chess-Williams, R., 1998. Endothelium-derived relaxing factor release on activation of NMDA receptors suggests role as intercellular messenger in the brain. Nature 336, 385–388.

Gibbs, S.M., Becker, A., Hardy, R.W., Truman, J.W., 2001. Soluble guanylate cyclase is required during development for visual system function in Drosophila. J. Neurosci. 21, 7705–7714.

Godfrey, E.W., Schwarte, R.C., 2010. Nitric oxide and cyclic GMP regulate early events in agrin signaling in skeletal muscle cells. Exp. Cell Res. 316, 1935–1945.

González-Hernández, T., Rustioni, A., 1999a. Expression of three forms of nitric oxide synthase in peripheral nerve regeneration. J. Neurosci. Res. 55, 198–207.

González-Hernández, T., Rustioni, A., 1999b. Nitric oxide synthase and growth-associated protein are coexpressed in primary sensory neurons after peripheral injury. J. Comp. Neurol. 404, 64–74.

Grossi, L., D'angelo, S., 2005. Sodium nitroprusside: mechanism of NO release mediated by sulfhydryl-containing molecules. J. Med. Chem. 48, 2622–2626.

Gu, Z., Kaul, M., Yan, B., Kridel, S.J., Cui, J., Strongin, A., et al., 2002. S-nitrosylation of matrix metalloproteinases: signaling pathway to neuronal cell death. Science 297, 1186–1190.

Guix, F., Uribesalgo, I., Coma, M., Munoz, F., 2005. The physiology and pathophysiology of nitric oxide in the brain. Prog. Neurobiol. 76, 126–152.

Hara, M.R., Thomas, B., Cascio, M.B., Bae, B.-I., Hester, L.D., Dawson, V.L., et al., 2006. Neuroprotection by pharmacologic blockade of the GAPDH death cascade. Proc. Natl. Acad. Sci. USA 103, 3887–3889.

Hayashi, T., Iguchi, A., 1998. Nipradilol: a β-adrenoceptor antagonist with nitric oxide-releasing action. Cardiovasc. Drug Rev. 16, 212–235.

Hunter, R.A., Storm, W.L., Coneski, P.N., Schoenfisch, M.H., 2013. Inaccuracies of nitric oxide measurement methods in biological media. Anal. Chem. 85, 1957–1963.

Imai, S., Nakahara, H., Nakazawa, M., Takeda, K., 1988. Adrenergic receptor effects and antihypertensive actions of β-adrenoceptor-blocking agents with ancillary properties. Drugs 36, 10–19.

Joslin, J.M., Damodaran, V.B., Reynolds, M.M., 2013. Selective nitrosation of modified dextran polymers. RSC Adv. 3, 15035–15043.

Kegeles, L.S., Martinez, D., Kochan, L.D., Hwang, D.R., Huang, Y., Mawlawi, O., et al., 2002. NMDA antagonist effects on striatal dopamine release: Positron emission tomography studies in humans. Synapse 43, 19–29.

Keilhoff, G., Fansa, H., Wolf, G., 2002. Differences in peripheral nerve degeneration/regeneration between wild-type and neuronal nitric oxide synthase knockout mice. J. Neurosci. Res. 68, 432–441.

Khan, M., Dhammu, T.S., Matsuda, F., Baarine, M., Dhindsa, T.S., Singh, I., et al., 2015. Promoting endothelial function by S-nitrosoglutathione through the HIF-1alpha/VEGF pathway stimulates neurorepair and functional recovery following experimental stroke in rats. Drug Des. Develop. Ther. 9, 2233–2247.

Khan, M., Sakakima, H., Dhammu, T.S., Shunmugavel, A., Im, W.-B., Gilg, A.G., et al., 2011. S-nitrosoglutathione reduces oxidative injury and promotes mechanisms of neurorepair following traumatic brain injury in rats. J. Neuroinflam. 8, 78.

Koriyama, Y., Yasuda, R., Homma, K., Mawatari, K., Nagashima, M., Sugitani, K., et al., 2009. Nitric oxide-cGMP signaling regulates axonal elongation during optic nerve regeneration in the goldfish in vitro and in vivo. J. Neurochem. 110, 890–901.

Krumenacker, J.S., Hanafy, K.A., Murad, F., 2004. Regulation of nitric oxide and soluble guanylyl cyclase. Brain Res. Bull. 62, 505–515.

Kuhn, D.M., Geddes, T.J., 2002. Reduced nicotinamide nucleotides prevent nitration of tyrosine hydroxylase by peroxynitrite. Brain Res. 933, 85–89.

Lin, M.T., Beal, M.F., 2006. Mitochondrial dysfunction and oxidative stress in neurodegenerative diseases. Nature 443, 787–795.

Lonart, G., Wang, J., Johnson, K.M., 1992. Nitric oxide induces neurotransmitter release from hippocampal slices. Eur. J. Pharmacol. 220, 271–272.

Lorrain, D.S., Hull, E.M., 1993. Nitric oxide increases dopamine and serotonin release in the medial pre-optic area. Neuroreport 5, 87–89.

Mancuso, C., 2004. Heme oxygenase and its products in the nervous system. Antioxid. Redox Signal. 6, 878–887.

Mancuso, C., Bonsignore, A., Di Stasio, E., Mordente, A., Motterlini, R., 2003. Bilirubin and S-nitrosothiols interaction: evidence for a possible role of bilirubin as a scavenger of nitric oxide. Biochem. Pharmacol. 66, 2355–2363.

Miklya, I., Göltl, P., Hafenscher, F., Pencz, N., 2014. The role of Parkin in Parkinson's disease. Off. J. Hung. Assoc. Psychopharmacol. 16, 67–76.

Miller, M.R., Megson, I.L., 2007. Recent developments in nitric oxide donor drugs. Br. J. Pharmacol. 151, 305–321.

Miranda, S., Opazo, C., Larrondo, L.F., Muñoz, F.J., Ruiz, F., Leighton, F., et al., 2000. The role of oxidative stress in the toxicity induced by amyloid β-peptide in Alzheimer's disease. Prog. Neurobiol. 62, 633–648.

Motterlini, R., Green, C.J., Foresti, R., 2002. Regulation of heme oxygenase-1 by redox signals involving nitric oxide. Antioxid. Redox Signal. 4, 615–624.

Nangle, M.R., Keast, J.R., 2007. Reduced efficacy of nitrergic neurotransmission exacerbates erectile dysfunction after penile nerve injury despite axonal regeneration. Exp. Neurol. 207, 30–41.

Nangle, M.R., Proietto, J., Keast, J.R., 2009. Impaired cavernous reinnervation after penile nerve injury in rats with features of the metabolic syndrome. J. Sex Med. 6, 3032–3044.

Naoi, M., Maruyama, W., 2001. Future of neuroprotection in Parkinson's disease. Parkinsonism Relat. Disord. 8, 139–145.

Pacher, P., Beckman, J.S., Liaudet, L., 2007. Nitric oxide and peroxynitrite in health and disease. Physiol. Rev. 87, 315–424.

Park, J.W., Qi, W.N., Cai, Y., Nunley, J.A., Urbaniak, J.R., Chen, L.E., 2005. The effects of exogenous nitric oxide donor on motor functional recovery of reperfused peripheral nerve. J. Hand Surg. Am. 30, 519–527.

Patschke, A., Bicker, G., Stern, M., 2004. Axonal regeneration of proctolinergic neurons in the central nervous system of the locust. Dev. Brain Res. 150, 73–76.

Prast, H., Tran, M., Fischer, H., Philippu, A., 1998. Nitric oxide-induced release of acetylcholine in the nucleus accumbens: role of cyclic GMP, glutamate, and GABA. J. Neurochem. 71, 266–273.

Rabinovich, D., Yaniv, S.P., Alyagor, I., Schuldiner, O., 2016. Nitric oxide as a switching mechanism between axon degeneration and regrowth during developmental remodeling. Cell 164, 170–182.

Rentería, R.C., Constantine-Paton, M., 1999. Nitric oxide in the retinotectal system: a signal but not a retrograde messenger during map refinement and segregation. J. Neurosci. 19, 7066–7076.

Riccio, A., Alvania, R.S., Lonze, B.E., Ramanan, N., Kim, T., Huang, Y., et al., 2006. A nitric oxide signaling pathway controls CREB-mediated gene expression in neurons. Mol. Cell 21, 283–294.

Ridnour, L.A., Thomas, D.D., Mancardi, D., Espey, M.G., Miranda, K.M., Paolocci, N., et al., 2004. The chemistry of nitrosative stress induced by nitric oxide and reactive nitrogen oxide species. Putting perspective on stressful biological situations. Biol. Chem. 385, 1–10.

Seidel, C., Bicker, G., 2000. Nitric oxide and cGMP influence axonogenesis of antennal pioneer neurons. Development 127, 4541–4549.

Sellak, H., Choi, C.S., Dey, N.B., Lincoln, T.M., 2013. Transcriptional and post-transcriptional regulation of cGMP-dependent protein kinase (PKG-I): pathophysiological significance. Cardiovasc. Res. 97, 200–207.

Sorolla, M.A., Reverter-Branchat, G., Tamarit, J., Ferrer, I., Ros, J., Cabiscol, E., 2008. Proteomic and oxidative stress analysis in human brain samples of Huntington disease. Free Radic. Biol. Med. 45, 667–678.

Stern, M., Bicker, G., 2008a. Nitric oxide regulates axonal regeneration in an insect embryonic CNS. Dev. Neurobiol. 68, 295–308.

Stern, M., Boger, N., Eickhoff, R., Lorbeer, C., Kerssen, U., Ziegler, M., et al., 2010. Development of nitrergic neurons in the nervous system of the locust embryo. J. Comp. Neurol. 518, 1157–1175.

Tamagnini, F., Barker, G., Warburton, E., Burattini, C., Aicardi, G., Bashir, Z.I., 2013. Nitric oxide-dependent long-term depression but not endocannabinoid-mediated long-term potentiation is crucial for visual recognition memory. J. Physiol. 591, 3963–3979.

Toshio, H., Iguchi, A., 1998. Nipradilol: A P-adrenoceptor antagonist with nitric oxide-releasing action. Cardiovasc. Drug Rev. 16, 212–235.

Uehara, T., Nakamura, T., Yao, D., Shi, Z.-Q., Gu, Z., Ma, Y., et al., 2006. S-nitrosylated protein-disulphide isomerase links protein misfolding to neurodegeneration. Nature 441, 513–517.

Van Wagenen, S., Rehder, V., 1999. Regulation of neuronal growth cone filopodia by nitric oxide. J. Neurobiol. 39, 168–185.

Watanabe, M., Tokita, Y., Yata, T., 2006. Axonal regeneration of cat retinal ganglion cells is promoted by nipradilol, an anti-glaucoma drug. Neuroscience 140, 517–528.

Yamamoto, T., Yuyama, K., Nakamura, K., Kato, T., Yamamoto, H., 2000. Kinetic characterization of the nitric oxide toxicity for PC12 cells: effect of half-life time of NO release. Eur. J. Pharmacol. 397, 25–33.

Yata, T., Nakamura, M., Sagawa, H., Tokita, Y., Terasaki, H., Watanabe, M., 2007. Survival and axonal regeneration of off-center retinal ganglion cells of adult cats are promoted with an anti-glaucoma drug, nipradilol, but not BDNF and CNTF. Neuroscience 148, 53–64.

Yuan, Q., Scott, D.E., So, K.-F., Lin, Z., Wu, W., 2009. The potential role of nitric oxide synthase in survival and regeneration of magnocellular neurons of hypothalamo-neurohypophyseal system. Neurochem. Res. 34, 1907–1913.

Zochodne, D.W., Levy, D., 2005. Nitric oxide in damage, disease and repair of the peripheral nervous system. Cell. Mol. Biol. 51, 225–267.

Zochodne, D.W., Misra, M., Cheng, C., Sun, H., 1997. Inhibition of nitric oxide synthase enhances peripheral nerve regeneration in mice. Neurosci. Lett. 228, 71–74.

CHAPTER 7

Nitric Oxide Donors as Antimicrobial Agents

Samuel K. Kutty, Kitty Ka Kit Ho and Naresh Kumar
School of Chemistry, University of New South Wales, Sydney, NSW, Australia

NITRIC OXIDE

Nitric oxide [nitrogen monoxide, •NO, IUPAC: oxidonitrogen(•)] is a colorless paramagnetic free radical gas formed in the troposphere during lightning or electrical storms. NO, initially suspected as a toxic, carcinogenic air pollutant produced from fuel burning, became the "molecule of the year" in 1992 on the cover story of *Science* (Culotta and Koshland, 1992) due to the discovery of its wide array of biological functions, including smooth muscle relaxation, inhibition of platelet aggregation, neurotransmission, and host defense mechanisms. The 1998 Nobel Prize in Physiology and Medicine was awarded to Ferid Murad, Robert F. Furchgott, and Louis Ignarro for their discovery of the signaling properties of NO (Sergio et al., 2008). NO, with a molecular mass of only $30 \, g \, mol^{-1}$, is the smallest, lightest, and the first gas known to act as a biological messenger in mammals. Thus, over the past decades, the recognition of the multiple biological roles of NO has transformed this molecule from a menace into a wonder in various areas of biological research.

BIOSYNTHESIS OF NO

NO is biosynthesized by the enzyme nitric oxide synthase (NOS) through the stepwise oxidation of L-arginine, via an N-hydroxyarginine intermediate, into NO, citrulline, and nicotinamide adenine dinucleotide phosphate (NADP) (Fig. 7.1) (Murad 1999).

NOSs are dimeric flavoproteins containing tetrahydrobiopterin, flavin adenine dinucleotide (FAD), flavin mononucleotide (FMN), and iron protoporphyrin IX (heme), and have homology with cytochrome P450 (Gorren and Mayer, 2007). There are three major distinct isoforms of NOS, including two constitutive forms present under physiological conditions in endothelial cells (eNOS or NOS-III) and in neurons (nNOS or NOS-I), and an inducible form (iNOS or NOS-II) expressed in macrophages, fibroblasts, and Kupffer cells (Crane et al., 1998; Groves and Wang, 2000). An additional isoform of NOS, mitochondrial NOS (mNOS), has been reported to help regulate mitochondrial energy production via NO signaling (Brown, 2003). The

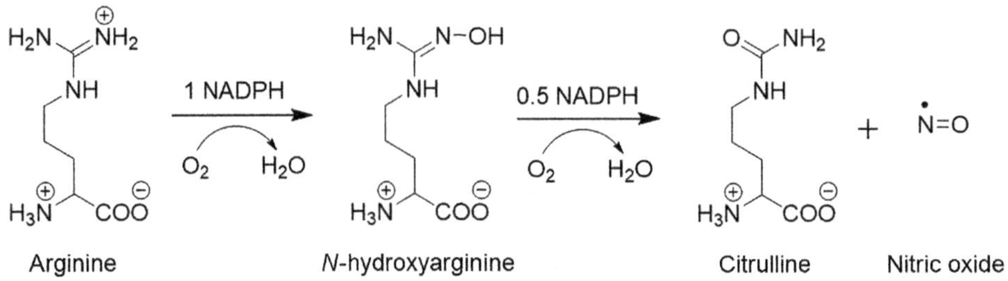

Figure 7.1 Biosynthesis of NO.

constitutive enzymes are activated by calcium-calmodulin and generate small amounts of NO, whereas iNOS produces much greater amounts of NO, especially during infection and inflammatory responses (Alderton et al., 2001; Pfeiffer et al., 1999).

NO INACTIVATION CHEMISTRY

NO is an unstable gas and reacts primarily with oxygen, superoxide, or redox metals. NO reacts with oxygen to form N_2O_4, which combines with water to produce a mixture of nitric and nitrous acids, or can be oxidized into nitrite or nitrate which are excreted through the urine (Eiserich et al., 1998). NO readily forms complexes with transition metals, such as the heme group of hemoglobin, leading to inactivation of NO. Low concentrations of NO are relatively stable in air. However, NO reacts rapidly with even low concentrations of superoxide anion (O_2^-) to produce the peroxynitrite anion ($ONOO^-$). The peroxynitrite anion is a strong oxidizing and nitrating species responsible for some of the toxic effects of NO, through free radical-mediated lipid peroxidation, sulfhydryl oxidation, tyrosine nitration, and DNA deamination (Eiserich et al., 1998; Pfeiffer et al., 1999; Stamler et al., 1992; Wink et al., 1991).

BIOLOGICAL EFFECTS OF NO

NO mediates many diverse physiological processes in the body, and its effects are largely concentration dependent (Coleman, 2001; Davis et al., 2001). At low concentrations, NO exerts direct effects on cellular systems, which are broadly orchestrated by two major biochemical signaling pathways.

One of the most important biochemical pathway via which NO exerts its effects is the activation of soluble guanylate cyclase (sGC), a heterodimeric enzyme present as distinct isoenzymes in vascular and nervous tissues. NO activates sGC by combining with its heme group to form a nitrosyl-heme complex, which results in the production of the secondary messenger cyclic guanosine monophosphate (cGMP) from guanosine triphosphate (GTP) (Ignarro 1999). This, in turn, leads to the activation of cGMP-dependent protein kinases, cGMP-regulated phosphodiesterases, and cyclic-nucleotide-gated ion channels, which ultimately promote the main biological functions of NO, including vasodilation, neurotransmission, inhibition of platelet aggregation, and smooth muscle relaxation (Bruckdorfer, 2005).

The second mechanism by which NO exerts its biological functions occurs through the reaction of NO with thiols and transition metals such as zinc. NO can also modify proteins, DNA, and lipids directly, without requiring enzymes involved with nitration or nitrosylation. S-nitrosylation of cysteine thiol groups is a reversible modification which regulates the function of many intracellular proteins and thus controlling cellular signaling pathways and activities (Handy and Loscalzo, 2006).

By comparison, higher concentrations of NO have an indirect effect on biomolecules, which is possibly mediated via the formation of reactive nitrogen species (RNS) from molecular O_2 or superoxide (O_2^-) (Eiserich et al., 1998). Large amounts of NO released following induction of NOS or excessive stimulation of N-methyl-D-aspartate (NMDA) receptors in the brain cause cytotoxic effects via peroxynitrite anions (Galla, 1993). NO contributes to inflammatory reactions and host defense against pathogens, but can also lead to tumor growth or neuronal destruction associated with overstimulation of NMDA receptors by glutamate (Coleman, 2001; Dawson and Dawson, 1996). Paradoxically, NO has also been reported to display cytoprotective effects under certain conditions (Bonthius et al., 2009).

In summary, NO exerts a variety of regulatory effects via multiple biological pathways in the cell and appears to play a vital role in the normal functioning and behavior of the human body.

NO REGULATION IN MICROBES

NO is a ubiquitous signaling molecule in nature crucial to humans and is gaining recognition as a key bacterial signaling molecule. In bacteria, NO plays multiple roles in the bacterial life cycle, and thus the regulation of NO levels in bacteria has received particular interest. NO appears to be involved in the regulation of denitrification, iron acquisition, and detoxification. Additionally, NO regulates diverse downstream processes in bacteria such as biofilm formation, motility and virulence. The levels of NO can be controlled by the interactions between different regulators (Spiro, 2007). To date, several NO-responsive regulatory networks have been identified in bacteria (Fig. 7.2).

NO is the reaction product and substrate of bacteria during the denitrification process, which involves dissimilatory reduction of different nitrogen oxide species such as nitrate to nitrite, and then further reduction to NO. Genes involved in denitrification are mainly regulated by two NO-responsive transcriptional activators: DNR (dissimilatory nitrate respiration regulator) and NNR (or NnrR; nitrite, nitric oxide reductase

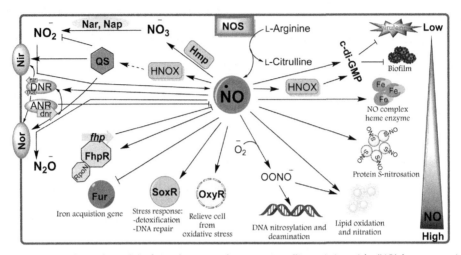

Figure 7.2 Hypothetical model of regulatory pathways controlling nitric oxide (NO) homeostasis in bacteria (*Pseudomonas aeruginosa*), left-hand side, and antimicrobial mechanism of NO against bacteria, right-hand side. Arrow-headed lines indicate activation and bar-headed lines indicate inhibition. *NOS*, NO synthase; *NO_3^-*, nitrate; *NO_2^-*, nitrite; *H-NOX*, heme-nitric oxide/oxygen binding domains; *Hmp*, flavohemoprotein; *QS*, quorum sensing; *DNR*, dissimilatory nitrate respiration regulator; *ANR*, anaerobic regulation; *c-di-GMP*, cyclic diguanosine monophosphate; *Fur*, Ferric uptake regulation protein. *Adapted from Barraud, N., 2007. Nitric oxide-mediated differentiation and dispersal in bacterial biofilms. Unpublished Ph.D. Thesis, The University of New South Wales, Sydney; Coleman, J.W., 2001. Nitric oxide in immunity and inflammation. Int. Immunopharmacol. 1, 1397–1406; and Davis, K.L., Martin, E., Turko, I.V., Murad, F., 2001. Novel effects of nitric oxide. Annu. Rev. Pharmacol. Toxicol. 41, 203–236.*

regulator). These two factors are responsible for maintaining steady-state concentrations of NO below cytotoxic levels (Cruz-Ramos et al., 2002; Korner et al., 2003; Rinaldo et al., 2006; Zumft, 2002). Another transcriptional activator in *Pseudomonas aeruginosa*, termed ANR (anaerobic regulation of arginine deiminase and nitrate reduction), is involved in the regulation of the denitrification pathway. ANR appears to function in a hierarchical fashion with DNR, with the expression of DNR being under the control of ANR (Arai et al., 1995, 1997). ANR activates different anaerobic pathways and induces the expression of DNR under low O_2 tension, and may also "shut down" the denitrification pathway in the presence of abnormal levels of NO (Giardina et al., 2008; Yoon et al., 2007).

DNR controls the activities of different denitrification enzymes, including membrane-bound nitrate reductase or a periplasmic nitrate reductase (Nar), which brings about reduction of nitrate (NO_3^-) to nitrite (NO_2^-), and nitrite reductase (Nir), which is responsible for the conversion of nitrite (NO_2^-) to NO. Nir can be further classified into two categories: the cytochrome nitrite reductase (cd_1-dNir) and the copper-containing nitrite reductase (Cu-dNir) (Ye et al., 1994). Another denitrification enzyme is NO reductase (Nor), a member of the heme-copper oxidase family with a high affinity for NO usually found as part of the membrane-bound cytochrome *bc* complex (Moura and Moura, 2001).

In *Escherichia coli* and *Bacillus subtilis*, NO was shown to inhibit DNA binding by the ferric uptake regulator (Fur), leading to upregulation of genes required for iron acquisition (D'Autréaux et al., 2002; Moore et al., 2004). In *P. aeruginosa*, microarray studies have revealed a large overlap in genes upregulation upon exposure to NO, and those expressed under iron-limiting conditions, indicating correlation between NO and iron-limiting processes (Firoved et al., 2004; Ochsner et al., 2002).

Anaerobic growth of *P. aeruginosa* PAO1 has been shown to be affected by quorum sensing (QS), a bacterial cell to cell communication system controlled by small signaling molecules called autoinducers. Deletion of genes that produces autoinducer signals resulted in an increase in denitrification activity, which was repressed by exogenous signaling molecules (Toyofuku et al., 2007, 2008). Similarly, anaerobic processes and NO levels in biofilms have also been found to be regulated by QS. Microarray studies have revealed that both *nirS* (nitrite reductase) and *norCB* (NO reductase) are highly expressed in biofilms compared to planktonic cells, and that their overexpression is QS dependent (Hentzer et al., 2005; Sauer et al., 2004; Wagner et al., 2003). Gene expression profiles in *P. aeruginosa* biofilms were shown to closely resemble those of bacteria in stationary phase (nutrient-limited), anaerobic and iron-limited modes of growth, or under QS regulatory control (Hassett et al., 2002; Hentzer et al., 2005; Schuster et al., 2003). NO and QS interconnection is further evident in *P. aeruginosa*, where NO derived from anaerobic respiration modulate diverse QS systems and virulence. This is also evident in bacterial systems such as in *Vibrio harveyi* and *V. fischeri*, where NO controls the QS circuit via heme NO/oxygen-binding (H-NOX) protein regulating motility, biofilm formation and iron uptake (Henares et al., 2012, Henares et al., 2013, Wang et al., 2010).

As described earlier, NO is intimately associated with bacterial survival under challenging environmental conditions and is also involved in different biofilm-relevant processes crucial for biofilm formation.

NO AS AN ANTIMICROBIAL AND ANTIBIOFILM AGENT

NO production from iNOS in mammals is stimulated by proinflammatory cytokines and is a critical component of the mammalian innate immune response. The cytotoxic and/or cytostatic effects of NO have been associated with primitive nonspecific host defense mechanisms against numerous pathogens, including viruses, bacteria, fungi, protozoa, and parasites. The importance of this is evidenced by the susceptibility of mice lacking the iNOS gene to infections as diverse as leishmaniasis, malaria, and tuberculosis (Burgner et al., 1999; Coleman, 2001).

The antimicrobial mechanisms of NO are complex and depend on the concentration of NO (Fig. 7.2). At higher concentrations, NO reacts to form reactive nitrogen and oxygen intermediates (RNOS) such as peroxynitrite or the hydroxyl radical, which are toxic due to its ability to induce nitrosation, deamination, or oxidation reactions with DNA, enzymes, proteins, and other macromolecules in microbial systems. Chemical alteration of DNA by RNOS is one of the main mechanisms of NO-mediated antimicrobial action. RNOS also causes nitrosylation of nucleic acids, thereby interfering with DNA replication of bacteria (Fang, 1997; Marcinkiewicz, 1997; Nathan and Hibbs-Jr, 1991; Richardson et al., 2011; Spiro, 2011; Wink et al., 1991). Additionally, NO inhibits DNA repair enzymes such as DNA alkyl transferases by *S*-nitrosation of enzyme cysteine and thereby inhibit the transfer of the alkyl group from guanine to the protein. NO may also exert its effects through complexation with thiols, other heme groups or the iron–sulfur centers of enzymes and proteins essential for cell functioning. For example, *S*-nitrosation of thiols has been implicated in inhibition of *Bacillus cereus* spores and oxidation of [Fe–S] protein aconitase by NO leads to anti-*Burkholderia mallei* activity (Jones-Carson et al., 2008). Membrane destruction via lipid peroxidation is another mechanism by which NO could show bactericidal effect.

Whereas NO at low concentration alters bacterial functions by selectively binding to sensor domains of signal transducing proteins or enzymes such as the heme-nitric oxide/oxygen (H-NOX) domains belonging to family of hemeprotein sensors. Bacterial H-NOX proteins are often found in the same operon as signaling proteins, such as histidine kinases or diguanylate cyclases, suggesting H-NOX's role as a sensor in prokaryotic NO signaling pathways. Binding of NO to H-NOX-diguanylate cyclase/phosphodiesterase protein complexes controls the levels of the bacterial secondary messenger cyclic diguanosine monophosphate (c-di-GMP). This secondary messenger brings about biofilm repression in *Legionella pneumophila* and *Shewanella woodyi* and controls expression of iron uptake genes in *Vibrio fischeri*. The NO binding to H-NOX is also known to be a crucial regulator for the transition between motile and sessile lifestyles in bacteria and the switch between chronic and acute infections (Barraud et al., 2009a; Liu et al., 2012; Plate and Marletta, 2012, 2013; Wang et al., 2010).

In biofilms, cell death and dispersal events are related to oxidative and/or nitrosative stress (Webb et al., 2003). In particular, NO or reactive species produced from NO can cause dispersion of the biofilm. For example, in *P. aeruginosa* biofilms enhanced levels of $ONOO^-$ (peroxynitrite, produced from the reaction of NO and O_2) were found in microcolonies of mature 7-day-old biofilms and were proposed to be responsible for cell death and dispersal. Moreover, a $\Delta nirS$ mutant strain (lacking nitrite reductase) without the ability to produce NO forms biofilms that fail to disperse, whereas a $\Delta norCB$ mutant (lacking NO reductase) produces large amounts of NO and shows enhanced biofilm dispersal (Barraud et al., 2006).

NO at low, nontoxic concentrations have also been shown to induce biofilm dispersal in a broad range of environmentally and clinically relevant bacterial species (Barraud et al., 2009a). Exogenous NO, at sublethal concentrations, is able to induce the transition of bacteria in biofilms from a sessile mode of growth to a free-swimming, planktonic mode. For instance, addition of NO donors was shown to induce dispersal in biofilms of many Gram-negative bacteria such as *P. aeruginosa*, *E. coli*, *Vibrio cholerae*, *Serratia marcescens*, *S. woodyi*, and *Neisseria gonorrhoeae* (Liu et al., 2012; Potter et al., 2009). Similarly in Gram-positive bacteria *Staphylococcus aureus*, exposure to nitrite inhibits biofilm formation and induces dispersal through generation of NO (Schlag et al., 2007). In *B. subtilis*, changes in endogenous production of NO by NOS resulted in dispersal of biofilm via transition from oxic to anoxic conditions (Schreiber et al., 2011). The molecular mechanism for biofilm dispersal at low NO concentrations is also expected to involve the secondary messenger c-di-GMP (Ha and Toole, 2015). NO can also induce dispersal in multispecies biofilms, for instance addition of low dose (20–500 nM) NO donors caused dispersal of multispecies microbial biofilms formed in drinking water pipes, recycled water systems, and on reverse osmosis water filtration membranes (Barraud et al., 2009b). Suspended biofilm aggregates in expectorated sputum from chronically infected cystic fibrosis (CF) patients were dispersed by using NO donors. Thus NO-mediated biofilm dispersal appears to be well conserved across bacterial species. NO at nonlethal concentrations has also been shown to increase the sensitivity of various biofilms to antimicrobial treatments such as chlorine (Barraud et al., 2006, 2009b, 2015).

Therefore, NO can be used to regulate different pathways in bacteria and thereby control pathogenicity and biofilm formation of bacteria.

NO GAS AS AN ANTIMICROBIAL AGENT

Gaseous NO (gNO) is bactericidal against various strains of bacteria, including both Gram-positive and Gram-negative organisms, fungi, mycobacteria, parasites, and viruses. In order to evaluate the topical use of gNO, different clinical isolates of *S. aureus*, methicillin-resistant *S. aureus*, *E. coli*, *Streptococcus* spp., *P. aeruginosa*, and *Candida albicans* were studied and showed potent bacteriocidal activity at 200 ppm gNO with an average of 4.1 ± 1.1 h to completely stop bacterial growth. This study suggests that exogenous gaseous NO has potent significant antibacterial properties that can be beneficial in reducing bacterial burden in infected wound in burn injuries or nonhealing ulcers without toxic effect on skin cells (Ghaffari et al., 2006, 2007). Exogenous gNO at 200 ppm in vitro showed complete bactericidal activity against pathogens derived from patients with pneumonia under the test conditions between 2 and 6 hours. Delivery of gNO to an infected tissue with specialized chamber allows controlled delivery of NO and prevents oxygen species from reacting with the NO. Continuous exposure to 80 ppm gNO under controlled environment inhibited growth of *P. aeruginosa* at 160 ppm, and showed bactericidal effect (McMullin et al., 2005). Studies with gNO with air and oxygen have also shown promising results, for example, two different research teams showed that exposure to 40 ppm NO in air or 10 ppm NO in 100% oxygen for 24 hours reduced *P. aeruginosa* infiltration and helped to clear lung infections in rats. In the case of Gram-positives, gNO at 200 ppm reduced *S. aureus* burden in a rabbit wound model without showing cytotoxicity to human fibroblast, keratinocyte, endothelial, monocyte, and macrophage cells in culture (Adler and Friedman, 2015; Schairer et al., 2012). A clinical trial study on disruption of *P. aeruginosa* biofilms in patients with cystic fibrosis using a low dose inhaled gNO in combination with intravenous ceftazidime and tobramycin antibiotic showed promising results. Study demonstrated that patients who received NO gas at 5–10 ppm (~200 nM NO) for 8 hours daily and standard ceftazidime and tobramycin treatments for a week showed significant reductions in the number of *Pseudomonas* biofilm aggregates and marginal improvement in lung function. These data suggest that using NO as adjunctive therapy may be highly beneficial for the treatment of CF-related infections (Barraud et al., 2015).

NO DONORS AS ANTIMICROBIALS

The discovery of the ever-increasing biological functions of NO has led to the development of a wide array of novel NO donors as research tools and as pharmaceuticals. Most of the NO donors known today were discovered or synthesized before their role as NO donors were fully understood, or were rediscovered over the past few decades. In addition, a wide range of novel molecules with NO-releasing properties have also been reported in the literature (Megson and Webb, 2002; Scatena et al., 2005).

NO donors can be classified into different types, based on either their chemical natures (Table 7.1) or their mechanisms of NO generation. The wide variety of structural motifs among NO-donating compounds results in remarkably varied chemical reactivities and NO-releasing mechanisms. One such mechanism is the spontaneous release of NO through hydrolytic, thermal, or photochemical self-decomposition, as exemplified by the diazeniumdiolates, *S*-nitrosothiols, or oximes. The second NO-releasing mechanism, exhibited by organic nitrates, nitrites, and sydnonimines, is via reaction with acids, alkalis, metals, or thiols. The third NO-releasing mechanism is through oxidation by enzymes such as cytochrome P450, xanthine oxidase, or NO synthases. For example, *N*-hydroxyguanidines require metabolic activation by NO synthases to release NO. Some NO donors may release NO by more than one route, such as the organic nitrates, which can generate NO by chemical reactions or by enzymatic catalysis (Sergio et al., 2008; Wang et al., 2002).

Of the NO donors described, nitrates, diazeniumdiolates, and *S*-nitrothiols are of particular interest due to their distinct advantages, and these three NO donors are discussed in greater detail below.

Table 7.1 Examples of NO donors (Keefer et al., 2001; Wang et al., 2002)

Entry	Name	Examples
1.	Organic nitrates (e.g., glyceryl trinitrate)	
2.	Organic nitrites (e.g., *tert*-pentyl nitrite)	
3.	Metal–NO complexes (e.g., sodium nitroprusside)	
4.	*N*-nitrosoamines (e.g., *N*-(2,5-dihydroxyphenyl)-*N*-methylnitrous amine)	
5.	*C*-nitrosoamines (e.g., 2-nitro-2-nitrosopropane)	
6.	*S*-nitrosoamines (e.g., *S*-nitroso-*N*-valerylpenicillamine)	
7.	Diazeniumdiolates (e.g., sodium 1-(pyrrolidin-1-yl) diazen-1-ium-1,2-diolate)	
8.	Furoxans and benzofuroxans (e.g., 4-phenyl-3-furoxancarbonitrile)	
9.	Sydnonimines (e.g., molsidomine)	
10.	Oximes (e.g., FK409)	
11.	Hydroxylamines (e.g., hydroxylamine)	
12.	*N*-hydroxyguanidines (e.g., (*S*)-2-amino-5-(3-hydroxyguanidino)pentanoic acid)	

Nitrates

The first therapeutic use of organic nitrates was discovered in 1867 by physician Launder Brunton, who used the inhalation of amyl nitrite vapors to relieve anginal pain (Rang et al., 2007). This led to development of nitrates such as glyceryl trinitrate (GTN) **1**, isosorbide dinitrate (ISDN) **2**, and isosorbide mononitrate (ISMN) **3**, which have been used as the main treatment of angina and other vasodilation-related cardiovascular diseases for over a century. Nitrates exert their effects via their metabolite NO, which activates guanylate cyclase leading to the formation of the secondary messenger cGMP, which in turn activates cGMP-dependent kinase to induce relaxation of vascular smooth muscle (Murad, 2006; Thatcher et al., 2004).

The mechanism of NO release from nitrates under physiological conditions is still a matter of considerable debate, but can be broadly classified into two major endogenous mechanisms, the enzymatic and nonenzymatic pathways. The enzymatic mechanism proceeds through activation of the nitrate by glutathione S-transferase, cytochrome P450, and/or xanthine oxidoreductase, or alternatively through catalytic activity mediated by mitochondrial aldehyde dehydrogenase (ALDH-2) (Li et al., 2005; Mayer, 2003; Needleman and Edmund Hunter, 1965). On the other hand, the nonenzymatic mechanism involves denitration by reaction with thiol (SH) groups to generate an S-nitrosothiol, which discharges NO upon decomposition (Ignarro, 1999).

Even though antimicrobial properties of nitrate-based NO donors has attracted less interest due to their potent cardiovascular effects, they are known to play role in body defense mechanism against pathogens. Nitrate in general is an inert compound, but conversion of nitrate by saliva bacteria to nitrite (NO_2) using nitrate reductase could lead to formation of NO and other RNS under acidic conditions (Tennyson and Lippard, 2011). This leads to beneficial gastric bactericidal functions and has a protective effect in the gastrointestinal system, such as the prevention of stress–induced injury in rats and inhibition of the *Salmonella* acid tolerance response (ATR) that enables enteropathogens to survive harsh acidity (Bourret et al., 2008). Bacterial infection of the urinary system also leads to NO_2 production and it has been observed that acidification of nitrite-enriched urine below pH 5.5 is lethal to all known urinary infectious bacteria. This strategy has been used to overcome urinary tract infection both in modern treatments and folk remedies and employs deliberate acidification via dietary habits such as cranberries and citrus fruits (Tennyson and Lippard, 2011). The role of acidified nitrate is also relevant to combating common cutaneous pathogens. NO can be generated from sweat nitrates in the acidic environment of the skin surface, and this is thought to protect against potential invasion by pathogens (Adler and Friedman, 2015; Schairer et al., 2012). Nitrite from endogenous or exogenous nitrate under acidic condition has also been shown to inhibit biofilm in both Gram-positive *S. aureus* and Gram-negative *P. aeruginosa* (Schlag et al., 2007).

Diazeniumdiolates

Diazeniumdiolates are anions containing the $[N(O)NO]^-$ functional group. The first report of diazeniumdiolates (NONOates) dates back to 1802, which described a sulfur-based diazeniumdiolate known as Pelouze's salt. Over the last two centuries, a wide range of diazeniumdiolates bound to oxygen (e.g., Angeli's salt), carbon (e.g., Traube product and cupferron) or nitrogen (e.g., Drago complexes (DEA/NO)) have been prepared. Thus diazeniumdiolates can be classified into different types based on the atom the $[N(O)NO]^-$ functional group is attached to, namely, the S-diazeniumdiolates, O-diazeniumdiolates, C-diazeniumdiolates, and N-diazeniumdiolates (Hrabie and Keefer, 2002; Keefer et al., 2001).

Of special interest are the N-bound diazeniumdiolates, which can release biologically significant amounts of NO (Fig. 7.3). In general, a diazeniumdiolate can

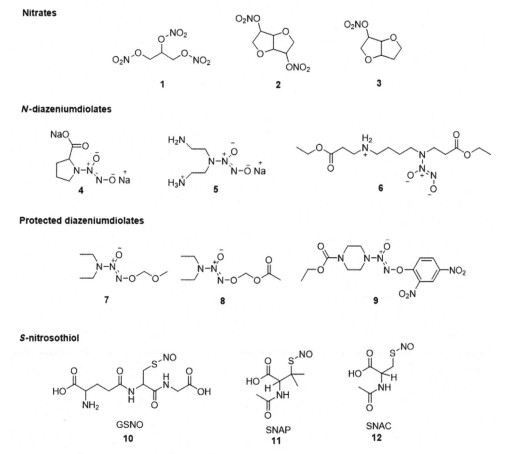

Figure 7.3 Different classes of NO donors.

be synthesized by the reaction of appropriate nucleophiles with NO. The synthesis of N-diazeniumdiolates was first reported by Drago et al. in the 1960s, and involved the reaction of an amine in a nonpolar solvent under a high pressure of NO (Drago and Karstetter, 1961; Longhi et al., 1962; Drago, 1960). A wide array of N-diazeniumdiolates have been synthesized based on amines, particularly secondary amines (Konter et al., 2007; Keefer et al., 2001). Each molecule of N-diazeniumdiolate can release two molecules of NO under normal physiological conditions and the half-life of an N-diazeniumdiolate molecule is structure dependent. The different rates of NO release of diazeniumdiolates could be leveraged to develop NO donors for very specific needs. For example, ultrafast short-burst NO donors such as proline-based N-diazeniumdiolate 4 with half-life of 2 seconds could be used to generate highly localized antiplatelet and vasodilatory effects, or longer NO-releasing donors such as the diethylenetriamine/NO (DETA/NONO) 5 and spermate derivatives 6 could be utilized for implantable biomedical devices and lung inhalation therapy (Hrabie et al., 1993; Rao and Smith, 1998; Saavedra et al., 1996).

The NO release mechanism from N-diazeniumdiolates is commonly considered to follow a pseudo first-order rate law and is pH dependent. It generally involves protonation of the amine nitrogen attached to the diazeniumdiolate, followed by the release of the parent amine and two molecules of NO (Davies et al., 2001; Dutton et al., 2004).

Diazeniumdiolates can be synthetically modified through substitution reactions at second oxygen (O^2) of the diazeniumdiolate group. Many O^2-protected diazeniumdiolates have been synthesized to modulate the stability or NO-releasing properties of the parent diazeniumdiolates. For example, the MOM-protected diazeniumdiolate 7 has an extended half-life of NO release (~17 days in pH 7.4 phosphate buffer) due to the need of the protecting group to be hydrolyzed prior to NO release (Saavedra et al., 1992). Similarly, O^2-substituted diazeniumdiolate, which requires activation by enzymes for NO release has been developed, e.g., O^2-diazeniumdiolate 8 is activated by esterase, and O^2-arylated diazeniumdiolate 9, prepared by nucleophilic aromatic substitution reactions is activated by glutathione and glutathione S-transferase (Saavedra et al., 2000, 2001).

The antimicrobial properties of diazeniumdiolates have been explored against different bacterial strains. DETA-NO a diazeniumdiolate of diethylenetriamine, a ubiquitous polyamine found in both eukaryotic and prokaryotic cells, is effective against Gram-negative and Gram-positive bacteria. Furthermore, out of the different diazeniumdiolates tested, DETA-NO was found to be the most suitable for examining NO activity against *Candida* sp. DETA-NO was active against six species of *Candida*, with minimum inhibitory concentration 50% (MIC 50%) in the range of 0.25–1.0 mg mL^{-1}. DETA-NO in combination with different azole antifungals such as fluconazole, ketoconazole, or miconazole showed synergistic activity against strains tested (McElhaney-Feser et al., 1998). Similarly, DETA/NO was evaluated against clinical isolates of multidrug-resistant extended-spectrum β-lactamase (ESBL)-producing uropathogenic *E. coli* and showed growth inhibition. Combination of miconazole and polymyxin B nonapeptide (PMBN), which increases the bacterial wall permeability, prolonged DETA/NO antimicrobial response for up to 24 hours (Bang et al., 2014). Diazeniumdiolate spermine NONOate and DETA NONOate showed bactericidal activity against *Burkholderia pseudomallei* in a concentration and time-dependent fashion. Additionally, rapidly growing *B. pseudomallei* are more prone to and readily killed upon exposure to the NO donor spermine NONOate (Jones-Carson et al., 2012).

S-nitrosothiols

The *S*-nitrosothiol (*S*-NO)-based compounds are an important class of NO donors. *S*-NO such as *S*-nitrosoglutathione (GSNO) **10** (Singh et al., 1996) have been discovered as naturally occurring molecules in animal and human airways (leading to bronchodilation), and also as *S*-nitroso proteins (in cysteine-containing proteins) in blood plasma. Thus *S*-NO has been considered a storage and transportation form of NO in plasma via the *S*-nitrosylation of serum albumin (Al-Sa'doni and Ferro, 2005; Al-Sa'doni and Ferro, 2000; Williams, 1999). Various *S*-NO derivatives have been synthesized and reported in the literature. Studies have indicated that an *S*-NO group attached to a tertiary carbon or a peptide thiol (SH), as in the tripeptide GSNO, is more stable compared to other types of *S*-NO compounds (Al-Sa'doni et al., 2000; Roy et al., 1994; Spivey et al., 2005). The *S*-NO derivative *S*-nitroso-*N*-acetyl-D,L-penicillamine (SNAP) **11**, obtained from D,L-penicillamine, is one of the most stable nonpeptide *S*-NOs reported (Field et al., 1978; Megson et al., 1999).

S-NO derivatives such as GSNO **10** and *S*-nitroso-*N*-acetylcysteine (SNAC) **12** have been reported to be bactericidal against common Gram-negative and Gram-positive pathogens. GSNO is bactericidal at concentrations safe to human cells—about twice the concentration of GSNO in human serum. GSNO, SNAC, and SNAP have antimicrobial activity and has also shown activity against parasites including Leishmania species, *Trypanosoma cruzi*, *Plasmodium falciparum*, and *Acanthamoeba castellanii* in vitro (Cariello et al., 2010; Venturini et al., 2000; Vespa et al., 1994). Cystine residues found in proteins can also be *S*-nitrosylated, e.g., *S*-nitrosylation of cystine residues found in albumin has shown to be effective against *S. typhimurium* in vitro. Human studies of *S*-NO are limited to case reports and one small study of 16 patients with cutaneous leishmaniasis. All patients treated with SNAP had complete healing of their ulcers while those treated with vehicle alone showed no clinical improvement after 1 month (Lopez-Jaramillo et al., 1998; Schairer et al., 2012).

DUAL ACTION NO DONORS AS ANTIMICROBIALS

As discussed earlier, NO is associated with controlling a wide range of activities through the modulation of multiple targets in different biological systems. This property of NO makes it a useful functionality to conjugate with existing drugs to increase the therapeutic efficacy of the drug, and/or to overcome an adverse side effect. Such hybrids have the potential to broaden the application of existing drugs for the treatment of other pathological conditions.

This strategy has been employed to develop important multitarget hybrid drugs that combine the "native" or existing mechanism of action of a drug with NO-releasing activities, to improve the therapeutic efficiency and/or reduce side effects of a particular drug. For example, the gastric toxicity of aspirin was alleviated by conjugation of aspirin with NO donors such as nitrate. Similar results have been obtained by conjugating other nonsteroidal antiinflammatory drugs (NSAIDs), Ca^{2+} channel blockers, 3-hydroxy-3-methyl-glutaryl-CoA (HMGCoA) reductase, and anticancer drugs with different NO donors such as nitrates, diazeniumdiolates, S-NO, and furoxans (Letts and Loscalzo, 2007). These hybrids were shown to display improved therapeutic effects, such as an increase in the antiplatelet activity of aspirin, or more potent antihypertensive activity of angiotensin receptor antagonists. Such hybrids have also broadened the application of existing drugs for new pathological conditions such as the use of an aspirin-based hybrid for the treatment of cancer (Rigas and Williams, 2008; Sergio et al., 2008). The increasing number of NO hybrid drugs for the treatment of inflammatory and cardiovascular diseases has led to an increase in interest of NO hybrids for treating infectious diseases (Bertinaria et al., 2003; Konter et al., 2008; Tang et al., 2003).

NO hybrid antimicrobials based on antibiotics

Antimicrobial hybrids consisting of NO donors such as nitrates, diazeniumdiolates, and furoxans combined with antibiotic such as cephalosporin or antifungal agents such as ketakonazole have been reported in the literature imparting value to existing drugs. Cephalosporin a β-lactam class of antibiotic and NO donors, cupferron and SIN-1 were conjugated to form hybrids to achieve site-specific delivery of NO in presence of β-lactamase. Evaluation for NO-releasing properties demonstrated promising beta–lactamase dependent NO-releasing ability for the cephalosporin conjugated with SIN-113 (Tang et al., 2003). Similarly, NO donor hybrids based on diazeniumdiolates and cephalosporin have been studied for antimicrobial properties such as biofilm dispersion for P. aeruginosa. The cephalosporin-3′-diazeniumdiolate due to its β-lactam unit provides a scaffold to prevent release of NO from NONOate donor until activated by a bacteria-specific enzyme β-lactamase. Cephalosporin was conjugated with secondary amine diazeniumdiolates derivative such as diethylamine diazeniumdiolate (DEA NONOate). The lead compound, DEA NONOate-cephalosporin prodrug (DEACP) **14** was found to be highly stable in solution and release NO upon incubation with commercially available β-lactamase penicillinase as well as bacterial whole cell extracts that produce β-lactamases. These compounds were effective at dispersing biofilms of several pathogenic species including mixed species biofilms. Co-administration experiments with known antibiotic

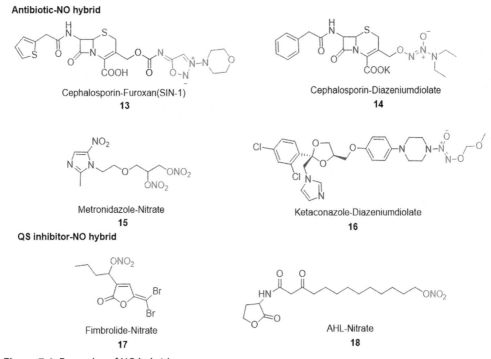

Figure 7.4 Examples of NO hybrids.

such as tobramycin and ciprofloxacin also showed improved antibiofilm outcome compared to antibiotic therapy alone (Barraud et al., 2012;Yepuri et al., 2013).

A series of hybrids by conjugation of metronidazole and NO donors, furoxan and nitrate were synthesized and evaluated in vitro for their activity against a number of *Helicobacter pylori* strains. All the synthesized hybrids (e.g., **15**) exhibited good anti-*H. pylori* activity, resulting in compounds more active than metronidazole. Ketoconazole–NO donor hybrid drugs, e.g., **16** were synthesized using different diazeniumdiolate and organic nitrate NO donors. Both classes of NO donor hybrids showed activity against a broad variety of fungal strains, and for the nitrate-based hybrids the in vitro efficacy was greatly influenced by the variation in structure compared to the NONOate hybrids (Konter et al. 2008) (Fig. 7.4).

Dual action NO antimicrobials based on QS inhibitors

There is growing evidence showing that NO and QS systems are inter-related, and NO influences QS in many different bacterial species. In *Vibrio harveyi*, and *V. fischeri* NO controls the QS circuit regulating motility and biofilm formation, and in *P. aeruginosa*, NO derived from anaerobic respiration modulates diverse QS systems and virulence (Henares, 2013; Plate and Marletta, 2013; Wang et al., 2010). QS deficient strains of *P. aeruginosa* were found to accumulate more NO, further highlighting the common molecular mechanism between NO and QS (Hassett et al., 2002; Toyofuku et al., 2007). This interrelation between NO and QS opens up avenues for interfering with multiple biofilm regulatory pathways using combinations of QS inhibitors and NO donors and thus controlling biofilms and bacterial virulence. The combination of fimbrolide analogs, a class of brominated furanones isolated from the marine red algae *Delisea pulchra* with potent QS inhibitory activity (de Nys et al., 1992; Wright et al., 2006) and NO can thus be used to control the biofilm life cycle and virulence without inhibiting the bacterial growth. Various hybrid molecules based upon fimbrolide derivatives and different NO donors such as nitrates and dizeniumdiolates have been synthesized. These hybrids particularly, fimbrolide nitrate **17**, retains potent QS inhibitory property and was found to be more potent than natural fimbrolide alone. Additionally, these compounds show NO release properties and inhibition of biofilm formation (Kutty et al., 2013).

Similarly, hybrids based on acylated homoserine lactones (AHL), the natural QS signaling molecule in Gram-negative bacteria and NO donors have been synthesized and show modulation of QS and virulence factor expression in bacteria. Hybrids with nitrate NO donors and the AHL containing 3-oxo substitution with longer alkyl chain, e.g., **18** were found to be the most potent QS inhibitors and NO donors. These hybrids also moderately inhibit pyocyanin and elastase production in *P. aeruginosa* (Kutty et al., 2015).

The success of NO donors and NO hybrids uncovers the potential of these new antimicrobial agents for biological applications. Additionally, delivery of these NO donors and hybrids through different delivery platforms, such as nanoparticles and polymeric materials, would further enhance the possibilities of using these molecules as antimicrobial and antibiofilm agents for clinical applications.

CONCLUSION

NO is an integral part of human innate immune response and plays a key role as signaling molecule in microbes. The signaling role of NO in regulating biofilm dispersal across microbial species offers an unprecedented opportunity to develop novel therapeutics to induce biofilm dispersal and improve treatment for bacterial infections. Such nonconventional mode of action that involves nontoxic activation of a signaling pathway, is unlikely to induce evolutionary stress and therefore is believed to decrease the opportunity of resistance development. Despite the promising activities of NO, the delivery and the ability to maintain a sustainable release of NO to target sites in body remain a challenge. Strategies including the use of novel carriers such as nanoparticles, dual action hybrid drugs, polymer coatings, and prodrugs specifically designed to release NO to biofilm infection sites are currently being investigated in order to uncover the full potential of NO as therapeutic agent.

REFERENCES

Adler, B.L., Friedman, A.J., 2015. Nitric oxide therapy for dermatologic disease. Future Sci. OA FSO37, 1.

Alderton, W.K., Cooper, C.E., Knowles, R.G., 2001. Nitric oxide synthases: structure, function and inhibition. Biochem. J. 357, 593–615.

Al-Sa'doni, H., Ferro, A., 2000. S-nitrosothiols: a class of nitric oxide-donor drugs. Clin. Sci. 98, 507–520.

Al-Sa'doni, H.H., Ferro, A., 2005. Current status and future possibilities of nitric oxide-donor drugs: focus on S-nitrosothiols. Mini-Rev. Med. Chem. 5 (3), 247–254.

Al-Sa'doni, H.H., Khan, I.Y., Poston, L., Fisher, I., Ferro, A., 2000. A novel family of S-nitrosothiols: chemical synthesis and biological actions. Nitric Oxide Biol. Chem. 4 (6), 550–560.

Arai, H., Igarashi, Y., Kodama, T., 1995. Expression of the nir and nor genes for denitrification of *Pseudomonas aeruginosa* requires a novel CRP/FNR-related transcriptional regulator, DNR, in addition to ANR. FEBS Lett 371, 73–76.

Arai, H., Kodama, T., Igarashi, Y., 1997. Cascade regulation of the two CRP/FNR-related transcriptional regulators (ANR and DNR) and the denitrification enzymes in *Pseudomonas aeruginosa*. Mol. Microbiol. 25, 1141–1148.

Bang, C.S., Kinnunen, A., Karlsson, M., Oennberg, A., Soederquist, B., Persson, K., 2014. The antibacterial effect of nitric oxide against ESBL-producing uropathogenic *E. coli* is improved by combination with miconazole and polymyxin B nonapeptide. BMC Microbiol. 14, 65/1–65/9. 9.

Barraud, N., 2007. Nitric oxide-mediated differentiation and dispersal in bacterial biofilms. Ph.D. Thesis, The University of New South Wales, Sydney.

Barraud, N., Hassett, D.J., Hwang, S.-H., Rice, S.A., Kjelleberg, S., Webb, J.S., 2006. Involvement of nitric oxide in biofilm dispersal of *Pseudomonas aeruginosa*. J. Bacteriol. 188 (21), 7344–7353.

Barraud, N., Kardak, B.G., Yepuri, N.R., Howlin, R.P., Webb, J.S., Faust, S.N., et al., 2012. Cephalosporin-3'-diazeniumdiolates: targeted NO-donor prodrugs for dispersing bacterial biofilms. Angew. Chem. Int. Ed. 51 (36), 9057–9060. S9057/1–S9057/21.

Barraud, N., Kelso, M.J., Rice, S.A., Kjelleberg, S., 2015. Nitric oxide: a key mediator of biofilm dispersal with applications in infectious diseases. Curr. Pharmaceut. Des. 21 (1), 31–42.

Barraud, N., Schleheck, D., Klebensberger, J., Webb, J.S., Hassett, D.J., Rice, S.A., et al., 2009a. Nitric oxide signaling in *Pseudomonas aeruginosa* biofilms mediates, phosphodiesterase activity, decreased cyclic Di-GMP levels and enhanced dispersal. J. Bacteriol. 191 (23), 7333–7342.

Barraud, N., Storey, M.V., Moore, Z.P., Webb, J.S., Rice, S.A., Kjelleberg, S., 2009b. Nitric oxide-mediated dispersal in single- and multi-species biofilms of clinically and industrially relevant microorganisms. Microb. Biotechnol. 2 (3), 370–378.

Bertinaria, M., Galli, U., Sorba, G., Fruttero, R., Gasco, A., Brenciaglia, M.I., et al., 2003. Synthesis and anti-*Helicobacter pylori* properties of NO-donor/metronidazole hybrids and related compounds. Drug Dev. Res. 60, 225–239.

Bonthius, D.J., Luong, T., Bonthius, N.E., Hostager, B.S., Karacay, B., 2009. Nitric oxide utilizes NF-kB to signal its neuroprotective effect against alcohol toxicity. Neuropharmacology 56, 716–731.

Bourret, T.J., Porwollik, S., McClelland, M., Zhao, R., Greco, T., Ischiropoulos, H., et al., 2008. Nitric oxide antagonizes the acid tolerance response that protects Salmonella against innate gastric defenses. PLoS ONE 3 (3).

Brown, G.C., 2003. NO says yes to mitochondria. Science 299, 838–839.

Bruckdorfer, R., 2005. The basics about nitric oxide. Mol. Aspects Med. 26, 3–31.

Burgner, D., Rockett, K., Kwiatkowski, D., 1999. Nitric oxide and infectious diseases. Arch. Dis. Child. 81, 185–188.

Cariello, A.J., de Souza, G.F.P., Foronda, A.S., Yu, M.C.Z., Hofling-Lima, A.L., de Oliveira, M.G., 2010. In vitro amoebicidal activity of S-nitrosoglutathione and S-nitroso-N-acetylcysteine against trophozoites of *Acanthamoeba castellanii*. J. Antimicrob. Chemother. 65 (3), 588–591.

Coleman, J.W., 2001. Nitric oxide in immunity and inflammation. Int. Immunopharmacol. 1, 1397–1406.

Crane, B.R., Arvai, A.S., Ghosh, D.K., Wu, C., Getzoff, E.D., Stuehr, D.J., et al., 1998. Structure of nitric oxide synthase oxygenase dimer with pterin and substrate. Science 279, 2121–2126.

Cruz-Ramos, H., Crack, J., Wu, G., Hughes, M.N., Scott, C., Thomson, A.J., et al., 2002. NO sensing by FNR: regulation of the *Escherichia coli* NO-detoxifying flavohaemoglobin, Hmp. EMBO J. 21, 3235–3244.

Culotta, E., Koshland, D.E., 1992. NO news is good news. Science 258, 1862–1863.

D'Autréaux, B., Touati, D., Bersch, B., Latour, J.M., Michaud-Soret, I., 2002. Direct inhibition by nitric oxide of the transcriptional ferric uptake regulation protein via nitrosylation of the iron. Proc. Natl. Acad. Sci. USA 99, 16619–16624.

Davies, K.M., Wink, D.A., Saavedra, J.E., Keefer, L.K., 2001. Chemistry of the diazeniumdiolates. 2. Kinetics and mechanism of dissociation to nitric oxide in aqueous solution. J. Am. Chem. Soc. 123 (23), 5473–5481.

Davis, K.L., Martin, E., Turko, I.V., Murad, F., 2001. Novel effects of nitric oxide. Annu. Rev. Pharmacol. Toxicol. 41, 203–236.

Dawson, V.L., Dawson, T.M., 1996. Nitric oxide neurotoxicity. J. Chem. Neuroanat. 10, 179–190.

de Nys, R., Coll, J.C., Bowden, B.F., 1992. *Delisea pulchra* (cf. *fimbriata*) revisited. The structural determination of two new metabolites from the red alga *Delisa pulchra*. Aust. J. Chem. 45, 1625–1632.

Drago, R.S., Karstetter, B.R., 1961. The reaction of nitrogen(II) oxide with various primary and secondary amines. J. Am. Chem. Soc. 83 (8), 1819–1822.

Drago, R.S., Paulik, F.E., 1960. The reaction of nitrogen(II) oxide with diethylamine. J. Am. Chem. Soc. 82 (1), 96–98.

Dutton, A.S., Fukuto, J.M., Houk, K.N., 2004. The mechanism of NO formation from the decomposition of dialkylamino diazeniumdiolates: density functional theory and CBS-QB3 predictions. Inorg. Chem. 43, 1039–1045.

Eiserich, J.P., Patel, R.P., O'Donnell, V.B., 1998. Pathophysiology of nitric oxide and related species: free radical reactions and modification of biomolecules. Molec. Aspects Med. 19, 221–357.

Fang, F.C., 1997. Mechanism of nitric oxide-related antimicrobial activity. J. Clin. Invest. 99, 2818–2825.

Field, L., Dilts, R.V., Ravichandran, R., Lenhert, P.G., Carnahan, G.E., 1978. An unusually stable thionitrite from N-acetyl-D,L-penicillamine; X-ray crystal and molecular structure of 2-(acetylamino)-2-carboxy-1,1-dimethylethyl thionitrite. JCS Chem. Commun., 249–250.

Firoved, A.M., Wood, S.R., Ornatowski, W., Deretic, V., Timmins, G.S., 2004. Microarray analysis and functional characterization of the nitrosative stress response in nonmucoid and mucoid *Pseudomonas aeruginosa*. J. Bacteriol. 186, 4046–4050.

Galla, H.-J., 1993. Nitric oxide, NO, an intercellular messenger. Angew. Chem. Int. Ed. 32, 378–380.

Ghaffari, A., Jalili, R., Ghaffari, M., Miller, C., Ghahary, A., 2007. Efficacy of gaseous nitric oxide in the treatment of skin and soft tissue infections. Wound Repair Regeneration 15 (3), 368–377.

Ghaffari, A., Miller, C.C., McMullin, B., Ghahary, A., 2006. Potential application of gaseous nitric oxide as a topical antimicrobial agent. Nitric Oxide 14 (1), 21–29.

Giardina, G., Rinaldo, S., Johnson, K.A., Matteo, A.D., Brunori, M., Cutruzzolà, F., 2008. NO sensing in *Pseudomonas aeruginosa*: structure of the transcriptional regulator DNR. J. Mol. Biol. 378, 1002–1015.

Gorren, A.C.F., Mayer, B., 2007. Nitric-oxide synthase: a cytochrome P450 family foster child. Biochim. Biophys. Acta 1770, 432–445.

Groves, J.T., Wang, C.C.-Y., 2000. Nitric oxide synthase: models and mechanisms. Curr. Opin. Chem. Biol. 4, 687–695.

Ha, D.-G., Toole, G.A., 2015. c-di-GMP and its effects on biofilm formation and dispersion: a *Pseudomonas aeruginosa* review. Microbiol. Spectr. 3, 2.

Handy, D.E., Loscalzo, J., 2006. Nitric oxide and posttranslational modification of the vascular proteome: S-nitrosation of reactive thiols. Arterioscler. Thromb. Vasc. Biol. 26, 1207–1214.

Hassett, D.J., Cuppoletti, J., Trapnell, B., Lymar, S.V., Rowe, J.J., Yoon, S.S., et al., 2002. Anaerobic metabolism and quorum sensing by *Pseudomonas aeruginosa* biofilms in chronically infected cystic fibrosis airways: rethinking antibiotic treatment strategies and drug targets. Adv. Drug. Deliv. Rev. 54, 1425–1443.

Henares, B.M., Higgins, K.E., Boon, E.M., 2012. Discovery of a nitric oxide responsive quorum sensing circuit in. Vibrio harveyi, ACS Chem. Biol 7 (8), 1331–1336.

Henares, B.M., Xu, Y., Boon, E.M., 2013. A nitric oxide-responsive quorum sensing circuit in *Vibrio harveyi* regulates flagella production and biofilm formation. Int. J. Mol. Sci. 14 (8), 16473–16484.

Hentzer, M., Eberl, L., Givskov, M., 2005. Transcriptome analysis of *Pseudomonas aeruginosa* biofilm development: anaerobic respiration and iron limitation. Biofilms 2, 37–61.

Hrabie, J.A., Keefer, L.K., 2002. Chemistry of the nitric oxide-releasing diazeniumdiolate ("Nitrosohydroxykamine") functional group and its oxygen substituted derivatives. Chem. Rev. 102, 1135–1154.

Hrabie, J.A., Klose, J.R., Wink, D.A., Keefer, L.K., 1993. New nitric oxide-releasing zwitterions derived from polyamines. J. Org. Chem. 58 (6), 1472–1476.

Ignarro, L.J., 1999. Nitric oxide: a unique endogenous signaling molecule in vascular biology (nobel lecture). Angew. Chem. Int. Ed. 38, 1882–1892.

Jones-Carson, J., Laughlin, J., Hamad, M.A., Stewart, A.L., Voskuil, M.I., Vazquez-Torres, A., 2008. Inactivation of [Fe–S] metalloproteins mediates nitric oxide-dependent killing of *Burkholderia mallei*. PLoS ONE, 3 (4).

Jones-Carson, J., Laughlin, J.R., Stewart, A.L., Voskuil, M.I., Vazquez-Torres, A., 2012. Nitric oxide-dependent killing of aerobic, anaerobic and persistent *Burkholderia pseudomallei*. Nitric Oxide 27 (1), 25–31.

Keefer, L.K., Flippen-Anderson, J.L., George, C., Shanklin, A.P., Dunams, T.M., Christodoulou, D., et al., 2001. Chemistry of the diazeniumdiolates. I. Structural and spectral characteristics of the [N(O)NO]$^-$ functional group. Nitric Oxide Biol. Chem. 5 (4), 377–394.

Konter, J., Abuo-Rahma, A.G.E.-D.A., El-Emam, A., Lehmann, J., 2007. Synthesis of diazen-1-ium-1,2-diolates monitored by the "NOtizer" apparatus: relationship between formation rates, molecular structure and the release of nitric oxide. Eur. J. Org. Chem., 616–624.

Konter, J., Möllmann, U., Lehmann, J., 2008. NO-donors. Part 17: synthesis and antimicrobial activity of novel ketoconazole–NO-donor hybrid compounds. Bioorg. Med. Chem. 16, 8294–8300.

Korner, H., Sofia, H.J., Zumft, W.G., 2003. Phylogeny of the bacterial superfamily of Crp-Fnr transcription regulators: exploiting the metabolic spectrum by controlling alternative gene programs. FEMS Microbiol. Rev. 27, 559–592.

Kutty, S.K., Barraud, N., Pham, A., Iskander, G., Rice, S.A., Black, D.S., et al., 2013. Design, synthesis, and evaluation of fimbrolide-nitric oxide donor hybrids as antimicrobial agents. J. Med. Chem. 56 (23), 9517–9529.

Kutty, S.K., Barraud, N., Ho, K.K.K., Iskander, G.M., Griffith, R., Rice, S.A., et al., 2015. Hybrids of acylated homoserine lactone and nitric oxide donors as inhibitors of quorum sensing and virulence factors in *Pseudomonas aeruginosa*. Org. Biomol. Chem. 13 (38), 9850–9861.

Letts, G., Loscalzo, J., 2007. Frontiers in nephrology: targeting inflammation using novel nitric oxide donors. J. Am. Soc. Nephrol. 18, 2863–2869.

Li, H., Cui, H., Liu, X., Zweier, J.L., 2005. Xanthine oxidase catalyzes anaerobic transformation of organic nitrates to nitric oxide and nitrosothiols. J. Biol. Chem. 280 (17), 16594–16600.

Liu, N., Xu, Y., Hossain, S., Huang, N., Coursolle, D., Gralnick, J.A., et al., 2012. Nitric oxide regulation of cyclic di-GMP synthesis and hydrolysis in *Shewanella woodyi*. Biochemistry 51 (10), 2087–2099.

Longhi, R., Ragsdale, R.O., Drago, R.S., 1962. Reactions of nitrogen(II) oxide with miscellaneous lewis bases. Inorg. Chem. 1 (4), 768–770.

Lopez-Jaramillo, P., Ruano, C., Rivera, J., Teran, E., Salazar-Irigoyen, R., Esplugues, J.V., et al., 1998. Treatment of cutaneous leishmaniasis with nitric-oxide donor. Lancet 351 (9110), 1176–1177.

Marcinkiewicz, J., 1997. Nitric oxide and antimicrobial activity of reactive oxygen intermediates. Immunopharmacology 37 (1), 35–41.

Mayer, B., 2003. Bioactivation of nitroglycerin–A new piece in the puzzle. Angew. Chem. Int. Ed. 42, 388–391.

McElhaney-Feser, G.E., Raulli, R.E., Cihlar, R.L., 1998. Synergy of nitric oxide and azoles against *Candida* species in vitro. Antimicrob. Agents Chemother. 42 (9), 2342–2346.

McMullin, B.B., Chittock, D.R., Roscoe, D.L., Garcha, H., Wang, L., Miller, C.C., 2005. The antimicrobial effect of nitric oxide on the bacteria that cause nosocomial pneumonia in mechanically ventilated patients in the intensive care unit. Respiratory Care 50 (11), 1451–1456.

Megson, I.L., Webb, D.J., 2002. Nitric oxide donors: current status future trends. Expert Opin. Investig. Drugs 11, 587–601.

Megson, I.L., Morton, S., Greig, I.R., Mazzei, F.A., Field, R.A., Butler, A.R., et al., 1999. *N*-substituted analogues of *S*-nitroso-*N*-acetyl-D,L-penicillamine: chemical stability and prolonged nitric oxide mediated vasodilatation in isolated rat femoral arteries. Br. J. Pharmacol. 126, 639–648.

Moore, C.M., Nakano, M.M., Wang, T., Ye, R.W., Helmann, J.D., 2004. Response of *Bacillus subtilis* to nitric oxide and the nitrosating agent sodium nitroprusside. J. Bacteriol. 186, 4655–4664.

Moura, I., Moura, J.J.G., 2001. Structural aspects of denitrifying enzymes. Curr. Opin. Chem. Biol. 5, 168–175.

Murad, F., 1999. Discovery of some of the biological effects of nitric oxide and its role in cell signaling (nobel lecture). Angew. Chem. Int. Ed. 38, 1856–1868.

Murad, F., 2006. Nitric oxide and cyclic GMP in cell signaling and drug development. N. Engl. J. Med. 355 (19), 2003–2011.

Nathan, C.F., Hibbs-Jr, J.B., 1991. Role of nitric oxide synthesis in macrophage antimicrobial activity. Curr. Opin. Immunol. 3, 65–70.

Needleman, P., Edmund Hunter Jr, F., 1965. The transformation of glyceryl trinitrate and other nitrates by glutathione-organic nitrate reductase. Mol. Pharmacol. 1, 77–86.

Ochsner, U.A., Wilderman, P.J., Vasil, A.I., Vasil, M.L., 2002. GeneChip expression analysis of the iron starvation response in *Pseudomonas aeruginosa*: identification of novel pyoverdine biosynthesis genes. Mol. Microbiol. 45, 1277–1287.

Pfeiffer, S., Mayer, B., Hemmens, B., 1999. Nitric oxide: chemical puzzles posed by a biological messenger. Angew. Chem. Int. Ed. 38, 1714–1731.

Plate, L., Marletta, Michael A., 2012. Nitric oxide modulates bacterial biofilm formation through a multicomponent cyclic-di-GMP signaling network. Mol. Cell 46 (4), 449–460.

Plate, L., Marletta, M.A., 2013. Nitric oxide-sensing H-NOX proteins govern bacterial communal behavior. Trends Biochem. Sci. 38 (11), 566–575.

Potter, A.J., Kidd, S.P., Edwards, J.L., Falsetta, M.L., Apicella, M.A., Jennings, M.P., et al., 2009. Thioredoxin reductase is essential for protection of *Neisseria gonorrhoeae* against killing by nitric oxide and for bacterial growth during interaction with cervical epithelial cells. J. Infect. Dis. 199 (2), 227–235.

Rang, H.P., Dale, M.M., Ritter, J.M., Flower, R.J., 2007. Rang & Dale's Pharmacology, 6 ed. Churchill Livingstone.

Rao, W., Smith, D.J., 1998. Spermic acid. Chem. Pharm. Bull. 46 (11), 1846–1847.

Richardson, A.R., Payne, E.C., Younger, N., Karlinsey, J.E., Thomas, V.C., Becker, L.A., et al., 2011. Multiple targets of nitric oxide in the tricarboxylic acid cycle of *Salmonella enterica* Serovar Typhimurium. Cell Host Microbes 10, 33–43.

Rigas, B., Williams, J.L., 2008. NO-donating NSAIDs and cancer: an overview with a note on whether NO is required for their action. Nitric Oxide 19, 199–204.

Rinaldo, S., Giardina, G., Brunori, M., Cutruzzolà, F., 2006. N-oxide sensing and denitrification: the DNR transcription factors. Biochem. Soc. Trans. 34, 185–187.

Roy, B., d'Hardemare, Ad.M., Fontecave, M., 1994. New thionitrites: synthesis, stability, and nitric oxide generation. J. Org. Chem. 59, 7019–7026.

Saavedra, J.E., Dunams, T.M., Flippen-Anderson, J.L., Keefer, L.K., 1992. Secondary amine/nitric oxide complex ions, R2N[N(0)NO]⁻ O-functionalization chemistry. J. Org. Chem. 57, 6134–6138.

Saavedra, J.E., Shami, P.J., Wang, L.Y., Davies, K.M., Booth, M.N., Citro, M.L., et al., 2000. Esterase-sensitive nitric oxide donors of the diazeniumdiolate family: in vitro antileukemic activity. J. Med. Chem. 43 (2), 261–269.

Saavedra, J.E., Southan, G.J., Davies, K.M., Lundell, A., Markou, C., Hanson, S.R., et al., 1996. Localizing antithrombotic and vasodilatory activity with a novel, ultrafast nitric oxide donor. J. Med. Chem. 39 (22), 4361–4365.

Saavedra, J.E., Srinivasan, A., Bonifant, C.L., Chu, J., Shanklin, A.P., Flippen-Anderson, J.L., et al., 2001. The secondary amine/nitric oxide complex ion R2N[N(O)NO]⁻ as nucleophile and leaving group in SNAr reactions. J. Org. Chem. 66, 3090–3098.

Sauer, K., Cullen, M.C., Rickard, A.H., Zeef, L.A., Davies, D.G., Gilbert, P., 2004. Characterization of nutrient-induced dispersion in *Pseudomonas aeruginosa* PAO1 biofilm. J. Bacteriol. 186, 7312–7326.

Scatena, R., Bottoni, P., Martorana, G.E., Giardina, B., 2005. Nitric oxide donor drugs: an update on pathophysiology and therapeutic potential. Expert Opin. Investig. Drugs 14, 835–846.

Schairer, D.O., Chouake, J.S., Nosanchuk, J.D., Friedman, A.J., 2012. The potential of nitric oxide releasing therapies as antimicrobial agents. Virulence 3 (3), 271–279.

Schlag, S., Nerz, C., Birkenstock, T.A., Altenberend, F., Gotz, F., 2007. Inhibition of staphylococcal biofilm formation by nitrite. J. Bacteriol. 189 (21), 7911–7919.

Schreiber, F., Beutler, M., Enning, D., Lamprecht-Grandio, M., Zafra, O., González-Pastor, J.E., et al., 2011. The role of nitric-oxide-synthase-derived nitric oxide in multicellular traits of *Bacillus subtilis* 3610: biofilm formation, swarming, and dispersal. BMC Microbiol. 11 (1), 1–12.

Schuster, M., Lostroh, C.P., Ogi, T., Greenberg, E.P., 2003. Identification, timing, and signal specificity of *Pseudomonas aeruginosa* quorum-controlled genes: a transcriptome analysis. J. Bacteriol. 185, 2066–2079.

Sergio, H., Sapna, C., Benjamin, B., 2008. Nitric oxide donors: novel cancer therapeutics (review). Int. J. Oncol. 33, 909–927.

Singh, S.P., Wishnok, J.S., Keshive, M., Deen, W.M., Tannenbaum, S.R., 1996. The chemistry of the S-nitrosoglutathione/glutathione system. Proc. Natl. Acad. Sci. USA 93, 14428–14433.

Spiro, S., 2007. Regulators of bacterial responses to nitric oxide. FEMS Microbiol. Rev. 31, 193–211.

Spiro, S., 2011. Another target for NO. Cell Host Microbe 10, 1–2.

Spivey, A.C., Colley, J., Sprigens, L., Hancock, S.M., Cameron, D.S., Chigboh, K.I., et al., 2005. The synthesis of water soluble decalin-based thiols and S-nitrosothiols—model systems for studying the reactions of nitric oxide with protein thiols. Org. Biomol. Chem. 3, 1942–1952.

Stamler, J.S., Singel, D.J., Loscalzo, J., 1992. Biochemistry of nitric oxide and its redox-activated forms. Science 258, 1898–1902.

Tang, X., Cai, T., Wang, P.G., 2003. Synthesis of beta-lactamase activated nitric oxide donors. Bioorg. Med. Chem. Lett. 13, 1687–1690.

Tennyson, Andrew G., Lippard, Stephen J., 2011. Generation, translocation, and action of nitric oxide in living systems. Chem. Biol. 18 (10), 1211–1220.

Thatcher, G.R.J., Nicolescu, A.C., Bennett, B.M., Toader, V., 2004. Nitrates and NO release: contemporary aspects in biological and medicinal chemistry. Free Radic. Biol. Med. 37, 1122–1143.

Toyofuku, M., Nomura, N., Fujii, T., Takaya, N., Maseda, H., Sawada, I., et al., 2007. Quorum sensing regulates denitrification in Pseudomonas aeruginosa PAO1. J. Bacteriol. 189, 4969–4972.

Toyofuku, M., Nomura, N., Kuno, E., Tashiro, Y., Nakajima, T., Uchiyama, H., 2008. Influence of the Pseudomonas quinolone signal on denitrification in Pseudomonas aeruginosa. J. Bacteriol. 190, 7947–7956.

Venturini, G., Salvati, L., Muolo, M., Colasanti, M., Gradoni, L., Ascenzi, P., 2000. Nitric oxide inhibits cruzipain, the major papain-like cysteine proteinase from Trypanosoma cruzi. Biochem. Biophys. Res. Commun. 270 (2), 437–441.

Vespa, G.N., Cunha, F.Q., Silva, J.S., 1994. Nitric oxide is involved in control of Trypanosoma cruzi-induced parasitemia and directly kills the parasite in vitro. Infect. Immun. 62 (11), 5177–5182.

Wagner, V.E., Bushnell, D., Passador, L., Brooks, A.I., Iglewski, B.H., 2003. Microarray analysis of Pseudomonas aeruginosa quorum sensing regulons: effects of growth phase and environment. J. Bacteriol. 185, 2080–2095.

Wang, P.G., Xian, M., Tang, X., Wu, X., Wen, Z., Cai, T., et al., 2002. Nitric oxide donors: chemical activities and biological applications. Chem. Rev. 102, 1091–1134.

Wang, Y., Dufour, Y.S., Carlson, H.K., Donohue, T.J., Marletta, M.A., Ruby, E.G., 2010. H-NOX–mediated nitric oxide sensing modulates symbiotic colonization by Vibrio fischeri. Proc. Natl. Acad. Sci. USA 107 (18), 8375–8380.

Webb, J.S., Thompson, L.S., James, S., Charlton, T., Tolker-Nielsen, T., Koch, B., et al., 2003. Cell death in Pseudomonas aeruginosa biofilm development. J. Bacteriol. 185, 4585–4592.

Williams, D.L.H., 1999. The chemistry of S-nitrosothiols. Acc. Chem. Res. 32, 869–876.

Wink, D.A., Kasprzak, K.S., Maragos, C.M., Elespuru, R.K., Misra, M., Dunams, T.M., et al., 1991. DNA deaminating ability and genotoxicity of nitric oxide and its progenitors. Science 254, 1001–1003.

Wright, A.D., Nys, Rd, Angerhofer, C.K., Pezzuto, J.M., Gurrath, M., 2006. Biological activities and 3D QSAR studies of a series of Delisea pulchra (cf. fimbriata) derived natural products. J. Natl. Prod. 69, 1180–1187.

Ye, R.W., Averill, B.A., Tiedje, J.M., 1994. Denitrification: production and consumption of nitric oxide. Appl. Environ. Microbiol. 60, 1053–1058.

Yepuri, N.R., Barraud, N., Mohammadi, N.S., Kardak, B.G., Kjelleberg, S., Rice, S.A., et al., 2013. Synthesis of cephalosporin-3′-diazeniumdiolates: biofilm dispersing NO-donor prodrugs activated by β-lactamase. Chem. Commun. 49 (42), 4791–4793.

Yoon, S.S., Karabulut, A.C., Lipscomb, J.D., Hennigan, R.F., Lymar, S.V., Groce, S.L., 2007. Two-pronged survival strategy for the major cystic fibrosis pathogen, Pseudomonas aeruginosa, lacking the capacity to degrade nitric oxide during anaerobic respiration. EMBO J. 26, 3662–3672.

Zumft, W.G., 2002. Nitric oxide signaling and NO dependent transcriptional control in bacterial denitrification by members of the FNR-CRP regulator family. J. Mol. Microbiol. Biotechnol. 4 (3), 277–286.

CHAPTER 8

Improving the Performance of Implantable Sensors with Nitric Oxide Release

Megan C. Frost
Department of Biomedical Engineering, Michigan Technological University, Houghton, MI, United States

INTRODUCTION

There are medical needs that could be better served if we were able to directly measure physiological analytes within the human body on a continuous, real-time basis. Examples that demonstrate the value of these measurements include care of patients in intensive care units (ICUs) and during surgery. Early detections of changes in oxygen, pH, and carbon dioxide levels in blood are indicators of respiratory distress. Tight control over glucose and lactate levels in critically ill patients has been linked to better patient outcomes (Frost and Meyerhoff, 2015; Frost et al., 2013; Gifford, 2013; Ganter and Zollinger, 2003). In addition to the hospital setting, reliable in vivo sensing of glucose levels, when used in conjunction with implantable insulin pumps, would allow diabetic patients to release insulin as needed, in an automatic manner that would in practice eliminate the functional outcome of diabetes (i.e., not a biological cure to the disease, but effectively eliminating the symptoms of the disease) and allow people to function without finger pricking and daily bolus injections of insulin. The technology for reliable sensors and insulin pumps has existed for decades. Part of the reason that closed loop systems have not become the standard treatment for diabetes is that the implantable sensors used to detect glucose levels in the body do not function reliably in the physiological environment. The problem of biological response to implanted sensors (both blood contacting and tissue contacting) is that the dynamic, rigorous host response to these devices eventually causes inaccurate sensor readings that cannot be used to make therapeutic decisions for patients. The potential benefits of continuous, real-time monitoring are completely erased if a medical professional questions the reliability of data obtained from implanted sensor reading. Controlling biological response toward the sensors is one of the largest, if not the single largest, technical obstacles remaining to achieve widespread use of implanted sensors. In this chapter, the basis of

Nitric Oxide Donors.

191

the biological response to implanted polymeric materials will be described for both blood contacting and tissue contacting sensors. The potential utility of nitric oxide (NO) as a means to facilitate control of biological response, NO donors that have been utilized to fabricate more reliable implanted sensors and future technical challenges that need to be addressed in the development of reliable implanted sensors will be presented.

BIOLOGICAL RESPONSE TO IMPLANTED SENSORS

Polymeric materials are used to fabricate or coat a wide variety of devices that treat medical conditions in the human body. Examples include a hugely diverse array of devices such as sutures, screws, meshes, tissue scaffolds, drug delivery patches, vascular grafts, access catheters, wound dressings, artificial skin, and sensors. These devices can be implanted for short-term temporary use, long-term permanent use, or may be intended to biodegrade over time. These different devices experience a wide range of different physical environments, depending on the location of implantation. As polymer science has advanced and our understanding of biological response to foreign materials has improved, we have reached a point of tremendous opportunity; the possibility exists of specifically matching materials properties and processing conditions to tailor materials to specific implantation situations, in theory allowing better matching of physical and chemical properties to the tissue in need of medical intervention. Although great strides have been made in this realm, a persistent problem with all implanted materials is the dynamic nature of the biological response toward implanted devices. The ability of the body to recognize foreign materials and work to isolate and remove these materials is tenaciously impressive. Universally, devices made from polymeric materials inspire some type of inflammatory response. Some level of this response is necessary to force wound healing to come to a resolution, but the uncontrolled, chronic inflammatory state that often occurs causes a great deal of damage at the implantation site, damaging healthy tissue surrounding the site and often leading to failure of devices such as sensors. Many research strategies have been developed and are currently under investigation that are aimed at resolving inflammation and wound healing in such a way that functional tissue develops, facilitating the implanted sensors ability to reach a predictable, stable state that will allow appropriate sensor calibration and reliable monitoring of analytes. The biological response to foreign materials will be discussed in the context of blood contacting materials and tissue contacting materials. Although there are common elements involved in both responses and often times the same device will interact with both blood and tissue, the primary biological responses will be described separately.

Biological response to polymers in blood

Nearly instantly upon contact with blood, protein will begin to adhere to a polymeric surface (Volger, 2012; Barnthip et al., 2009). The specific details of the polymer chemistry and surface topography will influence the protein layer that is adsorbed. Additionally, the specific details of the protein composition present in the blood will also play a significant role in determining the composition of the adsorbed protein layer. It is important to keep in mind that the adsorbed protein layer will change over time when it reaches a pseudo-state state (Barnthip et al., 2009). Thrombosis problems develop as platelets begin to adhere to the protein covered polymer. Platelet adhesion and activation on foreign surfaces is heavily influenced by the ability of platelets to recognize certain proteins that adsorb on the material surfaces after exposure to blood. Two critical proteins that mediate platelet adhesion are von Willebrand's Factor (vWF) and fibrinogen (Gorbet and Sefton, 2004). vWF is a circulating protein that readily adsorbs to collagen (the basement material of blood vessels) and foreign materials that do not have intact endothelial layer. Upon adsorption, vWF will bind with glycoprotein Ib (GPIb) on the surface of platelets to form the GPIb-V-IX complex, which allows platelet adhesion to the device surface (Dörmann et al., 2000). Once platelets adhere to a surface they will become activated, thereby releasing a host of factors such as prostaglandins (PGH_2 and PGG_2), Ca^{2+}, and thromboxane A_2 (Dee et al., 2002a) that promote more platelet adhesion and accelerate the coagulation cascade. Thomboxane A_2 promotes the presentation of GP IIb/IIIa on the surface of platelets that actively binds circulating fibrinogen, linking platelets together and locally converting soluble fibrinogen to insoluble fibrin (Dee et al., 2002a). The fibrin network and platelets then entangle red blood cells and form the thrombus that covers devices placed in blood. Coagulation factor XII and platelet factor 3 (which is released by adherent platelets) will bind directly to the foreign surface and also convert prothrombin to thrombin, which will further promote the conversion of fibrinogen to fibrin (Dee et al., 2002a). Often, the presence of an adsorbed protein layer only or protein and thrombus on a device can significantly interfere with the function of a device and can cause a local change in blood around the device by partially occluding the vessel. For example, the cross-sectional surface area of an artificial vascular graft can be considerably reduced by the adsorption of a layer of protein on the inner surface of the graft, particularly in small diameter vascular graft (<4 mm). If only a 1-mm layer of protein or thrombus is adsorbed onto the surface, this reduces the cross-sectional area by 36% (Dee et al., 2002b), thereby significantly altering blood flow through the graft. Additionally the presence of thrombus on the surface of this device creates the risk of emboli forming and increasing the risk of heart attack and stroke in patients.

In addition to activation of the coagulation cascade, the compliment system is also activated by foreign materials such as polymers in contact with blood. Although compliment activation is usually discussed in terms of increased efficiency in removing infectious microorganisms, it also plays a significant role in the overall biological response of blood contacting polymeric biomaterials. The compliment system is a system of approximately 30 membrane-bound and plasma proteins which collectively function to enhance the biological response to foreign substances by increasing the activation and release of proinflammatory mediators (Kazatchkine and Carreno, 1988; Moghimi et al., 2011; Szott and Horbert, 2011; Ekdahl et al., 2011). These mediators will affect the activation of many cell types, including leukocytes, mast cells, and smooth muscle cells. The compliment system functions by either causing direct lysis of marked cells or by mediating leukocyte function in inflammation and innate immunity (Kazatchkine and Carreno, 1988). There are three known mechanisms by which the compliment system is activated: classical pathway, lectin pathway, and the alternative pathway (AP). Both the classical and lectin pathways involve antibody binding and cleavage of compliment protein C3 and ultimately forming a protein assembly known as the membrane attack complex (MAC) that serves to induce cell lysis of microorganisms. Although biomaterials will not be affected by MAC in the same manner as microorganisms, adsorption of compliment proteins and activation of the AP can have a prolonged effect on biological response when materials are exposed to blood. Compliment factors that adsorb to the protein layer deposited on polymeric materials following exposure to blood can generate anaphylatoxin, activated fragments of C3, C4, and C5. These activated compliment factors (C3a and C5a) are responsible for mediating chemotaxis of specific cells to the foreign material, inflammation at the site of implantation, and vascular permeability near the implant. It has been reported (Kazatchkine and Carreno, 1988; Ekdahl et al., 2011) that compliment activation added approximately 25% of the mass of initially adsorbed proteins on a surface after 60 minutes of incubation with undiluted serum, consisting primarily of iC3b (Kazatchkine and Carreno, 1988). Surface-bound C3b is a major component that will accelerate the AP activation by forming more C3 convertase complexes (C3bBb) which activates more C3 into the active C3a fragment and the surface-bound C3b fragment (Ekdahl et al., 2011).

The activation of the coagulation cascade and compliment activation are directly linked, as described in more detail by Ekdahl et al. (2011). Specific coagulation factors such as Factor XIIa, thrombin, and plasmin are able to directly activate purified compliment factors and fragments in vitro (Ghebrehiwet et al., 1983; Thoman et al., 1984; Spath and Gabl, 1976; Goldberger et al., 1981; Lachmann et al., 1982). Additionally, other factors generated in the coagulation cascade such as XIa, Xa, and IXa are also thought to directly generate C3a and C5a (Amara et al., 2010). Additionally, C5a is able to increase the expression of tissue factor responsible for the activation of extrinsic coagulation. The activation of both of these systems leads to thrombosis and inflammation at the site of polymer implantation.

Biological response to polymers in subcutaneous tissue

When polymers are placed in subcutaneous tissue, the classically accepted sequence of events that take place begin with a cascade of reactions that are activated which seek to isolate the foreign material, destroy or remove it, and then prompt wound healing to move to completion. Upon implantation in tissue, both plasma and tissue proteins adsorb to the surface of the device such as albumin, fibrinogen, complement factors, globulins, fibronectin, and vitronectin (Anderson et al., 2008). Mast cells (MCs) present throughout subcutaneous tissue play a critical role in regulating cellular response toward the implanted sensor (Klueh et al., 2010; Thevenot et al., 2011). Upon tissue injury, MCs release stored proinflammatory agents such as vasoactive amines, chemotactic factors, proteases, and NO. This will imitate the recruitment of macrophages and lymphocytes to the site of injury. Upon continued stimulation, the MCs will also synthesize these proinflammatory agents to maintain a vigorous biological response. Continued stimulation could result from micromovement and macromovement of the device, infection, the accumulation of cellular debris, etc. Additionally, vascular endothelial cell growth factor (VEGF) is released that is important for angiogenesis at the site of injury, and cytokines that recruit and activate fibroblasts that will eventually participate in fibrous encapsulation will also be produced (Klueh et al., 2010). As these initial events are taking place, neutrophils are among the first cells recruited to the site of implantation will migrate toward the implanted sensor and begin attempting to phagocytose and degrade the foreign device (Bryers et al., 2012). Neutrophils dump lysosomal proteases, reactive oxygen species, and chemokines that serve as potent signals to recruit cells at the wound site, which further enhances the biological response at the device–tissue interface. Neutrophils have a lower capacity to remove bacteria and debris than macrophages, but they respond much more rapidly to the site of injury than the monocytes that exit the vasculature and mature into functional macrophages near the site of injury after tissue injury has occurred. The neutrophils will persist at the wound for 1–2 days before they are replaced by the matured macrophages and lymphocytes (Bryers et al., 2012).

Key in the biological response toward an in vivo sensor is the role macrophages play at the site of implantation. The conventional description of cellular activity resulting from activated macrophages indicates the release of IL-12 and IL-23, reactive nitrogen and oxygen species (including NO, superoxides, and H_2O_2), and proinflammatory cytokines initially occurs (Chang et al., 2008). They continue to release chemoattractants to recruit more monocytes/macrophages. It has been shown that this is an overly simplified description of the role macrophages play in inflammation and wound healing. There is heterogeneous activation of macrophages that is dependent on many factors that include the nature of the protein layer present and environmental signals that recruit monocytes from circulation (Anderson, 2001). There are two general extremes in the phenotypes of macrophages that will be expressed at the site of an implanted device (Badylak et al., 2008). In normal wound healing, during days 1–3, the proinflammatory macrophages (M1 phenotype or "classically activated" macrophages) will eventually be replaced by or exhibit a plasticity in phenotypic behavior to tissue repairing phenotype that promotes angiogenesis (M2 phenotype or "alternatively activated" macrophages) which allows productive wound healing to proceed after approximately day 3 (Velner et al., 2009). Although the two extremes of macrophage behavior are described as M1 and M2, the reality is the plasticity of expression indicates that a continuum of functions are played by the macrophages and it has not been fully elucidated what regulates the change from proinflammatory activation to the prohealing activation. It is clear that macrophages at the site of implantation possess a heterogeneous population of M1/M2 phenotypes that are highly regulated and both are essential for wound healing that returns the tissue to a normal state (Brown et al., 2012). If macrophages persist in an M1 phenotype and exhibit a state of chronic inflammation, adherent macrophages will fuse to form foreign body giant cells (FBGCs) that further attempt to biodegrade the implanted device. After approximately 1–2 weeks, this chronic inflammation causes local tissue damage and leads to the deposition of a collagen matrix by fibroblasts to encapsulate the implant and isolate it from native tissue (Helton et al., 2011). The nature of this collagen layer is partially dependent on factors such as the implant texture, implant stiffness, movement of the implant within tissue, the specific protein layer adhered to the implant, specific cells adhered to the protein layer, micro and macromovements of the device and how these factors change over time. It is important to recognize that the material properties, topography, and placement in vivo affect the biological response to the implant, but the biological response itself will further influence long-term tissue response (Helton et al., 2011).

THE ROLE NO COULD PLAY

Nitric oxide (NO) is a highly reactive, free radical gas that is endogenously produced by many cells in the body (Calabres et al., 2007; Lowenstein et al., 1994). It plays a major role in hemostasis and inhibiting thrombus formation. NO is continuously released from vascular endothelial cells that line all blood vessels and produce a steady-state level of NO that is estimated to be $0.5–4 \times 10^{-10}\,\mathrm{mol\,cm^{-2}\,min^{-1}}$, depending on the level of stimulation (i.e., caused by branch points in the vasculature, flow rate, etc.) (Vaughn et al., 1998; Radomski et al., 1987a). This NO diffuses in all directions from the endothelial cells and causes relaxation of the smooth muscle cells that surround vessels, thus maintaining normal blood pressure. It also diffuses into the lumen of the blood vessel and will directly interact with platelets to prevent activation along the inner surface of the vessels. NO is known to deactivate platelets by specifically binding to soluble guanyl cyclase (GC-S) and converting to guanosine $5'$ triphosphate (GTP).

GTP is in turn converted to guanosine 3′,5′-monophosphate (Radomski et al., 1987b, 1987c).

NO also plays a significant role in inflammation and recruitment of cells to sites of tissue injury (Sharma et al., 2007; Tripathi et al, 2007). Neutrophils and macrophages both produce NO at the interface of the foreign material as a means to destroy bacteria and pathogens at the site of injury. Additionally, NO has been implicated in longer term cell signaling in controlling MC stimulation and may play a role in shifting macrophage phenotype from the M1 to M2, thereby assisting in bringing inflammation to resolution/wound healing (Tabas, 2010). If the proper level of NO can be released at the sensor/blood or sensor/tissue interface for the proper duration with appropriate timing (i.e., high initial bolus, followed by low continuous release for 7–10 days or pulsed NO release for 3 days, etc.), then the biological response to implanted sensors may be controlled to the point that will allow reliable, in vivo performance of sensors.

NO DONORS AND GENERATORS APPLIED TO SENSORS

Because NO is a highly reactive, free radical gas that is toxic in high doses, it is typically delivered in vivo via donor molecules that will decompose to release NO or catalysts that will generate NO in situ under physiological conditions. Several classes of NO donors have been used and different approaches to release or generate NO have been employed. The ultimate goal of creating NO releasing/generating systems is to produce enough NO at the interface between the sensor and the biological environment that the NO is able to mediate the body's natural defensive responses such that the sensor is able to reproducibly detect the physiological analyte of interest such that therapeutic decisions can be made based on the sensor output. The extent and details of this mediation are dependent on the implant location of the sensor and its intended use (i.e., does the sensor need to perform for 6 hours in the highly controlled environment during surgery or function reliably for 6 months in an ambulatory patient?). Also critical to the application of NO release to improve sensor performance is the need to maintain NO production specifically at the sensor/tissue interface. If small molecules leach from a polymer used to construct a sensor and decompose even a few millimeters from the sensor surface, the NO produced may not be localized where it is most needed to control biological response on the sensing region of the device. Although this is not a comprehensive compilation of NO donors/generators, three major sources of localized NO production will be presented that have been used in such a manner that this sensor/interface NO production is maintained: diazeniumdiolates, S-nitrosothiols, and the catalytic generation of NO from S-nitrosothiols endogenously present in body fluids will be presented.

The development of diazeniumdiolates as NO donors for sensor coatings

One of the most studied moieties utilized for NO production in physiological systems are diazeniumdiolates (also called NONOates). Each mole of diazeniumdiolate releases 2 moles of NO either by means of thermal decomposition or proton-mediated decomposition (Keefer, 2011; Hrabie and Keefer, 2002; Wang et al., 2013; Reynolds et al., 2004). Most diazeniumdiolates used in sensor applications have been formed using a parent secondary amine that is reacted with NO under high pressure (5 atm) for 24–48 hours in the absence of oxygen. Fig. 8.1A shows the structure of (Z)-1-[N-butyl-N-[6-(N-methylammoniohexyl)amino]]-diazen-1-ium-1,2-diolate (DBHD/N_2O_2). The stability and NO release properties of diazeniumdiolates can be controlled by changing the molecular substitution of the parent molecule, by changing the water uptake of the polymer containing the diazeniumdiolate and/or by changing the pH of the polymer matrix containing the NO donor. These donors have been covalently linked to polymer backbones, used as small molecule donors blended within polymer matrices, and covalently linked to solid polymer fillers that can then be blended into various polymer matrices (all discussed later).

The first studies completed that combined NO release with functional sensors used hydrophobic polymer matrices as the outermost coatings for sensor fabrication. Initial work by Espadas-Torre et al. (1997) showed that the analytical performance of a classical membrane-type ion-selective electrode (ISE) for H^+ and K^+ functioned reliably in the presence of a continuous low level of NO released by the incorporation of the small molecule NO donor (Z)-1-[N-methyl-N-[6-(N-methylammoniohexyl)amino]]-diazen-1-ium-1,2-diolate (DMHD/N_2O_2) into a tecoflex polyurethane (PU) with the appropriate ionophores and additives (tridocey-lamine (TDDA) for H^+ detection or Valinomycin (VAL) for K^+ detection, and with potassium tertakis(4-chlorophenyl)borate (KTpClPB) and dioctyl sebacate (DOS) plasticizer). These films capable of sensing H^+ and K^+ and emitting a low level of NO had nearly identical potentiometric analytical performance as films that did not release NO (i.e., 66%PU/30%DOS/0.6%KTpClPB/2.5%VAL films with and without DMHD/N_2O_2 had slopes of 60.3 ± 0.1 and 60.6 ± 0.2 mV per decade and intercepts of −5.13 and −5.44, respectively). It is critically important that the NO releasing component of these films (including the original NO source, the NO that is released and the component of the donor remaining after NO is delivered/generated) do not interfere with the analytical performance of the sensor or the transduction methods of the sensors. Although these sensors were not tested in vivo, the authors suggest that this approach of blending a small molecule NO donor should be applicable to a wide range of materials used in sensors and blood/tissue contacting devices to improve biocompatibility of the devices fabricated from or coated with these polymeric materials.

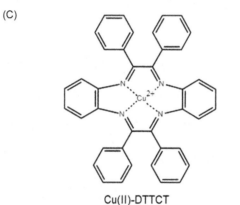

Figure 8.1 The structures of some NO donors used with polymeric materials: (A) (*Z*)-1-[*N*-butyl-*N*-[6-(*N*-methylammoniohexyl)amino]]-diazen-1-ium-1,2-diolate (DBHD/N$_2$O$_2$), (B) *S*-nitroso-*N*-acetyl-D-penicillamine covalently linked directly to PDMS (SNAP-PDMS), and (C) the lipophilic copper complex that entrapped Cu(II) in dibenzo[e,k]-2,3,8,9-tetraphenyl-1,4,7,10-tetraaza-cyclododeca-1,3,7,9-tetraene (Cu(II)-DTTCT) which is capable of catalytically generating NO.

The first in vivo testing of NO releasing sensors was then carried out by incorporating the same small molecule NO donor (DMHD/N$_2$O$_2$), but this time dispersed in polydimethylsiloxane (PDMS) that was used as the outermost coating on an intravascular needle–type amperometric oxygen sensor (Schoenfisch et al., 2000). This coating was freely diffusible to O_2 and simultaneously released a low level of NO. This NO releasing sensor was implanted in the carotid and femoral arteries of dogs with no systemic anticoagulation administered to the animal. The results showed an exciting improvement in blood compatibility toward the intravascular sensors that released NO compared to the blood compatibility of standard medical-grade silicone rubber sensors implanted in the same animal but only moderate improvement in the actual ability of the senor to track blood oxygen levels relative to standard blood gas analysis (i.e., 1 mL samples of blood withdrawn and injected into a standard discrete blood gas analyzer) was observed. Although the actual ability of the sensor to track blood oxygen levels was improved, it was not satisfactory for the demands of clinical needs. It was later determined that a major limitation of these initial studies was that the NO donor was leaching out of the polymer matrix, the sensor was coated with and actually decomposing to release NO in the bulk of flowing blood. This means that the NO was not at exactly the blood/sensor interface when it was released and in fact was probably washed slightly downstream from the sensor surface as a result of blood flow. Additionally, the diamine compound that was present after $DMHD/N_2O_2$ decomposed to release NO formed a detectable level of its corresponding nitrosamine by the reaction of NO and the diamine under physiological conditions. This nitrosamine is a known carcinogen and in general, it is not desirable to form any nitrosamine compounds during physiological applications.

Because of the very promising improvement in blood compatibility, a great deal of effort was put into developing materials that could release NO from the bulk of the polymer, ensuring that NO was present at precisely the blood/sensor interface and simultaneously preventing the formation of nitrosamines in the physiological environment. Three approaches were pursued to release NO at the sensor/tissue interface and prevent the release of by-products that could form toxic compounds. First, the NO donor/generating moiety could be covalently linked to the polymer backbone used for the sensor coating, thus allowing NO to be generated from within the polymer matrix used to fabricate the sensor but preventing the release of the parent diamine. This approach definitely keeps NO at the polymer/tissue interface and prevents leaching of by-products, however, it requires a fundamental change in polymer chemistry to achieve the NO donor system. The second approach is to increase the lipophilicity of the parent diamine compound used to form the diazeniumdiolate in order to prevent leaching of the NO donor out of the polymer matrix. The advantage of this approach is that the level and duration of NO release can be tuned by simply blending different amounts of the donor molecule into the polymer matrix. This approach has the potential to make nearly all polymer matrices NO releasing. The third approach investigated involved the derivatization of polymer fillers with NO release chemistry such that fillers routinely applied for controlling rheology could still perform this function but also impart NO release to the polymer composite. The derivatized fillers have the advantage of being blendable into many different base polymers like lipophilic NO donors, but due to their size relative to the polymer backbone, the fillers will definitively prevent leaching to the parent donor compounds from the polymer composite.

Zhang et al. (2002) covalently linked two different NO donors to the backbone of the PDMS. The strategy employed was to use a trimethoxy silane cross-linking agent that contained diamino or triamino as the fourth arm of the silane (see Fig. 8.2 for the synthetic scheme for diazeniumdiolated N-(6-aminohexyl)-3-aminopropyl-trimethoxysilane linked to 2000 cSt PDMS (DACA-6/N_2O_2)). The ultimate loading and release kinetics of NO could be controlled by varying the concentration of cross-linking agent used as well as the structure of the secondary amine upon which

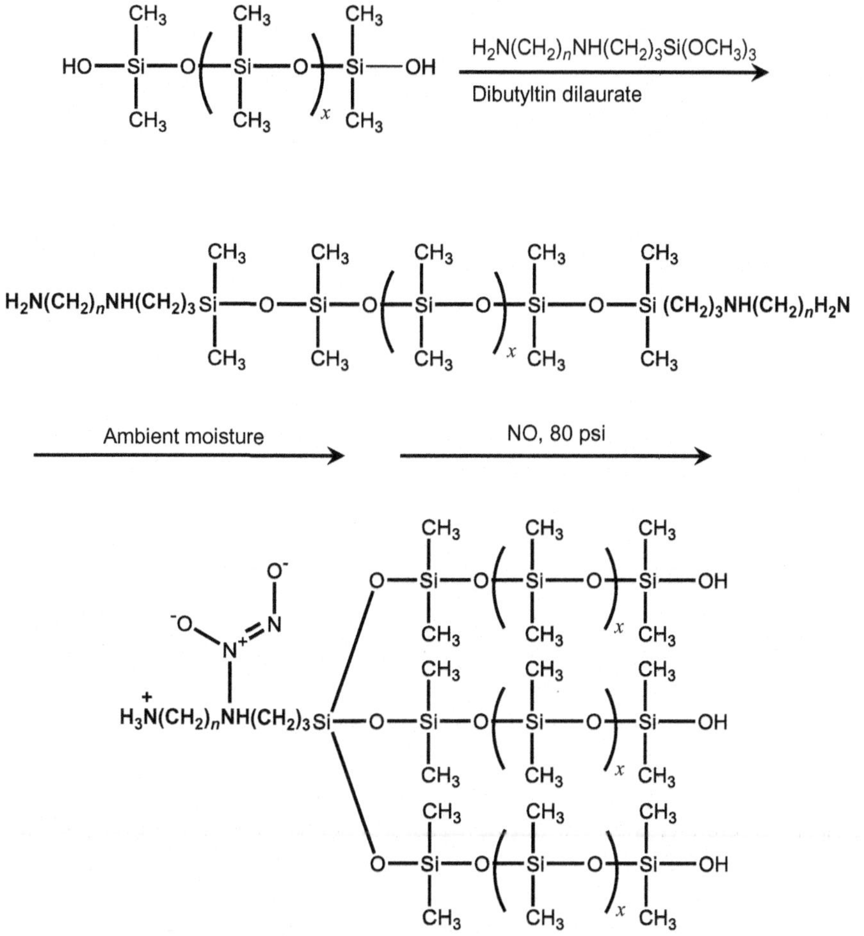

Figure 8.2 The synthetic scheme for diazeniumdiolated N-(6-aminohexyl)-3-aminopropyl-trimethoxysilane linked to 2000 cSt PDMS (DACA-6/N_2O_2). *Reproduced from Zhang, H., Annich, G.M., Miskulin, J., Osterholzer, K., Merz, S.I., Bartlett, R.H., et al., 2002. Nitric oxide releasing silicone rubbers with improved blood compatibility: preparation, characterization, and in vivo evaluation. Biomaterials 23, 1485–1494, with permission from Elsevier.*

the diazeniumdiolate was formed. Additionally the overall properties of the modified PDMS material were tunable by using different molecular weights of hydroxyl-terminated PDMS chains. These materials were cross-linked and cast or coated into their final cured state prior to exposure to 80 psi of NO in an oxygen-free environment for 24–120 hours to form the diazeniumdiolate moiety. Films (1-cm-diameter, 175-mm-thick) cast from DACA-6/N_2O_2, the most stable of the materials developed, released NO at a physiologically relevant level for up to 20 days. The total reservoir of NO available can be increased by simply increasing the thickness of the material present. One drawback of this particular material is that the polymer is not loaded with NO (i.e., is not capable of releasing NO) until it is cured in its final form. If the material is intended to be coated onto sensors (or other medical devices) then the entire device must be subjected to the high-pressure conditions used to form the diazeniumdiolate.

Frost et al. (2002) then used DACA-6/N_2O_2 to coat intravascular oxygen sensor that were tested in vivo and proved that with NO at the sensor/blood interface, hemocompatibility and analytical sensor performance were both radically improved. Clarke-style amperometric sensors were fabricated with and without final loading of NO and tested in the carotid and femoral arteries of swine for over 16 hours with no systemic anticoagulation treatment. The NO releasing sensors produced NO continuously at a level >1×10^{-10} mol cm^{-2} min^{-1}. The sensors were paired such that the carotid arteries and the femoral arteries each had an NO releasing sensor and a control sensor, respectively. The sensors monitored oxygen tension in blood continuously and both sensors were compared to periodic standard discrete blood gas analysis. All the NO releasing sensors ($N = 9$) showed excellent hemocompatibility as demonstrated by the lack of platelet adhesion and thrombus formation observed via scanning electron microscopy on the sensor surface, while the control sensors that did not release NO that were implanted in the same animals showed a great deal of platelet adhesion and activation (see Fig. 8.3). Furthermore, the NO releasing sensors were able to accurately monitor blood oxygen levels throughout the entire 16-hour duration of the experiments when compared to standard blood gas analysis. The oxygen level determined with the non-NO releasing sensors measured oxygen levels that were shown to be statistically different than the standard blood gas analysis. This work established a level of NO needed to achieve inhibition of platelet adhesion and activation (>1×10^{-10} mol cm^{-2} min^{-1}) and clearly indicated that NO present at the blood/polymer interface could indeed prevent thrombus formation. It is worth noting that this level was not established as the upper or lower limit of NO needed to ensure prevention of clot formation, but it was the first report of an empirical, in vivo experimental test that establish an effective level of NO that prevented platelet adhesion and activation. Prior to this, theoretical modeling had estimated NO production form endothelial cells lining healthy blood vessels and had suggested $0.5–4 \times 10^{-10}$ mol cm^{-2} min^{-1} was the surface flux produced endogenously (Vaughn et al., 1998; Radomski et al., 1987a).

(A)

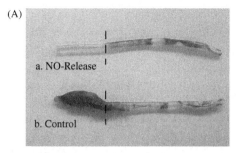

(B)

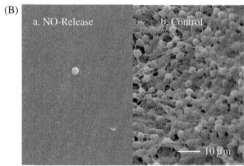

Figure 8.3 An example of an NO releasing oxygen sensor and its corresponding non-NO releasing sensor implanted in the same swine for 16 hours that showed excellent hemocompatibility (panel A) and further demonstrated by the lack of platelet adhesion and thrombus formation observed via scanning electron microscopy on the sensor surface (panel B) compared to the control which showed a great deal of platelet adhesion and activation. *Modified from Keefer, L.K., 2002. Progress toward clinical applications of the nitric oxide-releasing diazeniumdiolates. Annu. Rev. Pharmacol. Toxicol. 43, 585–607 (Keefer, 2002).*

This same DACA/N_2O_2 material was also applied to the fabrication of an optical, fluorescence-based oxygen sensing film (Schoenfisch et al., 2002). The sensor transduction method was based on quenching of fluorescence using the oxygen indicating pair pyrene/perylene as the electron donor/acceptor pair. The goal was to create oxygen sensing films that also simultaneously released NO. Two different NO donor systems were originally tested in single layer films that contained the pyrene/perylene sensing pair and the NO donor (either DMHD/N_2O_2 dispersed in PDMS or DACA/N_2O_2 PDMS). The single layer films suffered from two problems that lead to nonlinear Stern–Volmer plots, first DMHD/N_2O_2 is insoluble in the PDMS matrix, so a heterogeneous film resulted. Second, the excess amine sites present in the polymer matrices resulting from both NO donors quenched the fluorescent oxygen indicating pair. As NO release continued, the concentration of amines present in the polymer sensing film increased, creating an ever-changing level of background quenching that did not correspond to a change in oxygen levels present. In order to overcome the issue of both heterogeneous films and quenching from excess amine content, two layer sensors were fabricated, one layer consisted of DACA/N_2O_2 PDMS and was responsible for

NO release and the second layer cast over the NO release layer consisted of standard PDMS with pyrene/perylene homogenously dissolved in the polymer matrix. This separated the excess amine sites from the oxygen sensing pair and produced sensing films that were able to monitor oxygen while continuously releasing NO with identical calibration curves compared to non–NO releasing control films. This points out another critical issue in the design and fabrication of NO releasing sensors. The NO release moiety and all the resulting by-products present before and after NO release (i.e., parent compound, NO, amines present after NO release) cannot interfere with transduction method of the sensor. Fig. 8.4 shows the sensor configuration and Stern–Volmer plots obtained for the fluorescent oxygen films that monitored oxygen while continuously releasing NO.

In parallel, effort was also put into increasing the lipophilicity of small molecule NO donors that could be blended into a wide variety of polymers but that would not

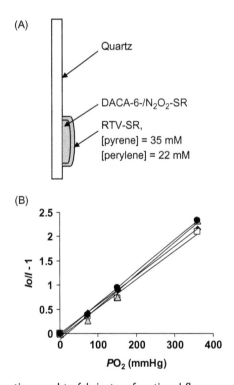

Figure 8.4 Sensor configuration used to fabricate a functional fluorescent oxygen sensor that continuously released NO. The sensing layer was a separate layer under the NO releasing layer to prevent the high amine content in the NO releasing polymer from interfering with the optical sensor. The corresponding Stern–Volmer plots show that the sensor maintains a linear response to oxygen even while releasing NO. *Modified from Schoenfisch, M.H., Zhang, H., Frost, M.C., Meyehoff, M.E., 2002. Nitric oxide-relesing fluoresce-based oxygen sensing polymeric films. Anal. Chem. 74, 5937–5941, with permission from American Chemical Society.*

leach from the polymer matrix into the physiological environment. The advantage of this approach over covalently linking the NO donor to the polymer is the reservoir of NO available for release can be adjusted by changing the amount of donor blended into the polymer cocktail used in sensor fabrication. Batchelor et al. (2003) systematically changed the substitution on dialkylhexamethylenediamine diazeniumdiolates with alkyl groups increasing in size (methyl, ethyl, propyl, butyl, pentyl, or dodecyl) in an effort to retain the NO donor within polymer matrices while still imparting NO release to the plasticized PVC films studied. The NO release kinetics and air stability of these compounds were profoundly influenced by the increasing size of the alkyl substituent. The calculated octanol/water partition coefficients ($\log P$) values varied from 0.97 for methyl alkyl substituents up to 12.6 for the dodecyl substituents. The hexyl and dodecyl substituted diazeniumdiolates were not air stable, indicating there is a limit to how lipophilic these dialkylhexamethylenediamine diazeniumdiolates can be made in this manner. Additionally, it was found that long-term sustained NO release required the addition of a lipophilic tetraphenyl borate to buffer effective proton concentration within the polymer matrix. Diazeniumdiolates are stabilized at elevated pH values and as the molecule decomposes to release NO, the pH of the polymer films increased due to the increased amine content that results from diazeniumiolate decomposition. With the addition of KTpClPB, NO release was sustained for 21 days while NO release significantly decreased after 1 hour without the buffering salt. Although this report did not use these dialkylhexamethylenediamine diazeniumdiolates specifically in sensor applications, (Z)-1-[N-butyl-N-[6-(N-methylammoniohexyl)amino]]-diazen-1-ium-1,2-diolate (DBHD/N$_2$O$_2$) was used as a coating on vascular graft that was tested for 21 days in a sheep model. The NO releasing grafts showed 95% surface area free of thrombus formation compared to controls in the same animal that showed only 42% surface area free of thrombus formation. Frost et al. (2003) published preliminary results using 5% (w/w) DBHD/N$_2$O$_2$ with KTpClPB as the NO donor in PDMS as the outer coating on catheter-type amperometric oxygen sensors that showed excellent sensor performance and hemocompatibility compared to non-NO releasing control sensors implanted in the same animal for 16 hours. Fig. 8.5 shows the oxygen level measured by the NO releasing sensor and the non-NO releasing control sensor both compared to standard discrete blood gas analysis as well as scanning electron micrographs of the senor surfaces after 16 hours of implantation. The NO releasing sensor determined oxygen levels that tracked well with the discrete analysis and the sensor surface showed greatly reduced cellular adhesion compared to the control sensor.

Rather than trying to make small molecule diazeniumdiolates more lipophilic, Zhang et al. (2003) tethered the small molecule NO donors to 7-10 nm sized silica particles. The approach limited leaching of the NO donors out of polymer matrices by mechanical entrapment of the supporting particles. The synthesis was accomplished by covalently linking alkylamines to the surface of the particles and then converting the

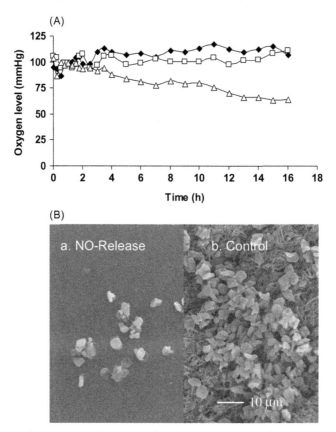

Figure 8.5 Intravascular oxygen sensors coated with PDMS doped with DBHD/N$_2$O$_2$ showed better analytical performance compared to conventional blood gas analysis than non-NO releasing control sensors placed in the same swine for 16 hours (panel A) and the corresponding scanning electron micrographs of the sensor surfaces demonstrate both improved analytical performance of the sensor with NO release and better hemocompatibility (panel B). *Reproduced from Frost, M.C., Batchelor, M.M., Lee, Y., Zhang, H., Kang, Y., Oh, B., et al., 2003. Preparation and characterization of implantable sensors with nitric oxide release coatings. Microchem. J. 74, 277–288, with permission from Elsevier.*

amine groups to their corresponding diazeniumdiolates by reaction with NO at 5 atm pressure with methoxide bases present under the exclusion of oxygen. The alkylamines used were 3-aminopropyltrimethoxysilane, *N*-methyl-3-aminopropyltrimethoxysilane, *N*-(2-aminoethyl)-3-aminopropyltrimethoxysilane, and *N*-(6-aminohexyl)-3-amino-propyltrimethoxysilane with added sodium methoxide or potassium methoxide bases. The addition of the sodium methoxide counter ion increased NO loading by twofold, and also increased cleavage of the silane bonds that tethered the amine compounds to the surface of the particles. This limited the amount of base that could be used to increase NO loading. *N*-(6-aminohexyl)-3-aminopropyltrimethoxysilane tethered to particles was able to store 0.28 ± 0.03 mmol NO per g of particles. Although these NO releasing particles were not tested in a sensor application, they were used in an in vivo/ex vivo extracorporeal circulation experiment. The particles were dispersed in a polyurethane coating (20 wt%) that was applied to the inner surface of a polyvinyl chloride tube and used in a 4-hour rabbit experiment. The NO flux that resulted from the inner surface of the coated tubing was 4.1×10^{-10} mol cm^{-2} min^{-1}. Thrombus formation and activation of adhered platelets was significantly reduced. To an certain extent, this approach shares the advantage of tailoring NO release by simply changing the mass of NO releasing particles in the polymer matrix, but is somewhat limited because the addition of particles has a significant impact on the mechanical properties of the cured polymer.

As the application of NO release to blood contacting sensors continued to show great promise and advances were made in terms of understanding and developing materials that retained the NO donors within the polymer matrix of the sensor coating and empirical measurements of NO inhibiting thrombus formation confirmed the validity of this approach for improving sensor performance, subcutaneous sensors were also developed with the hopes of controlling the inflammatory response around the implanted sensor. Gifford et al. (2005) tested the effectiveness of reducing inflammation around a needle-type enzyme-based electrochemical glucose sensor implanted in Sprague-Dawley rates for 6–48 hours. The NO release coating applied to these sensors was extremely thin and released NO continuously for 12–18 hours. The sensor signals from the NO releasing sensors were more stable and showed better sensitivity over the duration of implantation compared to the control sensors that did not release NO. This is important because the NO release sensors did not release NO for the entire duration of implantation, but appear to functionally perform better than control sensors implanted in the same animal. This points to NO mediating the initial biological response toward the implanted sensors, suggesting that NO will indeed positively influencing sensor performance. Histological examination showed a statistically significant reduced number of neutrophils infiltrating the site of implantation around the NO sensors compared to control sensors in the same animal after 24 hours. After 48 hours, the neutrophil infiltration was not significantly different between the NO releasing sensors and the control sensors, which may coincide with the complete exhaustion of NO release from this initial proof of principle coatings. Overall, this work clearly indicates that this may be a highly effective approach to improving subcutaneous sensor performance. Additionally, this work demonstrated that NO release has the potential to be compatible with enzyme-based sensor transduction methods.

Yan et al. (2011) published a report of intravascular glucose/lactate sensors using DBHD/N$_2$O$_2$ as the NO donor doped into poly(lactide-co-glycolide) (PLGA) as the active sensor coating. The outermost layer of the enzyme-based sensor was PurSil (polyurethane/dimethylsilxoane copolymer). The NO releasing layer (ca. 30 mm thick with 33 wt% DBHD/N$_2$O$_2$) was capable of releasing 1×10^{-10} mol cm^{-2} min^{-1} for

7 days without influencing the sensor performance (using either glucose oxidase or lactate oxidase) in the presence of ascorbic acid as an interferent. The release of NO from DBHD/N_2O_2 is proton meditated, previous work added a buffering salt (KTpClPB) to the polymer matrix to counteract the increasing pH that resulted from the accumulation of parent diamine compounds that resulted after NO release. In this case, the PGLA matrix itself will decompose to form lactic acid and glycolic acid, thus providing an acidic environment that allows proton-mediated NO release from the DBHD/N_2O_2 to continue. This demonstrates that controlling polymer degradation of the resulting microenvironment is a viable means to control NO release with diazeniumdiolate NO donors. The NO releasing glucose sensors were then implanted in the veins of rabbits for 8 hours to test hemocompatibility. These sensors showed greatly reduced thrombus formation on the sensor surface and sustained NO release resulting from the acidic environment created by the decomposing PLGA.

The huge promise of improved in vivo performance of implanted sensors constructed with hydrophobic polymers then lead to the development on NO releasing hydrophilic sol–gel derived materials. Marxer et al. (2004) developed an amperometric oxygen sensor that consisted of one of three aminosilane/ethyltrimethoxysilane (ETMOS3) hydrid xerogel doped with hydrophilic polyurethane (HPU). The material was coated onto a platinum working electrode and after it was fully cured, the NO donating diazeniumdiolate was formed in situ by placing the electrode under 5 atm of NO pressure for 3 days in the absence of air and water. Oxygen sensors fabricated with 20% N-(6-aminohexyl)aminopropyltrimethoxysilane (AHAP3) (balance ETMOS3) with 0.1%, 0.5%, and 1.0% HPU demonstrated sensitivities of ~0.3, 4.2, and 6.5 nA per mmHg, respectively, after 24 hours soaking in phosphate-buffered saline. The 1.0% HPU-doped coatings showed the greatest sensitivity and fastest hydration time of the three coatings tested. The sensors were characterized by rapid response time ($t_{90} < 30s$) and reproducible calibration curves over 24 hours. One concern with the materials is the potential for the xerogel to crack upon hydration. It was found that minimal fragmentation occurred with 20% AHAP3 and sufficient NO release was obtained to suggest this material may be a useful hydrophilic coating for sensors, although more development work was suggested by the authors to increase the NO loading of the material.

Shin et al. (2004) then built on this work and developed a sol–gel particle/polyurethane (PU) glucose sensor. The specific sensor configuration consisting of glucose oxidase (GOx) immobilized in a sol–gel layer protected with a PU coating, followed by diazeniumdiolated sol–gel particles suspended in PU and then an outer coating of PU as shown in Fig. 8.6. This configuration was required to counteract the drastic loss in sensitivity that was observed when the GOx was exposed to the harsh conditions necessary to form the NO donor. Additionally, it was found that exposure to NO irreversibly decreased the permeability of the sol–gel to hydrogen peroxide (H_2O_2),

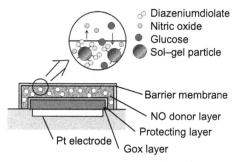

Figure 8.6 Schematic of the hybrid sol–gel/PU glucose sensor using NO donor-modified sol–gel particles dispersed in the PU supporting matrix. *Reproduced from Shin, J.H., Marxer, S.M., Schoenfisch, M.H., 2004. Nitric oxide-releasing sol–gel particles/polyurethane glucose biosensors. Anal. Chem. 76, 4543–4549, with permission from American Chemical Society.*

the analytical agent actually measured with GOx-based biosensors, regardless of the pressure used to form the diazeniumdiolate. The authors speculate that NO exposure enhances polycondensation of the residual silanol groups present. By using the preformed diazeniumdiolate sol–gel particles suspended in PU, the aggregate effect allowed simultaneous NO release and diffusion of H_2O_2 as well as functional GOx to remain in the sensor. The analytical performance of the hybrid NO releasing glucose biosensor remained stable for 18 days, after which the linear response range decreased from 0–60 to 0–20 mM glucose with a response time ($t_{95\%}$) increasing from <20 s to over 65 s, and after 22 days, the sensor performance completely degraded. This points out another critical design feature that should be noted in the development of NO releasing sensors. The fact that NO is a highly reactive free radical gas means that over time, the release of NO could cause continual changes in the sensors due to changes in material properties (i.e., NO itself increased cross-linking in the sol–gel material and changed the permeability to the analyte of interest, thereby influencing analytical performance of the sensor).

In order to overcome some of these limitations, a miniaturized glucose sensor was developed with a micropatterned outer layer that released NO in 5-μm-wide lines separated by 5 or 20 μm was reported by Oh et al. (2005). The idea was that most of the surface will remain unmodified, thereby allowing glucose sensing to take place, while NO generation is confined to specific locations that will allow sufficient diffusion of NO around the sensor surface to inhibit thrombus formation and bacterial adhesion. The sensor glucose response decreased by 64% when the modified xerogels were spaced 5 μm apart relative to a control sensor, while lines spaced 20 μm apart showed a 36% decrease in response compared to the control sensors. This was a vast improvement over xerogel films covering the entire sensor surface that saw >99% loss in sensitivity relative to control sensors. Bacterial adhesion to the micropatterned surfaces was tested and the NO releasing surfaces showed 70–80% reduction in the adhesion of *P. aeruginosa* compared to micropatterned surfaces that did not release NO. Similarly, the NO releasing microarrays showed >40% decrease in platelet adhesion in an in vitro assay while the control surfaces showed a high degree of a platelet adhesion. Although these micropatterned glucose biosensors had less than optimal sensor performance and showed some bacterial and platelet adhesion, it is a very interesting and potentially useful approach to begin trying to develop sensors that contain separate sensing region(s) integrated with NO generating region(s). Specifically, this points to the possibility that the entire sensor surface may not need to release NO to improve analytical performance and biocompatibility of in vivo sensors. If topographical features and processing conditions of the NO releasing microarrays could be optimized, a wider array of transduction methods may be able to be integrated with NO release strategies.

Because it was now clear that permeability of xerogels to both H_2O_2 and glucose was drastically reduced by the formation of the NO donor, poly(vinylpyrrolidone) (PVP) was added to the xerogel to enhance permeability of the NO-loaded material (Schoenfisch et al., 2006). The analytical response of the glucose biosensor and NO release characteristics were both present in the PVP-doped xerogel sensors. Several concentrations of PVP added to different xerogel compositions were systematically studied and it was found that a glucose biosensing membrane consisting of 20% (v/v) *N*-(2-aminoethyl)-3-aminopropyltrimethoxysilane/isobutyltrimethoxysilane xerogel doped with 20% (w/w) PVP, then treated with 5 atm of NO for 48 hours maintained sensor performance and stability for 17 days. Although H_2O_2 and glucose permeability were improved by the addition of PVP, the problem of noncovalently linked additives again presents the possibility of leaching membrane components into the physiological environment if these sensors were implanted. The composite membranes were soaked for up to 7 days and did in fact show leaching of silane compounds into the bathing solution (111 ± 10 and $171 \pm 15 \, pmol \, Si \, cm^{-2}$, respectively, for 4 and 7 days of soaking the 20% (v/v) *N*-(2-aminoethyl)-3-aminopropyltrimethoxysilane/isobutyltrimethoxysilane xerogel doped with 20% (w/w) PVP in PBS). In general, the films showed less leaching after exposure to NO, which is consistent with the increased cross-linking of silanol groups catalyzed by NO as previously hypothesized (Shin et al., 2004).

S-Nitrosothiols (RSNOs)

Another potentially useful class of NO donor that could be utilized to improve the biological response toward implanted sensors are *S*-nitrosothiols (RSNOs). *S*-nitrosothiols are a natural carrier of NO in biological systems (Zhang and Hogg, 2000). There are three published decomposition mechanisms for this class of NO donor (Williams, 1999). First, copper-mediated decomposition will rapidly catalyze the release of NO from RSNOs when Cu^{2+} is reduced to Cu^+, and Cu^+ will react with RSNOs to liberate NO, regenerate Cu^{2+} and ultimately form disulfide bonds. The second reaction mechanism to release NO from RSNOs is photolytic cleavage of the S–N bond of the RSNOs to produce NO and thiolate groups. The final reaction mechanism that has been reported to release NO from RSNOs is the direct reaction of high concentrations (>1 mm) ascorbate with RSNOs to produce disulfide bonds and NO. An inherent advantage of using RSNOs as an NO donor for implanted devices is that these carriers are in fact one of the ways the body naturally shuttles NO within living systems. Great care must be taken to eliminate all trace transition metal contaminants present in order to prevent premature NO release. Although Cu^{2+}, Fe^{2+}, Ag^+, and Hg^{2+} are all well known to release NO, other transition metals including Co^{2+}, Ni^{2+}, and Zn^{2+} were also recently reported to generate NO from RSNOs(McCarthy et al., 2016a).

Several early reports used RSNOs as the NO donor in a variety of materials that suggested the utility of this class of molecule for the creation of NO releasing materials (Shishido and de Oliveria, 2000; Etchenique et al., 2000; Shishido et al., 2003). The primary limitations of these early materials were the relatively small reservoir of NO and the lack of controlled release once NO generation was initiated. Additionally, the same concerns for potential leaching of NO donors exist when using RSNOs as the source of NO as were demonstrated with diazeniumdiolates. Since these reports, many new materials have been developed that utilize RSNOs as an NO donor (Bohl and West, 2000; Damodaran et al., 2012; Joslin et al., 2014; Lutzke et al., 2014; McCarthy et al., 2016b; Riccio et al., 2009; Yapr et al., 2015), although this is clearly not an exhaustive review of all RSNO containing materials, the following discussion will focus on materials that were designed for use with implanted sensors.

In order to both limit leaching of the NO donor and to create a system that will allow control over NO release, Frost and Meyerhoff (2004) reported the synthesis of a polymer filler with S-nitroso-N-acetyl-DL-penicillamine tethered to the surface of the fumed silica particles (SNAP-FS). These particles were then blended into PDMS and cast into trilayer films composed of the middle layer containing 20 wt% SNAP-PDMS sandwiched between two layers of plain PDMS. Because the NO donor was covalently linked to the particles mechanically entrapped in the polymer matrix, the donor was unable to leach from the cured film. The PDMS outer layers prevented transition metal ions from diffusing into the middle layer and thus prevented uncontrolled NO release from the RSNO. Controlled NO release was achieved by using light to trigger photolytic cleavage of the S-NO bond, thereby providing a means of controlling the level and duration of NO generated from the material. The higher the intensity of light used to illuminate the polymer composite, the higher the level of NO generated and when the light was turned off, NO generation ceased. The authors published full characterization of SNAP-FS as well as two structural analogs, S-nitrosocysteine and

S-nitroso-*N*-acetyl-L-cysteine each tethered to fumed silica particles (NO-Cys-FS and SNAC-FS, respectively) (Frost and Meyerhoff, 2005). All three sets of particles demonstrated different storage and release properties of NO when blended into PDMS and PU materials. Although these materials were not directly applied to the fabrication of sensors, the ability to control NO generation by the application of light provides a means to externally modulate after the system was fully cured and assembled. In all previous cases described, the release of NO was determined by how much donor was utilized prior to finished curing of the device and NO release could not be stopped or increased/decreased if it was no longer needed or was not at the optimal level required for the medical application.

Gierke et al. (2011) then developed *S*-nitroso-*N*-acetyl-D-penicillamine covalently linked directly to PDMS (SNAP-PDMS), shown in Fig. 8.1B. This polymer had a larger reservoir of NO available independently of changing the mechanical properties of the polymer as occurs with the addition of NO releasing fillers. Films of SNAP-PDMS (~100-µm-thick) were illuminated with 470 nm LED light and showed highly controlled NO release ranging from a surface flux of 0–3.5 × 10^{-10} mol cm^{-2} min^{-1}. This was the first time an inexpensive LED light source was shown to be effective at photolytic cleavage of the RSNO to release NO. This polymer was then applied to declad polymethylmethacrylate (PMMA) optical fibers and the illumination was controlled with a wirelessly controlled LED platform (Starrett et al., 2012). Fig. 8.7 shows the NO release generated from the SNAP-PDMS coated fibers. This clearly demonstrates the ability to control NO release from the material in a continuous manner external to the device. Although this system was not applied to a functional sensor, the authors suggest the possibility of including bundled optical fibers where some fibers are sensing specific analytes of interest while other fibers are generating NO to mediate the biological response to the implanted sensor. This system also provide a tool to allow careful investigation of exactly how much NO is necessary to result in the desired control over biological response. Questions can now be explored to determine if NO release needs to be continuous or could it be turned on and off, thereby extending the useful life time of the NO reservoir of an implanted device. Also, is it optimal to release NO continuously, at a steady state or would a higher NO surface flux initially help calm the inflammatory response triggered upon implantation and then a lower level may be suitable to maintain minimal biological response and speed wound healing around the sensor? A third significant question that can be answered is what is the long-term duration of NO needed? If the NO release is tailored to allow optimized wound healing, could a long-term sensor such as a glucose sensor be fabricated to release NO for 10 days and then no longer require NO because the sensor/tissue interface has been stabilized by completion of normal wound healing? The ability to turn NO release on and off and modulate the level of NO released during in vivo implantation makes it feasible to answer these questions and refine the design of NO releasing materials to specifically match the physiological site of implantation.

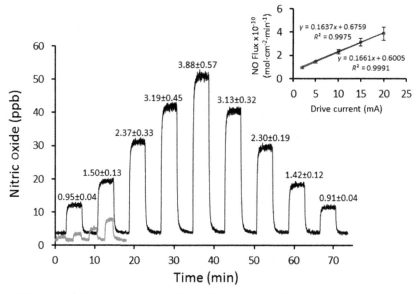

Figure 8.7 Demonstration of the controlled NO release achieved by illuminating a 500 µm poly(methyl methacrylate) (PMMA) optical fiber consisting of a declad region of the fiber coated with SNAP-PDMS. As the current applied to the illuminating LED was increased, the intensity of light increased and the corresponding surface flux of NO generated from the fiber increased. *Reproduced from Starrett, M.A., Nielsen, M., Smeenge, D.M., Romanowicz, G.E., Frost, M.C., 2012. Wireless platform for controlled nitric oxide releasing optical fibers for mediating biological response to implanted devices. Nitric Oxide 27, 228–234, with permission from Elsevier.*

Catalytic generation from endogenous RSNOs

Possibly the most significant limitation using NO release to improve sensor performance is the finite reservoir of NO available with all the materials developed that include a specific NO donor. In order to overcome this limitation, several groups have worked to develop materials that are capable of locally generating NO at the sensor/tissue interface using naturally occurring NO donor species present in physiological systems. The idea is to include a catalytic moiety that is able to repeatedly generate NO as long as a substrate such as an endogenous RSNO is present. Duan and Lewis (2002) first suggested this by covalently linking L-cysteine to the surface of PU and polyethylene terephthalate (PET). The approach relies on the natural transfer of NO from endogenously present RSNOs such as S-nitrosoalbumin (AlbSNO). The NO is transferred from the protein carrier to the thiol on the cysteine and then the unstable S-nitrosocysteine (CysNO) will decompose to release NO at the polymer/solution interface. Both cysteine-modified polymers tested showed NO generation at the surface and the ability to inhibit platelet adhesion on the polymer surface relating to control materials that did not contain the thiol group. In theory, as long as the patient was producing RSNOs that are part of normal metabolism, there should be an infinite supply of NO donor available to locally generate at specific artificial surfaces to improve sensor performance. Long-term studies were not performed with this material to assess effects of disulfide bond formation with the immobilized cysteine nor the effects of protein adsorption on the surface tested in vivo that has the potential to block the access of RSNOs to the thiol sites.

Oh and Meyerhoff developed a lipophilic copper complex that entrapped Cu(II) in dibenzo[e,k]-2,3,8,9-tetraphenyl-1,4,7,10-tetraaza-cyclododeca-1,3,7,9-tetraene (Cu(II)-DTTCT) (see Fig. 8.1C) doped into PVC plasticized with o-nitrophenyloctyl ether (o-NPOE) catalytically generated physiologically relevant levels of NO from nitrite (Oh and Meyerhoff, 2004) and S-nitrosoglutathione (GSNO) (Oh and Meyerhoff, 2003). The system requires the presence of a reducing agent such as ascorbic acid or free thiols such as glutathione to cycle Cu(II) to the active Cu(I) form. The catalytic center will continue to be active as long as reducing equivalents are present, which overcome one potential limitation with the tethered thiol system first tested by Daun and Lewis. However, there is a potential risk of protein fouling interfering with the ability of a reducing agent and an RSNO from reaching the Cu(I) active center at the surface of the doped polymer. To overcome this issue, Hwang and Meyerhoff (2008) tethered a Cu(II)–cyclen complex to the surface of PU. The complex was grafted onto the commercially available medical-grade polyurethane, Tecophilic SP-93A-100. Direct NO measurements showed that NO was generated when GSNO and CysNO at physiological pH in the presence of thiol reducing agents (GSH and Cys). Because GSNO and CysNO are both present in blood as well as the corresponding thiol reducing equivalents (GSH and Cys), the implication is that this polymer should generate NO in blood. The authors tested this material in fresh whole blood by using it as the outer membrane of an aerometric NO sensor. Hwang and Meyehorr (2006) also demonstrated that Cu(II)-cyclen could be covalently attached to cross-linked poly(2-hydroxyethyl methyl acrylate) (pHEMA) to create a high water uptake material that was capable of catalytically generating NO from naturally occurring RSNOs present in physiological fluids.

To expand the potential of catalytic NO generation, Cha and Mayerhoff (2007) developed an organoselenium species that is capable of generating NO from RSNOs to release NO and generate the corresponding free thiol group from the parent donor. The authors demonstrated that in the presence of sufficient reducing equivalents, GSNO, CysNO, AlbSNO, SNAC, and S-nitroso-penicillamine (SPA) all decomposed to release NO. Additionally, the organoselinium material continued to catalytically generate NO after being stored for up to 5 days in porcine plasma. This indicates that protein deposition on the material does not inhibit the catalytic generation of NO. Hwang and Meyerhoff (2007) then reported that organoditelluride (5,5′-ditelluro-2,2′-diohenecarboxylic acid and its polymeric derivative) catalytically generated NO from GSNO or CysNO in the presence of GSH or Cys.

The theory of in vivo generation of NO from catalytic NO generators was validated when it was demonstrated that intravascular electrochemical oxygen sensors monitored the partial pressure of oxygen in porcine arteries for 19–20 hours compared to non-NO generating controls (Wu et al., 2007). Copper particles (3 μ m and 80 nm) were embedded in a Tecoflex SG-80A polyurethane layer and coated on Clarke-style amperometric oxygen sensors. Although overall, the catalyst containing (copper particle) sensors had much less clot formation on the surfaces and monitored oxygen more accurately, some of the sensors did have particle thrombus formation on their surfaces. The authors suggest these could be cases where the animals may have had less RSNO circulating in their blood. This points out a critical limitation of this approach to local in situ generation of NO—the total NO production is limited by how much circulating RSNO is present. If a higher level of NO is needed to mediate biological response, it cannot be increased without the exogenous addition of RSNOs or reducing equivalents.

CONCLUSION AND FUTURE DIRECTIONS

To date, the combination of NO release at the interface between sensor and the biological environment has shown exciting promise as a means to allow real-time, continuous monitoring that has the potential to allow great advances in human health. Outstanding research has developed materials that can be applied to many medical devices. A broad range of NO donors have been shown to be capable of producing the general levels of NO needed to mediate biological response in blood and tissue. The strategies of using discrete donors and catalytically generating NO from endogenous sources from hydrophobic and hydrophilic polymers indicate that a range of NO release can be obtained from immediate bolus release and long-term generation of NO. This should provide tools to address a wide array physiological situations. Specifically in the case of NO releasing sensors, NO release has been shown to be compatible with electrochemical gas and ion sensors, enzyme-based sensors, and optical sensors. The next big challenge is translating this research to clinical utility to understand how much NO is needed and for what duration it should be delivered to ensure reliable sensor readings that allow therapeutic decisions to be made. This understanding also needs to be linked to different physiological states of patients who are already under physiological stress. For example, diabetic patients suffer from systemically poor circulation and will have altered wound healing response to implanted sensors compared to healthy patients. Other disease conditions may lower levels of circulating RSNOs, so if a sensor is relying on catalytic generation of NO from endogenous RSNOs, there needs to be clear understanding of what levels of NO can be obtained in vivo and if this will be sufficient to ensure proper sensor function. Ideally, our understanding will reach the level that will allow the necessary dose and duration of NO to be delivered that will bring disrupted tissue to the point of complete wound healing around an implanted sensor such that NO release or production will not be required for long-term functionality. This is a very challenging goal, but if it is accomplished, will revolutionize patient monitoring.

REFERENCES

Amara, U., Flierl, M.A., Rittirsch, D., Klos, A., Chen, H., Acker, B., et al., 2010. Molecular intercommunication between the complement and coagulation systems. J. Immunol. 185, 5628–5636.

Anderson, J.M., 2001. Biological responses to materials. Ann. Rev. Mater. Res. 31, 81–110.

Anderson, J.M., Rodriguez, A., Chang, D.T., 2008. Foreign body reaction to biomaterials. Semin. Immunol. 20, 86.

Badylak, S.F., Valentin, J.E., Ravindra, A.K., McCabe, G.P., Stewart-Akers, A.M., 2008. Macrophage phenotype as a determinant of biologic scaffold remodeling. Tissue Eng. A 14, 1835–1842.

Barnthip, N., Parhi, P., Golas, A., Volger, E.A., 2009. Volumetric interpretation of protein adsorption: kinetics of protein adsorption competition from binary solution. Biomaterials 30, 6495–6513.

Batchelorr, M.M., Reoma, S.L., Fleser, P.S., Nuthakki, V.K., Callahan, R.E., Shanley, C.J., et al., 2003. More lipophilic dialkyldiamine-based diazeniumdiolates: synthesis, characterization, and application in preparing thromboresistant nitric oxide release polymeric coatings. J. Med. Chem. 46, 5153–5161.

Bohl, K.S., West, J.L., 2000. Nitric oxide-generating polymers reduce platelet adhesion and smooth muscle cell proliferation. Biomaterials 21, 2273–2278.

Brown, B.N., Ratner, B.D., Goodman, S.B., Amar, S., Badylak, S.F., 2012. Macrophage polarization: an opportunity for improved outcomes in biomaterials and regenerative medicine. Biomaterials 33, 3792–3802.

Bryers, J.D., Giachelli, C.M., Ratner, B.D., 2012. Engineering biomaterials to integrate and heal: the biocompatibility paradigm shifts. Biotechnol. Bioeng. 109, 1898–1911.

Calabrese, V., Mancuso, C., Calvani, M., Rizzarello, E., Butterfield, D.A., Stella, A.M.G., 2007. Nitric oxide in the central nervous system: neuroprotection versus neurotoxicity. Nat. Rev. Neurosci. 8, 766–775.

Cha, W., Meyerhoff, M.E., 2007. Catalytic generation of nitric oxide from S-nitrosothiols using immobilized organoselenium species. Biomaterials 28, 19–27.

Chang, D.T., Jones, J.A., Meyerson, H., Colton, E., Kwon, I.K., Matsuda, T., et al., 2008. Lymphocyte/macrophage interactions: biomaterial surface-dependent cytokine, chemokine, and matrix protein production. J. Biomed. Mater. Res. 87A, 676–687.

Damodaran, V.B., Josiln, J.M., Wold, K.A., Lantvit, S.M., Reynolds, M.M., 2012. S-nitrosated biodegradable polymers for biomedical applications: synthesis, characterization and impact of thiol structure on the physiochemical properties. J. Mater. Chem. 22, 5990–6001.

Dee, K.C., Puleo, D.A., Bizios, R., 2002a. An Introduction to Tissue–Biomaterial Interactions. John Wiley & Sons, Inc, Hoboken, NJ.68–81

Dee, K.C., Puleo, D.A., Bizios, R., 2002b. An Introduction to Tissue–Biomaterial Interactions. John Wiley & Sons, Inc, Hoboken, NJ. p.185–189

Dörmann, D., Clemetson, K.J., Kehrel, B.E., 2000. The GPIb thrombin-binding site is essential for thrombin-induced platelet procoagulant activity. Blood 96, 2469–2478.

Duan, X., Lweis, R.S., 2002. Improved haemocompatibility of cysteine-modified polymers via endogenous nitric oxide. Biomaterials 23, 1197–1203.

Ekdahl, K.N., Lambris, J.D., Ewing, H., Ricklin, D., Nilsson, P.H., Teramura, Y., et al., 2011. Innate immunity activation on biomaterials surfaces: a mechanistic model and coping strategies. Adv. Drug Delivery 63 (12), 1042–1050.

Etchenique, R., Furman, M., Olabe, J.A., 2000. Photodelivery of nitric oxide from a nitrosothiol-derivatized surface. J. Am. Chem. Soc. 122, 3967–3968.

Espadas-Torre, C., Oklejas, V., Mowery, K., Meyerhoff, M.E., 1997. Thromboresistant chemical sensors using combined nitric oxide release/ion sensing polymeric films. J. Am. Chem. Soc. 119, 2321–2322.

Frost, M.C., Meyerhoff, M.E., 2004. Controlled photointiated release of nitric oxide from polymer films containing S-nitroso-N-acetyl-DL-penecillamine derivatized fumed silica filler. J. Am. Chem. Soc. 126, 1348–1349.

Frost, M.C., Meyerhoff, M.E., 2005. Synthesis, characterization, and controlled nitric oxide release from S-nitrisithiol-derivatized fumed silica polymer fillers. J. Biomed. Mater. Res. 72A, 409–419.

Frost, M.C., Meyerhoff, M.E., 2015. Real-time monitoring of critical care analytes in the bloodstream with chemical sensors: progress and challenges. Ann. Rev. Anal. Chem. 8, 171–192.

Frost, M.C., Batchelor, M.M., Lee, Y., Zhang, H., Kang, Y., Oh, B., et al., 2003. Preparation and characterization of implantable sensors with nitric oxide release coatings. Microchem. J. 74, 277–288.

Frost, M.C., Rudich, S.M., Zhang, H., Maraschio, M.A., Meyerhoff, M.E., 2002. In vivo biocompatibility and analytical performance of intravascular amperometric oxygen sensors prepared with improved nitric oxide-releasing silicone rubber coatings. Anal. Chem. 74, 5942–5947.

Frost, M.C., Wolf, A.K., Meyerhoff, M.E., 2013. In vivo sensors for continuous monitoring of blood gases, glucose and lactate: biocompatibility challenges and potential solutions. In: Clinical Diagnosis, P., Vadgama, Peteu, S. (Eds.), Detection Challenges. RSC Publishing, pp. 125–155.

Ganter, M., Zollinger, A., 2003. Continuous intravascular blood gas monitoring: development, current techniques, and clinical use of a commercial device. Brit. J. Anaesth. 91, 397–407.

Ghebrehiwet, B., Randazzo, B.P., Dunn, J.T., Silverberg, M., Kaplan, A.P., 1983. Mechanisms of activation of the classical pathways of complement by Hageman factor fragment. J. Clin. Invest. 71, 1450–1456.

Gierke, G.E., Nielsen, M., Frost, M.C., 2011. S-nitroso-N-acetyl-D-penicillamine covalently linked to polydimethylsiloxane (SNAP-PDMS) for use as a controlled photoinitiated nitric oxide release polymer. Sci. Technol. Adv. Mater. 12 55007 (5pp).

Gifford, R., 2013. Continuous glucose monitoring: 40 years, what we've learned and what's next. Chem. Phys. Chem. 14, 2032–2044.

Gifford, R., Batchelor, M.M., Lee, Y., Gokulrangan, G., Meyerhoff, M.E., Wilson, 2005. Mediation of in vivo glucose sensor inflammatory response via nitric oxide release. J. Biomed. Mater. Res. 75A, 755–766.

Goldberger, G., Thomas, M.L., Tack, B.F., Williams, J., Colten, H.R., Abraham, G.N., 1981. NH2-terminal structure and cleavage of guinea pig pro-C3, the precursor of the third compliment component. J. Biol. Chem. 256, 12617–12619.

Gorbet, M.B., Sefton, M.V., 2004. Biomaterials-associated thrombosis: roles of coagulation factors, complement, platelets and leukocytes. Biomaterials 25, 5681–5703.

Helton, K.L., Ratner, B.D., Wisniewski, N.A., 2011. Biomechanics of the sensor-tissue interface—effects of motion, pressure, and design on sensor performance and the foreign body response—Part I: theoretical framework. J. Diabetes Sci. Technol. 5, 632–646.

Hrabie, J.A., Keefer, L.K., 2002. Chemistry of the nitric oxide-releasing diazeniumdiolate ("nitrosohydroxylamine") functional group and its oxygen-substituted derivatives. Chem. Rev. 102, 1135–1154.

Hwang, S., Cha, W., Meyerhoff, M.E., 2006. Polymethylacrylates with covalently linked Cu^{II}-cyclen complex for in situ generation of nitric oxide from nitrosothiols in blood. Angew. Chem. Int. Ed. 45, 2745–2748.

Hwang, S., Meyerhoff, M.E., 2007. Organoditelluride-mediated catalytic S-nitrosothiol decomposition. J. Mater. Chem. 17, 1462–1465.

Hwang, S., Meyerhoff, M.E., 2008. Polyurethane with tethered copper(II)-cyclen complex: preparation, characterization and catalytic generation of nitric oxide from S-nitrosothiols. Biomaterials 29, 2443–2452.

Joslin, J.M., Neufeld, B.H., Reynolds, M.M., 2014. Correlating S-nitrosothiol decomposition and NO release from modified poly(lactic-co-glycolic acid) polymer films. RSC Adv 4, 42039–42043.

Kazatchkine, M.D., Carreno, M.P., 1988. Activation of the complement system at the interface between blood and artificial surfaces. Biomaterials 9, 30–35.

Keefer, L.K., 2002. Progress toward clinical applications of the nitric oxide-releasing diazeniumdiloates. Annu. Rev. Pharmacol. Toxicol. 43, 585–607.

Keefer, L.K., 2011. Fifty years of diazeniumdiolate research. From laboratory curiosity to broad-spectrum biomedical advances. ACS Chem. Biol. 6, 1147–1155.

Klueh, U., Kaur, M., Qiao, Y., Kreutzer, D.L., 2010. Critical role of tissue mast cells in controlling long-term glucose sensor function in vivo. Biomaterials 31, 4540–4551.

Lachmann, P.J., Pangburn, M.K., Oldroyd, R.G., 1982. Breakdown of C3 after complement activation. Identification of a new fragment C3g, using monoclonal antibodies. J. Exp. Med. 156, 205–216.

Lowenstein, C.J., Dinerman, J.L., Snyder, S.H., 1994. Nitric oxide: a physiologic messenger. Ann. Intern. Med. 120, 227–237.

Lutzke, A., Pegalajar-Jurado, A., Neufeld, B.H., Reynolds, M.M., 2014. Nitrix ocide-releasing S-nitrosalted derivatives of chitin and chitosan for biomedical applications. J. Mater. Chem. B 2, 7449–7458.

Marxer, S.M., Robbins, M.E., Schoenfisch, M.H., 2004. Sol–gel derived nitric oxide-releasing oxygen sensors. Analyst 130, 206–212.

McCarthy, C.W., Guillory, R.J., Goldman, J., Frost, M.C., 2016a. Transition-metal-mediated release of nitric oxide (NO) from S-nitroso-N-acetyl-D-penecillamine (SNAP): potential applications for endogenous release of NO at the surface of stents via corrosion products. ACS Appl. Mater. Interfaces 8, 10128–10135.

McCarthy, C.W., Goldman, J., Frost, M.C., 2016b. Synthesis and characterization of the novel nitric oxide (NO) donating compound, S-Nitroso-N-acteyl-D-peniecillamine derivatized cyclam (SNAP-Cyclam). ACS Appl. Mater. Interfaces 8, 5898–5905.

Moghimi, S.M., Andersen, A.J., Ahmadvand, D., Wibroe, P.P., Andresen, T.L., Hunter, A.C., 2011. Materials properties in complement activation. Adv. Drug Delivery Rev. 63, 1000–1007.

Oh, B.K., Meyerhoff, M.E., 2003. Spontaneous catalytic generation of nitric oxide from S-nitrosothiols at the surface of polymer films doped with lipophilic copper(II) complex. J. Am. Chem. Soc. 125, 9552–9553.

Oh, B.K., Meyerhoff, M.E., 2004. Catalytic generation of nitric oxide from nitrite at the interface of polymer films doped with lipophilic Cu(II)-complex: a potential route to the preparation of thromboresistant coatings. Biomaterials 24, 283–293.

Oh, B.K., Robbins, M.E., Nablo, B.J., Schoenfisch, M.H., 2005. Miniaturized glucose biosensor modified with a nitric oxide-releasing xerogel microarray. Biosens. Bioelectron. 21, 749–757.

Radomski, M.W., Palmer, R.M.J., Moncada, S., 1987a. The role of nitric oxide and cGMP in platelet adhesion to vascular endothelium. Biochem. Biophys. Res. Commun. 148, 1482–1489.

Radomski, M.W., Palmer, R.M., Moncada, S., 1987b. The anti-aggregating properties of vascular endothelium: interactions between prostacyclin and nitric oxide. Br. J. Pharmacol. 92, 639–646.

Radomski, M.W., Palmer, R.M., Moncada, S., 1987c. Endogenous nitric oxide inhibits human platelet adhesion to vascular endothelium. Lancet 2, 1057–1058.

Reynolds, M.M., Frost, M.C., Meyehofff, M.E., 2004. Nitric oxide-releasing hydrophobic polymers: preparation, characterization, and potential biomedical applications. Free Rad. Bio. Med. 37, 926–936.

Riccio, D.A., Dobmeier, K.P., Hetrick, E.M., Privett, B.J., Paul, H.S., Schoenfisch, M.H., 2009. Nitric oxide-releasing S-nitrosothiol-modified erogels. Biomaterials 30, 4494–4502.

Schoenfisch, M.H., Mowery, K.A., Rader, M.V., Baliga, N., Wahr, J.A., Meyerhoff, M.E., 2000. Improving the thromboresistivity of chemical sensors via nitric oxide release: fabrication and in vivo evaluation of NO-releasing oxygen-sensing catheters. Anal. Chem. 72, 1119–1126.

Schoenfisch, M.H., Rothrock, A.R., Shin, J.H., Polizzi, M.A., Brinkley, M.F., Dobmeier, K.P., 2006. Poly(vinylpyrrolidone)-doped nitric oxide-releasing xerogels as glucose biosensor membranes. Biosens. Bioelectron. 22, 306–312.

Schoenfisch, M.H., Zhang, H., Frost, M.C., Meyehoff, M.E., 2002. Nitric oxide-relesing fluoresce-based oxygen sensing polymeric films. Anal. Chem. 74, 5937–5941.

Sharma, J.N., Omran, A.A., Parvathy, S.S., 2007. Role of nitric oxide in inflammatory diseases. Inflammopharmacology 15, 252–259.

Shin, J.H., Marxer, S.M., Schoenfisch, M.H., 2004. Nitric oxide-releasing sol–gel particles/polyurethane glucose biosensors. Anal. Chem. 76, 4543–4549.

Shishido, S.M., de Oliveria, M.G., 2000. Polyethylene glycol matrix reduces the rates of photochemical and thermal release of nitric oxide from S-nitroso-N-acetylcysteine. Photochem. Photobiol. 71, 273–280.

Shishido, S.M., Seabra, A.B., Loh, W., de Oliveria, M.G., 2003. Thermal and photochemical nitric oxide release from S-nitrosothiols incorporated in Pluronic F127 gel: potential uses for local and controlled nitric oxide release. Biomaterials 24, 3543–3553.

Spath, P., Gabl, F., 1976. Critical role of the conversion of the third complement component C3 (beta 1C/beta 1A) for immunochemical quantitation. Clin. Chim. Acta 73, 171–175.

Starrett, M.A., Nielsen, M., Smeenge, D.M., Romanowicz, G.E., Frost, M.C., 2012. Wireless platform for controlled nitric oxide releasing optical fibers for mediating biological response to implanted devices. Nitric Oxide 27, 228–234.

Szott, L.M., Horbett, T.A., 2011. Protein interactions with surfaces: cellular responses, complement activation, and newer methods. Curr. Opin. Chem. Biol 15, 677–682.

Tabas, I., 2010. Macrophage death and defective inflammation resolution in atherosclerosis. Nat. Rev. Immunol. 10, 36–46.

Thevenot, P.T., Baker, D.W., Weng, H., Sun, M.W., Tang, L., 2011. The pivotal role of fibrocytes and mast cells in mediating fibrotic reactions to biomaterials. Biomaterials 32, 8394–8403.

Thoman, M.L., Meuth, J.L., Morgan, E.L., Weigle, W.O., Hugli, T.E., 1984. C3d-K. A kallikrein cleavage fragment of iC3b is a potent inhibitor of cellular proliferation. J. Immunol. 133, 2629–2633.

Tripathi, P., Tripathi, P., Kashyap, L., Singh, V., 2007. The role of nitric oxide in inflammaorty reactions. FEMS Immunol. Med. Microbiol. 51, 443–452.

Vaughn, M.W., Kuo, L., Liao, J.C., 1998. Estimation of nitric oxide production and reactionrates. Am. J. Physiol. Heart Circ. Physiol. 274, H2163–H2176.

Velnar, T., Bailey, T., Smrkolj, V., 2009. The wound helaing process: an overview of the cellular and molecular mechanisms. J. Int. Med. Res. 37, 1528–1542.

Volger, E.A., 2012. Protein adsorption in three dimensions. Biomaterials 33, 1201–1237.

Wang, Y.N., Collins, J., Holland, R.J., Keefer, L.K., Ivanic, J., 2013. Decoding nitric oxide release rates of amine-based diazeniumdiolates. J. Phys. Chem. A 117, 6671–6677.

Williams, D.L.H., 1999. The chemistry of S-nitrosothiols. Acc. Chem. Res. 32, 869–876.

Wu, Y., Rojas, A.P., Griffith, G.W., Skrzypchak, A.M., Lafayette, N., Bartlett, R.H., et al., 2007. Improving blood compatibility of intravascular oxygen sensors via catalytic decomposition. 121(1), 36–46.

Yan, Q., Major, T.C., Bartlett, R.H., Meyerhoff, M.E., 2011. Intravascualr glucose/lactate sensors prepared with nitric oxide releasing poly(lactide-co-glycolide)-based coatings for enhanced biocompatibility. Biosens. Bioelectron. 26, 4276–4282.

Yapr, J.P., Lutzke, A., Pegalajar-Jurado, A., Neufeld, B.H., Damodaran, V.B., Reynolds, M.M., 2015. Biodegradable citrate-based polyesters with S-nitrosothiol functional groups for nitric oxide release. J. Mater. Chem. B 3, 9233–9241.

Zhang, H., Annich, G.M., Miskulin, J., Osterholzer, K., Merz, S.I., Bartlett, R.H., et al., 2002. Nitric oxide releasing silicone rubbers with improved blood compatibility: preparation, characterization, and in vivo evaluation. Biomaterials 23, 1485–1494.

Zhang, H., Annich, G.M., Miskulin, J., Stankiewicz, K., Osterholzer, K., Merz, S.I., et al., 2003. Nitric oxide-releasing fumed silica particles: synthesis, characterization, and biomedical application. J. Am. Chem. Soc. 125, 5015–5024.

Zhang, Y., Hogg, N., 2000. Biological chemistry and clinical potential of S-nitrosothiols. Free Rad. Biol. Med. 28, 1478–1486.

CHAPTER 9

Strategies to Deliver Nitric Oxide Donors to Control Biofilms of Clinical and Industrial Interest

Massimiliano Marvasi[1] and Tania Henríquez[2]
[1]Middlesex University London, The Burroughs, London, United Kingdom
[2]Department of Microbiology and Mycology, Institute of Biomedical Sciences, University of Chile, Santiago, Chile

INTRODUCTION

Nitric oxide (NO), carbon monoxide, and hydrogen sulfide are a small family of gaseous molecules with an important role in intra- and extracellular signalings (Mustafa et al., 2009; Quinn et al., 2015). Among them, NO has attracted the attention during the last 30 years due to its capability as a singling molecule and to act as a bactericide (Barraud et al., 2015; Quinn et al., 2015).

NO is a small molecule that affects multiple processes and functions in eukaryote as well prokaryote organisms. In eukaryotes, NO controls vascular tonus, inhibition of platelet, leukocyte adhesion and aggregation, immune response, apoptosis, inflammation, would healing, neurotransmission, tissue repair, and angiogenesis (Howard et al., 2014; Kim et al., 2014; Seabra et al., 2015; Wang et al., 2005). NO has also been described as an anticancer agent, an efficient inhibitor of restenosis (Naghavi et al., 2013) and a key molecule involved in wound repair (Georgii et al., 2011). In prokaryotes, NO has a dual effect acting as an antimicrobial molecule and as a regulator of transcription (Arasimowicz-Jelonek and Floryszak-Wieczorek, 2014).

NO is very effective as gaseous molecule. Application of medical-grade gaseous form of NO (gNO) has shown consistent and controllable delivery of NO against a wide range of pathogens (*Staphylococcus aureus*, *Escherichia coli*, *Pseudomonas aeruginosa*, and *Candida albicans*), including antibiotic-resistant strains, demonstrating potent bactericidal activity (Ghaffari et al., 2006) (Fig. 9.1). However, the application of gNO has several safety concerns, NO is a gaseous radical specie, and its direct delivery, while possible, has significant limitations. Instead, it is safer and in some cases necessary to deliver NO using a reactive precursor (Quinn et al., 2015). Friedman and coworkers have highlighted the most important features of a NO delivery system. Ultimately the chemical should be easily applied and capable of delivering NO over a time interval (Friedman et al., 2008; Quinn et al., 2015) (Fig. 9.1).

Nitric Oxide Donors.

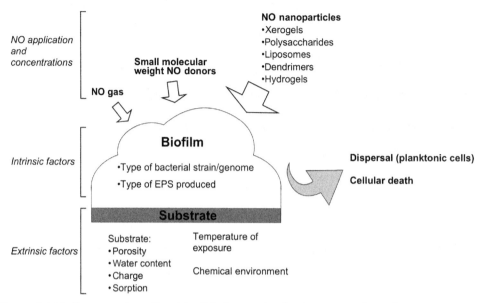

Figure 9.1 Biofilm control mediated by NO donors. Applications range from the NO gaseous to the delivery via donors. Response to the treatment varies according to intrinsic and extrinsic factors that ultimately constitute the environment. Examples of critical intrinsic factors are the type of microorganism(s) and the chemical structure of the extracellular polymeric substances (EPS). EPS can prevent NO diffusion within the biofilm. Extrinsic factors are more easily controlled than intrinsic factors and include the type of substrate (porosity, water content, charge, sorption, etc.), temperature of exposure, and chemical environment. With reference to the chemical environment, NO-scavengers can react with and neutralize NO (Pegalajar-Jurado et al., 2015). Intrinsic, extrinsic factors, type of NO application and donors concentrations all contribute to dispersal potential and/or biocidal activity.

Beside the bactericidal activity of NO, current research has identified NO as an effective molecule to control biofilms (Barraud et al., 2015) (Fig. 9.1). During this process, NO does not necessarily have bactericidal effect, in particular when is used at very low concentrations. In some cases, control and bactericidal effect overlap, however, in some of the strategies showed in this review, the donors serve as potent platform to control cellular adhesion, biofilm formation or dispersal, but they are not effective to kill bacteria. Although planktonic bactericidal assays are helpful in determining antibacterial agent's potential and its biomedical utility, most bacteria establish biofilms as a protective mechanism against therapeutics agents or to cope with environmental fluctuations (Corcoran et al., 2014; Lu et al., 2014; Zhang and Mah, 2008). Biofilms prevail in natural and man-made environments on all surfaces. They play a beneficial role in water treatment, waste management, and are important reservoirs of pathogens involved in fouling of medical and industrial equipment (Flemming and Wingender, 2010).

Despite the fact that biofilms can be resistant to antimicrobial agents (Zarena and Gopal, 2013), several disinfectants, antibiotics, and messenger molecules have been developed for their ability to dislodge existing biofilms (Barraud et al., 2006, 2009b, 2012).

Of these, NO appears to be very promising. NO is effective as a biofilm dispersant, functioning as a messenger rather than a generic poison (Barraud et al., 2006, 2009b, 2012). Current research focuses on different application of NO, ranging from food safety to medicine. NO technology has been applied in food contact materials (Marvasi et al., 2014), orthopedic implants (Charville et al., 2008; Nablo et al., 2005), in reverse osmosis membrane (Barnes et al., 2014), silicone rubber coatings (Cai et al., 2015), preparation of polyester mesh (Fernandez-Moure et al., 2014; Moretti et al., 2012), for coating of blood-contacting artificial materials (Seabra et al., 2010), and liposomal NO treatments (Jardeleza et al., 2015), to give some examples.

Recently, nontraditional antibiotic agents have attracted tremendous interest, substances against which microbial pathogens may not be able to develop resistance (Huh and Kwon, 2011). Among these groups, nanoparticle-based systems, which enable the sustained exogenous delivery of NO, have been subject of considerable investigation (Huh and Kwon, 2011).

Several authors have extensively reviewed a reasonable volume of current literature focused on NO releasing compounds within different platforms (Eroy-Reveles and Mascharak, 2009; Jen et al., 2012; Quinn et al., 2015; Riccio and Schoenfisch, 2012; Roveda and Wagner, 2013; Tfouni et al., 2012). Examples of NO releasing platforms are: nanoparticles (Friedman et al., 2008; Seabra and Duran, 2010), silica gel (Doro et al., 2007; Zanichelli et al., 2006), xerogels and dendrimers (Benini et al., 2008; Lu et al., 2011; Stasko et al., 2008; Stasko and Schoenfisch, 2006).

Here, in this chapter, current strategies for NO delivery to control biofilms of clinical and industrial interest are reviewed. Also, it highlights strategies and perspectives to improve the effectiveness of NO donors in several biotechnological applications.

DUAL FUNCTION OF NO: ANTIBACTERIAL AGENT AND GENETIC REGULATOR IN BACTERIA

NO is effective as an antibacterial agent by acting on different routes (McMullin et al., 2005; Quinn et al., 2015): (1) NO has been shown to have a direct chemical effect on DNA, deaminating cytosine, guanine, and adenine through the formation of nitrosating intermediates such as N_2O_3. (2) NO causes S-nitrosation of the SH moiety in the cysteine residues. (3) Elevated concentrations of NO cause formation of secondary species such as peroxynitrite and nitrogen dioxide, which can facilitate lipid peroxidation (Quinn et al., 2015). In addition to these effects, NO also is a signaling regulator.

NO affects several routes that allow transition from sessile biofilms to free-swimming bacteria, and vice versa (Plate and Marletta, 2012). The mechanisms are not entirely clear but cellular adhesion may provide a protection for bacteria against reactive and damaging NO (Barraud et al., 2006, 2009b; Plate and Marletta, 2012). Microarray studies have revealed that *P. aeruginosa* PAO1 genes involved in adherence are downregulated upon exposure to NO (Firoved et al., 2004), and the involvement of this compound in regulating biofilm formation and dispersal was also supported by other studies (Barraud et al., 2006, 2009b; Darling and Evans, 2003; Van Alst et al., 2007). Genes involved in the production and perception of NO have been characterized in *P. aeruginosa* PAO1. For example, the chemotaxis protein BdlA is involved in biofilm dispersal of *P. aeruginosa*: biofilms formed by the *bdlA* mutant do not detach when exposed to low doses of NO in continuous-flow cultures (Barraud et al., 2006, 2009a; Petrova and Sauer, 2012). *P. aeruginosa nirS* and *norCB* encode a NO_2^- and NO reductases, respectively. A mutation in *nirS* led to a reduced biofilm dispersal (Barraud et al., 2006), while biofilms formed by a NO reductase-deficient strain Δ*norCB* did not shift to the planktonic state when exposed to endogenous NO. No homologs of *nirS* and *norCB* are found in *Salmonella*. However, when grown anaerobically with nitrate, *Salmonella* is capable of generating NO after nitrite addition, likely via products of *fnr* and *hmp* genes (Gilberthorpe and Poole, 2008).

When a *P. aeruginosa* mutant strain lacking the phosphodiesterase *rdbA* was exposed to a class of core cross-linked star polymers releasing NO, biofilm dispersal was reduced up to 50% less when compared to the wild type. These results indicate that the released NO was likely involved in stimulating phosphodiesterase activity, thereby maintaining low intracellular concentrations of the bacterial intracellular messenger cyclic diguanosine monophosphate (c-di-GMP) and maintaining *P. aeruginosa* cells in planktonic form (Duong et al., 2014; Quinn et al., 2015). c-di-GMP is involved in biofilm formation and it is also known that modulate swarming motility in *P. aeruginosa* (Ryan et al., 2009).

In *Shewanella oneidensis*, NO stimulates biofilm scaffolding by controlling the levels of c-di-GMP (Plate and Marletta, 2012). A feed-forward loop between response regulators with phosphodiesterase domains and phosphorylation-mediated activation regulates c-di-GMP levels leading to biofilm dispersal (Barraud et al., 2015; Liu et al., 2012; Plate and Marletta, 2012) (Fig. 9.2).

NO affects many other metabolic cascades in prokaryotes. For example, in *Deinococcus radiodurans*, nitric oxide synthase (bNOS) is related to the recovery from UV radiation damage (Crane, 2008). bNOS also seems to be an important regulator of virulence in *Bacillus anthracis* since endogenous NO protects it from oxidative immune response (Shatalin et al., 2007). In *Staphylococcus aureus*, the NO-dependent NOS is involved in the response to oxidative stress (H_2O_2) (Sapp et al., 2014). In *Bacillus subtilis*, NO protects bacterial cells from reactive oxygen species by using two independent mechanisms (Gusarov et al., 2009; Gusarov and Nudler, 2005): first, transiently suppresses the enzymatic reduction of free cysteine, and second, directly reactivates

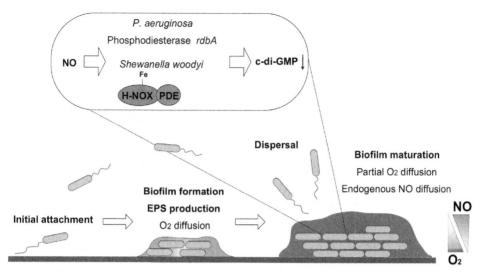

Figure 9.2 Effect of NO signaling on biofilm dispersal. The biofilm life cycle. Oxygen (O₂) and nutrients gradients are present in mature biofilm leading to production of NO signals that trigger cell death and dispersal. In *Shewanella woodyi*, direct binding of NO to an heme NO/oxygen binding sensor (H-NOX) stimulates phosphodiesterase (PDE) activity, which leads to decreased c-di-GMP levels and enhances dispersal. In *P. aeruginosa*, NO was likely involved in stimulating phosphodiesterase activity (*rdbA*), thereby maintaining low intracellular concentrations of c-di-GMP *(Barraud et al., 2015; Quinn et al., 2015; Duong et al., 2014; Liu et al. 2012).*

catalase, a major antioxidant enzyme that is inhibited by endogenous cysteine (Gusarov et al., 2009; Gusarov and Nudler, 2005).

Finally, in plant pathogens NO has been related to virulence suggesting that NO produced by pathogens could be associated to phytotoxins activation (Arasimowicz-Jelonek and Floryszak-Wieczorek, 2014).

Genes involved in NO perception are regulated in both Gram-positive and Gram-negative bacteria showing a universal effect on bacteria (Firoved et al., 2004; Xiong and Liu, 2010). Therefore, it is important to deeply understand the role of NO in signaling and gene regulation of bacterial pathogens. NO seems to regulate different functions apart from biofilm development, as stress resistance and virulence. The combined effect of chemicals and regulative mechanisms makes NO a potent antimicrobial agent which may be employed either alone or in concert with antibiotics (Quinn et al., 2015).

The knowledge of the genetics and physiology of bacteria is substantial in developing NO delivering strategies. Such knowledge would allow identifying substrates, environmental conditions and temperatures that effectively trigger bacterial response in favor of an efficient NO response.

NO DONOR PLATFORMS TO CONTROL BIOFILM OF CLINICAL INTEREST

Donors release NO according to their physical–chemical surface properties, temperature, pH, light, molecule half-life, and chemical environment (Wang et al., 2005). From the biofilm standpoint, it is well known that environmental conditions ultimately dictate the key properties of biofilms such as porosity, density, water content, charge, sorption and ion exchange properties, hydrophobicity, and mechanical stability (Wingender et al., 1999) (Fig. 9.1). So it is not unusual that biofilms attached to different substrate can develop different extracellular polymeric substances (EPS) leading to diverse dispersal potentials (Davey and O'Toole, 2000; Marvasi et al., 2010). For example, *Salmonella* biofilm preformed on polypropylene or polystyrene showed a different dispersal potential when treated with the same donor diethylamine NONOate diethylammonium. Such molecule was less effective in dispersing biofilms formed on polystyrene when compared with polypropylene (Marvasi et al., 2014).

NO is effective at low concentrations. Studies by Barraud et al. (2006) showed that dispersal of *P. aeruginosa* biofilm was achieved with low, sublethal concentrations ranging 25–500 nmol L^{-1} of the NO donor sodium nitroprusside (SNP). Other studies showed that donors such as molsidomine, MAHMA NONOate, diethylamine NONOate diethylammonium, PROLI NONOate were effective up to 10 pmol L^{-1} (Barnes et al., 2015; Marvasi et al., 2014).

Different temperatures also contribute to the dispersal potential, which is related to the molecular structure of the donor (Fig. 9.1). Dispersal of the preformed *Salmonella* Typhimurium 14028 biofilms exposed to MAHMA NONOate was effective at 22°C but less effective at 4°C on polystyrene or polypropylene plastics. However, certain donors are very sensitive to temperature and 1°C variation can result in a wide change in the NO releasing rate (Marvasi et al., 2014, 2015; Wingender et al., 1999). Different strains also react differently to NO exposure, for example, a quaternary ammonium-NO compound showed that the performance of such particles against *P. aeruginosa* was more effective when compared with *S. aureus* (Carpenter et al., 2012).

When NO donors are tested to measure their potential to control biofilm, researcher should consider the origin of the strains used. It is well known that laboratory strains can produce different or reduced EPS composition. For example, the laboratory strain *B. subtilis* 168, widely used in genetic studies, is defective or attenuated in EPS production. Several of the biosynthetic pathways are not functional because of the domestication processes (i.e., mutations that allow easier genetic manipulations coupled with repeated growth under artificial settings) (Aguilar et al., 2007; Stein, 2005). In these attenuated strains, the effectiveness of the donors could be enhanced when compared with the environmental homologous strains.

According to these examples, beside the chemical composition of the NO donors, other factors need to be taken in account when planning strategies to improve the effectiveness of donors in biotechnological applications. As previously mentioned, the genetics of the target(s) is certainly determinant.

NO delivery using nanoparticles

This section will focus on strategies currently used to reduce bacterial adhesion, biofilm formation, and to promote biofilm dispersal on biofilms of clinical interest by using NO releasing nanoparticles. The current state of preparation of the nanoparticles, postassembly modification to incorporate the NO delivering moiety and bactericidal activity have been recently reviewed by Quinn et al. (2015) and Seabra et al. (2015). The vehicles for exogenous delivery of NO can be categorized in sol–gel based silica particles, functionalized metal/metal oxide nanoparticles, polymer-coated metal nanoparticles, dendrimers, micelles, and star polymers (Huh and Kwon, 2011; Quinn et al., 2015) (Fig. 9.1).

Sol gels physical and chemical properties can be easily altered by changing reaction constituents and conditions (e.g., solvent type and amounts, pH, catalyst, etc.). Such characteristics make sol–gel chemistry a versatile and powerful method for coating materials (Nablo et al., 2001; Nablo and Schoenfisch, 2003, 2004).

A NO sol–gel film composed by 40% N-aminohexyl-N-aminopropyltrimethoxysilane and 60% isobutyltrimethoxysilane was used as potential antibacterial coating for orthopedic devices medical-grade stainless steel (Nablo et al., 2005). NO was released by converting the diamine groups in these films to diazeniumdiolate NO donors resulting in NO releasing profile temperature dependent. Sol–gel films incubated at 25°C showed a lower NO release but five times longer flux over the first 24 hours when compared to those at 37°C. The effectiveness to prevent biofilm formation of the sol–gel and NO releasing sol–gel-coated steel was tested by using bacterial suspensions of *P. aeruginosa*, *S. aureus*, and *S. epidermidis* at 25°C and 37°C. NO releasing surfaces showed significantly less bacterial adhesion for all species and temperatures investigated. Bacterial adhesion was semiquantitatively measured by using fluorescent staining. At 37°C, *P. aeruginosa* coverage on the NO releasing film was 12% of the entire surface, noticeably less than the controls (44% of the entire surface) (Nablo et al., 2005; Nablo and Schoenfisch, 2003).

Similarly, perspectives for avoiding bacterial colonization of human pathogens on polyurethane have been recently described by Seabra and collaborators. Poly(sulfhydrylated polyester) (PSPE) was synthesized by the polyesterification reaction of mercaptosuccinic acid with 3-mercapto-1,2-propanediol and blended with poly-(methyl methacrylate) (PMMA) from solution, leading to solid PSPE/PMMA films with three different PSPE:PMMMA mass ratios. These films were S-nitrosated through the immersion in acidified nitrite solution, yielding poly(nitrosated) polyester/PMMA (PNPE/PMMA) films. A polyurethane intravascular catheter coated with PNPE/PMMA was shown to exert a potent dose- and time-dependent antimicrobial activity against *S. aureus* and multidrug-resistant *P. aeruginosa* strains (Seabra and Duran, 2010).

In experiments carried out by Nablo and coworkers, poly(vinyl chloride) (PVC) over coated sol–gel (diazeniumdiolate-modified sol–gel film) provided controlled surface chemistry for systematic studies on the effect of NO release on bacterial adhesion. According to the researchers, *P. aeruginosa* adhesion was proportionately reduced according to the NO flux. A linear relationship was noted over NO fluxes of $5-20 \, pmol \, cm^{-2} s^{-1}$ (Fig. 9.3).

A sustained NO flux above $20 \, pmol \, cm^{-2} s^{-1}$ is required to prevent 75% of the bacterial adhesion to PVC (Fig. 9.3). Above $20 \, pmol \, cm^{-2} s^{-1}$, the adhesion dependence shallows resulting in a minimal reduction in relative coverage to $0.15 \, pmol \, cm^{-2} s^{-1}$ (Fig. 9.3). When the experiments were repeated by measuring the adhesion of *Proteus mirabilis*, it did not appear to be correlated with NO releasing profile (Nablo and Schoenfisch, 2004). The studies conducted with *P. mirabilis* demonstrated that low surface NO fluxes inhibit cellular adhesion, but the adhesion is not flux dependent. These results suggested that localized release of NO (pmol) is a potent method for reducing bacterial adhesion, but it also seems to be specie specific (see intrinsic factors, Fig. 9.1). Interestingly, in agreement with previous literature, concentration in the order of $pmol \, L^{-1}$ has been showed to be effective in controlling biofilm dispersal (Marvasi et al., 2014).

NO-xerogels also represent a unique class of materials that form porous, glass-like materials (Nablo and Schoenfisch, 2003). Coating strategies may consist in preparing low molecular weight PVC spin coated onto NO releasing xerogels (Marxer et al.,

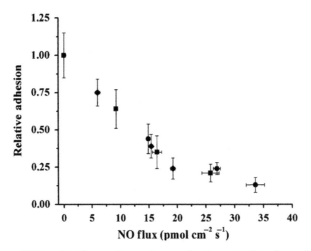

Figure 9.3 Influence of NO surface flux on *P. aeruginosa* adhesion tested on three different PVC-coated NO releasing sol gels: (●) 3-trimethoxy-silylpropyl-diethylenetriamine (DET3); (■) *N*-(6-aminohexyl)-aminopropyltrimethoxysilane (AHAP3); (♦) (aminoethylaminomethyl)-phenylethyltrimethoxysilane (AEMP3). *Reproduced from Nablo, B.J., Schoenfisch, M.H., 2004. Poly(vinyl chloride)-coated sol gels for studying the effects of nitric oxide release on bacterial adhesion. Biomacromolecules 5 (5), 2034–2041, with permission from American Chemical Society.*

2003). PVC coated with a thin ($15\pm4\,\mu m$) NO-xerogels overlayer was used to reduce *S. aureus* and *E. coli* cells adhesion on fibrinogen (Charville et al., 2008). The reduced adhesion was NO flux dependent from 0 to $30\,pmol\,cm^{-2}\,s^{-1}$. A similar dependence for *S. epidermidis* was demonstrated from 18 to $30\,pmol\,cm^{-2}\,s^{-1}$. At an NO flux of $30\,pmol\,cm^{-2}\,s^{-1}$, surface coverage of *S. aureus*, *S. epidermidis*, and *E. coli* was reduced by 96%, 48%, and 88%, respectively, compared to not-NO releasing controls. Polymeric NO release was thus demonstrated to be an effective approach for significantly reducing fibrinogen-mediated adhesion of both Gram-positive and Gram-negative bacteria in vitro, thereby illustrating the advantage of active NO release as a strategy to inhibit bacterial adhesion in presence of preadsorbed protein (Charville et al., 2008).

Other example about the use of coating NO releasing xerogels in reducing bacteria adhesion is the application of isobutyltrimethoxysilane (BTMOS) and *N*-(6-aminohexyl) aminopropyltrimethoxysilane (AHAP3)—AHAP3/BTMOS xerogel films (Hetrick et al., 2007). Bacterial adhesion was measured by using glass slides coated with xerogel films in a flow chamber. The NO releasing xerogels reduced bacterial adhesion up to 90% in a flow-dependent manner. Interestingly, the treatment was also able to kill the cells within 7 hours of application (Hetrick et al., 2007).

NO releasing silica nanoparticles have also been proposed as an alternative strategy for delivering NO to pathogenic biofilms. *P. aeruginosa*, *E. coli*, *S. aureus*, *S. epidermidis*, and *C. albicans* biofilms were preformed on class VI medical-grade silicone rubber (SiR) treated with NO releasing silica nanoparticles. The treatment was able to kill 99% of cells from each type of biofilm with reduced fibroblast cytotoxicity (Hetrick et al., 2009).

Current novel strategies also aim to control the NO efflux of the top coating. NO releasing films with a bilayer configuration were recently synthetized by doping dibutyhexyldiamine diazeniumdiolate (DBHD/N_2O_2) in a poly(lactic-*co*-glycolic acid) (PLGA) layer and further encapsulating this base layer with a silicone rubber top coating (Cai et al., 2012). Interestingly, in this exercise, PLGA acts as both a promoter and controller of NO release from the coating. Experiments performed at 37°C and room temperature indicated that all the NO releasing films show considerable antibiofilm properties exhibiting about 98.4% reduction in biofilm biomass of *S. aureus* and 99.9% reduction for *E. coli* at 37°C (Cai et al., 2012).

NO release can be also controlled via light-induced manner. Metal nitrosyl incorporated within the columnar pores of biocompatible Al-MCM-41 type material was able to eradicate *C. albicans* in a dose-dependent way under exposure of light (Heilman et al., 2013).

NO releasing polysaccharides

Biofilms are more resistant to antibiotics and biocides due to the difficult penetration of such molecules through EPS (Flemming and Wingender, 2010; Marvasi et al., 2010).

NO releasing polysaccharides have been explored for their ability to penetrate into the EPS layer and for their capability to retain bacterial activity for long time after application, contributing to prevent biofouling by supporting the inability of the bacteria to recover. In fact, water solubility and charge may enable polysaccharides penetration across the extracellular matrix (Lu et al., 2014).

The donor *S*-nitrosated dextran-cysteamine was tested on *E. coli*, *Acinetobacter baumannii*, and *S. aureus*, commonly responsible for healthcare associated infections. Even though the experiments were conducted in liquid culture, treatments showed 8-log reduction in bacterial growth within the first 24 hours and absence of bacteria growth after 72 hours of exposure to NO, illustrating the ability of this water-soluble compound for treating healthcare associated infections (Pegalajar-Jurado et al., 2015).

Water-soluble NO releasing chitosan oligosaccharides have been described to be effective against *P. aeruginosa* biofilm, notoriously known to be a strong biofilm former (Lu et al., 2014). The chitosan oligosaccharides were synthesized via grafting 2-methyl aziridine from the primary amines on chitosan oligosaccharides, followed by reaction with NO gas under basic conditions to yield *N*-diazeniumdiolate NO donors. Chitosan oligosaccharides were effective to eradicate biofilm. Eradication was dependent on both the molecular weight and ionic characteristics. Lu and collaborators tested different types of chitosan oligosaccharides which differ in terms of NO storage and release. Among them, chitosan 1/NO-5k, 2/NO-5k, 2/NO-10k, and 3/NO-5k were further used to test to what extent the donors behave differently in terms of biofilm eradication. Chitosan 2/NO-5k exhibited the greatest antibiofilm efficacy, a likely result due to both increased NO storage/release and rapid association with the negatively charged bacteria. *P. aeruginosa* biofilm treated with 2/NO-5k (0.10 μmol NO mL−1) showed 5-log reduction of biofilm biomass. Biofilm reduction was measured as leftover of sonicated and vortexed biofilms followed by plating suspensions (Lu et al., 2014). The authors hypothesized that the decreased antibiofilm efficacy of chitosan 3/NO-5k (when compared with chitosan 2/NO-5k) resulted from the shielding of the amine moieties by the neutral functional group surrounding the *N*-diazeniumdiolates (e.g., polyethyleneglycol (PEG) chains), thus impeding association with the negatively charged exterior of the bacteria or EPS. Association of chitosan 2/NO-5k and chitosan 3/NO-5k with *P. aeruginosa* biofilm was evaluated using confocal microscopy. In an elegant experiment carried out by Lu et al. (2014), rhodamine B isothiocyanate (RITC) was used to label chitosan oligosaccharides (showing red fluorescence). The association of RITC-labeled chitosan oligosaccharides with *P. aeruginosa* biofilms was measured via confocal microscopy. Biofilms exposed to chitosan 2/NO-5k exhibited more intense red fluorescence compared to chitosan 3/NO-5k, confirming the enhanced association of the positively charged chitosan 2/NO-5k with the bacteria (Fig. 9.4). On the opposite, the less effective chitosan 3/NO-5k was not associated with the negatively charged exterior of the bacteria or EPS (Fig. 9.4).

Such experimental approach had the great advantage to visualize the association of polymers with the biofilm and helps to develop polymers able to properly associate and penetrate into the EPS layer. The chitosan oligosaccharides exhibited rapid association with bacteria throughout the entire biofilm, leading to enhanced biofilm dispersal and elicited no significant cytotoxicity to mouse fibroblast L929 cells in vitro (Lu et al., 2014).

Recently, novel NO releasing S-nitrosothiol-modified silica/chitosan core–shell nanoparticles have been developed with a prolonged NO release property (Chang et al., 2015). Hydrogels are water-swollen and cross-linked polymeric network (including polysaccharides) produced by the simple reaction of one or more monomers (Ahmed, 2015). The industrial applications of hydrogels are extremely heterogeneous, ranging from agriculture, drug delivery systems, food additives, biomedical applications such as implants, and separation of biomolecules. Hydrogels composed of cellulose

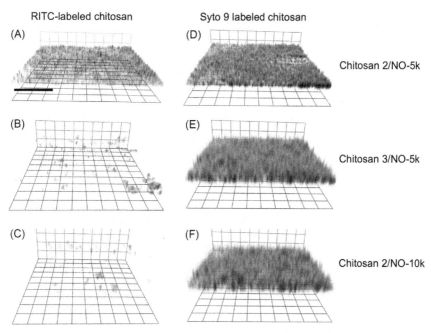

Figure 9.4 Confocal fluorescence images of rhodamine B isothiocyanate (RITC)-labeled chitosan oligosaccharide in association with *P. aeruginosa* in biofilms: (A) Chitosan 2/NO-5k; (B) Chitosan 3/NO-5k; (C) Chitosan 2-10k. Images of Syto 9 (green fluorescent nucleic acid stain) labeled biofilms incubated with (D) Chitosan 2/NO-5k; (E) Chitosan 3/NO-5k; and (F) Chitosan 2/NO-10k. Scale bar: 40 mm. The green fluorescence of Syto 9 indicates that the *P. aeruginosa* bacteria are embedded within the biofilms. Red fluorescence of RITC indicates the association of RITC-labeled chitosan oligosaccharides with *P. aeruginosa* in biofilms. *Modified from Lu, Y., Slomberg, D.L., Schoenfisch, M.H., 2014. Nitric oxide-releasing chitosan oligosaccharides as antibacterial agents. Biomaterials 35 (5), 1716–1724, with permission from Elsevier.*

nanocrystals (CNC) have also attracted a lot of interest due to high strength/stiffness, optical transparency, biocompatibility, biodegradability, highly porous structural network (Klemm et al., 2011; Siro and Plackett, 2011) and the ability to carry bactericidal molecules (Azizi et al., 2013; Drogat et al., 2011; Feese et al., 2011; George et al., 2011). CNC has high crystallinity, high water holding capacity, and excellent mechanical and thermal properties (George et al., 2011). CNC has an ionic-charged surface that could be appropriate for binding outer EPS (George et al., 2011; Marvasi et al., 2010, 2014, 2015). MAHMA NONOate was therefore encapsulated with CNC forming MAHMA NONOate-CNC. At 25°C for a composite of 0.1 µM MAHMA NONOate-CNC, NO diffuses up to 500 µm from the hydrogel surface, with flux decreasing according to Fick's Law. 60% of NO was released from the hydrogel composite during the first 23 minutes. Encapsulation of the NO donors MAHMA NONOate within CNC has shown a synergistic effect in dispersing 1-week-old biofilms when compared with the two chemicals tested alone. After 2 hours of exposure to 0.1 µM MAHMA NONOate-CNC moderate but significant dispersion was measured on preformed *Salmonella enterica* biofilms. After 6 hours of exposure, the number of cells transitioning from the biofilm to the planktonic state was up to 0.6 $\log_{10}$ higher when compared with nontreated *Salmonella* biofilms. These data suggest that the combined treatments with NO donor and hydrogels may allow for new sustainable cleaning strategies (Marvasi et al., 2015).

It is worth mentioning that the improved ability of MAHMA NONOate-CNC association to disperse biofilm may be also associated to the low-moderate antimicrobial activity of CNC (Azizi et al., 2013). In the literature, other molecules have been associated with CNC to improve its antimicrobial effectiveness: for example, CNC stabilized with ZnO-Ag exhibited greater bactericidal activity against *S. enterica* and *S. aureus* compared to cellulose-free ZnO-Ag heterostructure nanoparticles of the same particle size (Azizi et al., 2013). Association of CNC with porphyrin (Feese et al., 2011) and silver nanoparticles (Drogat et al., 2011) also showed antimicrobial effectiveness.

NO liposomes

Liposomal formulations enriched with NO donors have been recently proposed as a novel strategy to deliver NO. Anionic liposomes were enriched with the NO donor isosorbide mononitrate (ISMN), encapsulated into different anionic liposomal formulations based on particle size (unilamellar ULV, multilamellar MLV), and tested against *S. aureus* biofilms in vitro (Jardeleza et al., 2014). Reduction of biofilm biomass from the treatments and controls was compared and measured via confocal imaging. After 24 hours of exposure, 3 and 60 mg mL^{-1} of ISMN-ULV liposomes had a significant reduction of the biomass compared to the untreated control. ULV alone also had significant biomass reduction at both 24 hours and 5 minutes exposure. Just after

5 minutes of exposure, $60 \, mg \, mL^{-1}$ ISMN-MLV liposomes appeared to have greater antibiofilm effects compared to pure ISMN or ULV particles alone. The effectiveness of the NO liposomal preparations was improved by increasing liposomal lipid content of both MLV and ULVs (Jardeleza et al., 2014). As liposome-encapsulated NO showed high effectiveness in eradicating *S. aureus* biofilms in vitro, the application was repeated in vivo for treating an established rhinosinusitis in sheep. Interestingly, the NO donor liposome demonstrated a dose–response effect. At 3 and $60 \, mg \, mL^{-1}$, liposomal NO donor showed a significant antibiofilm effect compared to untreated control after 24 hours of exposure. Interestingly, the treatment showed reduced inflammation and significant ciliary preservation when compared with the control group (Jardeleza et al., 2014, 2015).

NO releasing dendrimers

Other novel strategy in controlling biofilm has been recently developed by using NO releasing dendrimers (Backlund et al., 2016). Dendrimer, synthetic polymers with branching structure, relies on supramolecular properties which are new dimension in targeting biofilms featuring drug encapsulation, binding, and delivery to the target site (Astruc et al., 2010; Zarena and Gopal, 2013). Several authors have been extensively reviewing dendrimers as a platform for NO transport (Roveda and Wagner, 2013; Seabra et al., 2015). According to the type of NO donor conjugated with the dendrimer, the mechanism of NO release can be modified in terms of releasing rate, concentration, and continuity of the flow. The mechanism involved in the NO release can be chosen according to the purpose, from spontaneous to triggered (Roveda and Wagner, 2013).

NO releasing dendrimers technology has been successfully applied to control biofilm responsible for dental caries (Backlund et al., 2016). In these experiments, Backlund and coworkers used a pH-dependent NO releasing strategy by taking advantage of the natural pH that occurs for dental caries. At the lower pH associated with dental caries (pH ~6.4), the molecules also had a more pronounced antibacterial effect. The authors measured the biocidal effect and biofilm disruption of NO releasing dendrimers on Gram-positive *Streptococcus mutans* at pH 7.4 and 6.4. Improved bactericidal efficacy at pH 6.4 (when compared with pH 7.4) was attributed to increased scaffold surface charge than enhanced dendrimer–bacteria association, ensuing membrane damage. Octyl- and dodecyl-modified dendrimers were the most effective for eradicating *S. mutans* biofilms (Backlund et al., 2016).

Similarly, antibacterial activity of a series of amphiphilic NO releasing poly(amidoamine) (PAMAM) dendrimers with different exterior functionalities has been recently reported by Lu et al. (2013). The hydrophobicity of the exterior functionality was tuned by varying the ratio of propylene oxide (PO) and 1,2-epoxy-9-decene (ED) grafted onto the dendrimers. Both the size and exterior functionalization of dendrimer proved to be important to a number of parameters including dendrimer–bacteria association, NO delivery efficiency, bacteria membrane disruption, and toxicity to mammalian cells. *P. aeruginosa* biofilms were preformed on medical-grade silicone rubber assembled in a biofilm reactor. Biofilm eradication was measured by vortexing the remained biofilms and by counting free dispersed aggregates. As complementary approach, the group observed cellular viability by using confocal vital staining techniques. The authors showed that amphiphilic dendrimers were more effective in reducing bacterial biofilms than the hydrophilic dendrimer due to the membrane disruption properties of the amphiphilic structures (Lu et al., 2013).

Recently, Worley et al. (2015) described the efficacy of dual-action NO releasing alkyl chain modified poly(amidoamine) dendrimers against bacterial biofilms. In this work, the authors reported that the antibiofilm action of the dendrimer biocides was dependent on alkyl chain length, bacterial Gram-class and dendrimer generation (sizes) with the most effective biofilm eradication occurring when antibacterial agents were capable of efficient biofilm infiltration (Worley et al., 2015). In addition, they found that the enhancement of NO release markedly increased antibiofilm effect of dendrimers that were unable to effectively penetrate the biofilm (Worley et al., 2015).

DELIVERY OF NO VIA SMALL MOLECULAR WEIGHT DONORS

Over 105 NO donors have been characterized, only few have been currently tested as biofilm dispersants (examples of molecular types are showed in Fig. 9.5).

Among these, the most studied are SNP (sodium nitroprusside) (Barraud et al. 2006, 2009a; Charville et al., 2008; Marvasi et al., 2014), molsidomine, diethylamine NONOate diethylammonium, NONOate diethylammonium, MAHMA NONOate (Marvasi et al., 2014), and NORS (sodium nitrite citric acid) (Regev-Shoshani et al., 2013). Studies by Barraud et al. (2006) showed the potential dispersal of biofilms preformed by *P. aeruginosa* by using SNP. Dispersal was induced with low, sublethal concentrations (25–500 nmol L^{-1}) of SNP (Fig. 9.5). With reference to the bacterial adhesion, the donor SNP can produce a flux of NO of 30 pmol cm^{-2} s^{-1}, and this can reduce adhesion of *S. aureus*, *S. epidermidis*, and *E. coli* by 96%, 48%, and 88%, respectively (Charville et al., 2008). Other studies showed the dispersal potential of donors such as molsidomine, MAHMA NONOate, diethylamine NONOate diethylammonium, PROLI NONOate (Barnes et al., 2015; Marvasi et al., 2014) (Fig. 9.5). Among these, MAHMA NONOate showed interesting dispersal potential. It spontaneously dissociates in a pH/temperature-dependent manner, in serum it has a half-life of 1 minute at 37°C and 3 minutes at 22–25°C, and at pH 7.4 liberates 2 moles of NO per mole of parent compound (Hrabie et al., 1993; Keefer et al., 1996; Wang et al., 2005). Interesting, when MAHMA NONOate was used to disperse *S. enterica* and *E. coli* biofilms (24-hour-old biofilm), up to 50% of *Salmonella* 14028 and a cocktail of six

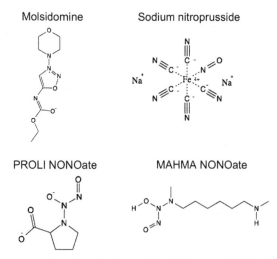

Figure 9.5 Four different types of small molecular weight NO donors. Molsidomine: *N*-(Ethoxycarbonyl)-3-(4-morpholino)sydnone imine; sodium nitroprusside: sodium nitroferricyanide(III) dehydrate; PROLI NONOate: 1-(hydroxy-NNO-azoxy)-ʟ-proline, disodium salt; MAHMA NONOate: 6-(2-hydroxy-1-methyl-2-nitrosohydrazino)-*N*-methyl-1-hexanamine.

Salmonella outbreak strains biofilms were dispersed when incubated for 6 and 24 hours at 22°C. About 40% of dispersal was also measured for the pathogenic *E. coli* O157:H7 when exposed at room temperature for 6 and 24 hours (Marvasi et al., 2014).

NO donors can be associated with disinfectants obtaining a synergistic effect in terms of biofilm dispersal. For example, a 20-fold increase in efficiency was observed when NO was applied together with chlorine in removing multispecies biofilms (Barraud et al., 2009b). MAHMA NONOate and molsidomine were also able to increase up to 15% of *Salmonella* biofilm dispersal when associated with the industrial disinfectant SaniDate 12.0 (Marvasi et al., 2014).

NO donor PROLI NONOate was effective on the dispersal of *P. aeruginosa* PAO1 wild-type biofilm and it was used to prove that EPS contributes to the degree of biofilm dispersal upon NO exposure. Alginate is a capsular polysaccharide produced by *P. aeruginosa* covering the exterior of one or more cells (Barnes et al., 2014). Overproduction of alginate by a *P. aeruginosa* PDO300 mutant increased the biofouling rate up to 59% when compared with the wild type, highlighting the ability of this polysaccharide to promote biofilm adherence on reverse osmosis membrane (Barnes et al., 2014). To further prove that EPS contributes to dispersal during exposure of NO, preformed biofilms were exposed to 40 µM PROLI NONOate for 1 hour and 24 hours. *P. aeruginosa* PAO1 biofilms (48 hours old) showed dispersal of the entire biomass up to 27% and 25%, respectively. In comparison, the alginate overproducing PDO300 mutant (24-hour-old biofilm) only dispersed by 5% over the same experimental period and the 48-hour-old biofilm showed no dispersal in the presence of PROLI NONOate. Alginate thus appears to play an important role in NO-induced dispersal of PAO1 biofilms (Barnes et al., 2014). The study stressed how the mucoid layer of bacteria led to a reduced responsiveness to NO-mediated dispersal, whereas a loss of mucoid layer production may enhance this response. Further strategies in controlling biofilm with NO should be oriented in identifying materials or chemicals that in synergistic effect with NO, may reduce EPS formation by biofilms.

CONCLUSION

In this chapter, current strategies for NO delivery to control biofilms of clinical and industrial interest were reviewed. Donors serve as a potent platform to control cellular adhesion, biofilm formation, or dispersal and they may be not effective to kill pathogenic bacteria or fungi (Fig. 9.1). With reference to NO signaling, NO in bacteria affects several metabolic routes and many of these are related to the transition from sessile biofilm to free-swimming bacteria, indicating a fine genetic regulation. Further studies should be focused in identifying regulatory routes used by NO to control biofilm attachment/dispersal in different bacteria.

A reasonable volume of the current research has been focused on including NO releasing compounds to different platforms. In this chapter, we presented the use of nanoparticles, as well as small donor molecules in controlling biofilms.

NO donors have been tested as free molecules or by coatings different surfaces such as sol–gel-coated stainless steel, glass slides, or plastics such as polyurethane intravascular catheter and poly(vinyl chloride) (Hetrick et al., 2007; Nablo et al., 2005; Nablo and Schoenfisch, 2004, Seabra and Duran, 2010). All these experiments adopted different coating NO donors, finding possible applications in medicine and food industry. Nanoparticles are of great interests, e.g., NO liposomal preparations appear to be effective just after 5 minutes of applications in eradicating *S. aureus* biofilm (Jardeleza et al., 2014). Water-soluble NO-chitosan polysaccharides also showed interesting properties as molecules able to deeply penetrate within the biofilm due to the water solubility and charge (Lu et al., 2014; Pegalajar-Jurado et al., 2015).

However, perspectives to develop additional strategies should consider: (1) studies on synergistic effects with detergents and biocides in disassembling biofilm structures and facilitate dislodging of attached cells. (2) Additional evidences for the efficiency of NO donors against biofilms formed on a wide range of abiotic and biotic surfaces (e.g., plants). (3) When NO donors are tested to measure their potential to control biofilms, researchers should consider the origin of the strains. It is well known that laboratory strains can produce a different response when compared with the environmental ones.

Before the application of NO donors could take place on the market, further studies must be done to: (1) test the effectiveness of combined products on actual industrial environments which may have multiple strains and very strong biofilms; (2) identify methods to control the release of NO; (3) to assess the neutralization/toxicity of the donors once depleted from the NO.

REFERENCES

Aguilar, C., Vlamakis, H., Losick, R., Kolter, R., 2007. Thinking about *Bacillus subtilis* as a multicellular organism. Curr. Opin. Microbiol. 10, 638–643.

Ahmed, E.M., 2015. Hydrogel: preparation, characterization, and applications. J. Adv. Res 6, 105–121.

Arasimowicz-Jelonek, M., Floryszak-Wieczorek, J., 2014. Nitric oxide: an effective weapon of the plant or the pathogen? Mol. Plant Pathol. 15, 406–416.

Astruc, D., Boisselier, E., Ornelas, C., 2010. Dendrimers designed for functions: from physical, photophysical, and supramolecular properties to applications in sensing, catalysis, molecular electronics, photonics, and nanomedicine. Chem. Rev. 110 (4), 1857–1959.

Azizi, S., Ahmad, M.B., Hussein, M.Z., Ibrahim, N.A., 2013. Synthesis, antibacterial and thermal studies of cellulose nanocrystal stabilized ZnO–Ag heterostructure nanoparticles. Molecules 18 (6), 6269–6280.

Backlund, C.J., Worley, B.V., Schoenfisch, M.H., 2016. Anti-biofilm action of nitric oxide-releasing alkyl-modified poly(amidoamine) dendrimers against *Streptococcus mutans*. Acta Biomater. 29, 198–205.

Barnes, R.J., Bandi, R.R., Chua, F., Low, J.H., Aung, T., Barraud, N., et al., 2014. The roles of *Pseudomonas aeruginosa* extracellular polysaccharides in biofouling of reverse osmosis membranes and nitric oxide induced dispersal. J. Membr. Sci. 466, 161–172.

Barnes, R.J., Low, J.H., Bandi, R.R., Tay, M., Chua, F., Aung, T., et al., 2015. Nitric oxide treatment for the control of reverse osmosis membrane biofouling. Appl. Environ. Microbiol. 81 (7), 2515–2524.

Barraud, N., Hassett, D.J., Hwang, S.H., Rice, S.A., Kjelleberg, S., Webb, J.S., 2006. Involvement of nitric oxide in biofilm dispersal of *Pseudomonas aeruginosa*. J. Bacteriol. 188, 7344–7353.

Barraud, N., Kardak, B.G., Yepuri, N.R., Howlin, R.P., Webb, J.S., Faust, S.N., et al., 2012. Cephalosporin-3′-diazeniumdiolates: targeted NO-donor prodrugs for dispersing bacterial biofilms. Angew. Chem. Int. Ed. Engl. 51, 9057–9060.

Barraud, N., Kelso, M.J., Scott, A.Ra, 2015. Nitric oxide: a key mediator of biofilm dispersal with applications in infectious diseases. Curr. Pharm. Des 21, 31–42.

Barraud, N., Schleheck, D., Klebensberger, J., Webb, J.S., Hassett, D.J., Rice, S.A., et al., 2009a. Nitric oxide signaling in *Pseudomonas aeruginosa* biofilms mediates phosphodiesterase activity, decreased cyclic di-GMP levels, and enhanced dispersal. J. Bacteriol. 191, 7333–7342.

Barraud, N., Storey, M.V., Moore, Z.P., Webb, J.S., Rice, S.A., Kjelleberg, S., 2009b. Nitric oxide-mediated dispersal in single- and multi-species biofilms of clinically and industrially relevant microorganisms. Microb. Biotechnol. 2, 370–378.

Benini, P.G.Z., McGarvey, B.R., Franco, D.W., 2008. Functionalization of PAMAM dendrimers with [RuIII(edta)(H₂O)]−. Nitric Oxide 19 (3), 245–251.

Cai, W., Wu, J., Xi, C., Meyerhoff, M.E., 2012. Diazeniumdiolate-doped poly(lactic-*co*-glycolic acid)-based nitric oxide releasing films as antibiofilm coatings. Biomaterials 33 (32), 7933–7944.

Cai, W., Tang, Z., Li, J., 2015. Removal of nitric oxide from simulated gas by the corona discharge combined with cobalt ethylenediamine solution. Fuel Process. Technol. 140, 82–87.

Carpenter, A.W., Worley, B.V., Slomberg, D.L., Schoenfisch, M.H., 2012. Dual action antimicrobials: nitric oxide release from quaternary ammonium-functionalized silica nanoparticles. Biomacromolecules 13 (10), 3334–3342.

Chang, W., Peng, K., Hu, T., Chiu, S., Liu, Y., 2015. Nitric oxide-releasing S-nitrosothiol-modified silica/chitosan core–shell nanoparticles. Polymer 57, 70–76.

Charville, G.W., Hetrick, E.M., Geer, C.B., Schoenfisch, M.H., 2008. Reduced bacterial adhesion to fibrinogen-coated substrates via nitric oxide release. Biomaterials 29, 4039–4044.

Corcoran, M., Morris, D., De Lappe, N., O'Connor, J., Lalor, P., Dockery, P., et al., 2014. Commonly used disinfectants fail to eradicate *Salmonella enterica* biofilms from food contact surface materials. Appl. Environ. Microbiol. 80, 1507–1514.

Crane, B., 2008. The enzymology of nitric oxide in bacterial pathogenesis and resistance. Biochem. Soc. Trans. 36 (6), 1149–1154.

Darling, K.E., Evans, T.J., 2003. Effects of nitric oxide on *Pseudomonas aeruginosa* infection of epithelial cells from a human respiratory cell line derived from a patient with cystic fibrosis. Infect. Immun. 71, 2341–2349.

Davey, M.E., O'Toole, G.A., 2000. Microbial biofilms: from ecology to molecular genetics. Microbiol. Mol. Biol. Rev. 64, 847–867.

Doro, F.G., Rodrigues-Filho, U.P., Tfouni, E., 2007. A regenerable ruthenium tetraammine nitrosyl complex immobilized on a modified silica gel surface: preparation and studies of nitric oxide release and nitrite-to-NO conversion. J. Colloid Interface Sci. 307 (2), 405–417.

Drogat, N., Granet, R., Sol, V., Memmi, A., Saad, N., Klein Koerkamp, C., et al., 2011. Antimicrobial silver nanoparticles generated on cellulose nanocrystals. J. Nanopart. Res. 13 (4), 1557–1562.

Duong, H.T.T., Jung, K., Kutty, S.K., Agustina, S., Adnan, N.N.M., Basuki, J.S., et al., 2014. Nanoparticle (star polymer) delivery of nitric oxide effectively negates *Pseudomonas aeruginosa* biofilm formation. Biomacromolecules 15 (7), 2583–2589.

Eroy-Reveles, A., Mascharak, P.K., 2009. Nitric oxide-donating materials and their potential in pharmacological applications for site-specific nitric oxide delivery. Future Med. Chem. 1 (8), 1497–1507.

Feese, E., Sadeghifar, H., Gracz, H.S., Argyropoulos, D.S., Ghiladi, R.A., 2011. Photobactericidal porphyrin-cellulose nanocrystals: synthesis, characterization, and antimicrobial properties. Biomacromolecules 12 (10), 3528–3539.

Fernandez-Moure, J.S., Van Eps, J.L., Haddix, S., Bryan, N.S., Olsen, R., Cabrera, F., et al., 2014. Polyester mesh functionalization with nitric oxide releasing silica nanoparticles prevents MRSA colonization and biofilm formation in vitro and in vivo. J. Am. Coll. Surg. 219 (3), S65.

Firoved, A.M., Wood, S.R., Ornatowski, W., Deretic, V., Timmins, G.S., 2004. Microarray analysis and functional characterization of the nitrosative stress response in nonmucoid and mucoid *Pseudomonas aeruginosa*. J. Bacteriol. 186, 4046–4050.

Flemming, H., Wingender, J., 2010. The biofilm matrix. Nat. Rev. Micro. 8 (9), 623–633.

Friedman, A.J., Han, G., Navati, M.S., Chacko, M., Gunther, L., Alfieri, A., et al., 2008. Sustained release nitric oxide releasing nanoparticles: characterization of a novel delivery platform based on nitrite containing hydrogel/glass composites. Nitric Oxide 19 (1), 12–20.

George, J., Ramana, K.V., Bawa, A.S., Siddaramaiah, 2011. Bacterial cellulose nanocrystals exhibiting high thermal stability and their polymer nanocomposites. Int. J. Biol. Macromol. 48 (1), 50–57.

Georgii, J.L., Amadeu, T.P., Seabra, A.B., de Oliveira, M.G., Monte-Alto-Costa, A., 2011. Topical *S*-nitrosoglutathione-releasing hydrogel improves healing of rat ischaemic wounds. J. Tissue Eng. Regener. Med. 5 (8), 612–619.

Ghaffari, A., Miller, C.C., McMullin, B., Ghahary, A., 2006. Potential application of gaseous nitric oxide as a topical antimicrobial agent. Nitric Oxide 14 (1), 21–29.

Gilberthorpe, N.J., Poole, R.K., 2008. Nitric oxide homeostasis in *Salmonella* typhimurium: roles of respiratory nitrate reductase and flavohemoglobin. J. Biol. Chem. 283, 11146–11154.

Gusarov, I., Nudler, E., 2005. NO-mediated cytoprotection: instant adaptation to oxidative stress in bacteria. PNAS 102 (39), 13855–13860.

Gusarov, I., Shatalin, K., Starodubtseva, M., Nudler, E., 2009. Endogenous nitric oxide protects bacteria against a wide spectrum of antibiotics. Science 325 (5946), 1380–1384.

Heilman, B.J., Tadle, A.C., Pimentel, L.R., Mascharak, P.K., 2013. Selective damage to hyphal form through light-induced delivery of nitric oxide to *Candida albicans* colonies. J. Inorg. Biochem. 123, 18–22.

Hetrick, E.M., Prichard, H.L., Klitzman, B., Schoenfisch, M.H., 2007. Reduced foreign body response at nitric oxide-releasing subcutaneous implants. Biomaterials 28 (31), 4571–4580.

Hetrick, E.M., Shin, J.H., Paul, H.S., Schoenfisch, M.H., 2009. Anti-biofilm efficacy of nitric oxide-releasing silica nanoparticles. Biomaterials 30 (14), 2782–2789.

Howard, M.D., Hood, E.D., Zern, B., Shuvaev, V.V., Grosser, T., Muzykantov, V.R., 2014. Nanocarriers for vascular delivery of anti-inflammatory agents. Annu. Rev. Pharmacol. Toxicol. 54 (1), 205–226.

Hrabie, J.A., Klose, J.R., Wink, D.A., Keefer, L.K., 1993. New nitric oxide-releasing zwitterions derived from polyamines. J. Org. Chem. 58, 1472–1476.

Huh, A.J., Kwon, Y.J., 2011. Nanoantibiotics: a new paradigm for treating infectious diseases using nanomaterials in the antibiotics resistant era. J. Control. Release 156 (2), 128–145.

Jardeleza, C., Rao, S., Thierry, B., Gajjar, P., Vreugde, S., Prestidge, C.A., et al., 2014. Liposome-encapsulated ISMN: a novel nitric oxide-based therapeutic agent against *Staphylococcus aureus* biofilms. PLoS ONE 9 (3), e92117. <http://dx.doi.org/10.1371/journal.pone.0092117> (accessed 14.07.16).

Jardeleza, C., Thierry, B., Rao, S., Rajiv, S., Drilling, A., Miljkovic, D., et al., 2015. An in vivo safety and efficacy demonstration of a topical liposomal nitric oxide donor treatment for *Staphylococcus aureus* biofilm-associated rhinosinusitis. Transl. Res. 166 (6), 683–692.

Jen, M.C., Serrano, M.C., van Lith, R., Ameer, G.A., 2012. Polymer-based nitric oxide therapies: recent insights for biomedical applications. Adv. Funct. Mater. 22 (2), 239–260.

Keefer, L.K., Nims, R.W., Davies, K.M., Wink, D.A., 1996. "NONOates" (1-substituted diazen-1-ium-1,2-diolates) as nitric oxide donors: convenient nitric oxide dosage forms. Methods Enzymol 268, 281–293.

Kim, J., Saravanakumar, G., Choi, H.W., Park, D., Kim, W.J., 2014. A platform for nitric oxide delivery. J. Mater. Chem. B 2 (4), 341–356.

Klemm, D., Kramer, F., Moritz, S., Lindstrom, T., Ankerfors, M., Gray, D., et al., 2011. Nanocelluloses: a new family of nature-based materials. Angew. Chem. Int. Ed. Engl 50, 5438–5466.

Liu, N., Xu, Y., Hossain, S., Huang, N., Coursolle, D., Gralnick, J.A., et al., 2012. Nitric oxide regulation of cyclic di-GMP synthesis and hydrolysis in *Shewanella woodyi*. Biochemistry 51 (10), 2087–2099.

Lu, Y., Sun, B., Li, C., Schoenfisch, M.H., 2011. Structurally diverse nitric oxide-releasing poly(propylene imine) dendrimers. Chem. Mater. 23 (18), 4227–4233.

Lu, Y., Slomberg, D.L., Shah, A., Schoenfisch, M.H., 2013. Nitric oxide-releasing amphiphilic poly(amidoamine) (PAMAM) dendrimers as antibacterial agents. Biomacromolecules 14 (10), 3589–3598.

Lu, Y., Slomberg, D.L., Schoenfisch, M.H., 2014. Nitric oxide-releasing chitosan oligosaccharides as antibacterial agents. Biomaterials 35 (5), 1716–1724.

Marvasi, M., Visscher, P.T., Casillas Martinez, L., 2010. Exopolymeric substances (EPS) from *Bacillus subtilis*: polymers and genes encoding their synthesis. FEMS Microbiol. Lett. 313, 1–9.

Marvasi, M., Chen, C., Carrazana, M., Durie, I.A., Teplitski, M., 2014. Systematic analysis of the ability of nitric oxide donors to dislodge biofilms formed by *Salmonella enterica and Escherichia coli* O157:H7. AMB Express 4. <http://amb-express.springeropen.com/articles/10.1186/s13568-014-0042-y> (accessed 14.07.16).

Marvasi, M., Durie, I.A., McLamore, E.S., Vanegas, D.C., Chaturvedi, P., 2015. *Salmonella enterica* biofilm-mediated dispersal by nitric oxide donors in association with cellulose nanocrystal hydrogels. AMB Express 5, 28. <http://amb-express.springeropen.com/articles/10.1186/s13568-015-0114-7> (accessed 14.07.16).

Marxer, S.M., Rothrock, A.R., Nablo, B.J., Robbins, M.E., Schoenfisch, M.H., 2003. Preparation of nitric oxide (NO)-releasing sol-gels for biomaterial applications. Chem. Mater. 15 (22), 4193–4199.

McMullin, B.B., Chittock, D.R., Roscoe, D.L., Garcha, H., Wang, L., Miller, C.C., 2005. The antimicrobial effect of nitric oxide on the bacteria that cause nosocomial pneumonia in mechanically ventilated patients in the intensive care unit. Respir. Care 50 (11), 1451–1456.

Moretti, A.I.S., Souza Pinto, F.J.P., Cury, V., Jurado, M.C., Marcondes, W., Velasco, I.T., et al., 2012. Nitric oxide modulates metalloproteinase-2, collagen deposition and adhesion rate after polypropylene mesh implantation in the intra-abdominal wall. Acta Biomater. 8 (1), 108–115.

Mustafa, A.K., Gadalla, M.M., Snyder, S.H., 2009. Signaling by gasotransmitters. Sci. Signal. 2 (68) re2-re2. <http://dx.doi.org/10.1126/scisignal.268re2> (accessed 14.07.16).

Nablo, B.J., Chen, T., Schoenfisch, M.H., 2001. Sol–gel derived nitric-oxide releasing materials that reduce bacterial adhesion. JACS 123 (39), 9712–9713.

Nablo, B.J., Schoenfisch, M.H., 2003. Antibacterial properties of nitric oxide-releasing sol gels. J. Biomed. Mater. Res. A 67A (4), 1276–1283.

Nablo, B.J., Schoenfisch, M.H., 2004. Poly(vinyl chloride)-coated sol gels for studying the effects of nitric oxide release on bacterial adhesion. Biomacromolecules 5 (5), 2034–2041.

Nablo, B.J., Rothrock, A.R., Schoenfisch, M.H., 2005. Nitric oxide-releasing sol gels as antibacterial coatings for orthopedic implants. Biomaterials 26 (8), 917–924.

Naghavi, N., De Mel, A., Alavijeh, O.S., Cousins, B.G., Seifalian, A.M., 2013. Nitric oxide donors for cardiovascular implant applications. Small 9 (1), 22–35.

Pegalajar-Jurado, A., Wold, K.A., Joslin, J.M., Neufeld, B.H., Arabea, K.A., Suazo, L.A., et al., 2015. Nitric oxide-releasing polysaccharide derivative exhibits 8-log reduction against *Escherichia coli*, *Acinetobacter baumannii* and *Staphylococcus aureus*. J. Control. Release 217, 228–234.

Petrova, O.E., Sauer, K., 2012. Dispersion by *Pseudomonas aeruginosa* requires an unusual posttranslational modification of BdlA. PNAS 109, 16690–16695.

Plate, L., Marletta, M., 2012. Nitric oxide modulates bacterial biofilm formation through a multicomponent cyclic-di-GMP signaling network. Mol. Cell. 46 (4), 449–460.

Quinn, J.F., Whittaker, M.R., Davis, T.P., 2015. Delivering nitric oxide with nanoparticles. J. Control. Release 205, 190–205.

Regev-Shoshani, G., Crowe, A., Miller, C.C., 2013. A nitric oxide-releasing solution as a potential treatment for fungi associated with tinea pedis. J. Appl. Microbiol. 114 (2), 536–544.

Riccio, D.A., Schoenfisch, M.H., 2012. Nitric oxide release: part I. Macromolecular scaffolds. Chem. Soc. Rev. 41 (10), 3731–3741.

Roveda, A.C., Wagner, D., 2013. Nitric oxide releasing-dendrimers: an overview. Braz. J. Pharm. Sci. 49, 1–14.

Ryan, R.P., Lucey, J., O'Donovan, K., McCarthy, Y., Yang, L., Tolker-Nielsen, T., et al., 2009. HD-GYP domain proteins regulate biofilm formation and virulence in *Pseudomonas aeruginosa*. Environ. Microbiol 11 (5), 1126–1136.

Sapp, A.M., Mogen, A.B., Almand, E.A., Rivera, F.E., Shaw, L.N., Richardson, A.R., et al., 2014. Contribution of the *nos-pdt* operon to virulence phenotypes in methicillin-sensitive *Staphylococcus aureus*. PLoS ONE 9 (10), e108868. <http://dx.doi.org/10.1371/journal.pone.0108868> (accessed 14.07.16).

Seabra, A.B., Duran, N., 2010. Nitric oxide-releasing vehicles for biomedical applications. J. Mater. Chem. 20 (9), 1624–1637.

Seabra, A.B., Martins, D., Simões, M.M.S.G., Da Silva, R., Brocchi, M., De Oliveira, M.G., 2010. Antibacterial nitric oxide-releasing polyester for the coating of blood-contacting artificial materials. Artif. Organs 34 (7), E204–E214.

Seabra, A.B., Justo, G.Z., Haddad, P.S., 2015. State of the art, challenges and perspectives in the design of nitric oxide-releasing polymeric nanomaterials for biomedical applications. Biotechnol. Adv. 33 (6), 1370–1379.

Shatalin, K., Gusarov, I., Avetissova, E., Shatalina, Y., McQuade, L.E., Lippard, S.J., et al., 2007. *Bacillus anthracis*-derived nitric oxide is essential for pathogen virulence and survival in macrophages. PNAS 105 (3), 1009–1013.

Siro, I., Plackett, D., 2011. Microfibrillated cellulose and new nanocomposite materials: a review. Cellulose 17, 459–494.

Stasko, N.A., Schoenfisch, M.H., 2006. Dendrimers as a scaffold for nitric oxide release. JACS 128 (25), 8265–8271.

Stasko, N.A., Fischer, T.H., Schoenfisch, M.H., 2008. *S*-nitrosothiol-modified dendrimers as nitric oxide delivery vehicles. Biomacromolecules 9 (3), 834–841.

Stein, T., 2005. *Bacillus subtilis* antibiotics: structures, syntheses and specific functions. Mol. Microbiol. 56, 845–857.

Tfouni, E., Truzzi, D.R., Tavares, A., Gomes, A.J., Figueiredo, L.E., Franco, D.W., 2012. Biological activity of ruthenium nitrosyl complexes. Nitric Oxide 26 (1), 38–53.

Van Alst, N.E., Picardo, K.F., Iglewski, B.H., Haidaris, C.G., 2007. Nitrate sensing and metabolism modulate motility, biofilm formation, and virulence in *Pseudomonas aeruginosa*. Infect. Immun. 75, 3780–3790.

Wang, P., Cai, T., Taniguchi, N., 2005. Nitric Oxide Donors. Wiley-VCH, Weinheim.

Wingender, J., Neu, T.R., Flemming, H.C., 1999. Microbial Extracellular Polymeric Substances: Characterization, Structure, and Function. Springer.

Worley, B.V., Schilly, K.M., Schoenfisch, M.H., 2015. Anti-biofilm efficacy of dual-action nitric oxide-releasing alkyl chain modified poly(amidoamine) dendrimers. Mol. Pharmaceut. 12 (5), 1573–1583.

Xiong, Y., Liu, Y., 2010. Biological control of microbial attachment: a promising alternative for mitigating membrane biofouling. Appl. Microbiol. Biotechnol. 86, 825–837.

Zanichelli, P.G., Sernaglia, R.L., Franco, D.W., 2006. Immobilization of the ruII(edta)NO$^+$ ion on the surface of functionalized silica gel. Langmuir 22 (1), 203–208.

Zarena, A.S., Gopal, S., 2013. Dendrimer a new dimension in targeting biofilms. Mini. Rev. Med. Chem. 13 (10), 1448–1461.

Zhang, L., Mah, T.F., 2008. Involvement of a novel efflux system in biofilm-specific resistance to antibiotics. J. Bacteriol. 190, 4447–4452.

CHAPTER 10

Developing New Organic Nitrates for Treating Hypertension

Camille M. Balarini[1,2], Josiane C. Cruz[2], José L.B. Alves[1], Maria S. França-Silva[2] and Valdir A. Braga[2]

[1]Health Sciences Center, Federal University of Paraiba, João Pessoa, Paraíba, Brazil
[2]Biotechnology Center, Federal University of Paraíba, João Pessoa, Paraíba, Brazil

INTRODUCTION

Nitric oxide (NO) is a free-radical diatomic gas, presenting a shot half-life. It is considered as one of the most important mediators of intra- and extracellular regulating processes (Wever et al., 1998). NO is produced endogenously by oxidation of the precursor amino acid L-arginine to L-citrulline, in a reaction catalyzed by the enzyme NO synthase (NOS). This enzyme presents two constitutive isoforms: endothelial (eNOS) and neuronal (nNOS), and an inducible isoform (iNOS) (Rudolph and Freeman, 2009). The effective activity of eNOS requires that the enzyme is in its dimeric form and the participation of different cofactors such as calcium, calmodulin, nicotinamide adenine dinucleotide phosphate in the reduced form (NADPH) and tetrahydrobiopterin (BH_4) (Förstermann and Sessa, 2012; Vanhoutte, 2009).

Once it is released by endothelial cells, NO diffuses to subendothelial space where induces the relaxation of vascular smooth muscle cells (VSMCs) and inhibits VSMC proliferation and remodeling (Lincoln et al., 2001; Mitchell et al., 2008). Intracellular action mechanisms of NO includes, but is not limited to, the activation of soluble guanylate cyclase (sGC) due to the conformational alteration induced by the formation of the iron–nitrosyl intermediate in the heme prosthetic group of this enzyme (Rudolph and Freeman, 2009). The activation of sGC leads to the production of cyclic guanosine monophosphate (cGMP), which regulates several effector systems such as cGMP-dependent protein kinases, ion channels, and phosphodiesterases (Follmann et al., 2013).

The NO/cGMP/protein kinase G (PKG) is considered one of the most important pathways in the regulation of vascular tone (Fig. 10.1). In VSMC, it is involved in the decrease of intracellular calcium, calcium desensitization, and thin filament regulation (Lincoln et al., 2001). The decrease in calcium concentration is related to removal of this ion through calcium pumps, inhibition of voltage-gated calcium channels; inhibition of receptor/G-protein coupling; increase in calcium-activated potassium channels,

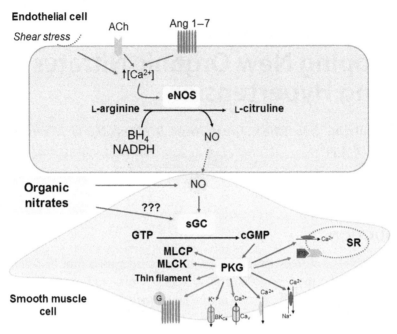

Figure 10.1 Schematic representation of physiological NO production and its action on vascular smooth muscle cells. Different stimuli such as shear stress and binding of ligands to receptors lead to increase in calcium content on endothelial cells. Calcium/calmodulin complex activates eNOS, which in the presence of cofactors, catalyzes the conversion of L-arginine to L-citrulline and NO release. NO diffuses to subjacent VSMC and activates sGC. The increase in cGMP activates PKG, the main protein kinase responsible for NO actions on the vascular tone. PKG increases calcium uptake by sarcoplasmic endoplasmic reticulum calcium pump (SERCA) and favors calcium elimination by membrane pumps and sodium/calcium exchanger. PKG also inhibits voltage-dependent calcium channels (Ca_V) and activates calcium-dependent potassium channels (BK_{Ca}). It inhibits receptor/G-protein coupling and negatively regulates thin and thick filaments, resulting in relaxation of VSMC. The *red arrows* indicate "inhibition" while the *blue arrows* indicate "activation." MLCK, myosin light chain kinase; MLCP, myosin light chain phosphatase.

leading to hyperpolarization; regulation of inositol 1,4,5-triphosphate (IP3) receptor and decreasing calcium release from sarcoplasmic reticulum (SR) (Lincoln et al., 2001; Münzel et al., 2014). The action of PKG on myosin light chain phosphatase (MLCP) contributes to calcium desensitization. In addition, PKG participates in thin filament regulation through two regulatory proteins (VASP and HSP20) (Lincoln et al., 2001). It is important to highlight that recently it has been shown that organic nitrates also perform epigenetic regulation of smooth muscle relaxation. It was described that nitroglycerine (NTG), a well-known organic nitrate that acts as NO donor, increases histone acetylase activity and that *N*-lysine acetylation of contractile proteins influences the vascular responses to this drug (Münzel et al., 2014).

Reduced bioavailability of NO is involved in the pathophysiology of many cardio-vascular disorders (Vasquez et al., 2016). Therefore the use of NO donor drugs, including organic nitrates, appears to be an effective alternative to replace the deficient endoge-nous NO, mimic the role of this molecule in the body, and consequently contribute to the treatment of cardiovascular disorders (Pörsti and Paakkari, 1995; França-Silva et al., 2012a,b; Queiroz et al., 2014; Mendes-Júnior et al., 2015; Porpino et al., 2016).

HISTORICAL PERSPECTIVE

The identity of NO was only revealed in the 1980s, but its vasodilatory properties have been used in pharmacotherapy for more than 150 years. NTG was discovered by Sobrero in 1847, who observed a violent headache after the used of the newly dis-covered explosive (Marsh and Marsh, 2000). Alfred Nobel also worked with Sobrero's compound and discovered a way to transform the explosive substance in a more con-trollable formulation, which he called "dynamite" and patented it in 1867 (Follmann et al., 2013). Physicians at that time observed that organic nitrates were effective in relieving chest pain due to angina pectoris (Murrel, 1879; Follmann et al., 2013). The phenomenon of nitrate tolerance was also described. Brunton noted that if a nitrate was used for a long time, the dose needed to be increased before the effect was repro-duced (Brunton, 1870; Marsh and Marsh, 2000). Ironically, Nobel's physician recom-mended NTG for his angina pectoris, which he refuses to use (Marsh and Marsh, 2000; Follmann et al., 2013).

Although the exact mechanism of action for organic nitrates was not yet established, Brunton suggested in 1870 that nitrates produced vasodilation due to a direct action in the vessel wall (Brunton, 1870). Murad et al. (1978) proposed in the 1970s that organic nitrates like glyceryl trinitrate and sodium nitroprusside (SNP) could stimulate sGC, producing relaxation due to increase in cGMP content. Considering that inhaled NO also increased sGC activity, they suggest that cGMP activation was due to NO for-mation from nitrates (Marsh and Marsh, 2000). At that time, Furchgott and Zawadzki observed that acetylcholine (ACh)-induced vasodilation *in vitro* was dependent on the presence of functional endothelium. Using the elegant experiment called "sandwich" (Fig. 10.2), where a transverse strip of aorta without endothelium was placed in con-tact with the intimal surface of a longitudinal strip with functional endothelium, they proved that the endothelium was capable of releasing a diffusible relaxing substance in response to ACh stimulation (Furchgott et al., 1981; Furchgott, 1999). Although they knew that NO could activate sGC, the link was not immediate and the substance was first identified as endothelium-derived relaxing factor (EDRF) (Marsh and Marsh, 2000). It was already demonstrated that there was a positive relationship between an increase in cGMP and relaxation, which lead the idea that EDRF could elicit activation of sGC (Furchgott, 1999). In 1987, Ignarro and Moncada groups concluded that EDRF

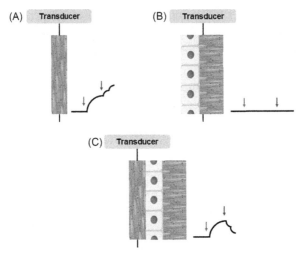

Figure 10.2 Sandwich experiment proposed by Furchgott and Zawadzzki. (A) Preparation with only a transverse strip of aorta without functional endothelium. In response to norepinephrine (NE, *blue arrow*) a contraction is observed. ACh (*red arrow*) fails to promote relaxation and further increased the contraction. (B) Preparation with only a longitudinal strip with functional endothelium. This preparation does not respond to NE or ACh. (C) Sandwich preparation: the endothelial surface of longitudinal strip is placed in contact with intimal surface of a transverse strip. Under these experimental conditions, ACh produces relaxation to NE-induced contractile tone.

was either NO or a chemically related radical (Ignarro et al., 1987; Palmer et al., 1987). In 1998, Furchgott, Murad, and Ignarro received the Nobel Prize in Physiology and Medicine for their discoveries concerning NO as a signaling molecule in the cardiovascular system. Unfortunately, Moncada was excluded from the Prize.

After the introduction of NTG as a therapeutic agent for the treatment of angina pectoris, other nitro compounds with similar chemical properties were developed. Most recent studies show that, in addition to angina, NTG and other organic nitrates such as isosorbide dinitrate (ISDN) and isosorbide 5-mononitrate (ISMN) are able to improve left ventricular function in patients with congestive heart failure and pulmonary hypertension. Also, they show favorable effects on left ventricular remodeling after myocardial infarction and silent ischemia, while reducing blood pressure alone or in combination with other drugs (França-Silva et al., 2014).

IMPORTANT ORGANIC NITRATES USED IN CLINIC

Organic nitrates are nitric acid esters of mono- and polyhydric alcohols, representing the oldest class of NO donors that have been clinically applied. The main representatives of this class are NTG, ISDN, ISMN, and pentaerythritol tetranitrate (PETN) (Wang et al., 2002; França-Silva et al., 2014). Those drugs can be administered as sublingual tablets, capsules, sprays, patches, or ointments (Münzel et al., 2014). All these compounds present a similar molecular structure, with the nitrate ester bond ($R-O-NO_2$) as an essential feature. This chemical group confers unique biological properties to this class of compounds, characterized by NO release in biological media (Al-Sa'Doni and Ferro, 2000; Miller and Megson, 2007).

The donation of NO by organic nitrates and subsequently activation of NO/sGC/cGMP pathway is described as the main mechanism of action of these drugs (Münzel et al., 2014). Organic nitrates are potential NO donors in biological systems. In general, they do not exert their effect in cell-free systems and act as prodrugs that need to be bioactivated by either enzymatic or nonenzymatic pathways to release NO (Mayer and Beretta, 2008). Nonenzymatic activation of NTG involves the reaction with reduced thiols in cysteine or its derivatives, resulting in activation of sGC. Additionally, it was shown that ascorbate or an intermediate of ascorbate autoxidation can react with NTG in a nonenzymatic manner to release NO or NO-related species (Kollau et al., 2009; Mayer and Beretta, 2008). Enzymatic bioactivation of high-potency nitrates like NTG and PETN (tri- and tetranitrates, respectively) depends on the activity of cytosolic and/or mitochondrial aldehyde dehydrogenase (ALDH-2), which converts them into nitrite and the denitrated metabolite. In contrast, low-potency nitrates such as ISMN and ISDN (mono- and dinitrates, respectively) undergo activation through a mechanism independent of ALDH-2, which usually occurs in endoplasmic reticulum via P450 enzymes (Mayer and Beretta, 2008; Münzel et al., 2011; Münzel and Gori, 2013; França-Silva et al., 2014).

In the past few years, a new concept has emerged suggesting that vascular effects of organic nitrates, more specifically NTG, can be unassociated with NO releasing capability. It has been described that NO is present only when NTG concentrations considerably exceeded the plasma levels reached during clinical dosing (Núñez et al., 2005). Kleschyov et al. (2003) found a marked dissociation between the potent vascular effects of NTG and its poor NO donor properties. Although both NTG, ISDN and A23187 (calcium ionophore, an endothelium-dependent vasodilator) elicited almost full relaxation in experimental preparations of aorta precontracted with phenylephrine, NTG did not increased basal vascular NO production, in contrast to ISDN or A23187. Authors suggest that NTG can activate vascular sGC/cGMP/PKG pathway independently of NO release. Confirming this hypothesis, Núñez et al. (2005) compared intracellular levels of NO and relaxation capability in response to NO donors, ACh and NTG. Authors found that, although NTG produced important vascular relaxation, there was no increase in NO bioavailability. This suggests that NTG does not release free NO but different molecules that can activate sGC. The effects of different organic nitrates on myocytes confirm that NTG and PETN release different vasoactive molecules. NTG leads to the expression of cardiotoxic genes and inhibition of cardioprotective genes while PETN enhanced the expression of beneficial genes (Pautz et al., 2009).

Organic nitrates and other NO donors have been used for many years in the treatment of cardiovascular disorders such as angina pectoris, arterial hypertension (AH), heart failure, preeclampsia, and pulmonary hypertension (Follmann et al., 2013; Kalidindi et al., 2012; Maul et al., 2003; Miller and Wadsworth, 2009). Despite the benefits elicited by organic nitrates to the cardiovascular system, the prolonged use of these drugs may cause tolerance (Münzel et al., 2005). Next, we will discuss the use of organic nitrates as a therapeutic approach to treat AH, some mechanisms involved in the development of nitrate tolerance and present further information about new insights on this class of drugs, especially involving new molecules developed and tested in preclinical trials by our research group.

ORGANIC NITRATES AND ARTERIAL HYPERTENSION

Cardiovascular diseases are the major cause of deaths worldwide according to the World Health Organization (WHO). AH is an important risk factor for the development of cardiovascular diseases, such as coronary heart disease, stroke, and renal failure, as well as death. Approximately 40% of adults aged 25 and above had been diagnosed with hypertension worldwide. In Brazil, the prevalence of hypertension is of approximately 30% and around 33% of deaths are caused by cardiovascular diseases (Hedner et al., 2012; Minelli et al., 2016; Ribeiro et al., 2016). In this context, understanding the fundamental mechanisms underlying the etiology of essential hypertension and also developing effective treatments to reduce its potential clinical, social, and economic consequences is becoming progressively urgent.

AH is a chronic condition characterized by high levels of systolic blood pressure greater than 140 mmHg and/or diastolic blood pressure greater than 90 mmHg. The regulation of arterial pressure is one of the most complex physiological functions and depends on integrated actions of cardiovascular, renal, neural, and endocrine systems, all of which act on different time scales and with different means of maintaining the blood pressure in a physiological range. Alterations in reflex control of blood pressure are involved in the development and maintenance of AH. However, the underlying cause of AH has been difficult to identify due to its multifactorial nature, which involves a complex interaction between numerous factors such as genetic, physiological, and environmental characteristics.

It is well recognized that the endothelium plays essential homeostatic functions such as regulation of the vascular tone, circulation of blood cells, inflammation, and platelet activity, and consequently helps maintain suitable blood pressure levels. Studies conducted in experimental models of hypertension and hypertensive patients revealed that AH is linked to alterations related to endothelium-dependent vasorelaxation mechanisms, and these alterations are jointly known as endothelial dysfunction (Tousoulis et al., 2014). The mechanisms associated with endothelial dysfunction include: the reduction in the release of EDRF such as NO, EDHF, and/or prostacy-clin; reduced bioavailability of these factors, mainly NO; decreased bioavailability of L-arginine; changes in signal transduction; decreased sensitivity of vascular smooth muscle to the factors involved in vasorelaxation; increased production of endothelium-derived constrictor factors such as endothelin-1, PGH_2, thromboxane; increased oxidative stress, leading to increased degradation of NO due to its reaction with superoxide anion (Förstermann and Sessa, 2012; Thakali et al., 2006).

Currently, it has been well established that the decreased bioavailability of NO contributes in the pathogenesis, maintenance, and progression of many cardiovascular disorders, such as essential hypertension, stroke, atherosclerosis, coronary artery disease, platelet aggregation after percutaneous transluminal coronary angioplasty, ischemia/reperfusion injury, and other systemic diseases. Therefore, drugs which mimic the role of this molecule in the body replacing or augmenting endogenous NO are known to act in the treatment of several cardiovascular diseases, mainly in arterial hypertension and coronary artery disease. The beneficial effects of the NO donors on the cardiovascular dysfunctions are mainly related to the (1) improvement of endothelial function (Yeo et al., 2007; França-Silva et al., 2012a), (2) prevention of hypoxia-induced vasoconstriction (Bednov et al., 2015), (3) improvement of ischemia (Engel et al., 2014), and (4) reduction of cardiac hypertrophy and fibrosis in hypertensive animals (Chang et al., 2005; Porpino et al., 2016).

Organic nitrates, such as NTG, ISMN, ISDN, and PETN, are able to release NO and promote potent vasodilator effects improving symptoms in patients with AH. The mechanism responsible for vasodilation mediated by organic nitrates comprises releasing NO, which in turn activates the enzyme sGC, potassium channels, or other targets involved in the reduction of cytosolic calcium and inactivation of myosin light chain kinase (MLCK) (Ledoux et al., 2006; Mendes-Júnior et al., 2015). In addition, NO can cause vasodilation by sGC-independent mechanisms (Bolotina et al., 1994; Mistry and Garland, 1998).

Preclinical studies demonstrated that organic nitrates modulate vascular tone, reduce the blood pressure, and promote beneficial effects on hypertension (Katsumi et al., 2007; Levine et al., 2012). The treatment with PETN ($15\,mg\,kg^{-1}$ per day for 7 days) improved the endothelial function in angiotensin II-treated rats (Schuhmacher et al., 2010) and SHR (Dovinová et al., 2009). In addition, clinical trials have evaluated the effects of organic nitrates in patients with AH. ISMN significantly decreased systolic blood pressure in chronically treated elderly hypertensive patients. Those effects were potentiated when treatment with ISMN was associated with L-arginine (Stokes et al., 2003). In this same way, a randomized, placebo-controlled, double-blind, cross-over study performed in patients with resistant hypertension on their usual antihypertensive treatment verified that a combination of single oral doses of phosphodiesterase inhibitor (50 mg) and ISMN (10 mg) acutely produced additional blood pressure reduction in these patients (Oliver et al., 2010). Oral treatment with ISMN improved the control of systolic blood pressure in patients with isolated systolic hypertension (Abad-Pérezz et al., 2013). AH is commonly found in patients with chronic kidney diseases. Recently, an prospective open-label, randomized controlled study performed in patients with chronic continuous ambulatory peritoneal dialysis demonstrated that oral treatment with ISMN for 24 weeks effectively improved left ventricular hypertrophy and reduced blood pressure at the end of the treatment (Han Li and Shixiang Wang, 2013).

The use of organic nitrate to restore NO to the vasculature in hypertensive individuals or hypertensive rodents began about a century ago. However, due to the rapid development of tolerance and acute vasorelaxant response (Stewart, 1905), which are the main problems also associated to organic nitrates, today its use is limited to the treatment of hypertensive crisis and their long-term use for blood pressure reduction is not considered a feasible first-line option.

For example, long-term treatment with molsidomine ($100\,mg\,kg^{-1}$ per day) or PETN ($200\,mg\,kg^{-1}$ per day) during 6 weeks did not alleviate cardiovascular dysfunction in SHR (Kristek et al., 2003), neither resulted in any improvement in arterial blood pressure in this hypertensive model (Dovinová et al., 2009). Similarly, NTG has its effect attenuated during long-term treatment; it also promotes tolerance to other nitrates, a phenomenon known as cross-tolerance (Kosmicki, 2009). Taken together, these data show an important limitation of the existing NO donors used in the long-term treatment of AH, which might potentially be solved by the development and synthesis of new NO donors.

Our research group demonstrated that the 2-nitrate-1,3-dibuthoxypropan (NDBP), an organic nitrate obtained from the synthetic route of glycerin in the biodiesel production process, exhibits an convincing and promising hypotensive activity. Initially, França-Silva et al. (2012a, 2012b) demonstrated that NDBP-induced vasorelaxation in preconstricted resistance vessels and acutely promoted hypotension associated with bradycardia in normotensive and spontaneously hypertensive rats, suggesting a therapeutical potential for the treatment of essential hypertension. Surprisingly, a recent study performed in a well-established model of hypertension, in which blood pressure progressively increases during chronic infusion of angiotensin II, we demonstrated that the chronic treatment with NDBP ($40\,mg\,kg^{-1}$) was not associated with tolerance and effectively attenuated the development of hypertension on this model over the course of 14 days. In addition, we observed a robust reduction in NADPH oxidase activity with NDBP in our study, which was associated with attenuated blood pressure elevation in a model of AH caused by oxidative stress. Interestingly, similar effects on both NADPH oxidase activity and angiotensin II-induced hypertension have been observed with inorganic nitrate and nitrite. Therefore the mechanism of action of NDBP could possibly be dependent on metabolism, or release of nitrite from the molecule, which is then reduced by xantine oxidase to NO, rather than rapid metabolism to NO with subsequent oxidation to nitrite. However, future studies are warranted to elucidate the mechanisms and also compare the effects elicited by NDBP with existing antihypertensive drugs and with inorganic nitrate/nitrite. Taken together, our *in vivo* and *in vitro* findings using NDBP for the treatment of experimental hypertension suggest absence of tolerance, and a novel mechanism where NDBP-mediated NO release inhibits NADPH oxidase-derived oxidative stress. We believe that results from this study will help to advance the field toward clinical trials, considering that currently used organic nitrates exhibit several undesirable effects with frequent dosing (Porpino et al., 2016). These exciting findings demonstrate that the development of new organic nitrates devoided of the undesirable side effects, such as tolerance at long term, may represent an attractive strategy to future preclinical and clinical applications for this class of drugs in essential hypertension. In this way, a recent study evaluating the cardiovascular effects of NO donor cyclohexane nitrate demonstrated that intravenous acute administration of cyclohexane nitrate $(1-20\,mg\,kg^{-1})$ induced, in a dose-dependent fashion, hypotension and bradycardia in normotensive and renovascular hypertensive rats. Interestingly, subchronic oral administration with the cyclohexane nitrate $(10\,mg\,kg^{-1}$ per day) for 7 days produced a significant antihypertensive effect also in renovascular hypertensive rats (Mendes-Júnior et al., 2015). Considering their hypotensive properties at short and long terms and the absence of undesirable effects such as tolerance, future clinical trial studies will be required to evaluate the effects induced by the promising NDBP and cyclohexane nitrates on the blood pressure of patients with AH.

NITRATE TOLERANCE

Sustained therapy or acute high doses of organic nitrate induce a tachyphylaxis response, where NTG decreases its efficacy for augmented doses, inducing a phenomenon called tolerance, which organic nitrate lost its hemodynamic, antiischemic, and clinical effects (Thomas et al., 2007; Klemenska and Beręsewicz 2009; Tarkin and Kaski, 2016). The first description of nitrate tolerance was in 1888, when Dr. Stewart, an American physician, observed that a patient with angina under chronic treatment with NTG required a higher dose of the nitrate over the time to induce the same antianginal effects observed after the first dose. In supporting of Dr. Stewart findings, different clinical studies have shown development of tolerance in patients with acute myocardial infarction and chronic congestive heart failure (Jugdutt and Warnica, 1989; Elkayam et al., 1987). In addition, nitrate tolerance was also observed in experimental studies using rodents, both using *in vivo* or *in vitro* techniques (Bennett et al., 1998; Heitzer et al., 1998; Münzel et al., 1995; Parker et al., 1991; França-Silva et al., 2014; Porpino et al., 2016). Of note, tolerance has been observed after chronic treatment with ISDN and ISMN (Thomas et al., 2007; Tarkin and Kaski, 2016). Studies by

Thomas et al. (2007) observed a significant reduction in the ACh or L-MNNA (NOS isoforms inhibitor) induced forearm blood flow responses in healthy volunteers chronically treated with ISMN.

Nitrate tolerance is a complex and distinct phenomenon, which involves neurohormonal and intravascular volume expansion (defined as pseudotolerance) and a decrease in the vascular responsiveness (defined as true tolerance). Despite several studies about nitrate tolerance, mechanisms underlying this phenomenon remain completely unknown. There are several hypothesis trying to explain the mechanisms underlying nitrate tolerance: one of them supports the impairment in NTG bioactivation (Needleman et al. 1976; Chen et al., 2002). In fact, studies by Chen et al. (2002) showed that NTG higher concentration inhibits ALDH-2 dehydrogenase activity, an enzyme responsible for NTG bioactivation. However, this hypothesis does not explain the remarkable endothelial dysfunction observed in NTG tolerance. Furthermore, other studies proposed the oxidative stress hypothesis, suggesting that increase in the vascular reactive oxygen species (ROS) takes place during organic nitrate treatment, inducing the so-called true tolerance (Münzel et al., 1995, 2005, 2011; Sydow et al., 2004). In this way, Münzel et al. (1995) reported that NTG treatment increases vascular superoxide anion formation, consequently decreasing NO bioavailability in rabbit aortas by the formation of peroxynitrite. In addition, studies by Sydow et al. (2004) reported a significant decrease in vascular ALDH-2 activity, NTG biotransformation, as well as cGMP-dependent kinase (cGK-I) activity in rats treated with NTG. Furthermore, the same study observed a significant increase in ROS production in isolated mitochondria exposed to a higher concentration of NTG. These findings suggest that NTG tolerance is mediated by inhibition of vascular ALDH-2 and mitochondrial dysfunction. Moreover, further studies showed that mitochondrial ROS overproduction induces increases in vascular NADPH oxidase-derived ROS, suggesting that a cross-talk between mitochondrial and vascular NADPH-derived ROS is involved in the development of nitrate tolerance (Wenzel et al., 2008). The NADPH oxidase in a higher concentration reacts with NO (derived from NTG and from endothelium) producing the oxidant peroxynitrite ($ONOO^-$), which decreases NO bioavailability (Beckman and Koppenol, 1996). Together, all these studies strongly suggest that mitochondrial and endothelial oxidative stresses are part of the intrinsic mechanisms involved in the vascular development of nitrate tolerance.

The oxidative stress hypothesis becomes more attractive supported by evidence showing that application of antioxidants/reductants in tolerant vessels decreased mitochondrial-induced ROS production and restored ALDH-2 activity (Sydow et al. 2004; Daiber et al., 2004, 2005). Interestingly, there are several clinical and experimental evidences proposing combined therapies to prevent nitrate tolerance; coadministration of nitrate and antioxidant such as hydralazine (an efficient ROS scavenger), vitamin C, vitamin E, and folic acid (Daiber et al., 2005, 2008; Mayer and Beretta,

2008; Münzel et al., 2005; Thomas et al., 2007; Wenzel et al., 2008). Other alternatives to explore the clinical use of organic nitrate are the development of new organic nitrates (França-Silva et al., 2012a, 2012b, 2014; Santos, 2009). In fact, oxidative stress is a limitation for the clinical use of the existing organic nitrates, all these strategies to prevent nitrate tolerance indicate how it is important to fully understand the complex intrinsic mechanisms involved in how the organic nitrate induces tolerance and endothelial dysfunction. Recently our group developed a new organic nitrate obtained from glycerin, NDBP, which appears not to induce oxidative stress or tolerance.

NEW INSIGHTS ON DRUG DEVELOPMENT

Due to tolerance caused by organic nitrates, the search for new compounds of this class that are unable to induce this undesirable effect increased in the last years. Recently, our research group evaluated new organic nitrates obtained from glycerin, which is generated as a byproduct of the biodiesel production (França-Silva et al., 2012a, 2012b, 2014; Santos, 2009). Kirk-Othmer (2007) reported that the production of $90\,m^3$ of biodiesel generates about $10\,m^3$ of pure glycerin. The excess of this material causes environmental problems due to its accumulation in nature. Therefore, the extra volume of glycerin challenges the pharmaceutical and biotechnology companies to include it in large-scale applications. In this context, four compounds were obtained from glycerin, whose structures are shown in Fig. 10.3.

Firstly, we evaluated if these compounds were able to induce vascular relaxation, the main feature of organic nitrates. It was observed that all the tested organic nitrates caused vasorelaxation in superior mesenteric artery rings (SMAR) isolated from rats. The nitrates-induced relaxation was concentration dependent and endothelium

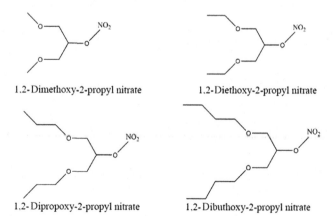

1.2-Dimethoxy-2-propyl nitrate   1.2-Diethoxy-2-propyl nitrate

1.2-Dipropoxy-2-propyl nitrate   1.2-Dibuthoxy-2-propyl nitrate

Figure 10.3 Structural formula of the organic nitrates obtained from glycerin as previously described by Santos (2009).

Table 10.1 Maximum relaxing effect (ME) and negative logarithm of the EC_{50} (pD_2) values for each organic nitrate tested

Compounds	Intact endothelium		Denuded endothelium	
	(%) ME ± SEM	pD_2 ± SEM	(%) ME ± SEM	pD_2 ± SEM
1,2-Dimethoxy-2-propyl nitrate	88.5 ± 11.2	4.7 ± 0.13	93.8 ± 11.7	4.4 ± 0.07
1,2-Diethoxy-2-propyl nitrate	94.1 ± 6.7	4.6 ± 0.08	108.8 ± 5.4	4.8 ± 0.06
1,2-Dipropoxy-2-propyl nitrate	96.4 ± 8.3	5.5 ± 0.10	111.1 ± 8.5	5.4 ± 0.08
1,2-Dibuthoxy-2-propyl nitrate	89.5 ± 3.4	5.8 ± 0.10	105.4 ± 2.7	5.9 ± 0.06

independent, as shown by França-Silva et al. (2012a). The maximum relaxing effect (ME) and negative logarithm of the EC50 (pD2) values for each compound are listed in Table 10.1. Data were previously published by França-Silva et al. (2014).

Although no significant differences were found in ME and pD2 values between the components studied, we observed a tendency toward increasing the vasodilator response as the number of carbon atoms in the carbonic chain increased. Thus we choose NDBP to further test its mechanism of action. Our laboratory had shown that NDBP-induced vasodilatation was attenuated after preincubation with the NO scavenger hydroxocobalamin (Kruszyna et al., 1998) or sGC inhibitor ODQ (1-H-[1,2,4] oxadiazolo-[4,3-a] quinoxalin-1-one) (Garthwaite et al., 1995), showing the participation of NO and soluble cyclase guanylyl enzyme in the effect induced by NDBP.

The pretreatment with KCl (20 mM), a K^+ efflux modulator (Campbell et al., 1996), significantly attenuated the vasorelaxant response promoted by NDBP. Moreover, the selective K^+ channels blockade with charybdotoxin, a blocker for large conductance calcium-sensitive potassium channels; glibenclamide, an ATP-sensitive potassium channel blocker, and 4-aminopyridine-4-AP, a voltage-operated potassium channel, also blunted the NDBP-induced vasodilatation. These data revealed that NDBP evoked potent vasorelaxation through NO generation and activation of the sCG/cGMP/PKG pathway with opening of K^+ channels (França-Silva et al., 2012a).

The kinetic of NO release was similar to other NO donors such as NTG and SNP, confirming this molecule as a new organic nitrate that acts as an NO donor. In agreement with these findings, *in vivo* preclinical studies proved that the administration of different doses of NDBP (1, 5, 10, 15, and 20 mg kg^{-1}) elicited dose-dependent hypotension in normotensive and SHR (Queiroz et al., 2014; França-Silva et al., 2012b). The blood pressure reduction was associated with intense bradycardia that showed to be dependent on NO and, more than that, dependent on an action of the compound in the central nervous system, since in vagotomized animals, the reduction in heart rate was significantly attenuated, a differential feature of the NDBP when compared to other NO donors such as SNP and NTG that induce reflex tachycardia (Needleman, 1976; Caputi et al., 1980). Besides to promote vasorelaxation by NO release, the

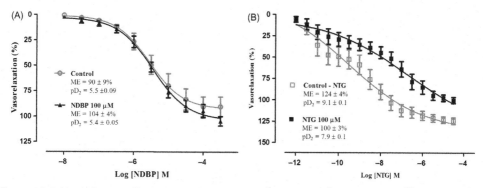

Figure 10.4 Vasorelaxant effect induced by NDBP (10^{-8} to 3×10^{-4} M) or NTG (10^{-12} to 3×10^{-5}) in mesenteric arteries isolated from rats subjected to prior incubation with NDBP or NTG 100 μM compared to the group not exposed to preincubation (Control) ($n = 7$ per group). The vasorelaxant effect was expressed as a percentage of relaxation of phenylephrine-induced contraction and data were expressed as mean ± SE, *$p < 0.05$ versus specific control (Porpino et al., 2016).

compound was able to act in brain areas and to stimulate the vagal activity directed to the heart. This effect reduces cardiac output and consequently the blood pressure.

In addition to the cardiovascular effects, NDBP is also able to induce changes in respiratory frequency. NDBP when administered in the doses of 1, 5, 10, 15, and 20 mg kg^{-1} was able to induce bradypnoea such as NTG and this effect was attenuated by hydroxocobalamin, showing the involvement of NO (Queiroz et al., 2014). All the actions induced by NDBP need to be further studied to assess the physiological relevance and the side effects of this molecule as medicine for treating cardiovascular disorders, including hypertension.

Considering that tolerance is one of the most important undesirable effect evoked by organic nitrates, we performed experimental trials to evaluate the ability of NDBP to induce tolerance. Experimental protocols were conducted according to Daiber et al. (2004). The cumulative addition of NDBP (10^{-8} to 10^{-4} M) in SMAR isolated from rats promoted endothelium-independent vasodilation. Previous exposition of SMAR to NDBP (100 μM) for 60 minutes did not alter the vasorelaxant response induced by cumulative addition of the NDBP (10^{-8} to 10^{-4} M), suggesting that NDBP did not induce tolerance to vasodilatation *in vitro*, unlike NTG under the same experimental conditions (Porpino et al., 2016), as shown in Fig. 10.4. Taken together, these data confirm the promising role of NDBP as an NO donor with potential to be used clinically on cardiovascular system.

In order to better understand the mechanisms by which NDBP induces cardiovascular changes without tolerance, a recent preclinical study was conducted by our research group using a combination of *ex vivo*, *in vitro*, and *in vivo* techniques in a model of hypertensive mice under chronic angiotensin II infusion. It was demonstrated that treatment with NDBP promotes vasorelaxation in mesenteric resistance arteries, and dose-dependent and sustained NO generation in liver and kidney. Furthermore, NDBP reduced NADPH oxidase activity in liver and prevented angiotensin II-induced activation of NADPH oxidase in kidney. Interestingly, *in vivo* studies showed that NDBP attenuated hypertension induced by angiotensin II infusion, which was accompanied by reduction in cardiac hypertrophy, NADPH oxidase-derived oxidative stress, and fibrosis in kidney and heart. These findings suggest a novel molecular mechanism: NDBP-mediated NO release inhibits NADPH oxidase-derived oxidative stress, attenuating angiotensin II-induced hypertension (Porpino et al., 2016). These preclinical studies confirm that this molecule has antihypertensive effects and this effect is also due to the ability of the NDBP to reduce oxidative stress through NO release.

Probably the decrease in oxidative stress induced by NDBP explains the absence of tolerance induced by the compound, since oxidative stress is closely related to tolerance to organic nitrates as aforementioned. However, further studies *in vitro* and *in vivo* regarding potential of NDBP to induce tolerance need to be performed.

Another study by our research group evaluated the effects of another organic nitrate, the cyclohexane nitrate. It was synthesized from cyclohexanol as explained by Mendes-Júnior et al. (2015) (Fig. 10.5). Our data show that cyclohexane nitrate increases NO levels in endothelium-denuded mesenteric arteries, promoting vasorelaxation by NO release and activation of sGC. However, unlike NDBP, the vasorelaxant effect induced by cyclohexane nitrate was dependent of only ATP-sensitive K^+ channels (KATP), since blockade of other K^+ channels did not alter the vasodilatation induced by cyclohexane nitrate.

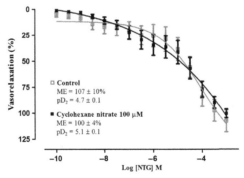

Figure 10.5 Vasorelaxant effect induced by cyclohexane nitrate (10^{-10} to 10^{-3}M) in mesenteric arteries isolated from rats subjected to prior incubation with cyclohexane nitrate (100 μM) compared to the group not exposed to preincubation (Control) (n = 7 per group). The vasorelaxant effect was expressed as a percentage of relaxation of phenylephrine-induced contraction and data were expressed as mean ± SE, *p < 0.05 versus specific control (unpublished data).

Furthermore, cyclohexane nitrate promotes hypotension in normotensive sham and with renovascular hypertensive rats. The renovascular hypertension model (2-kidney-1-cip, 2K1C), developed for the first time by Goldblatt et al. (1934), is featured by an implantation of an U-shaped silver clip over the renal artery to reduce renal blood flow. Reduced blood perfusion stimulates the secretion of renin by the juxtaglomerular cells from afferent arteriole, the increase in endogenous renin–angiotensin system activity, and consequently activation of AT1 receptors contributes to establishment of hypertension (Navar et al., 1998; Fazan Jr et al., 2001; Cervenka et al., 2008).

High blood pressure is accompanied by ROS generation in the central nervous system and periphery, with endothelial dysfunction, vascular remodeling, and increased sympathetic activity (Intengan and Schiffrin, 2001; Oliveira-Sales et al., 2008; Costa et al., 2009). ROS may directly alter vascular function or cause changes in vascular tone by several mechanisms including altered NO bioavailability or signaling (Schulz et al., 2011). In rats, the increase of blood pressure reaches the maximum at sixth week after implantation of the silver clip. We observed that oral administration of cyclohexane nitrate for 7 days, from fifth week after clip implantation, induced antihypertensive effect in rats with renovascular hypertension, showing promising effects of this nitrate against the AH.

Experimental assays to evaluate the ability of cyclohexane nitrate in inducing tolerance to vasodilatation *in vitro* were performed. The previous incubations with cyclohexane nitrate ($100\,\mu$M) for 60 minutes did not alter the vasorelaxant response induced by cumulative addition of the nitrate (10^{-10} to 10^{-3} M) such as NDBP (unpublished data), as shown in Fig. 10.5.

In the battle against cardiovascular disease, and its adverse complications, new strategies to increase NO production or bioavailability have gained a lot of interest. This chapter highlights the potential and limitations of the organic nitrates as NO donors. Our data indicate a new way to go with new compounds belonging to the class of NO donors named organic nitrates, currently sidelined due to their undesirable side effects. These compounds may help us to understand about nitrate tolerance mechanisms and possibly can play a role as a new therapeutic alternatives for treating cardiovascular disorders in substitution to current available NO donors.

ACKNOWLEDGMENTS

This work was funded by the Conselho Nacional de Desenvolvimento Científico e Tecnológico (CNPq) and Coordenação de Aperfeiçoamento de Pessoal de Ensino Superior (CAPES).
Author Contributions: All authors were involved in drafting and reviewing of the manuscript.
Conflicts of Interest: The authors declare no conflict of interest.

REFERENCES

Abad-Pérez, D., Novella-Arribas, B., Rodríguez-Salvanés, F.J., Sánchez-Gómez, L.M., García-Polo, I., Verge-González, C., et al., 2013. Effect of oral nitrates on pulse pressure and arterial elasticity in patients aged over 65 years with refractory isolated systolic hypertension: study protocol for a randomized controlled trial. Trials 14, 388. http://dx.doi.org/10.1186/1745-6215-14-388.

Al-Sa'Doni, H., Ferro, A., 2000. S-Nitrosothiols: a class of nitric oxide-donor drugs. Clin. Sci. (Lond.) 98, 507–520. http://dx.doi.org/10.1042/cs0980507.

Bednov, A., Espinoza, J., Betancourt, A., Vedernikov, Y., Belfort, M., Yallampalli, C., 2015. L-Arginine prevents hypoxia-induced vasoconstriction in dual-perfused human placental cotyledons. Placenta. 36(11), 1254-1259. http://dx.doi.org/10.1016/j.placenta.2015.08.012.

Beckman, J.S., Koppenol, W.H., 1996. Nitric oxide, superoxide, and peroxynitrite: the good, the bad, and ugly. Am. J. Physiol 271 (5 Pt 1), C1424–C1437. Nov.

Bennett, B.M., Schroder, H., Hayward, L.D., Waldman, S.A., Murad, F., 1998. Effect of in vitro organic nitrate tolerance on relaxation, cyclic GMP accumulation, and guanylate cyclase activation by glyceryl trinitrate and the enantiomers of isoidide dinitrate. Circ. Res. 63, 693–701. http://dx.doi.org/10.1161/01.RES.63.4.693.

Bolotina, V.M., Najibi, S., Palacino, J.J., Pagano, P.J., Cohen, R.A., 1994. Nitric oxide directly activates calcium-dependent potassium channels in vascular smooth muscle. Nature 368, 850–853. http://dx.doi.org/10.1038/368850a0.

Brunton, T.L., 1870. The Action of Nitrate of Amyl on the Circulation. J. Anat. Physiol. 5 (Pt 1), 92–101. Nov.

Campbell, W.B., Gebremedhin, D., Prait, P.F., Herder, D.R., 1996. Identification of epoxyeicosatrienoic acids as endothelium-derived hyperpolarizing factors. Circ. Res. 78, 415–423.

Caputi, A.P., Rossi, F., Carney, K., Brezenoff, H.E., 1980. Modulatory effect of brain acetylcholine on reflex-induced bradycardia and tachycardia in conscious rats. JPET 215, 309–316.

Cervenka, L., Vanièková, I., Husková, Z., Vaòourková, Z., Erbanová, M., Thumová, M., et al., 2008. Pivotal role of AT_{1A} receptors in the development of two-kidney, one-clip hypertension: study in AT1A receptor knockout mice. J. Hypertens. 26, 1379–1389. http://dx.doi.org/10.1097/HJH.0b013e3282.

Chang, H., Wu, C.Y., Hsu, Y.H., Chen, H.I., 2005. Reduction of ventricular hypertrophy and fibrosis in spontaneously hypertensive rats by L-arginine. Chin. J. Physiol. 48, 15–22.

Chen, Z., Zhang, J., Stamler, J.S., 2002. Identification of the enzymatic mechanism of nitroglycerin bioactivation. Proc. Natl. Acad. Sci. USA 99, 8306–8311. http://dx.doi.org/10.1073/pnas.122225199.

Costa, C.A., Amaral, T.A., Carvalho, L.C., Ognibene, D.T., da Silva, A.F., Moss, M.B., et al., 2009. Antioxidant treatment with tempol and apocynin prevents endothelial dysfunction and development of renovascular hypertension. Am. J. Hypertens. 22, 1242–1249. http://dx.doi.org/10.1038/ajh.2009.186.

Daiber, A., Mulsch, A., Hink, U., Mollnau, H., Warnholtz, A., Oelze, M., et al., 2005. The oxidative stress concept of nitrate tolerance and the antioxidant properties of hydralazine. Am. J. Cardiol. 96, 25I–36I. http://dx.doi.org/10.1016/j.amjcard.2005.07.030.

Daiber, A., Oelze, M., Coldewey, M., Bachschmid, M., Sydow, K., Wendt, M., et al., 2004. Oxidative stress and mitochondrial aldehyde dehydrogenase activity: a comparison of pentaerythritol tetranitrate with other organic nitrates. Mol. Pharmacol. 66, 1372–1382. http://dx.doi.org/10.1124/mol.104.002600.

Daiber, A., Wenzel, P., Oelze, M., Munzel, T., 2008. New insights into bioactivation of organic nitrates, nitrate tolerance and cross-tolerance. Clin. Res. Cardiol. 97, 12–20.

Dovinová, I., Cacányiová, S., Fáberová, V., Kristek, F., 2009. The effect of an NO donor, pentaerythrityl tetranitrate, on biochemical, functional, and morphological attributes of cardiovascular system of spontaneously hypertensive rats. Gen. Physiol. Biophys. 1, 86–93.

Elkayam, U., Kulick, D., McIntosh, N., Roth, A., Hsueh, W., Rahimtoola, S.H., 1987. Incidence of early tolerance tohemodynamic effects of continuous infusion of nitroglycerin in patients withcoronary artery disease and heart failure. Circulation 76 (3), 577–584. http://dx.doi.org/10.1161/01.CIR.76.3.577.

Engel, H., Friedrich, S., Schleich, C., Heyer, M., Gebhardt, M.M., Gross, W., et al., 2014. Enhancing nitric oxide bioavailability via exogen nitric oxide synthase and L-arginine attenuates ischemia-reperfusion-induced microcirculatory alterations. Ann. Plast. Surg 29, 2014. http://dx.doi.org/10.1097/SAP.0000000000000148.

Fazan Jr, R., Silva, V.J.D., Salgado, H.C., 2001. Modelos de hipertensão arterial. Rev. Bras. Hypertens. 8 (1), 19–29.

Follmann, M., Griebenow, N., Hahn, M.G., Hartung, I., Mais, F.J., Mittendorf, J., et al., 2013. The chemistry and biology of soluble guanylate cyclase stimulators and activators. Angew. Chem. Int. Ed. Engl 52, 9442–9462. http://dx.doi.org/10.1002/anie.201302588.

Förstermann, U., Sessa, W.C., 2012. Nitric oxide synthases: regulation and function. Eur. Heart J. 33, 829a–837d. http://dx.doi.org/10.1093/eurheartj/ehr304.

França-Silva, M.S., Balarini, C.M., Cruz, J.C., Khan, B.A., Rampelotto, P.H., Braga, V.A., 2014. Organic nitrates: past, present and future. Molecules. 19, 15314–15323.

França-Silva, M.S., Luciano, M.N., Ribeiro, T.P., Silva, J.S., Santos, A.F., França, K.C., et al., 2012a. The 2-nitrate-1,3-dibuthoxypropan, a new nitric oxide donor, induces vasorelaxation in mesenteric arteries of the rat. Eur. J. Pharmacol. 690, 170–175. http://dx.doi.org/10.1016/j.ejphar.2012.06.043.

França-Silva, M.S., Monteiro, M.M.O., Queiroz, T.M., Santos, A.F., Athayde-Filho, P.F., Braga, V.A., 2012b. The new nitric oxide donor 2-nitrate-1,3-dibuthoxypropan alters autonomic function in spontaneously hypertensive rats. Autonom. Neurosci. Basic Clin. 171, 28–35. http://dx.doi.org/10.1016/j.autneu.2012.10.002.

Furchgott, R.F., 1999. Endothelium-derived relaxing factor: discovery, early studies, and identification as nitric oxide. Biosci. Rep. 19 (4), 235–251.

Furchgott, R.F., Zawadzki, J.V., Cherry, P.D., 1981. Role of endothelium in the vasodilator response to ACh. In: Vanhoutte, P., Leusen, I. (Eds.), Vasodilatation. Raven Press, New York, pp. 49–66.

Garthwaite, J., Southam, E., Boulton, C.L., Nielsen, E.B., Schmidt, K., Mayer, B., 1995. Potent and selective inhibition of nitric oxide-sensitive guanylyl cyclase by 1-H-[1,2,4]oxadiazo-lo[4,3-a]quinoxalin-1-one. Mol. Pharmacol. 48, 184–188.

Goldblatt, B.Y.H., Lynch, J., Ramon, F., Summerville, W.W., 1934. Studies on experimental hypertension. The production of persistent elevation of systolic blood pressure by means of renal ischemia.. J. Exp. Med. 59, 347–379. http://dx.doi.org/10.1084/jem.59.3.347.

Hedner, T., Kjeldsen, S.E., Narkiewicz, K., 2012. State of global health—hypertension burden and control. Blood Press. 21 (Suppl 1), 1–2. http://dx.doi.org/10.3109/08037051.2012.704786.

Heitzer, T., Just, H., Brockhoff, C., Meinertz, T., Olschewski, M., Münzel, T., 1998. Long-term nitroglycerin treatment is associated with supersensitivity to vasoconstrictors in men with stable coronary artery disease: prevention by concomitant treatment with captopril. J. Am. Coll. Cardiol. 31, 83–88. http://dx.doi.org/10.1016/S0735-1097(97)00431-2.

Ignarro, L.J., Byrns, R.E., Buga, G.M., Wood, K.S., 1987. Endothelium-derived relaxing factor from pulmonary artery and vein possesses pharmacologic and chemical properties identical to those of nitric oxide radical. Circ. Res. 61, 866–879.

Intengan, H.D., Schiffrin, E.L., 2001. Vascular remodeling in hypertension: roles of apoptosis, inflammation, and fibrosis. Hypertension 38 (3Pt2), 581–587. http://dx.doi.org/10.1161/hy09t1.096249.

Jugdutt, B.I., Warnica, J.W., 1989. Tolerance with low dose intravenous nitroglycerin therapy in acute myocardial infarction. Am. J. Cardiol. 64 (10), 581–587. http://dx.doi.org/10.1016/0002-9149(89)90482-7.

Kalidindi, M., Velauthar, L., Khan, K., Aquilina, J., 2012. The role of nitrates in the prevention of preeclampsia: an update. Curr. Opin. Obstet. Gynecol. 24, 361–367. http://dx.doi.org/10.1097/GCO.0b013e32835a31de.

Katsumi, H., Nishikawa, M., Hashida, M., 2007. Development of nitric oxide donors for the treatment of cardiovascular diseases. Cardiovasc. Hematol. Agents Med. Chem. 5, 204–208. http://dx.doi.org/10.2174/187152507781058735.

Kirk-Othmer, E.T., 2007. Glycerol American Society of Chemistry. Encyclopedia of Chemical Technology. John Wiley, New York.

Klemenska, E., Bereşewicz, A., 2009. Bioactivation of organic nitrates and the mechanism of nitrate tolerance. Cardiol. J. 16, 11–19.

Kleschyov, A.L., Oelze, M., Daiber, A., Huang, Y., Mollnau, H., Schulz, E., et al., 2003. Does nitric oxide mediate the vasodilator activity of nitroglycerin? Circ. Res. 93 (9), e104–e112.

Kollau, A., Beretta, M., Russwurm, M., Koesling, D., Keung, W.M., Schmidt, K., et al., 2009. Mitochondrial nitrite reduction coupled to soluble guanylate cyclase activation: lack of evidence for a role in the bioactivation of nitroglycerin. Nitric Oxide 20 (1), 53–60.

Kosmicki, M.A., 2009. Long-term use of short- and long-acting nitrates in stable angina pectoris. Curr. Clin. Pharmacol 4, 132–141.

Kristek, F., Fáberová, V., Varga, I., 2003. Long-term effect of molsidomine and pentaerythrityl tetranitrate on cardiovascular system of spontaneously hypertensive rats. Physiol. Res. 52 (6), 709–717.

Kruszyna, H., Magyar, J.S., Rochelle, L.G., Russel, M.A., Smith, R.P.E., Wilcox, D.E., 1998. Spectroscopic studies of Nitric Oxide (NO) interations with Cobalamins: reaction of NO with superoxocobalamin(Iii) likely accounts for Cobalamin reversal of the biological effects of NO. J. Pharmacol. Exp. Ther. 285, 665–671.

Ledoux, J., Werner, M.E., Brayden, J.E., Nelson, M.T., 2006. Calcium-activated potassium channels and the regulation of vascular tone. Physiology 21, 69–78. http://dx.doi.org/10.1152/physiol.00040.2005.

Levine, A.B., Punihaole, D., Levine, T.B., 2012. Characterization of the role of nitric oxide and its clinical applications. Cardiology. 122, 55–68. http://dx.doi.org/10.1159/000338150.

Li, H., Wang, S., 2013. Organic nitrates favor regression of left ventricular hypertrophy in hypertensive patients on chronic peritoneal dialysis. Int. J. Mol. Sci. 14, 1069–1079. http://dx.doi.org/10.3390/ijms14011069.

Lincoln, T.,M., Dey, N., Sellak, H., 2001. Invited review: cGMP-dependent protein kinase signaling mechanisms in smooth muscle: from the regulation of tone to gene expression. J. Appl. Physiol. (1985) 91 (3), 1421–1430.

Marsh, N., Marsh, A., 2000. A short history of nitroglycerine and nitric oxide in pharmacology and physiology. Clin. Exp. Pharmacol. Physiol. 27 (4), 313–319.

Maul, H., Longo, M., Saade, G.R., Garfield, R.E., 2003. Nitric oxide and its role during pregnancy: from ovulation to delivery. Curr. Pharm. Des. 9, 359–380. http://dx.doi.org/10.2174/1381612033391784.

Mayer, B., Beretta, M., 2008. The enigma of nitroglycerin bioactivation and nitrate tolerance: news, views and troubles. Br. J. Pharmacol. 155, 170–184. http://dx.doi.org/10.1038/bjp.2008.263.

Mendes-Júnior, L.G., Guimarães, D.D., Gadelha, D.D., Diniz, T.F., Brandão, M.C., Athayde-Filho, P.F., et al., 2015. The new nitric oxide donor cyclohexane nitrate induces vasorelaxation, hypotension, and anti-hypertensive effects via NO/cGMP/PKG pathway. Front. Physiol. 6, 243. http://dx.doi.org/10.3389/fphys.2015.00243.

Miller, M.R., Megson, I.L., 2007. Recent developments in nitric oxide donor drugs. Br. J. Pharmacol. 151, 305–321. http://dx.doi.org/10.1038/sj.bjp.0707224.

Miller, M.R., Wadsworth, R.M., 2009. Understanding organic nitrates—a vein hope? Br. J. Pharmacol. 157, 565–567. http://dx.doi.org/10.1111/j.1476-5381.2009.00193.x.

Minelli, C., Borin, L.A., Trovo Mde, C., Dos Reis, G.C., 2016. Hypertension prevalence, awareness and blood pressure control in Matao, Brazil: a pilot study in partnership with the Brazilian Family Health Strategy Program. J. Clin. Med. Res. 8 (7), 524–530. http://dx.doi.org/10.14740/jocmr2582w.

Mistry, D.I., Garland, C.J., 1998. Nitric oxide (NO)-induced activation of large conductance Ca2-dependent K_ channels (BKCa) in smooth muscle cells isolated from the rat mesenteric artery. Br. J. Pharmacol. 124, 1131–1140. http://dx.doi.org/10.1038/sj.bjp.0701940.

Mitchell, J.A., Ali, F., Bailey, F., Moreno, L., Harrington, L.S., 2008. Role of nitric oxide and prostacyclin as vasoactive hormones released by the endothelium. Exp. Physiol. 93 (1), 141–147.

Moncada, S., Higgs, A., 1993. Mechanisms of disease: the L-arginine-nitric oxide pathway. N. Engl. J. Med. 329 (27), 2002–2012.

Münzel, T., Daiber, A., Gori, T., 2011. Nitrate therapy: new aspects concerning molecular action and tolerance. Circulation 123, 2132–2144. http://dx.doi.org/10.1161/CIRCULATIONAHA.

Münzel, T., Daiber, A., Mülsch, A., 2005. Explaining the phenomenon of nitrate tolerance. Circ. Res. 97, 618–628. http://dx.doi.org/10.1161/01.RES.0000184694.03262.6d.

Münzel, T., Gori, T., 2013. Nitrate therapy and nitrate tolerance in patients with coronary artery disease. Curr. Opin. Pharmacol. 13, 251–259.

Münzel, T., Sayegh, H., Freeman, B.A., Tarpey, M.M., Harisson, G.D., 1995. Evidence for enhanced vascular superoxide anion production in nitrate tolerance. A novel mechanism underlying tolerance and cross-tolerance. J. Clin. Invest. 9, 187–194. http://dx.doi.org/10.1172/JCI117637.

Münzel, T., Steven, S., Daiber, A., 2014. Organic nitrates: update on mechanisms underlying vasodilation, tolerance and endothelial dysfunction. Vascul. Pharmacol. 63 (3), 105 113.

Murad, F., Mittal, C.K., Arnold, W.P., Katsuki, S., Kimura, H., 1978. Guanylate cyclase activation by azide, nitro compounds, nitric oxide, and hydroxyl radical and inhibition by hemoglobin and myoglobin. Adv. Cycl. Nucl. Res. 9, 145–158.

Murrel, W., 1879. Nitro-glycerine as a remedy for angina pectoris. Lancet 113 (2890), 80–81.

Navar, L.G., Zou, L., Von Thun, A., Tarng Wang, C., Imig, J.D., Mitchell, K.D., 1998. Unraveling the mystery of goldblatt hypertension. News Physiol. Sci. 13, 170–176.

Needleman, P., 1976. Organic nitrate metabolism. Annu. Rev. Pharmacol. Toxicol 16, 81–93. http://dx.doi.org/10.1146/annurev.pa.16.040176.000501.

Núñez, C., Víctor, V.M., Tur, R., Alvarez-Barrientos, A., Moncada, S., Esplugues, J.V., et al., 2005. Discrepancies between nitroglycerin and NO-releasing drugs on mitochondrial oxygen consumption, vasoactivity, and the release of NO. Circ. Res. 97 (10), 1063–1069.

Oliveira-Sales, E.B., Dugaich, A.P., Carillo, B.A., Abreu, N.P., Boim, M.A., Martins, P.J., et al., 2008. Oxidative stress contributes to renovascular hypertension. Am. J. Hypertens. 21, 98–104. http://dx.doi.org/10.1038/ajh.2007.12.

Oliver, J.J., Hughes, V.E., Dear, J.W., Webb, D.J., 2010. Clinical potential of combined organic nitrate and phosphodiesterase type 5 inhibitor in treatment-resistant hypertension. Hypertension. 56, 62–67. http://dx.doi.org/10.1161/HYPERTENSIONAHA.109.147686.

Palmer, R.M., Ferrige, A.G., Moncada, S., 1987. Nitric oxide release accounts for the biological activity of endothelium-derived relaxing factor. Nature 327, 524–526.

Parker, J.D., Farrell, B., Fenton, T., Cohanim, M., Parker, J.O., 1991. Counter-regulatory responses to continuous and intermittent therapy with nitroglycerin. Circulation 84, 2336–2345. http://dx.doi.org/10.1161/01.CIR.84.6.2336.

Pautz, A., Rauschkolb, P., Schmidt, N., Art, J., Oelze, M., Wenzel, P., et al., 2009. Effects of nitroglycerin or pentaerithrityl tetranitrate treatment on the gene expression in rat hearts: evidence for cardiotoxic and cardioprotective effects. Physiol. Genomics 38 (2), 176–185.

Porpino, S.K., Zollbrecht, C., Paleli, M., Montenegro, M.F., Brandão, M.C., Athayde-Filho, P.F., et al., 2016. Nitric oxide generation by the organic nitrate NDBP attenuates oxidative stress and angiotensin II-mediated hypertension. Br. J. Pharmacol. http://dx.doi.org/10.1111/bph.13511.

Pörsti, I., Paakkari, I., 1995. Nitric oxide-based possibilities for pharmacotherapy. Ann. Med. 27, 407–420.

Queiroz, T.M., Mendes-Júnior, L.G., Guimarães, D.D., França-Silva, M.S., Eugene, N., Braga, V.A., 2014. Cardiorespiratory effects induced by 2-nitrate-1,3-dibuthoxypropan are reduced by nitric oxide scavenger in rats Autonomic Neuroscience. Basic Clin. 181, 31–36. http://dx.doi.org/10.1016/j.autneu.2013.12.012.

Ribeiro, A.L., Duncan, B.B., Brant, L.C., Lotufo, P.A., Mill, J.G., Barreto, S.M., 2016. Cardiovascular health in Brazil: trends and perspectives. Circulation 133 (4), 422–433. http://dx.doi.org/10.1161/CIRCULATIONAHA.114.008727.

Rudolph, V., Freeman, B.A., 2009. Cardiovascular consequences when nitric oxide and lipid signaling converge. Circ. Res. 105 (6), 511–522.

Santos, A.F., 2009. Novas Perspectivas da Glicerina – Síntese de Novos Nitratos com Propriedades Farmacológicas e Melhoradores de Cetano. Dissertação de Mestrado. Universidade Federal da Paraíba, João Pessoa.

Schuhmacher, S., Wenzel, P., Schulz, E., Oelze, M., Mang, C., Kamuf, J., et al., 2010. Pentaerythritol tetranitrate improves angiotensin II-induced vascular dysfunction via induction of heme oxygenase-1. Hypertension 55 (4), 897–904. http://dx.doi.org/10.1161/HYPERTENSIONAHA.109.149542.

Schulz, E., Gori, T., Münzel, T., 2011. Oxidative stress and endothelial dysfunction in hypertension. Hypertens. Res. 34, 665–673. http://dx.doi.org/10.1038/hr.2011.39.

Stewart, D.D., 1905. Remarkable tolerance to nitroglycerin, Philadelphia Polyclinic, p. 172. From Stewart, DD. Tolerance to nitroglycerin. JAMA 44, 1678–1679. Accessed on November 2015.

Stokes, G.S., Barin, E.S., Gilfillan, K.L., Kaesemeyer, W.H., 2003. Interactions of L-arginine, isosorbide mononitrate, and angiotensin II inhibitors on arterial pulse wave. Am. J. Hypertens. 16, 719–724. http://dx.doi.org/10.1016/S0895-7061(03)00979-8.

Sydow, K., Daiber, A., Oelze, M., Chen, Z., August, M., Wendt, M., et al., 2004. Central role of mitochondrial aldehyde dehydrogenase and reactive oxygen species in nitroglycerin toleranceand cross-tolerance. J. Clin. Invest. 113 (3), 482–489. http://dx.doi.org/10.1172/JCI200419267.

Tarkin, J.M., Kaski, J.C., 2016. Vasodilator therapy: nitrates and nicorandil. Cardiovasc. Drugs Ther., 16. http://dx.doi.org/10.1007/s10557-016-6668-z10.

Thakali, K.M., Lau, Y., Fink, G.D., Galligan, J.J., Chen, A.F., Watts, S.W., 2006. Mechanisms of hypertension induced by nitric oxide (NO) deficiency: focus on venous function. J. Cardiovasc. Pharmacol. 47, 742–750. http://dx.doi.org/10.1097/01.fjc.0000211789.37658.e4.

Thomas, G.R., DiFabio, J.M., Gori, T., Parker, J.D., 2007. Once daily therapy with isosorbide-5-mononitrate causes endothelial dysfunction in humans: evidence of a free-radical-mediated mechanism. J. Am. Coll. Cardiol. 49 (12), 1289–1295. http://dx.doi.org/10.1016/j.jacc.2006.10.074.

Tousoulis, D., Simopoulou, C., Papageorgiou, N., Oikonomou, E., Hatzis, G., Siasos, G., et al., 2014. Endothelial dysfunction in conduit arteries and in microcirculation. Novel therapeutic approaches. Pharmacol. Ther. 144, 253–267. http://dx.doi.org/10.1016/j.pharmthera.2014.06.003.

Vanhoutte, P.M., 2009. Endothelial dysfunction—the first step toward coronary arteriosclerosis. Circ. J. 73 (4), 595–601.

Vasquez, E.C., Gava, A.L., Graceli, J.B., Balarini, C.M., Campagnaro, B.P., Pereira, T.M., et al., 2016. Novel therapeutic targets for phosphodiesterase 5 inhibitors: current state-of-the-art on systemic arterial hypertension and atherosclerosis. Curr. Pharm. Biotechnol. 17 (4), 347–364.

Wang, P.G., Xian, M., Tang, X., Wu, X., Wen, Z., Cai, T., et al., 2002. Nitric oxide donors: chemical activities and biological applications. Chem. Rev. 102, 1091–1134. http://dx.doi.org/10.1021/cr000040l.

Wenzel, P., Mollnau, H., Oelze, M., Schulz, E., Wickramanayake, J.M., Müller, J., et al., 2008. First evidence for a crosstalk between mitochondrial and NADPH oxidase-derived reactive oxygen species in nitroglycerin-triggered vascular dysfunction. Antioxid. Redox Signal 10, 1435–1447.

Wever, R., Stroes, E., Rabelink, T.J., 1998. Nitric oxide and hypercholesterolemia: a matter of oxidation and reduction? Atherosclerosis 137, S51–S60.

Yeo, T.W., Lampah, D.A., Gitawati, R., Tjitra, E., Kenangalem, E., McNeil, Y.R., et al., 2007. Impaired nitric oxide bioavailability and L-arginine reversible endothelial dysfunction in adults with falciparum malaria. J. Exp. Med. 204, 2693–2704. http://dx.doi.org/10.1084/jem.20070819.

CHAPTER 11

Nitric Oxide Donors in Brain Inflammation

Marisol Godínez-Rubí[1] and Daniel Ortuño-Sahagún[2]
[1]Departamento de Microbiología y Patología, Universidad de Guadalajara, Jalisco, México
[2]Instituto de Investigación en Ciencias Biomédicas, Departamento de Biología Molecular y Genómica.
Universidad de Guadalajara, Jalisco, México

INTRODUCTION

It has now been clearly established that the central nervous system (CNS) is able to raise an immune response against the majority of threatening stimuli, whereby resident cells generate inflammatory mediators that include cytokines, prostaglandins, free radicals, complementary chemokines, and adhesion molecules, which activate glia and microglia and recruit immune cells (reviewed in Ahmad and Graham, 2010; Ceulemans et al., 2010; Downes and Crack, 2010; Zierath et al., 2010).

Control of early CNS inflammation is a careful balancing act, because both too much and too little inflammation will lead to a decrease or a delay in recovery. Whether the inflammation is neurotoxic or protective may depend upon the context and the location of the inflammatory mediator in relation to an injury or affliction, and may also depend on the timing of the inflammatory response, which may determine its outcome (see Table 1 in Downes and Crack, 2010).

ROLE OF NITRIC OXIDE IN BRAIN PHYSIOLOGY

Nitric oxide (NO) is a gas that may be present as a free nitrogen radical or a nitrous ion, depending on the redox state of the cell. NO is synthesized by nitric oxide synthase (NOS), for which there are three known isoforms: neuronal NOS (nNOS or NOS1), inducible NOS (iNOS or NOS2), and endothelial NOS (eNOS or NOS3). NOS requires three substrates to synthesize NO: L-arginine, oxygen, and nicotinamide adenine dinucleotide phosphate (NADPH). nNOS and eNOS are expressed constitutively and are calcium-/calmodulin dependent, while iNOS acts in a Ca^{2+}-independent manner (Guix et al., 2005). Baseline concentration of NO in the brain is mainly due to nNOS activity and secondarily to eNOS. iNOS is normally not expressed under physiological conditions (Forstermann and Sessa, 2012; Zhou and Zhu, 2009).

NO participates in numerous normal biological processes, including regulation of vascular tone, immunomodulation, and neurotransmission, among others (Guix et al., 2005), being an extremely important signaling molecule responsible for maintaining vascular homeostasis by promoting vasodilatation and inhibiting platelet aggregation and leukocyte adhesion. As a free radical, NO becomes less reactive than the majority of oxygen free radicals (Chiueh and Rauhala, 1999). However, peroxynitrite (product of interaction between NO and O_2^-) is a lipoperoxidant agent and a nitrosylator of proteins.

The majority of NO biological activities is exerted via at least four different mechanisms: (1) through activation of classic cyclic guanosine monophosphate/protein kinase G (cGMP/PKG) pathway by interaction with the heme group of soluble guanylate cyclase (Toda et al., 2009; Wobst et al., 2015); (2) via oxidative and nitrosative effects via peroxynitrite generation (Cherian et al., 2004; Garry et al., 2015); (3) via nitrosative posttranslational modifications of proteins such as S-nitrosylator and tyrosine nitration (Heinrich et al., 2013), and (4) via S-nitrosoglutathione (GSNO), formed by the reaction of NO and glutathione. GSNO is an endogenous carrier of NO and the most abundant S-nitrosothiol in human and animals. cGMP-independent actions are more frequently related with the pathological effects of NO (Corti et al., 2014; Garry et al., 2015).

eNOS is a membrane-bound enzyme. In the brain, eNOS is mainly produced by the vascular endothelium and the choroid plexus (Stanarius et al., 1997). Although eNOS-NO production comprises a minor part of total brain NOS activity, this enzyme is critical for the regulation of cerebrovascular hemodynamic (Toda et al., 2009) and for the protection of endothelium integrity from inflammatory, oxidative, and procoagulant stimuli. It has been demonstrated that eNOS-derived NO scavenges reactive oxygen species (ROS) (Kuhlencordt et al., 2004) and inhibits the expression of cellular adhesion molecules (Nabah et al., 2005), platelet aggregation (Moore et al., 2011), and leukocyte adhesion (Hossain et al., 2012).

nNOS is cytosolic enzyme expressed in both immature and mature neurons (Zhou and Zhu, 2009). However, its expression has been also documented in astrocytes, the adventitia of brain blood vessels and cardiac myocytes (Mercanoglu et al., 2015; Munoz et al., 2015; Zhou and Zhu, 2009). NO derived from nNOS is a critical molecule in mediating synaptic plasticity and neuronal signaling related with memory formation, regulation of the CNS blood flow, transmission of pain signals, neurotransmitter release, and neurogenesis (Zhou and Zhu, 2009; Zhu et al., 2016). However, the NO-nNOS derived changes into a neurotoxic factor when an excessive amount of NO is produced (Ito et al., 2010; Liu et al., 2009; Yang et al., 2008; Yu et al., 2008).

iNOS sources in brain comprise microglia (Khan et al., 2005), astrocytes (Gibson et al., 2005), endothelial cells (Niwa et al., 2001), and infiltrated leukocytes (Suzuki et al., 2002). The amount of iNOS-NO derived is 100–1000 times more than that produced by constitutive NOS. iNOS expression can be stimulated by inflammation and oxidative stress (OS) (Pannu and Singh, 2006). Once activated, iNOS continuously produces NO until the enzyme is degraded (MacMicking et al., 1997).

NEUROINFLAMMATION AND NO IN DAMAGED BRAIN

Several factors can affect the brain and its functioning. Here we analyze three quite different etiologies for brain damage: one caused by external mechanical force; a second caused by blood flux reperfusion after an ischemic event; and a third, a multietiological disease, Alzheimer's disease (AD). All of these have in common the presentation of brain inflammation; therefore, at least at cellular and molecular levels, we can establish certain parallelisms between some of the mechanisms involved, mainly regarding NO and the participation of nitrosative derivatives in all of these.

Traumatic brain injury

Traumatic brain injury (TBI) is defined as damage to the brain resulting from an external mechanical force, including accelerating, decelerating, and rotating forces (Chen et al., 2014). TBI causes physical disruption of the neuronal tissue directly affected by the trauma and secondarily affects the surrounding tissue. Although there are many factors contributing to establishing the damage in TBI, OS, endothelial dysfunction, acute inflammation, and reduction in cerebral blood flow (CBF) appear to be the most significant, similarly to the physiopathology of cerebral ischemia (Ahn et al., 2004; Bayir et al., 2007).

In each of these events, NO plays multiple and complex roles and, as in the other entities covered in this chapter, the neurotoxic or neuroprotective effect depends on the microenvironment, NO concentration, and oxidative balance. In the first minutes after the trauma, there is a temporary and very short rise of NO. It then decreases below baseline levels for several hours (Ahn et al., 2004; Cherian et al., 2004), and finally increases again to baseline levels or even above these after 6–12 hours of impact (Cherian et al., 2004; Wada et al., 1998).

CBF is markedly reduced immediately afterward and during the first few hours after trauma, especially in the pericontusional region (Engel et al., 2008). There is evidence that early reduction of CBF is related with a reduction in the amount of NO available after trauma (Ahn et al., 2004). This may be due to the dysfunction of endothelial NO production. In transgenic rodent models (eNOS knockout mice) undergoing controlled cortical impact, removing eNOS expression leads to a significant reduction in CBF and neuronal damage (Cruz Navarro et al., 2014; Hlatky et al., 2003), whereas administration of L-arginine, a substrate of eNOS, reversed this effect (Cherian et al., 2003; Cherian and Robertson, 2003). Similarly, pharmacological overactivation of eNOS is associated with a reduction in tissue damage, OS, and proinflammatory cytokines, and an improvement of neurological score (Chen et al., 2014; Cruz Navarro et al., 2014). This appears to be mediated by overactivation of protein kinase B (Akt), whose phosphorylation induces eNOS activation. This suggests that eNOS-derived NO plays a major role in regulating the vascular dynamics that control CBF after TBI, the inflammatory response, and oxidative environment.

Hypoperfusion in turn contributes to the excessive production of reactive oxygen/ nitrogen species (ROS/NOS), not only in the brain cells but also in the endothelial and glial cells (Cherian and Robertson, 2003; Bao and Liu, 2004; Liu et al., 1998). As described in other sections of this chapter, the interaction of superoxide with NO generates peroxynitrite, a powerful nitrosylator agent that is highly injurious. Production of peroxynitrite is an upstream, early posttraumatic event that appears to lead to mitochondrial damage, intracellular calcium overload, and neurodegeneration following TBI (Deng et al., 2007; Khan et al., 2011, 2016). It is regarded as the cause of lipid peroxidation, protein carbonylation, and the nitration of protein tyrosine residues (Deng et al., 2007).

The elevated concentration of peroxynitrite after trauma exerts an influence on the oxidative balance of the microenvironment. It has been described, in humans and animals, that following TBI, nitrosylation inhibits the function of manganese superoxide dismutase (MnSOD) (Bayir et al., 2007), an antioxidant enzyme that reduces the amount of superoxide and whose action is associated with protection after acute brain injury (Gunther et al., 2015), while its inhibition aids in establishing damage. Such inhibition appears to be related with the activity of nNOS and its indirect production of peroxynitrite, but not with eNOS or iNOS activity (Bayir et al., 2007). Inhibition of nNOS has been shown to be neuroprotective in TBI, indicating a harmful role of nNOS activity after cerebral trauma (Wada et al., 1998).

Similarly to what takes place in an ischemic stroke, endothelial dysfunction, OS, and neuronal death lead to the expression of inflammatory mediators after TBI. The neuroinflammatory response can magnify the secondary brain injury strikingly after trauma. This phenomenon is characterized by microglia and astrocytic activation, increased levels of cyclooxygenase-2 (COX-2), prostaglandins, adhesion molecules, and proinflammatory cytokines (Ahn et al., 2004; Chen et al., 2014; Knoblach and Faden, 2002).

A pivotal player in the regulation of these molecules is the Nuclear Factor kappa B (NF-κB) family. NF-κB is a key regulator of innate immunity, inflammation, and of cell survival and proliferation (Mankan et al., 2009). This inducible transcription factor is comprised of two subunits. There are five subunits that can be combined to yield homo- or heterodimers of NF-κB as follows: p50, p52, c-Rel, p65 (RelA), and RelB (Sarkar et al., 2008). C-Rel containing dimer activation increases neuron resistance to ischemia (Sarnico et al., 2009). Moreover, the prevalent heterodimer during cerebral ischemia and reperfusion is formed by p50- and p65-inducible subunits, and its activation contributes to the pathogenesis of postischemic injury (Campelo et al., 2012; Sarnico et al., 2009). NF-κB is maintained in latent form in the cytoplasm of cells bound to inhibitory IκB proteins. Phosphorylation of IκB releases NF-κB by permitting its translocation into the nucleus, its binding with NF-κB motifs, and the subsequent activation of its target genes. There is, in turn, an enzymatic complex responsible for IκB phosphorylation in specific serine residues, the so-called IκB kinases (IKK). Activation of IKK is essential to induce NF-κB activity (Ridder and Schwaninger, 2009).

NF-κB expression can be triggered by OS, bacterial endotoxin, or cytokines, and it subsequently activates the genes encoding cytokines, ICAM-1, and iNOS, among others (Sarkar et al., 2008; Sarnico et al., 2009; Blais and Rivest, 2001). NF-κB is upregulated in the injured tissue after TBI in a biphasic manner (Hu et al., 2014). There is a first increase 12 hours after trauma, and then a second and higher rise at 72 hours. Concomitant to the upregulation of NF-κB, there is an increase in proinflammatory cytokines: tumor necrosis factor (TNF-α), interleukin (IL)-6, and ICAM-1 (Hang et al., 2005). Upregulation of NF-κB and its proinflammatory effect has been associated with damage in several models of brain injury (Zhang et al., 2005; Kolev et al., 2003; Sarkar et al., 2008; Sarnico et al., 2009; Blais and Rivest, 2001; Cummins et al., 2006; Greco et al., 2011). In the late phase after TBI, the rise in NO is mainly attributed to overactivation of iNOS in microglia, macrophages, and endothelial cells, which are also transcriptionally regulated by NF-κB (Katsuyama et al., 1998; Sarkar et al., 2008; Trickler et al., 2005). The activity of this enzyme and the excessive production of NO are well known to cause brain damage by multiple mechanisms, including neuroinflammation, apoptosis, excitotoxicity, and ROS and NOS production (Guix et al., 2005; Pannu and Singh, 2006; Suzuki et al., 2002; Jafarian-Tehrani et al., 2005). Consequently, its inhibition is involved in neuroprotection (Jafarian-Tehrani et al., 2005; Louin et al., 2006; Terpolilli et al., 2009) (Fig. 11.1).

Ischemia–reperfusion injury

When the brain blood flow is interrupted, it results in deprivation of oxygen and nutrients to the cells; this situation constitutes an ischemic stroke. Restoration of the flux, or reperfusion, can reduce the damage, and its efficacy is restricted by secondary injuries, mainly by OS and an inflammatory reaction, which lead to cell death by apoptosis (Lakhan et al., 2009). In contrast to the known vulnerability of neurons and astrocytes, endothelial cells tend to be more resistant to ischemic or oxidative injury (Gertz, 2008). Hence, to be successful, stroke therapies should be widely effective and must protect the neurovascular unit, namely, all of the neuronal, glial, and endothelial components in the brain (Lee et al., 2010).

After focal ischemia, primary neuronal death appears rapidly, and one of the first events is the rapid decline of adenosine triphosphate (ATP) reserves, which increases energy consumption and promotes Ca^{2+} influx into the cells. The increase in intracellular Ca^{2+} in neurons and glial cells initiates a set of events that produce deep brain tissue damage: Ca^{2+} mitochondrial overload (which compromises the already affected ATP production and promotes the opening of the mitochondrial transition

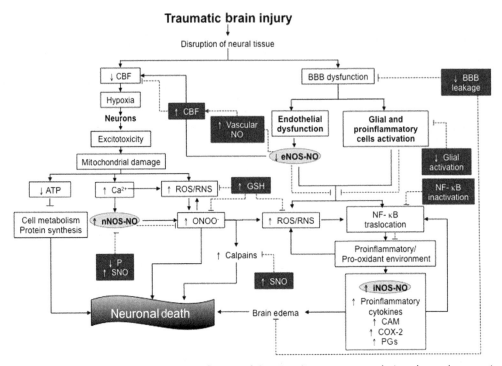

Figure 11.1 Schematic representation of some of the signaling events regulating the pathogenesis of traumatic brain injury and the effect of NODs. Events common to the three pathologies discussed in this chapter are in *white boxes*; NO synthases are identified with *blue ovals*; the effects of NODs are highlighted with *blue squares and blue dashed lines*. TBI causes a neural tissue disruption that leads to a significant reduction in CBF and BBB dysfunction. Reduction in CBF triggers a cascade of events that cause mitochondrial damage, reduction in ATP synthesis, intracellular Ca²⁺ overload, and production of ROS and RNS which in turn induce oxidative and nitrosative stress. Ca²⁺ overload triggers nNOS activation and a rise in its NO production, which contributes to nitrosative stress by production of peroxynitrite. BBB damage is characterized by endothelial dysfunction and glial activation. Endothelial dysfunction implies a reduction in NO-eNOS derived that affects CBF and contributes to oxidative stress and neuroinflammation. In endothelial and glial cells, nuclear translocation of the transcription factor NF-κB promotes expression of several molecules involved in neuroinflammation and oxidative/nitrosative stress, including iNOS. Together all these events lead to neuronal death via apoptosis and necrosis. NODs can act at different levels in these pathways: (1) NODs improve CBF by increasing NO at vascular level; (2) reduce BBB dysfunction and brain edema; (3) block nNOS activity by decreasing its phosphorylation and enhance nNOS nitrosylation, hence, its NO and peroxynitrite production; (4) prevent the nuclear translocation of NF-κB, therefore the expression of their target genes; (5) inhibit calpain activity by SNO. See the text for further explanation. ATP, adenosine triphosphate; BBB, blood brain barrier; CAM, cell adhesion molecules; CBF, cerebral blood flow; COX-2, cyclooxygenase-2; eNOS, endothelial nitric oxide synthase; GSH, glutathione; iNOS, inducible nitric oxide synthase; NF-κB, nuclear factor kappa B; NODs, nitric oxide donors; nNOS, neuronal nitric oxide synthase; NO, nitric oxide; P, phosphorylation; PGs, Prostaglandins; RNS, reactive nitrogen species; ROS, reactive oxygen species; SNO, S-nitrosylation; TBI, traumatic brain injury.

pore), the increase in OS, and the activation of a number of Ca^{2+}-dependent enzymes. Additionally, increased intracellular Ca^{2+} also promotes the production of NO from constitutive synthases that, together with acidosis and peri-infarct depolarization, contribute to the initiation of damage. Later, inflammation and activation of apoptotic phenomena contribute to increased injury (Mehta et al., 2007; Ahmad and Graham, 2010; Barber et al., 2003).

OS is a major mechanism implicated in stroke and in a variety of neurodegenerative diseases. It is during reperfusion that OS paradoxically induces the excessive generation of ROS, such as superoxide anion radical (O_2^-), hydroxyl radical (OH'), hydrogen peroxide (H_2O_2), and NO, which contribute to increased neuronal death by oxidizing proteins, damaging DNA, and inducing lipid peroxidation (Barber et al., 2003).

Reperfusion-induced ROS contribute to a decrease of the NO availability responsible for postischemic endothelial dysfunction. During the ischemic period, reduction in O_2 availability reduces the activity of NO synthase, producing O_2^- instead of NO; later, during reperfusion, the arrival of O_2 increases NO synthase activity. These can exert an injurious effect by promoting nitrosative stress and diminishing the availability of NO for preserving endothelial integrity (Forstermann and Sessa, 2012; De Pascali et al., 2014).

Thus NO plays a dual role, i.e., neuroprotection and neurotoxicity, in the pathophysiology of cerebral ischemia–reperfusion injury, depending mainly on NO concentration, its source (endogenous or exogenous), the compartment in which is located, the enzyme and cell producing it, and the redox environment (Godinez-Rubi et al., 2013). During ischemia, NO concentration decreases due to oxygen deficiency (Yang et al., 2008). However, immediately after reperfusion, the biosynthesis of this molecule is triggered mainly by overactivation of nNOS, as evidenced in nNOS null mice (Ito et al., 2010) or with specific inhibitors such as 7-Nitroindazole (7-NI) (Liu et al., 2009; Yang et al., 2008). Glutamate-induced Ca^{2+} overload in ischemic neurons is responsible for the rise of nNOS-derived NO (Zhou and Zhu, 2009). The concentration of NO returns to physiological levels approximately 1 hour after reperfusion (Ito et al., 2010; Liu et al., 2009; Yang et al., 2008) and increases again due to iNOS expression between 12 hours and up to 8 days later (Danielisova et al., 2004; Khan et al., 2005). iNOS sources at this stage comprise microglia (Khan et al., 2005), astrocytes (Gibson et al., 2005), endothelial cells (Niwa et al., 2001), and infiltrated leukocytes (Suzuki et al., 2002).

On the other hand, NO derived from eNOS plays a neuroprotective role. Total suppression of eNOS activity in knockout mice ($eNOS^{-/-}$) renders them hypertensive (Ito et al., 2010) and more susceptible to ischemia–reperfusion injury, with larger infarcts compared with those of controls and a more severe reduction in CBF (Asahi et al., 2005; Ito et al., 2010; Wei et al., 1999). Conversely, overactivation of eNOS by flavonoids induces a protective effect (Duarte et al., 2014). In contrast to eNOS, infarct volume and neuronal death are consistently decreased by nNOS gene deficiencies or by nNOS inhibition (Greco et al., 2011; Ito et al., 2010; Pei et al., 2008; Sun et al., 2009; Khan et al., 2015; Liu et al., 2013). nNOS abolition also reduces excitotoxicity (Khan et al., 2015; Rameau et al., 2007), nitrosative stress (Hu et al., 2012; Liu et al., 2013), and O_2^- production (Gursoy-Ozdemir et al., 2004) and downregulates calpain and caspase-3 in ischemic lesion (Pei et al., 2008; Sun et al., 2009).

Additionally, during reperfusion, iNOS-produced NO contributes to brain injury (Danielisova et al., 2004, 2011). iNOS expression is, as mentioned earlier, transcriptionally regulated by NF-κB secondarily to an inflammatory stimulus, such as TNF-α (Trickler et al., 2005) and interleukin (IL)-1β (Katsuyama et al., 1998), and also by oxidative radicals (Sarkar et al., 2008). Due to the large amount of NO produced by iNOS, this enzyme is related with high peroxynitrite production and significant nitrosative damage of biological molecules (Danielisova et al., 2004). Consequently, nNOS mediates early neuronal injury, while iNOS contributes to late neuronal injury, whereas eNOS activity might be protective (Ito et al., 2010; Liu et al., 2009; Yang et al., 2008).

Hence, whether the effects of NO are beneficial or harmful depends on the cellular compartment in which NO is generated, on its concentration, on the environment's redox state, and on the evolutive stage of ischemic brain injury (Ito et al., 2010; Liu et al., 2009; Yang et al., 2008). According to the dual role of NO in brain ischemia, there is a rationale for the use of NO for promoting treatments shortly after the occurrence of focal cerebral ischemia as neuroprotective strategies (Godinez-Rubi et al., 2013). The neurovascular protective mechanism of eNOS-NO derived suggests that intervention with NO may be most effective when delivered at an optimal amount by a suitable source at the correct time in an appropriate environment (Gertz, 2008; Godinez-Rubi et al., 2013) (Fig. 11.2).

Alzheimer's disease

AD is the main cause of dementia and decreased quality of life (QOL) among elderly persons. The two histological features that define AD are neurofibrillary tangles and extracellular amyloid beta (Aβ) deposits within senile plaques. Aβ deposition and tau protein cause loss of synaptic function, mitochondrial damage, activation of microglia, and finally neuronal death (Lowe, 2008). Aβ is produced by the sequential cleavages of amyloid precursor protein (APP) by the beta-site APP cleaving enzyme (BACE) 1 and gamma-secretase. BACE1 activity is the most important limiting factor in the synthesis of Aβ (Zhang and Xu, 2007).

The precise cause of the sporadic form of AD, which is the most common form of presentation, is unknown. However, virtually all risk factors for vascular disease are also those for AD. Age is the major risk factor for the development of dementia and

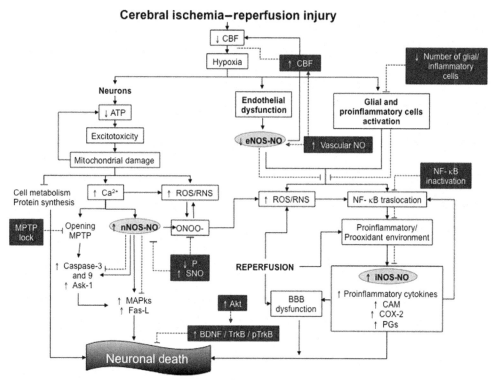

Figure 11.2 Schematic representation of some of the signaling events regulating the pathogenesis of cerebral ischemia–reperfusion injury and the effect of NODs. Events common to the three pathologies discussed in this chapter are in *white boxes*; NO synthases are identified with *blue ovals*; the effects of NODs are highlighted with *blue squares and blue dashed lines*. The sudden reduction of CBF triggers a cascade of events which are similar to those described in TBI. Similarly, endothelial dysfunction and glial activation are characterized by deprivation of NO-eNOS derived and an activation of a proinflammatory and pro-oxidant profile. Reperfusion brings with it an excess of ROS and RNS and a second wave of inflammatory molecules. The evidence shows that NODs are able to interfere with these signaling pathways as follows: (1) NODs improve CBF by increasing NO at vascular level; (2) block nNOS activity by decreasing its phosphorylation and enhance nNOS nitrosylation, hence, (3) its NO and peroxynitrite production; (4) inhibition of synthesis and activation of molecules associated with neuronal death and which activation depends on nNOS, as MAPK, caspases, Ask-1 and Fas-L; (5) prevent the nuclear translocation of NF-κB, therefore the expression of their target genes and neuroinflammation; (6) enhance neurotrophic factors like BDNF and TrkB. See the text for further explanation. Ask-1, apoptosis signal-regulating kinase-1; ATP, adenosine triphosphate; BDNF, brain-derived neurotrophic factor; BBB, blood brain barrier; CAM, cell adhesion molecules; CBF, cerebral blood flow; COX-2, cyclooxygenase-2; eNOS, endothelial nitric oxide synthase; Fas-L, Fas-ligand; iNOS, inducible nitric oxide synthase; MAPK, mitogen-activated protein kinases; MTPT, mitochondrial permeability transition pore; NF-κB, nuclear factor kappa B; NODs, nitric oxide donors; nNOS, neuronal nitric oxide synthase; NO, nitric oxide; P, phosphorylation; PGs, prostaglandins; RNS, reactive nitrogen species; ROS, reactive oxygen species; SNO, *S*-nitrosylation; STAT, signal transducer and activator of transcription; TBI, traumatic brain injury; TrkB, tyrosine receptor kinase B.

AD (Love and Miners, 2015; Lowe, 2008). The presence of cardiovascular disease such as diabetes mellitus, obesity, atherosclerosis, coronary heart disease, and ischemic stroke has also been linked with an increased risk of developing dementia and AD (Carcaillon et al., 2015; de Bruijn and Ikram, 2014; Knopman et al., 2016). It is relevant that the link between these entities and AD comprises OS, inflammation, and endothelial dysfunction (Jeynes and Provias, 2009; Park and Park, 2015; Zlokovic, 2011).

At the vascular level, a major histopathological feature of AD is amyloid deposition within the walls of arteries and arterioles in the subarachnoid space and the cerebral and cerebellar cortex, denominated cerebral amyloid angiopathy (CAA). It is present to some degree in nearly all cases of AD and is also associated with endothelial dysfunction (Brenowitz et al., 2015; Lowe, 2008). It is now well established that Aβ deposition can induce vascular damage and, at the same time, endothelial dysfunction causes Aβ deposition, resulting in a vicious cycle in which vascular damage and Aβ accumulation interact synergistically to increase endothelial dysfunction. Once the precise balance is disturbed, it may result in Aβ deposition, microbleeds, and neuroinflammatory disturbances. In transgenic models of AD, vascular modifications precede the deposit of Aβ and changes in memory (Austin et al., 2013b; Han et al., 2008; Iadecola, 2004).

In addition to the CAA, other inflammatory vascular changes have been described in AD. Among these can be found the activation of endothelial cells, perivascular astrocytes, and microglia, leading in conjunction to dysfunction of the blood brain barrier (BBB). Cerebral hypoperfusion is another feature that is observed frequently in AD and a potential trigger of the neuroinflammatory response (Attems and Jellinger, 2014; Gupta and Iadecola, 2015). These changes contribute to maintaining chronic perfusion deficits, reduced NO synthesis by eNOS, and a proinflammatory and pro-oxidant environment (Brenowitz et al., 2015; Tong et al., 2005).

Therefore, evidence suggests that vascular disease with endothelial dysfunction plays a critical role in establishing the long-term damage in AD. Endothelial dysfunction is characterized by a reduction in the bioavailability of NO synthesized by endothelial NOS. Hence, NO plays a major role in the vascular pathology that characterizes AD. Both of these, its lacks and its excess, can contribute to the development of the disease. The contribution of aberrant activity of NO synthases in AD pathology is extensively described and is intimately associated with cerebral hypoperfusion and oxidative/nitrosative stress, as well with neuroinflammation (Aliev et al., 2009).

Endothelial NO deficiency, or its absence, is directly related with AD pathology. In brains from patients with AD, eNOS levels in capillary endothelium are reduced compared with controls, and there is a significant inverse correlation between eNOS-positive microvessels and the presence of neurofibrillary tangles, amyloid deposition in senile plaques, and amyloid-positive capillaries (Jeynes and Provias, 2009; Provias and Jeynes, 2008). Abnormal eNOS activity in the cerebrovascular endothelium was identified a very long time before the development of AD pathology and the appearance of detectable cognitive decline (Iadecola, 2004). This suggests a dysfunction at the vascular level in AD brains long before disturbances in memory appear.

Pharmacological or genetic inactivation of eNOS in human brain microvascular endothelium increased expression of APP and BACE1, thus favoring the amyloidogenic process, via the cGMPC pathway (Austin et al., 2010). The same effect is observed in a late-middle age model with eNOS knockout mice, which proved the increased APP expression and amyloidogenic processing in the hippocampus, associated with an impaired performance in a memory test (Austin et al., 2013b). In an aging rat model of chronic brain hypoperfusion that mimics human mild cognitive impairment, eNOS inhibition triggered heavy Aβ deposition in hippocampal capillaries and perivascular regions, and impaired memory tasks (de la Torre and Aliev, 2005). These findings indicate that loss of NO promotes the production and release of amyloid beta, and also demonstrated that endothelial NO is a critical modulator of APP processing in vivo.

At the same time, excessive amyloidogenic processing of APP and an elevated local concentration of Aβ cause endothelial dysfunction in the cerebral blood vessels via OS and neuroinflammation (Iadecola, 2004). Recently, a study in cultured vascular endothelial cells demonstrated that amyloid beta interferes with eNOS-NO production by inhibiting its phosphorylation. This effect was achieved by stimulation of the interaction between eNOS and its regulatory partner HSP90, and by diminishing eNOS interaction with Akt, the kinase responsible for eNOS phosphorylation. The trigger of this reaction was the production of ROS (superoxide and hydroxyl radicals) by Aβ (Lamoke et al., 2015). However, whether loss of endothelial NO may affect the initiation and progression of cerebral amyloid deposition in humans is unknown and remains to be determined (Katusic and Austin, 2014).

In addition to the impact at the vascular level, experimental studies favor the concept that eNOS/NO/cGMP signaling comprises a central mechanism required for memory formation (Blackshaw et al., 2003; Doreulee et al., 2003). Thus it is likely that the loss of endothelial NO caused by prolonged exposure to vascular risk factors might result in impairment of neuronal function and cognitive decline (Bon and Garthwaite, 2003; Katusic and Austin, 2014).

The relationship between ischemic brain damage and AD is well documented. In fact, alterations in CBF have turned out to be a robust predictor of AD (Iadecola, 2004; Love and Miners, 2015). Evidence suggests that ischemic lesions in the brain leads to rapid deposition of Aβ in the brain parenchyma and in the vessels surrounding the infarcted area, via activation of proinflammatory and pro-oxidant reactions (Garcia-Alloza et al., 2011). Furthermore, partial eNOS deficiency elicits spontaneous cerebral infarctions that increase with age, leading to progressive CAA and cognitive impairments (Tan et al., 2015). This suggests that cerebrovascular injury is at least one of the many contributors to Aβ deposition, and that both factors maintain a continuous cycle in which ischemia and amyloid deposition exacerbate each other and, at the same time, both subsequently contribute to neuroinflammation and to oxidative damage (Love and Miners, 2015).

Several lines of evidence indicate that Aβ is a source of ROS and lipid peroxidation (Butterfield et al., 2013; Minjarez et al., 2016; Qu et al., 2016). Aβ itself is able to produce hydrogen peroxide, which in turn causes accumulation of hydroxyl radical (Ill-Raga et al., 2010; Tong et al., 2005). This produces uncoupling of the mitochondrial respiratory chain in neurons and subsequently generates superoxide anion accumulation. Additionally, H_2O_2 influences the function of calcium channels, increasing intracellular Ca^{2+} stores. That activates nNOS, with consequent NO production. Overexpression of nNOS has been documented in neurons with neurofibrillary tangles and senile plaques in the entorhinal cortex and hippocampus of patients with AD (Thorns et al., 1998).

Subsequently, NO reacts with superoxide anion to form peroxynitrite anion, which nitrosylates cell proteins. Activation of mitochondrial NADPH oxidase could also cause enhanced superoxide production (Ill-Raga et al., 2010; Malinski, 2007). In the brains of patients with AD, nitrosylation of proteins is a hallmark of tissue damage (Aliev et al., 2009) and is particularly elevated in hippocampus and cerebral cortex. Amyloid-induced changes in NO production and mitochondrial activity lead to neuronal apoptosis (Keil et al., 2004; Valerio et al., 2006). Therefore, ROS/RNS production can be observed both as a consequence of and as a driving force in neuroinflammatory cycles.

The events previously described are sufficient to trigger the proinflammatory profile of glia and endothelial cells via NF-κB activation (Granic et al., 2009; Shi et al., 2016). Essentially, almost every pathological hallmarks of AD, described previously, are able to lead to an inflammatory response. In this regard, chronic activation of microglia has a central role in AD pathology (Shi et al., 2016; Wang et al., 2015) because it is capable of responding against many harmful stimuli by releasing neurotoxic molecules, such as cytokines and toxic oxygen and nitrogen species, which are harmful to the neurons. Even more so, APP and Aβ can directly activate microglia and initiate a local inflammatory response via NF-κB translocation (Malinski, 2007; McCarthy et al., 2016; Shi et al., 2016).

Activated microglia can damage neurons by generating peroxynitrite via iNOS. Levels of iNOS are increased in AD cortex relative to normal brains in cortical capillaries and within senile plaques (where there is active microglial participation); thus microglial and endothelial cells are probably the main source of NO-iNOS derived in AD brains (Jeynes and Provias, 2009). The nitro-OS elicited by iNOS overactivation can initiate a cascade of reactions that trigger apoptosis and evoke cytotoxic effects and degeneration in neurons and endothelial cells (Malinski, 2007), maintaining a feedback loop of irreversible inflammation and neurodegeneration.

Proinflammatory cytokines also play a direct role in promoting neuronal death and the progression of AD pathology. Elevated markers of inflammation have been described in both human AD brains (Wood et al., 2015) and in animal models (Shi et al., 2016). In Aβ-stimulated primary microglia, levels of IL-1β, TNF-α, prostaglandin E2 (PGE2), and COX-2 were substantially increased (Shi et al., 2016). TNF-α and IL-1β are pleiotropic proinflammatory cytokines that are raised in the brains and plasma of patients with AD, and are proximal to amyloid plaques on autopsy. Overexpression of both enhance intracellular Aβ, inflammation, and Tau pathology, and eventually lead to neuronal cell death (Kitazawa et al., 2011; McAlpine et al., 2009; Sheng et al., 2000). COX-2 expression is elevated in AD brain (Fujimi et al., 2007; Ho et al., 2001; Rao et al., 2011) and its expression in hippocampus is related to disease progression (Ho et al., 2001). PGE2 is likewise a well-known inflammatory mediator that is synthesized by cyclooxygenases. Both COX-2 and PGE2 induced BACE1 expression and upregulate the levels of IL-1β in vitro, which in turn established a pro-amyloidogenic and proinflammatory milieu by reciprocal regulation of IL-1β and Aβ between glial and neuron cells, all of this regulated by COX-2 (Wang et al., 2014).

Expression of iNOS, TNF-α, IL-1β, and COX-2 is regulated by NF-κB (Sarkar et al., 2008). As discussed previously, NF-κB activation comprises a central event of inflammation and is a common feature of many neurodegenerative diseases, as well as a primary regulator of target proinflammatory genes. In the brains of patients with AD, activated NF-κB was found predominantly in neurons and glial cells in areas surrounding Aβ plaques of hippocampus and temporal cortex (Chen et al., 2012; Granic et al., 2009). In primary rat and human neurons, Aβ activates NF-κB transcription factor composed of p50 and p65 subunits, and decreases neuronal viability by apoptosis. In turn, NF-κB promotes Aβ42 intracellular accumulation (Valerio et al., 2006), at least in part by upregulation of BACE1 activity (Chen et al., 2012). This finding supports that regulation of NF-κB-dependent gene transcription may mediate some of the neurodegenerative effects described in AD pathology.

Hence, AD is a complex pathological process that involves many pathways and factors. Therefore, a multipurpose treatment strategy that could be able to simultaneously modulate OS, inflammatory response, endothelial dysfunction, and NF-κB activation, or the majority of these aspects, might be an ideal therapy for patients with AD (Fig. 11.3).

NO DONORS AS NEUROPROTECTIVE AGENTS IN BRAIN INFLAMMATION

NO donors in cerebral ischemia–reperfusion injury

NO donors (NODs) exerted a neuroprotective effect against cerebral ischemia–reperfusion injury at different levels by influencing CBF, cellular oxidative status,

Alzheimer's disease

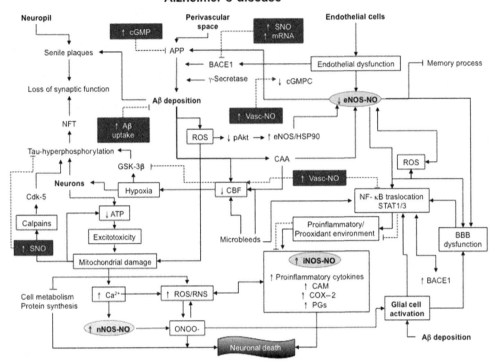

Figure 11.3 Schematic representation of some of the signaling events regulating the pathogenesis of Alzheimer's disease and the effect of NODs. Events common to the three pathologies discussed in this chapter are in *white boxes*; NO synthases are identified with *blue ovals*; the effects of NODs are highlighted with *blue squares and blue dashed lines*. Aβ and tau-hyperphosphorylation deposits are both hallmarks of AD pathology. Aβ is deposited in senile plaques in the extracellular space and in the perivascular space, namely CAA. Aβ is synthesized from APP through BACE1 and γ-secretase activities. The sole presence of Aβ is capable of generating ROS production that induces neuronal and endothelial oxidative stress with eNOS dysfunction. CAA is also associated with vascular damage and endothelial dysfunction with ablation of NO-eNOS derived. At the same time, eNOS-NO deprivation is able to increase APP accumulation and BACE1 activities via inhibition of cGMP, thus, increasing Aβ deposition. Failure of eNOS functioning leads to ROS production, BBB dysfunction, and NF-κB translocation, which together elicit a neuroinflammatory response. eNOS-NO deprivation affects as well the memory process that depends on NO/cGMP pathway. Aβ deposition also induces mitochondrial damage and glial activation. Mitochondrial damage leads to ROS production, Ca²⁺ overload, and nNOS activation, which produces NO that is deleterious to the neurons because it generates peroxynitrite by interaction with superoxide anion. Meanwhile, glial activation is also associated with NF-κB translocation, with ROS/RNS production, and expression of its target genes. Finally, vascular damage and BBB dysfunction generate a reduction in CBF and hypoxia, with the consequences described extensively in the text. The evidence suggests that NODs can alter some of the process described previously as follows: (1) NODs can blocks BACE1 activity by SNO and by inhibition of its transcription, with the corresponding decrease in Aβ deposition; (2) the boost on cGMP production is related to an increase in NO available at vascular level and with a decrease in APP generation; this also is related with an (3) amelioration of CBF; (4) NODs modulate activity of Aβ-transporters, thereby

inflammatory response, and neuronal death. Hemodynamically, sodium nitroprusside (SNP), *S*-nitrosoglutathione (GSNO), *S*-nitroso-*N*-acetylpenicillamine (SNAP), and 11H-indeno[1,2-b]quinoxalin-11-one (IQ-1S, a novel NOD) increase CBF in the penumbral region when administered at the onset of reperfusion (Atochin et al., 2016; Khan et al., 2006; Zhang and Iadecola, 1994; Zhang et al., 1994) (Fig. 11.2).

SNP and SPERMINE NONOate are able to reduce infarct size after transient, focal cerebral ischemia when administered early (Salom et al., 2000). Likewise, *S*-nitrosothiols (GSNO and SNAP) (Khan et al., 2006; Parent et al., 2015; Sakakima et al., 2012) and 3-morpholinosydnonimine (SIN-1) (Coert et al., 2002; Zhang and Iadecola, 1994) additionally reduced infarct volume and improved neurological performance.

In addition, pre- and postischemic administration of SNP and GSNO attenuates the ischemia-induced increase of proteins related with death, such as caspase-3 (Pei et al., 2008; Sakakima et al., 2012; Yin et al., 2013), caspase-9 (Yin et al., 2013), and the apoptosis signal-regulating kinase-1 (Ask-1) (Liu et al., 2013). The consequence is downregulation of neuronal apoptosis by inhibiting the increased phosphorylation of p38 (Qi et al., 2013), JNK, c-Jun (Pei et al., 2008; Yu et al., 2008), Fas (Yin et al., 2013), and MLK-3 (Hu et al., 2012). All of these are proteins expressed downstream of signaling pathways related with apoptosis. The common mechanism by which this effect is attained is through nitrosylation of nNOS, which restricts its phosphorylation and decreases its NO production and neuronal peroxynitrite levels (Khan et al., 2015; Yin et al., 2013), indicating that the activity of nNOS is vulnerable to exogenous NO.

In focal cerebral ischemia, SNP and *S*-nitrosothiols decrease lipid peroxidation and nitrotyrosine formation in plasma, which is associated with less oxidative and nitrosative stress, neuroprotection, and fewer antiinflammatory effects (Khan et al., 2006). In an in vitro model with brain slices under hypoxia, preischemic treatment with NOC-18, another NOD, was able to reduce tissue necrosis. This effect is achieved by inhibiting the opening of the mitochondrial permeability transition pore (MPTP) through the intervention of protein kinase G and protein kinase C. The opening of MPTP

Figure 11.3 (Continued)   improving its clearance from neuropil and perivascular space; (5) by SNO, NODs inhibit the activity of calpain and kinases (Cdk5 and GSK-3β) found downstream and related to tau-hyperphosphorylation. See the text for further explanation. Aβ, amyloid beta; APP, amyloid precursor protein; Ask-1, apoptosis signal-regulating kinase-1; ATP, adenosine triphosphate; BACE1, beta-site APP cleaving enzyme 1; BBB, blood brain barrier; CAM, cell adhesion molecules; CBF, cerebral blood flow; Cdk5, cyclin-dependent kinase 5; cGMP, cyclic guanosine monophosphate; COX-2, cyclooxygenase-2; eNOS, endothelial nitric oxide synthase; GSK-3β, glycogen synthase kinase 3 beta; iNOS, inducible nitric oxide synthase; mRNA, messenger ribonucleic acid; NFT, neurofibrillary tangles; NF-κB, nuclear factor kappa B; NODs, nitric oxide donors; nNOS, neuronal nitric oxide synthase; NO, nitric oxide; PGs, prostaglandins; RNS, reactive nitrogen species; ROS, reactive oxygen species; SNO, *S*-nitrosylation; STAT, signal transducer and activator of transcription.

leads to the release of cytochrome c and the subsequent inhibition of mitochondrial respiration and apoptosis (Arandarcikaite et al., 2015).

NODs not only attenuate signaling pathways associated with neuronal death but also stimulate the expression of neurotrophins that influence neuronal proliferation, survival, and differentiation. Such is the case of the GSNO that, in a model of focal cerebral ischemia, demonstrated a neuroprotective effect by stimulating the expression of neurorepair mediators such as BDNF/TrkB/pTrkB via activation of Akt (Sakakima et al., 2012). The effects of NODs on ischemia–reperfusion injury are also related with modulation of the inflammatory response, and these effects are probably the neuroprotective effects with greatest impact after cerebral ischemia and the reperfusion of this drug type.

Nevertheless, to shield the nerve tissue from ischemic reperfusion injury is not sufficient to protect the brain parenchyma, but the integrity of the BBB also must be preserved. Thus the cerebral vascular endothelium is essential in the control of vascular inflammatory and oxidative responses, leukocyte migration, and the production of inflammatory mediators capable of spreading to nerve tissue (Ceulemans et al., 2010). Under physiological conditions, eNOS-NO is responsible for maintaining the integrity of the vascular endothelium. But under ischemic conditions, endothelial dysfunction could be offset by mimicking eNOS-derived NO neuroprotective functions by intravascular administration of an NOD (Godinez-Rubi et al., 2013). Control of endothelial inflammatory and oxidative responses in turn allows restriction of their impact on resident brain cells, particularly on those with an inflammatory phenotype such as microglia and astrocytes. Therefore, effective neuroprotection should include protection of the BBB and of the elements within it (Tajes et al., 2013; Yasuda et al., 2011; Yilmaz and Granger, 2008).

In vivo, high expression of TNF-α, IL-1β, and iNOS in microglia and astrocytes after focal cerebral ischemia is reduced by GSNO. Likewise, GSNO induces a reduction in microglial and macrophage cells in the penumbral region, which is associated with less expression of cellular adhesion molecules such as ICAM-1 in endothelial cells (Suzuki et al., 2002). Decreased expression of adhesion molecules (ICAM-1 and E-selectin) was also demonstrated with SNP and SNAP in the same model (Khan et al., 2006).

In vitro studies have elucidated some of the mechanisms involved in the antiinflammatory effect of NOD. It is well documented that cerebral ischemia, and particularly reperfusion, leads to nuclear translocation of NF-κB into the core and ischemic penumbra (Campelo et al., 2012; Zhang et al., 2005), as well as into the microvessels of the affected region (Kolev et al., 2003; Liu et al., 2006). In the ischemic brain, a wide range of stimuli may trigger activation of NF-κB including, among others, the following: hypoxia (Cummins et al., 2006); IL-1 and TNF-α (Blais and Rivest, 2001); OS (Hu et al., 2012); glutamate (Xu et al., 2005); and NOS activity such as nNOS and iNOS

(Greco et al., 2011). Overactivation of NF-κB after ischemia has been documented in neurons (Campelo et al., 2012; Zhang et al., 2005), astrocytes (Khan et al., 2005), microglia (Webster et al., 2009), and endothelial cells (Kolev et al., 2003). Although in some hippocampal neurons, NF-κB has a constitutive action related with neuronal survival (Engelmann et al., 2014), overactivation of the p50/p65 heterodimer in neurons and in glia and endothelial cells due to ischemia appears to contribute to acute neurodegeneration. In neurons, NF-κB translocation has been associated with apoptosis (Zhang et al., 2005), while in glia and in vascular endothelium, NF-κB activates a proinflammatory phenotype (Khan et al., 2005; Kolev et al., 2003; Webster et al., 2009). Therefore, blocking the inflammatory phenotype activation of NF-κB could disrupt the cascade of events that culminate in proinflammatory brain tissue destruction.

In a rat model of focal cerebral ischemia, preconditioning with cis-[Ru(bpy)$_2$(SO$_3$)(NO)]PF$_6$) (Rut-bpy, a NOD) causes a reduction in infarct volume, and this effect was associated with reduced expression of NF-κB in the hippocampus (Campelo et al., 2012). In human endothelial cells, the addition of exogenous NO through GSNO and SNP limits activation of NF-κB and its target gene VCAM-1 (Mohan et al., 2003). This is also achieved by SNP upon stimulation with IL-1α, IL-1β, IL-4, and LPS (De Caterina et al., 1995). NF-κB inhibition is in fact sustained by the constitutive activity of eNOS, because its inhibition, without an inflammatory stimulus, triggers nuclear translocation of NF-κB (Blais and Rivest, 2001; De Caterina et al., 1995), while its overexpression or substitution by an NOD prevents NF-κB activation (Grumbach et al., 2005). Additionally, by inhibiting the activation of this transcription factor, NO effectively blocks monocyte adhesion, as well as the expression of the proinflammatory target genes of NF-κB, such as iNOS, TNF-α, IL-6, iNOS, V-CAM, ICAM-1, E-selectin, and COX-2 (Blais and Rivest, 2001; De Caterina et al., 1995; Garrean et al., 2006; Katsuyama et al., 1998; Mohan et al., 2003; Park et al., 2005).

In astrocytes and microglial cells, NOD also exhibits an antiinflammatory profile through downregulation of NF-κB. In primary rat astrocytes and in a microglial cell line, GSNO mitigated iNOS production by inhibiting the ability of NF-κB to bind to DNA (Khan et al., 2005). Therefore, NODs are capable not only of regulating NF-κB at the vascular level, but they also possess the capability of influencing glial cell reactivity and limiting their production of iNOS, proinflammatory cytokines, and other molecules and pro-oxidant enzymes that are potentially harmful to neuronal cells (Tajes et al., 2013).

In the brain ischemic environment, activation of NF-κB occurs, at least in part, via ROS (Song et al., 2010). Former researches have shown that one of the most significant sources of ROS in the ischemic brain is through the metabolism of arachidonic acid by COX (Candelario-Jalil et al., 2003; Tabassum et al., 2015). COX-2 expression is increased in brain tissue after global (Llorente et al., 2013) and focal (Ahmad et al., 2009; Tabassum et al., 2015) cerebral ischemia. ROS are produced by the peroxidase step of the COX reaction in which prostaglandin G2 is converted into prostaglandin H2 (Im et al., 2006; Speed and Blair, 2011). Hence, reducing COX-2 activity reduces oxidative damage of the ischemic brain (Ahmad et al., 2009; Tabassum et al., 2015). NODs GSNO and SNP are able to downregulate LPS-induced COX-2 protein expression via inhibition of NF-κB DNA-binding activity in murine monocytes (D'Acquisto et al., 2001). Therefore, both drugs may be candidates as neuroprotective antioxidants in cerebral ischemia. In addition, NO is a superoxide scavenger; hence, NO may inhibit NF-κB by scavenging superoxide anion (De Caterina et al., 1995).

Furthermore, SNP is capable of interfering directly with the ability of NF-κB to translocate into the nucleus. Specifically, SNP inactivates NF-κB by nitration of the p65 subunit at Tyr-66 and Tyr-152. This protein modification suppresses iNOS messenger RNA (mRNA) expression and prevents the activation of NF-κB target genes by TNF-α stimulation (Park et al., 2005).

With respect to IκB-α, exogenous NO increases mRNA IκB levels and stabilizes the complex formed with NF-κB (D'Acquisto et al., 2001). This stabilization is related with S-nitrosylation of the Cys-179 of IKKß, which decreases its ability to phosphorylate IκB (Reynaert et al., 2004). Additionally, NO interferes with the transient degradation of IκB-α induced by cytokines (Katsuyama et al., 1998). Together, these three actions induce negative regulation of NF-κB DNA-binding activity by NOD.

NODs in TBI

Because the pathophysiology of cerebral ischemia and TBI is similar, the neuroprotective effect of NOD and its therapeutic potential is also replicated in different models of TBI. As in brain ischemia–reperfusion injury, the exogenous NO released by NOD appears to replicate the actions of eNOS-derived NO, without contributing to neurotoxicity (Garry et al., 2015).

In that regard, NODs, such as SNP, GSNO, SNAP, and iQ-1S, have demonstrated the ability to improve CBF in the penumbral area in models of cerebral ischemia (Atochin et al., 2016; Khan et al., 2005, 2006; Zhang and Iadecola, 1994; Zhang et al., 1994). This effect is also achieved in a model of controlled cortical impact with SNAP, by reversing the attenuation of the hypercapnia cerebral vasoconstriction associated with TBI. Conversely, the increase in CBF was not achieved by papaverine (Zhang et al., 2002), another vasodilator whose hemodynamic effects are independent of NO, suggesting that, as in ischemic stroke, regulation of CBF after acute brain injury depends mainly on NO (Fig. 11.1).

GSNO is the most studied NOD in TBI. Within the context of acute brain injury, it has demonstrated that GSNO is not only a neuroprotective and antiinflammatory agent, but is also an endothelial protecting agent, in that it confers protection on the BBB. In a rodent model of controlled cortical impact, administration of GSNO after TBI reduced BBB leakage and brain edema, and also decreased the expression of

ICAM-1, a mediator of endothelial cell activation leading to BBB disruption (Khan et al., 2009, 2011). Blood outflow into the brain parenchyma through impaired BBB contributes further to inflammation and OS, disrupting cellular defense and antioxidant mechanisms (Khan et al., 2005, 2009, 2011).

GSNO is also capable of exerting an impact on oxidative and nitrosative stress after TBI. This NOD reduced peroxynitrite production by two main mechanisms that involved the regulation of nNOS and iNOS. On the one hand, GSNO is able to decrease nNOS activity by decreasing its phosphorylation, and in turn reducing peroxynitrite levels associated with the enzyme activity (Khan et al., 2016). Additionally, the NO donor also downregulated iNOS expression in microglia and macrophages in the injured area, thereby attenuating the production of peroxynitrite (Khan et al., 2009, 2011, 2016). Because iNOS expression is transcriptionally regulated by NF-κB and GSNO is able to regulate the expression of this transcription factor (Khan et al., 2005; Tajes et al., 2013), GSNO may reduce the expression of iNOS via inhibition of NF-κB.

Calpains are proteases associated with neurodegeneration and axonal loss within the context of acute brain injury (Liu et al., 2008), and are affected by nitrosative stress. The TBI-induced neuronal cytoskeleton degradation leading to neurodegeneration is mediated prominently by calpain activities. In humans, calpain is expressed after TBI in all resident brain cells: neurons, glial and endothelial cells, from 5 hours after the trauma until 10 days, with a peak at 72 hours (Bralic and Stemberga, 2012). Neuronal peroxynitrite production in TBI causes activation of calpain, but in addition, the activity of the enzyme is inhibited by NO/S-nitrosylation of cysteine at the active site of calpain. GSNO inhibits the nNOS/peroxynitrite/calpain system via the reversible mechanism of S-nitrosylation of both nNOS and calpain, thus reducing not only the levels of nNOS-derived peroxynitrite but also TBI-induced calpain activity, leading to reduced neurodegeneration and improved neurological functions (Khan et al., 2016).

In relation to OS, GSNO increases levels of reduced glutathione. Accordingly, the drug also decreases lipid peroxidation products in plasma and in traumatic penumbra (Khan et al., 2011). Thus it impacts positively on the oxidative balance of the microenvironment damaged by trauma.

The consequences of the previously described actions of GSNO include neuroprotection and promotion of functional recovery, expressed as a reduction in neuronal apoptotic cell death and neurodegeneration, reduction in contusion size, improvement synaptic plasticity, expression of neurotrophic factors such as brain-derived neurotrophic factor (BDNF) and its receptor (tyrosine receptor kinase B, TrkB), and the improvement of neurobehavioral functions (Khan et al., 2009, 2011, 2016).

GSNO is not the only NO donor that possesses neurorepair abilities in acute brain injury. In a model of controlled cortical impact, DETA/NONOate demonstrated that is able to significantly enhance the proliferation, survival, migration, and differentiation of neural progenitor cells in the hippocampus, subventricular zone, striatum, corpus callosum, and the boundary zone of the injured cortex, as well as in the contralateral hemisphere. This effect was also associated with improved neurological functional outcomes (Lu et al., 2003) and replicated in a model of brain ischemia. These effects were apparently achieved by an increase in cGMP (Zhang et al., 2001).

NODs in AD

Sporadic AD is a complex and multifactorial disease with the involvement of several genetic and environmental risk factors. In previous sections, we presented evidence that reported the implication of OS and neuroinflammation in the development and progression of AD, and also the possible implication of endothelium-derived NO deficiency in the pathophysiology of this degenerative pathology. This leads one to consider that restoration of NO bioavailability or NO-mediated vascular function may be beneficial in the attenuation of the cerebrovascular complications of AD, thus, attenuation of the imminent progress of sporadic AD (Fig. 11.3). Thereby, several NOS have been applied in vitro and in vivo models of AD with the results presented later.

As in cerebral ischemia and TBI, the most studied NODs in AD pathology are the S-nitrosothiol donors, including GSNO, SNAP, and S-nitrosocysteine (SNOC), in that all three have exhibited effective protection via several mechanisms (Kwak et al., 2011; Won et al., 2013, 2015) without the injurious effects of other NODs such as excessive release of NO and the production of peroxynitrite.

Experimentally, in vivo and in vitro, S-nitrosothiol donor-mediated mechanisms may regulate extracellular Aβ homeostasis by reduction of Aβ production in neurons and in endothelial and microglial cells without any significant cytotoxicity. Aβ production may be modulated by GSNO and SNOC via inhibition of BACE1. On the one hand, at low concentrations, GSNO and SNOC are capable of inhibiting BACE1 transcription via a cGMP-dependent pathway. Also, at higher concentrations, GSNO and SNOC induced S-nitrosylation of BACE1, leading to inactivation of the enzyme (Kwak et al., 2011; Won et al., 2013). These findings are in relation to the studies of Austin and colleagues, who demonstrated that lack of NO produced by eNOS is associated with increased neuronal and vascular Aβ deposition and with increased activity of BACE1 (Austin et al., 2010, 2013a, 2013b). In this regard, the same group demonstrated that in the cerebral microvasculature of eNOS knockout mice, supplementation of NO with nitroglycerine significantly reduced APP and BACE1 levels via cGMP (Austin et al., 2013a, 2013b). Thus the modulation of the endothelial NO/cGMPC pathway may provide novel therapeutic opportunities for preventing sporadic AD.

GSNO is also capable of interfering with Aβ clearance by diminishing its accumulation. Recently, it was acknowledged that levels of Aβ are determined by the balance between its production and clearance, rendering dysfunction in Aβ clearance a crucial event for the accumulation of Aβ in AD brains (Wang et al., 2006). In vitro assays of endothelial and microglial cells under GSNO treatment demonstrated that this NOD is able to enhance Aβ uptake in both cells by *S*-nitrosylation of a protein (dynamin-2) involved in Aβ transporting (Won et al., 2013). This data suggests that GSNO-mediated mechanisms may play a functional role in Aβ clearance in the brain and in microvasculature.

Another of the mechanisms described in AD pathology is tau hyperphosphorylation, which forms neurofibrillary tangles in the neurons of AD brains. Hyperphosphorylation of tau is triggered by protein kinases, among which the most important are Cdk5 and GSK-3β (Cruz et al., 2003; Llorens-Martin et al., 2014). Cdk5 is aberrantly activated under AD conditions, mainly by calpain, and this is related with neurodegeneration and neurofibrillary tangle deposition (Czapski et al., 2016; Gao et al., 2013). NO is able to inhibit calpain activity reversibly via *S*-nitrosylation (Samengo et al., 2012) and by decreasing its glutamate- and calcium-induced activation. Inhibition of calpain by GSNO is related in turn with a decreased activation of Cdk5 (Annamalai et al., 2015; Won et al., 2015). Calpain and Cdk5 inhibition are also associated with NF-κB inhibition in mice hippocampus (Czapski et al., 2016). Therefore, GSNO benefits achieved in this manner may be related with inhibition of tau hyperphosphorylation and with antiinflammatory activity.

GSK-3β activity is controlled by phosphorylation (Llorens-Martin et al., 2014). Chronic oligoemia induces increased GSK-3β activity and phosphorylation, but treatment with GSNO reduces its phosphorylation and activity (Annamalai et al., 2015). Abrogation of both kinases (Cdk5 and GSK-3β) is associated with a decline in tau hyperphosphorylation. These effects of GSNO are also related with an attenuation of neuronal degeneration and with an improved spatial learning performance (Annamalai et al., 2015; Won et al., 2015).

In addition to the previously described mechanisms, GSNO also possesses potential antiinflammatory activity in rodents subjected to chronic oligoemia. In this model, exogenous supplementation of NO through GSNO decreased the transcriptional activity and protein levels of vascular inflammatory markers such as ICAM-1 and VCAM-1. Moreover, in in vitro assays with endothelial and microglial cells, GSNO reproduced the same effects. This was associated with complete abrogation of NF-κB and STAT1/3 activities (Won et al., 2013), both considered key inflammatory transcription factors (Kim et al., 2014; Sarkar et al., 2008). The antiinflammatory effect of GSNO has been also proved in other models of neuroinflammation such as like ischemia–reperfusion injury (Khan et al., 2005, 2006, 2015), TBI (Khan et al., 2009, 2011, 2016), and experimental autoimmune encephalomyelitis (Prasad et al., 2007).

Thus, according to the evidence presented, it can be concluded that NODs, particularly *S*-nitrosothiol donors, are able to displace a set of mechanisms that situate them as potential therapeutic options for providing neuroprotection and improving cognitive functions under AD conditions.

CONCLUDING REMARKS

NO is a gas that is naturally produced by some cells; it is a simple but important molecule. In the brain, its concentration is generally low; however, under certain circumstances, it can rise to a higher, compartmentalized concentration. This could, in turn, elicit a neurotoxic effect. A common feature of neurodegenerative diseases, such as AD and other brain affections, such as TBI or ischemic reperfusion damage, is mediated by NF-κB nuclear translocation, which activates an inflammatory response. Originated by different agents, these three pathologies share many cellular and molecular mechanisms in common. One of these is the involvement of NO, which can be harmful under certain circumstances, but which also can function as a neuroprotective agent when it is exogenous administered at the right time and in the right place and at the right concentrations. Consequently, NODs can mitigate iNOS production by downregulating NF-κB nuclear translocation, therefore also limiting the inflammatory response in the brain. Understanding the interaction between the CNS and the immune system will provide greater insight into several different pathologies that involve CNS inflammation and the increase in the number of potential pharmacological targets for treating them such as NOD.

REFERENCES

Ahmad, M., Graham, S.H., 2010. Inflammation after stroke: mechanisms and therapeutic approaches. Transl. Stroke Res. 1, 74–84.

Ahmad, M., Zhang, Y., Liu, H., Rose, M.E., Graham, S.H., 2009. Prolonged opportunity for neuroprotection in experimental stroke with selective blockade of cyclooxygenase-2 activity. Brain Res. 1279, 168–173.

Ahn, M.J., Sherwood, E.R., Prough, D.S., Lin, C.Y., Dewitt, D.S., 2004. The effects of traumatic brain injury on cerebral blood flow and brain tissue nitric oxide levels and cytokine expression. J. Neurotrauma 21, 1431–1442.

Aliev, G., Palacios, H.H., Lipsitt, A.E., Fischbach, K., Lamb, B.T., Obrenovich, M.E., et al., 2009. Nitric oxide as an initiator of brain lesions during the development of Alzheimer disease. Neurotox. Res. 16, 293–305.

Annamalai, B., Won, J.S., Choi, S., Singh, I., Singh, A.K., 2015. Role of S-nitrosoglutathione mediated mechanisms in tau hyperphosphorylation. Biochem. Biophys. Res. Commun. 458, 214–219.

Arandarcikaite, O., Jokubka, R., Borutaite, V., 2015. Neuroprotective effects of nitric oxide donor NOC-18 against brain ischemia-induced mitochondrial damages: role of PKG and PKC. Neurosci. Lett. 586, 65–70.

Asahi, M., Huang, Z., Thomas, S., Yoshimura, S., Sumii, T., Mori, T., et al., 2005. Protective effects of statins involving both eNOS and tPA in focal cerebral ischemia. J. Cereb. Blood Flow Metab. 25, 722–729.

Atochin, D.N., Schepetkin, I.A., Khlebnikov, A.I., Seledtsov, V.I., Swanson, H., Quinn, M.T., et al., 2016. A novel dual NO-donating oxime and c-Jun N-terminal kinase inhibitor protects against cerebral ischemia-reperfusion injury in mice. Neurosci. Lett. 618, 45–49.

Attems, J., Jellinger, K.A., 2014. The overlap between vascular disease and Alzheimer's disease—lessons from pathology. BMC Med. 12, 206.

Austin, S.A., D'uscio, L.V., Katusic, Z.S., 2013a. Supplementation of nitric oxide attenuates AbetaPP and BACE1 protein in cerebral microcirculation of eNOS-deficient mice. J. Alzheimers Dis. 33, 29–33.

Austin, S.A., Santhanam, A.V., Hinton, D.J., Choi, D.S., Katusic, Z.S., 2013b. Endothelial nitric oxide deficiency promotes Alzheimer's disease pathology. J. Neurochem. 127, 691–700.

Austin, S.A., Santhanam, A.V., Katusic, Z.S., 2010. Endothelial nitric oxide modulates expression and processing of amyloid precursor protein. Circ. Res. 107, 1498–1502.

Bao, F., Liu, D., 2004. Hydroxyl radicals generated in the rat spinal cord at the level produced by impact injury induce cell death by necrosis and apoptosis: protection by a metalloporphyrin. Neuroscience 126, 285–295.

Barber, P.A., Demchuk, A.M., Hirt, L., Buchan, A.M., 2003. Biochemistry of ischemic stroke. Adv. Neurol. 92, 151–164.

Bayir, H., Kagan, V.E., Clark, R.S., Janesko-Feldman, K., Rafikov, R., Huang, Z., et al., 2007. Neuronal NOS-mediated nitration and inactivation of manganese superoxide dismutase in brain after experimental and human brain injury. J. Neurochem. 101, 168–181.

Blackshaw, S., Eliasson, M.J., Sawa, A., Watkins, C.C., Krug, D., Gupta, A., et al., 2003. Species, strain and developmental variations in hippocampal neuronal and endothelial nitric oxide synthase clarify discrepancies in nitric oxide-dependent synaptic plasticity. Neuroscience 119, 979–990.

Blais, V., Rivest, S., 2001. Inhibitory action of nitric oxide on circulating tumor necrosis factor-induced NF-kappaB activity and COX-2 transcription in the endothelium of the brain capillaries. J. Neuropathol. Exp. Neurol. 60, 893–905.

Bon, C.L., Garthwaite, J., 2003. On the role of nitric oxide in hippocampal long-term potentiation. J. Neurosci. 23, 1941–1948.

Bralic, M., Stemberga, V., 2012. Calpain expression in the brain cortex after traumatic brain injury. Coll. Antropol. 36, 1319–1323.

Brenowitz, W.D., Nelson, P.T., Besser, L.M., Heller, K.B., Kukull, W.A., 2015. Cerebral amyloid angiopathy and its co-occurrence with Alzheimer's disease and other cerebrovascular neuropathologic changes. Neurobiol. Aging 36, 2702–2708.

Butterfield, D.A., Swomley, A.M., Sultana, R., 2013. Amyloid beta-peptide (1-42)-induced oxidative stress in Alzheimer disease: importance in disease pathogenesis and progression. Antioxid. Redox Signal. 19, 823–835.

Campelo, M.W., Oria, R.B., Lopes, L.G., Brito, G.A., Santos, A.A., Vasconcelos, R.C., et al., 2012. Preconditioning with a novel metallopharmaceutical NO donor in anesthetized rats subjected to brain ischemia/reperfusion. Neurochem. Res. 37, 749–758.

Candelario-Jalil, E., Gonzalez-Falcon, A., Garcia-Cabrera, M., Alvarez, D., Al-Dalain, S., Martinez, G., et al., 2003. Assessment of the relative contribution of COX-1 and COX-2 isoforms to ischemia-induced oxidative damage and neurodegeneration following transient global cerebral ischemia. J. Neurochem. 86, 545–555.

Carcaillon, L., Plichart, M., Zureik, M., Rouaud, O., Majed, B., Ritchie, K., et al., 2015. Carotid plaque as a predictor of dementia in older adults: the Three-City Study. Alzheimers Dement. 11, 239–248.

Ceulemans, A.G., Zgavc, T., Kooijman, R., Hachimi-Idrissi, S., Sarre, S., Michotte, Y., 2010. The dual role of the neuroinflammatory response after ischemic stroke: modulatory effects of hypothermia. J. Neuroinflammation 7, 74.

Coert, B.A., Anderson, R.E., Meyer, F.B., 2002. Effects of the nitric oxide donor 3-morpholinosydnonimine (SIN-1) in focal cerebral ischemia dependent on intracellular brain pH. J. Neurosurg. 97, 914–921.

Corti, A., Franzini, M., Scataglini, I., Pompella, A., 2014. Mechanisms and targets of the modulatory action of S-nitrosoglutathione (GSNO) on inflammatory cytokines expression. Arch. Biochem. Biophys. 562, 80–91.

Cruz, J.C., Tseng, H.C., Goldman, J.A., Shih, H., Tsai, L.H., 2003. Aberrant Cdk5 activation by p25 triggers pathological events leading to neurodegeneration and neurofibrillary tangles. Neuron 40, 471–483.

Cruz Navarro, J., Pillai, S., Ponce, L.L., Van, M., Goodman, J.C., Robertson, C.S., 2014. Endothelial nitric oxide synthase mediates the cerebrovascular effects of erythropoietin in traumatic brain injury. Front Immunol. 5, 494.

Cummins, E.P., Berra, E., Comerford, K.M., Ginouves, A., Fitzgerald, K.T., Seeballuck, F., et al., 2006. Prolyl hydroxylase-1 negatively regulates IkappaB kinase-beta, giving insight into hypoxia-induced NFkappaB activity. Proc. Natl. Acad. Sci. USA 103, 18154–18159.

Czapski, G.A., Gassowska, M., Wilkaniec, A., Chalimoniuk, M., Strosznajder, J.B., Adamczyk, A., 2016. The mechanisms regulating cyclin-dependent kinase 5 in hippocampus during systemic inflammatory response: the effect on inflammatory gene expression. Neurochem. Int. 93, 103–112.

Chen, C.H., Zhou, W., Liu, S., Deng, Y., Cai, F., Tone, M., et al., 2012. Increased NF-kappaB signalling up-regulates BACE1 expression and its therapeutic potential in Alzheimer's disease. Int. J. Neuropsychopharmacol. 15, 77–90.

Chen, W., Qi, J., Feng, F., Wang, M.D., Bao, G., Wang, T., et al., 2014. Neuroprotective effect of allicin against traumatic brain injury via Akt/endothelial nitric oxide synthase pathway-mediated anti-inflammatory and anti-oxidative activities. Neurochem. Int. 68, 28–37.

Cherian, L., Chacko, G., Goodman, C., Robertson, C.S., 2003. Neuroprotective effects of L-arginine administration after cortical impact injury in rats: dose response and time window. J. Pharmacol. Exp. Ther. 304, 617–623.

Cherian, L., Hlatky, R., Robertson, C.S., 2004. Nitric oxide in traumatic brain injury. Brain Pathol. 14, 195–201.

Cherian, L., Robertson, C.S., 2003. L-Arginine and free radical scavengers increase cerebral blood flow and brain tissue nitric oxide concentrations after controlled cortical impact injury in rats. J. Neurotrauma 20, 77–85.

Chiueh, C.C., Rauhala, P., 1999. The redox pathway of S-nitrosoglutathione, glutathione and nitric oxide in cell to neuron communications. Free Radic. Res. 31, 641–650.

D'Acquisto, F., Maiuri, M.C., De Cristofaro, F., Carnuccio, R., 2001. Nitric oxide prevents inducible cyclo-oxygenase expression by inhibiting nuclear factor-kappa B and nuclear factor-interleukin-6 activation. Naunyn Schmiedebergs Arch. Pharmacol. 364, 157–165.

Danielisova, V., Burda, J., Nemethova, M., Gottlieb, M., 2011. Aminoguanidine administration ameliorates hippocampal damage after middle cerebral artery occlusion in rat. Neurochem. Res. 36, 476–486.

Danielisova, V., Nemethova, M., Burda, J., 2004. The protective effect of aminoguanidine on cerebral ischemic damage in the rat brain. Physiol. Res. 53, 533–540.

de Bruijn, R.F., Ikram, M.A., 2014. Cardiovascular risk factors and future risk of Alzheimer's disease. BMC Med. 12, 130.

De Caterina, R., Libby, P., Peng, H.B., Thannickal, V.J., Rajavashisth, T.B., Gimbrone Jr., M.A., et al., 1995. Nitric oxide decreases cytokine-induced endothelial activation. Nitric oxide selectively reduces endothelial expression of adhesion molecules and proinflammatory cytokines. J. Clin. Invest. 96, 60–68.

de La Torre, J.C., Aliev, G., 2005. Inhibition of vascular nitric oxide after rat chronic brain hypoperfusion: spatial memory and immunocytochemical changes. J. Cereb. Blood Flow Metab. 25, 663–672.

De Pascali, F., Hemann, C., Samons, K., Chen, C.A., Zweier, J.L., 2014. Hypoxia and reoxygenation induce endothelial nitric oxide synthase uncoupling in endothelial cells through tetrahydrobiopterin depletion and S-glutathionylation. Biochemistry 53, 3679–3688.

Deng, Y., Thompson, B.M., Gao, X., Hall, E.D., 2007. Temporal relationship of peroxynitrite-induced oxidative damage, calpain-mediated cytoskeletal degradation and neurodegeneration after traumatic brain injury. Exp. Neurol. 205, 154–165.

Doreulee, N., Sergeeva, O.A., Yanovsky, Y., Chepkova, A.N., Selbach, O., Godecke, A., et al., 2003. Cortico-striatal synaptic plasticity in endothelial nitric oxide synthase deficient mice. Brain Res. 964, 159–163.

Downes, C.E., Crack, P.J., 2010. Neural injury following stroke: are Toll-like receptors the link between the immune system and the CNS? Br. J. Pharmacol. 160, 1872–1888.

Duarte, J., Francisco, V., Perez-Vizcaino, F., 2014. Modulation of nitric oxide by flavonoids. Food Funct. 5, 1653–1668.

Engel, D.C., Mies, G., Terpolilli, N.A., Trabold, R., Loch, A., de Zeeuw, C.I., et al., 2008. Changes of cerebral blood flow during the secondary expansion of a cortical contusion assessed by 14C-iodoantipyrine autoradiography in mice using a non-invasive protocol. J. Neurotrauma 25, 739–753.

Engelmann, C., Weih, F., Haenold, R., 2014. Role of nuclear factor kappa B in central nervous system regeneration. Neural Regen. Res. 9, 707–711.

Forstermann, U., Sessa, W.C., 2012. Nitric oxide synthases: regulation and function. Eur. Heart J. 33, 829–837. 837a–837d.

Fujimi, K., Noda, K., Sasaki, K., Wakisaka, Y., Tanizaki, Y., Iida, M., et al., 2007. Altered expression of COX-2 in subdivisions of the hippocampus during aging and in Alzheimer's disease: the Hisayama study. Dement. Geriatr. Cogn. Disord. 23, 423–431.

Gao, L., Tian, S., Gao, H., Xu, Y., 2013. Hypoxia increases Abeta-induced tau phosphorylation by calpain and promotes behavioral consequences in AD transgenic mice. J. Mol. Neurosci. 51, 138–147.

Garcia-Alloza, M., Gregory, J., Kuchibhotla, K.V., Fine, S., Wei, Y., Ayata, C., et al., 2011. Cerebrovascular lesions induce transient beta-amyloid deposition. Brain 134, 3697–3707.

Garrean, S., Gao, X.P., Brovkovych, V., Shimizu, J., Zhao, Y.Y., Vogel, S.M., et al., 2006. Caveolin-1 regulates NF-kappaB activation and lung inflammatory response to sepsis induced by lipopolysaccharide. J. Immunol. 177, 4853–4860.

Garry, P.S., Ezra, M., Rowland, M.J., Westbrook, J., Pattinson, K.T., 2015. The role of the nitric oxide pathway in brain injury and its treatment—from bench to bedside. Exp. Neurol. 263, 235–243.

Gertz, K.A.E.M., 2008. eNOS and stroke: prevention, treatment and recovery. Future Neurol. 3, 537–550.

Gibson, C.L., Coughlan, T.C., Murphy, S.P., 2005. Glial nitric oxide and ischemia. Glia 50, 417–426.

Godinez-Rubi, M., Rojas-Mayorquin, A.E., Ortuno-Sahagun, D., 2013. Nitric oxide donors as neuroprotective agents after an ischemic stroke-related inflammatory reaction. Oxid. Med. Cell Longev. 2013, 297357.

Granic, I., Dolga, A.M., Nijholt, I.M., Van Dijk, G., Eisel, U.L., 2009. Inflammation and NF-kappaB in Alzheimer's disease and diabetes. J. Alzheimers Dis. 16, 809–821.

Greco, R., Mangione, A.S., Amantea, D., Bagetta, G., Nappi, G., Tassorelli, C., 2011. IkappaB-alpha expression following transient focal cerebral ischemia is modulated by nitric oxide. Brain Res. 1372, 145–151.

Grumbach, I.M., Chen, W., Mertens, S.A., Harrison, D.G., 2005. A negative feedback mechanism involving nitric oxide and nuclear factor kappa-B modulates endothelial nitric oxide synthase transcription. J. Mol. Cell Cardiol. 39, 595–603.

Guix, F.X., Uribesalgo, I., Coma, M., Munoz, F.J., 2005. The physiology and pathophysiology of nitric oxide in the brain. Prog. Neurobiol. 76, 126–152.

Gunther, M., Davidsson, J., Plantman, S., Norgren, S., Mathiesen, T., Risling, M., 2015. Neuroprotective effects of N-acetylcysteine amide on experimental focal penetrating brain injury in rats. J. Clin. Neurosci. 22, 1477–1483.

Gupta, A., Iadecola, C., 2015. Impaired Abeta clearance: a potential link between atherosclerosis and Alzheimer's disease. Front. Aging Neurosci. 7, 115.

Gursoy-Ozdemir, Y., Can, A., Dalkara, T., 2004. Reperfusion-induced oxidative/nitrative injury to neurovascular unit after focal cerebral ischemia. Stroke 35, 1449–1453.

Han, B.H., Zhou, M.L., Abousaleh, F., Brendza, R.P., Dietrich, H.H., Koenigsknecht-Talboo, J., et al., 2008. Cerebrovascular dysfunction in amyloid precursor protein transgenic mice: contribution of soluble and insoluble amyloid-beta peptide, partial restoration via gamma-secretase inhibition. J. Neurosci. 28, 13542–13550.

Hang, C.H., Shi, J.X., Li, J.S., Wu, W., Yin, H.X., 2005. Concomitant upregulation of nuclear factor-kB activity, proinflammatory cytokines and ICAM-1 in the injured brain after cortical contusion trauma in a rat model. Neurol. India 53, 312–317.

Heinrich, T.A., da Silva, R.S., Miranda, K.M., Switzer, C.H., Wink, D.A., Fukuto, J.M., 2013. Biological nitric oxide signalling: chemistry and terminology. Br. J. Pharmacol. 169, 1417–1429.

Hlatky, R., Lui, H., Cherian, L., Goodman, J.C., O'brien, W.E., Contant, C.F., et al., 2003. The role of endothelial nitric oxide synthase in the cerebral hemodynamics after controlled cortical impact injury in mice. J. Neurotrauma 20, 995–1006.

Ho, L., Purohit, D., Haroutunian, V., Luterman, J.D., Willis, F., Naslund, J., et al., 2001. Neuronal cyclooxygenase 2 expression in the hippocampal formation as a function of the clinical progression of Alzheimer disease. Arch. Neurol. 58, 487–492.

Hossain, M., Qadri, S.M., Liu, L., 2012. Inhibition of nitric oxide synthesis enhances leukocyte rolling and adhesion in human microvasculature. J. Inflamm. (Lond.) 9, 28.

Hu, S.Q., Ye, J.S., Zong, Y.Y., Sun, C.C., Liu, D.H., Wu, Y.P., et al., 2012. S-nitrosylation of mixed lineage kinase 3 contributes to its activation after cerebral ischemia. J. Biol. Chem. 287, 2364–2377.

Hu, Y.C., Sun, Q., Li, W., Zhang, D.D., Ma, B., Li, S., et al., 2014. Biphasic activation of nuclear factor kappa B and expression of p65 and c-Rel after traumatic brain injury in rats. Inflamm. Res. 63, 109–115.

Iadecola, C., 2004. Neurovascular regulation in the normal brain and in Alzheimer's disease. Nat. Rev. Neurosci. 5, 347–360.

Ill-Raga, G., Ramos-Fernandez, E., Guix, F.X., Tajes, M., Bosch-Morato, M., Palomer, E., et al., 2010. Amyloid-beta peptide fibrils induce nitro-oxidative stress in neuronal cells. J. Alzheimers Dis. 22, 641–652.

Im, J.Y., Kim, D., Paik, S.G., Han, P.L., 2006. Cyclooxygenase-2-dependent neuronal death proceeds via superoxide anion generation. Free Radic. Biol. Med. 41, 960–972.

Ito, Y., Ohkubo, T., Asano, Y., Hattori, K., Shimazu, T., Yamazato, M., et al., 2010. Nitric oxide production during cerebral ischemia and reperfusion in eNOS- and nNOS-knockout mice. Curr. Neurovasc. Res. 7, 23–31.

Jafarian-Tehrani, M., Louin, G., Royo, N.C., Besson, V.C., Bohme, G.A., Plotkine, M., et al., 2005. 1400W, a potent selective inducible NOS inhibitor, improves histopathological outcome following traumatic brain injury in rats. Nitric Oxide 12, 61–69.

Jeynes, B., Provias, J., 2009. Significant negative correlations between capillary expressed eNOS and Alzheimer lesion burden. Neurosci. Lett. 463, 244–468.

Katsuyama, K., Shichiri, M., Marumo, F., Hirata, Y., 1998. NO inhibits cytokine-induced iNOS expression and NF-kappaB activation by interfering with phosphorylation and degradation of IkappaB-alpha. Arterioscler. Thromb. Vasc. Biol. 18, 1796–1802.

Katusic, Z.S., Austin, S.A., 2014. Endothelial nitric oxide: protector of a healthy mind. Eur. Heart J. 35, 888–894.

Keil, U., Bonert, A., Marques, C.A., Scherping, I., Weyermann, J., Strosznajder, J.B., et al., 2004. Amyloid beta-induced changes in nitric oxide production and mitochondrial activity lead to apoptosis. J. Biol. Chem. 279, 50310–50320.

Khan, M., Dhammu, T.S., Matsuda, F., Singh, A.K., Singh, I., 2015. Blocking a vicious cycle nNOS/peroxynitrite/AMPK by S-nitrosoglutathione: implication for stroke therapy. BMC Neurosci. 16, 42.

Khan, M., Dhammu, T.S., Matsuda, F., Annamalai, B., Dhindsa, T.S., Singh, I., et al., 2016. Targeting the nNOS/peroxynitrite/calpain system to confer neuroprotection and aid functional recovery in a mouse model of TBI. Brain Res. 1630, 159–170.

Khan, M., Im, Y.B., Shunmugavel, A., Gilg, A.G., Dhindsa, R.K., Singh, A.K., et al., 2009. Administration of S-nitrosoglutathione after traumatic brain injury protects the neurovascular unit and reduces secondary injury in a rat model of controlled cortical impact. J. Neuroinflammation 6, 32.

Khan, M., Jatana, M., Elango, C., Paintlia, A.S., Singh, A.K., Singh, I., 2006. Cerebrovascular protection by various nitric oxide donors in rats after experimental stroke. Nitric Oxide 15, 114–124.

Khan, M., Sakakima, H., Dhammu, T.S., Shunmugavel, A., Im, Y.B., Gilg, A.G., et al., 2011. S-nitrosoglutathione reduces oxidative injury and promotes mechanisms of neurorepair following traumatic brain injury in rats. J Neuroinflammation 8, 78.

Khan, M., Sekhon, B., Giri, S., Jatana, M., Gilg, A.G., Ayasolla, K., et al., 2005. S-Nitrosoglutathione reduces inflammation and protects brain against focal cerebral ischemia in a rat model of experimental stroke. J. Cereb. Blood Flow Metab. 25, 177–192.

Kim, J., Won, J.S., Singh, A.K., Sharma, A.K., Singh, I., 2014. STAT3 regulation by S-nitrosylation: implication for inflammatory disease. Antioxid. Redox Signal. 20, 2514–2527.

Kitazawa, M., Cheng, D., Tsukamoto, M.R., Koike, M.A., Wes, P.D., Vasilevko, V., et al., 2011. Blocking IL-1 signaling rescues cognition, attenuates tau pathology, and restores neuronal beta-catenin pathway function in an Alzheimer's disease model. J. Immunol. 187, 6539–6549.

Knoblach, S.M., Faden, A.I., 2002. Administration of either anti-intercellular adhesion molecule-1 or a nonspecific control antibody improves recovery after traumatic brain injury in the rat. J. Neurotrauma 19, 1039–1050.

Knopman, D.S., Gottesman, R.F., Sharrett, A.R., Wruck, L.M., Windham, B.G., Coker JR., L., et al., 2016. Mild cognitive impairment and dementia prevalence: the Atherosclerosis Risk in Communities Neurocognitive Study (ARIC-NCS). Alzheimers Dement. (Amst.) 2, 1–11.

Kolev, K., Skopal, J., Simon, L., Csonka, E., Machovich, R., Nagy, Z., 2003. Matrix metalloproteinase-9 expression in post-hypoxic human brain capillary endothelial cells: H_2O_2 as a trigger and NF-kappaB as a signal transducer. Thromb. Haemost. 90, 528–537.

Kuhlencordt, P.J., Rosel, E., Gerszten, R.E., Morales-Ruiz, M., Dombkowski, D., Atkinson Jr., W.J., et al., 2004. Role of endothelial nitric oxide synthase in endothelial activation: insights from eNOS knockout endothelial cells. Am. J. Physiol. Cell Physiol. 286, C1195–C1202.

Kwak, Y.D., Wang, R., Li, J.J., Zhang, Y.W., Xu, H., Liao, F.F., 2011. Differential regulation of BACE1 expression by oxidative and nitrosative signals. Mol. Neurodegener. 6, 17.

Lakhan, S.E., Kirchgessner, A., Hofer, M., 2009. Inflammatory mechanisms in ischemic stroke: therapeutic approaches. J. Transl. Med. 7, 97.

Lamoke, F., Mazzone, V., Persichini, T., Maraschi, A., Harris, M.B., Venema, R.C., et al., 2015. Amyloid beta peptide-induced inhibition of endothelial nitric oxide production involves oxidative stress-mediated constitutive eNOS/HSP^{90} interaction and disruption of agonist-mediated Akt activation. J. Neuroinflammation 12, 84.

Lee, B.J., Egi, Y., Van Leyen, K., Lo, E.H., Arai, K., 2010. Edaravone, a free radical scavenger, protects components of the neurovascular unit against oxidative stress in vitro. Brain Res. 1307, 22–27.

Liu, D., Sybert, T.E., Qian, H., Liu, J., 1998. Superoxide production after spinal injury detected by micro-perfusion of cytochrome c. Free Radic. Biol. Med. 25, 298–304.

Liu, D.H., Yuan, F.G., Hu, S.Q., Diao, F., Wu, Y.P., Zong, Y.Y., et al., 2013. Endogenous nitric oxide induces activation of apoptosis signal-regulating kinase 1 via S-nitrosylation in rat hippocampus during cerebral ischemia-reperfusion. Neuroscience 229, 36–48.

Liu, H.Q., Wei, X.B., Sun, R., Cai, Y.W., Lou, H.Y., Wang, J.W., et al., 2006. Angiotensin II stimulates inter-cellular adhesion molecule-1 via an AT1 receptor/nuclear factor-kappaB pathway in brain microvas-cular endothelial cells. Life Sci. 78, 1293–1298.

Liu, J., Liu, M.C., Wang, K.K., 2008. Calpain in the CNS: from synaptic function to neurotoxicity. Sci. Signal. 1 re1.

Liu, K., Li, Q., Zhang, L., Zheng, X., 2009. The dynamic detection of NO during stroke and reperfusion in vivo. Brain Inj. 23, 450–458.

Louin, G., Marchand-Verrecchia, C., Palmier, B., Plotkine, M., Jafarian-Tehrani, M., 2006. Selective inhibi-tion of inducible nitric oxide synthase reduces neurological deficit but not cerebral edema following traumatic brain injury. Neuropharmacology 50, 182–190.

Love, S., Miners, J.S., 2016. Cerebrovascular disease in ageing and Alzheimer's disease. Acta Neuropathol 131(5), 645–658.

Lowe J, M.S., Thyman, B., Dickson, D.W., 2008. Ageing and dementia. In: Love, S., Louis, D., Ellison, D. (Eds.), Greenfield's Neuropathology. Hodder Arnold, London.

Lu, D., Mahmood, A., Zhang, R., Copp, M., 2003. Upregulation of neurogenesis and reduction in func-tional deficits following administration of DEtA/NONOate, a nitric oxide donor, after traumatic brain injury in rats. J. Neurosurg. 99, 351–361.

Llorens-Martin, M., Jurado, J., Hernandez, F., Avila, J., 2014. GSK-3beta, a pivotal kinase in Alzheimer dis-ease. Front. Mol. Neurosci. 7, 46.

Llorente, I.L., Perez-Rodriguez, D., Burgin, T.C., Gonzalo-Orden, J.M., Martinez-Villayandre, B., Fernandez-Lopez, A., 2013. Age and meloxicam modify the response of the glutamate vesicular transporters (VGLUTs) after transient global cerebral ischemia in the rat brain. Brain Res. Bull. 94, 90–97.

MacMicking, J., Xie, Q.W., Nathan, C., 1997. Nitric oxide and macrophage function. Annu. Rev. Immunol. 15, 323–350.

Malinski, T., 2007. Nitric oxide and nitroxidative stress in Alzheimer's disease. J. Alzheimers Dis. 11, 207–218.

Mankan, A.K., Lawless, M.W., Gray, S.G., Kelleher, D., Mcmanus, R., 2009. NF-kappaB regulation: the nuclear response. J. Cell. Mol. Med. 13, 631–643.

McAlpine, F.E., Lee, J.K., Harms, A.S., Ruhn, K.A., Blurton-Jones, M., Hong, J., et al., 2009. Inhibition of soluble TNF signaling in a mouse model of Alzheimer's disease prevents pre-plaque amyloid-associated neuropathology. Neurobiol. Dis. 34, 163–177.

McCarthy, R.C., Lu, D.Y., Alkhateeb, A., Gardeck, A.M., Lee, C.H., Wessling-Resnick, M., 2016. Characterization of a novel adult murine immortalized microglial cell line and its activation by amy-loid-beta. J. Neuroinflammation 13, 21.

Mehta, S.L., Manhas, N., Raghubir, R., 2007. Molecular targets in cerebral ischemia for developing novel therapeutics. Brain Res. Rev. 54, 34–66.

Mercanoglu, G., Safran, N., Ahishali, B.B., Uzun, H., Yalcin, A., Mercanoglu, F., 2015. Nitric oxide mediated effects of nebivolol in myocardial infarction: the source of nitric oxide. Eur. Rev. Med. Pharmacol. Sci. 19, 4872–4889.

Minjarez, B., Calderon-Gonzalez, K.G., Rustarazo, M.L., Herrera-Aguirre, M.E., Labra-Barrios, M.L., Rincon-Limas, D.E., et al., 2016. Identification of proteins that are differentially expressed in brains with Alzheimer's disease using iTRAQ labeling and tandem mass spectrometry. J. Proteomics 139, 103–121.

Mohan, S., Hamuro, M., Sorescu, G.P., Koyoma, K., Sprague, E.A., Jo, H., et al., 2003. IkappaBalpha-dependent regulation of low-shear flow-induced NF-kappa B activity: role of nitric oxide. Am. J. Physiol. Cell Physiol. 284, C1039–C1047.

Moore, C., Sanz-Rosa, D., Emerson, M., 2011. Distinct role and location of the endothelial isoform of nitric oxide synthase in regulating platelet aggregation in males and females in vivo. Eur. J. Pharmacol. 651, 152–158.

Munoz, M.F., Puebla, M., Figueroa, X.F., 2015. Control of the neurovascular coupling by nitric oxide-dependent regulation of astrocytic Ca(2+) signaling. Front. Cell Neurosci. 9, 59.

Nabah, Y.N., Mateo, T., Cerda-Nicolas, M., Alvarez, A., Martinez, M., Issekutz, A.C., et al., 2005. L-NAME induces direct arteriolar leukocyte adhesion, which is mainly mediated by angiotensin-II. Microcirculation 12, 443–453.

Niwa, M., Inao, S., Takayasu, M., Kawai, T., Kajita, Y., Nihashi, T., et al., 2001. Time course of expression of three nitric oxide synthase isoforms after transient middle cerebral artery occlusion in rats. Neurol. Med. Chir. (Tokyo) 41, 63–72. discussion 72-3.

Pannu, R., Singh, I., 2006. Pharmacological strategies for the regulation of inducible nitric oxide synthase: neurodegenerative versus neuroprotective mechanisms. Neurochem. Int. 49, 170–182.

Parent, M., Boudier, A., Perrin, J., Vigneron, C., Maincent, P., Violle, N., et al., 2015. In situ microparticles loaded with S-nitrosoglutathione protect from stroke. PLoS ONE 10, e0144659.

Park, K.H., Park, W.J., 2015. Endothelial dysfunction: clinical implications in cardiovascular disease and therapeutic approaches. J. Korean Med. Sci. 30, 1213–1225.

Park, S.W., Huq, M.D., Hu, X., Wei, L.N., 2005. Tyrosine nitration on p65: a novel mechanism to rapidly inactivate nuclear factor-kappaB. Mol. Cell Proteomics 4, 300–309.

Pei, D.S., Song, Y.J., Yu, H.M., Hu, W.W., Du, Y., Zhang, G.Y., 2008. Exogenous nitric oxide negatively regulates c-Jun N-terminal kinase activation via inhibiting endogenous NO-induced S-nitrosylation during cerebral ischemia and reperfusion in rat hippocampus. J. Neurochem. 106, 1952–1963.

Prasad, R., Giri, S., Nath, N., Singh, I., Singh, A.K., 2007. GSNO attenuates EAE disease by S-nitrosylation-mediated modulation of endothelial-monocyte interactions. Glia 55, 65–77.

Provias, J., Jeynes, B., 2008. Neurofibrillary tangles and senile plaques in Alzheimer's brains are associated with reduced capillary expression of vascular endothelial growth factor and endothelial nitric oxide synthase. Curr. Neurovasc. Res. 5, 199–205.

Qi, S.H., Hao, L.Y., Yue, J., Zong, Y.Y., Zhang, G.Y., 2013. Exogenous nitric oxide negatively regulates the S-nitrosylation p38 mitogen-activated protein kinase activation during cerebral ischaemia and reperfusion. Neuropathol. Appl. Neurobiol. 39, 284–297.

Qu, M., Jiang, Z., Liao, Y., Song, Z., Nan, X., 2016. Lycopene prevents amyloid [beta]-induced mitochondrial oxidative stress and dysfunctions in cultured rat cortical neurons. Neurochem. Res 41(6), 1354–1364.

Rameau, G.A., Tukey, D.S., Garcin-Hosfield, E.D., Titcombe, R.F., Misra, C., Khatri, L., et al., 2007. Biphasic coupling of neuronal nitric oxide synthase phosphorylation to the NMDA receptor regulates AMPA receptor trafficking and neuronal cell death. J. Neurosci. 27, 3445–3455.

Rao, J.S., Rapoport, S.I., Kim, H.W., 2011. Altered neuroinflammatory, arachidonic acid cascade and synaptic markers in postmortem Alzheimer's disease brain. Transl. Psychiatry 1, e31.

Reynaert, N.L., Ckless, K., Korn, S.H., Vos, N., Guala, A.S., Wouters, E.F., et al., 2004. Nitric oxide represses inhibitory kappaB kinase through S-nitrosylation. Proc. Natl. Acad. Sci. USA 101, 8945–8950.

Ridder, D.A., Schwaninger, M., 2009. NF-kappaB signaling in cerebral ischemia. Neuroscience 158, 995–1006.

Sakakima, H., Khan, M., Dhammu, T.S., Shunmugavel, A., Yoshida, Y., Singh, I., et al., 2012. Stimulation of functional recovery via the mechanisms of neurorepair by S-nitrosoglutathione and motor exercise in a rat model of transient cerebral ischemia and reperfusion. Restor. Neurol. Neurosci. 30, 383–396.

Salom, J.B., Orti, M., Centeno, J.M., Torregrosa, G., Alborch, E., 2000. Reduction of infarct size by the NO donors sodium nitroprusside and spermine/NO after transient focal cerebral ischemia in rats. Brain Res. 865, 149–156.

Samengo, G., Avik, A., Fedor, B., Whittaker, D., Myung, K.H., Wehling-Henricks, M., et al., 2012. Age-related loss of nitric oxide synthase in skeletal muscle causes reductions in calpain S-nitrosylation that increase myofibril degradation and sarcopenia. Aging Cell 11, 1036–1045.

Sarkar, F.H., Li, Y., Wang, Z., Kong, D., 2008. NF-kappaB signaling pathway and its therapeutic implications in human diseases. Int. Rev. Immunol. 27, 293–319.

Sarnico, I., Lanzillotta, A., Boroni, F., Benarese, M., Alghisi, M., Schwaninger, M., et al., 2009. NF-kappaB p50/RelA and c-Rel-containing dimers: opposite regulators of neuron vulnerability to ischaemia. J. Neurochem. 108, 475–485.

Sheng, J.G., Zhu, S.G., Jones, R.A., Griffin, W.S., Mrak, R.E., 2000. Interleukin-1 promotes expression and phosphorylation of neurofilament and tau proteins in vivo. Exp. Neurol. 163, 388–391.

Shi, S., Liang, D., Chen, Y., Xie, Y., Wang, Y., Wang, L., et al., 2016. Gx-50 reduces beta-amyloid-induced TNF-alpha, IL-1beta, NO, and PGE2 expression and inhibits NF-kappaB signaling in a mouse model of Alzheimer's disease. Eur. J. Immunol. 46, 665–676.

Song, Y.S., Kim, M.S., Kim, H.A., Jung, B.I., Yang, J., Narasimhan, P., et al., 2010. Oxidative stress increases phosphorylation of IkappaB kinase-alpha by enhancing NF-kappaB-inducing kinase after transient focal cerebral ischemia. J. Cereb. Blood Flow Metab. 30, 1265–1274.

Speed, N., Blair, I.A., 2011. Cyclooxygenase- and lipoxygenase-mediated DNA damage. Cancer Metastasis Rev. 30, 437–447.

Stanarius, A., Topel, I., Schulz, S., Noack, H., Wolf, G., 1997. Immunocytochemistry of endothelial nitric oxide synthase in the rat brain: a light and electron microscopical study using the tyramide signal amplification technique. Acta Histochem. 99, 411–429.

Sun, M., Zhao, Y., Gu, Y., Xu, C., 2009. Inhibition of nNOS reduces ischemic cell death through down-regulating calpain and caspase-3 after experimental stroke. Neurochem. Int. 54, 339–346.

Suzuki, M., Tabuchi, M., Ikeda, M., Tomita, T., 2002. Concurrent formation of peroxynitrite with the expression of inducible nitric oxide synthase in the brain during middle cerebral artery occlusion and reperfusion in rats. Brain Res. 951, 113–120.

Tabassum, R., Vaibhav, K., Shrivastava, P., Khan, A., Ahmed, M.E., Ashafaq, M., et al., 2015. Perillyl alcohol improves functional and histological outcomes against ischemia-reperfusion injury by attenuation of oxidative stress and repression of COX-2, NOS-2 and NF-kappaB in middle cerebral artery occlusion rats. Eur. J. Pharmacol. 747, 190–199.

Tajes, M., Ill-Raga, G., Palomer, E., Ramos-Fernandez, E., Guix, F.X., Bosch-Morato, M., et al., 2013. Nitro-oxidative stress after neuronal ischemia induces protein nitrotyrosination and cell death. Oxid. Med. Cell Longev. 2013, 826143.

Tan, X.L., Xue, Y.Q., Ma, T., Wang, X., Li, J.J., Lan, L., et al., 2015. Partial eNOS deficiency causes spontaneous thrombotic cerebral infarction, amyloid angiopathy and cognitive impairment. Mol. Neurodegener. 10, 24.

Terpolilli, N.A., Zweckberger, K., Trabold, R., Schilling, L., Schinzel, R., Tegtmeier, F., et al., 2009. The novel nitric oxide synthase inhibitor 4-amino-tetrahydro-L-biopterine prevents brain edema formation and intracranial hypertension following traumatic brain injury in mice. J. Neurotrauma 26, 1963–1975.

Thorns, V., Hansen, L., Masliah, E., 1998. nNOS expressing neurons in the entorhinal cortex and hippocampus are affected in patients with Alzheimer's disease. Exp. Neurol. 150, 14–20.

Toda, N., Ayajiki, K., Okamura, T., 2009. Cerebral blood flow regulation by nitric oxide: recent advances. Pharmacol. Rev. 61, 62–97.

Tong, X.K., Nicolakakis, N., Kocharyan, A., Hamel, E., 2005. Vascular remodeling versus amyloid beta-induced oxidative stress in the cerebrovascular dysfunctions associated with Alzheimer's disease. J. Neurosci. 25, 11165–11174.

Trickler, W.J., Mayhan, W.G., Miller, D.W., 2005. Brain microvessel endothelial cell responses to tumor necrosis factor-alpha involve a nuclear factor kappa B (NF-kappaB) signal transduction pathway. Brain Res. 1048, 24–31.

Valerio, A., Boroni, F., Benarese, M., Sarnico, I., Ghisi, V., Bresciani, L.G., et al., 2006. NF-kappaB pathway: a target for preventing beta-amyloid (Abeta)-induced neuronal damage and Abeta42 production. Eur. J. Neurosci. 23, 1711–1720.

Wada, K., Chatzipanteli, K., Busto, R., Dietrich, W.D., 1998. Role of nitric oxide in traumatic brain injury in the rat. J. Neurosurg. 89, 807–818.

Wang, P., Guan, P.P., Wang, T., Yu, X., Guo, J.J., Wang, Z.Y., 2014. Aggravation of Alzheimer's disease due to the COX-2-mediated reciprocal regulation of IL-1beta and Abeta between glial and neuron cells. Aging Cell 13, 605–615.

Wang, W.Y., Tan, M.S., Yu, J.T., Tan, L., 2015. Role of pro-inflammatory cytokines released from microglia in Alzheimer's disease. Ann. Transl. Med. 3, 136.

Wang, Y.J., Zhou, H.D., Zhou, X.F., 2006. Clearance of amyloid-beta in Alzheimer's disease: progress, problems and perspectives. Drug Discov. Today 11, 931–938.

Webster, C.M., Kelly, S., Koike, M.A., Chock, V.Y., Giffard, R.G., Yenari, M.A., 2009. Inflammation and NFkappaB activation is decreased by hypothermia following global cerebral ischemia. Neurobiol. Dis. 33, 301–312.

Wei, G., Dawson, V.L., Zweier, J.L., 1999. Role of neuronal and endothelial nitric oxide synthase in nitric oxide generation in the brain following cerebral ischemia. Biochim. Biophys. Acta 1455, 23–34.

Wobst, J., Kessler, T., Dang, T.A., Erdmann, J., Schunkert, H., 2015. Role of sGC-dependent NO signalling and myocardial infarction risk. J. Mol. Med. (Berl.) 93, 383–394.

Won, J.S., Annamalai, B., Choi, S., Singh, I., Singh, A.K., 2015. S-nitrosoglutathione reduces tau hyperphosphorylation and provides neuroprotection in rat model of chronic cerebral hypoperfusion. Brain Res. 1624, 359–369.

Won, J.S., Kim, J., Annamalai, B., Shunmugavel, A., Singh, I., Singh, A.K., 2013. Protective role of S-nitrosoglutathione (GSNO) against cognitive impairment in rat model of chronic cerebral hypoperfusion. J. Alzheimers Dis. 34, 621–635.

Wood, L.B., Winslow, A.R., Proctor, E.A., Mcguone, D., Mordes, D.A., Frosch, M.P., et al., 2015. Identification of neurotoxic cytokines by profiling Alzheimer's disease tissues and neuron culture viability screening. Sci. Rep. 5, 16622.

Xu, L., Sun, J., Lu, R., Ji, Q., Xu, J.G., 2005. Effect of glutamate on inflammatory responses of intestine and brain after focal cerebral ischemia. World J. Gastroenterol. 11, 733–736.

Yang, Y., Ke-Zhou, L., Ning, G.M., Wang, M.L., Zheng, X.X., 2008. Dynamics of nitric oxide and peroxynitrite during global brain ischemia/reperfusion in rat hippocampus: NO-sensor measurement and modeling study. Neurochem. Res. 33, 73–80.

Yasuda, Y., Shimoda, T., Uno, K., Tateishi, N., Furuya, S., Tsuchihashi, Y., et al., 2011. Temporal and sequential changes of glial cells and cytokine expression during neuronal degeneration after transient global ischemia in rats. J. Neuroinflammation 8, 70.

Yilmaz, G., Granger, D.N., 2008. Cell adhesion molecules and ischemic stroke. Neurol. Res. 30, 783–793.

Yin, X.H., Yan, J.Z., Hou, X.Y., Wu, S.L., Zhang, G.Y., 2013. Neuroprotection of S-nitrosoglutathione against ischemic injury by down-regulating Fas S-nitrosylation and downstream signaling. Neuroscience 248, 290–298.

Yu, H.M., Xu, J., Li, C., Zhou, C., Zhang, F., Han, D., et al., 2008. Coupling between neuronal nitric oxide synthase and glutamate receptor 6-mediated c-Jun N-terminal kinase signaling pathway via S-nitrosylation contributes to ischemia neuronal death. Neuroscience 155, 1120–1132.

Zhang, F., Iadecola, C., 1994. Reduction of focal cerebral ischemic damage by delayed treatment with nitric oxide donors. J. Cereb. Blood Flow Metab. 14, 574–580.

Zhang, F., Sprague, S.M., Farrokhi, F., Henry, M.N., Son, M.G., Vollmer, D.G., 2002. Reversal of attenuation of cerebrovascular reactivity to hypercapnia by a nitric oxide donor after controlled cortical impact in a rat model of traumatic brain injury. J. Neurosurg. 97, 963–969.

Zhang, W., Potrovita, I., Tarabin, V., Herrmann, O., Beer, V., Weih, F., et al., 2005. Neuronal activation of NF-kappaB contributes to cell death in cerebral ischemia. J. Cereb. Blood Flow Metab. 25, 30–40.

Zhang, F., White, J.G., Iadecola, C., 1994. Nitric oxide donors increase blood flow and reduce brain damage in focal ischemia: evidence that nitric oxide is beneficial in the early stages of cerebral ischemia. J. Cereb. Blood Flow Metab. 14, 217–226.

Zhang, Y.W., Xu, H., 2007. Molecular and cellular mechanisms for Alzheimer's disease: understanding APP metabolism. Curr. Mol. Med. 7, 687–696.

Zhang, R., Zhang, L., Zhang, Z., Wang, Y., Lu, M., Lapointe, M., et al., 2001. A nitric oxide donor induces neurogenesis and reduces functional deficits after stroke in rats. Ann. Neurol. 50, 602–611.

Zhu, X., Dong, J., Shen, K., Bai, Y., Chao, J., Yao, H., 2016. Neuronal nitric oxide synthase contributes to pentylenetetrazole-kindling-induced hippocampal neurogenesis. Brain Res. Bull. 121, 138–147.

Zhou, L., Zhu, D.Y., 2009. Neuronal nitric oxide synthase: structure, subcellular localization, regulation, and clinical implications. Nitric Oxide 20, 223–230.

Zierath, D., Thullbery, M., Hadwin, J., Gee, J.M., Savos, A., Kalil, A., et al., 2010. CNS immune responses following experimental stroke. Neurocrit. Care 12, 274–284.

Zlokovic, B.V., 2011. Neurovascular pathways to neurodegeneration in Alzheimer's disease and other disorders. Nat. Rev. Neurosci. 12, 723–738.

CHAPTER 12

Synergistic Activities of Nitric Oxide and Various Drugs

Govindan Ravikumar and Harinath Chakrapani
Department of Chemistry, Indian Institute of Science Education and Research Pune, Maharashtra, India

INTRODUCTION

The importance of nitric oxide (NO) in mediating numerous cellular processes is by now well established. The relationship between NO and cancer is complex and is largely dependent on the location, concentration, duration of release as well as the presence or lack thereof of other reactive entities (Heinrich et al., 2013; Ridnour et al., 2008; Sharma and Chakrapani, 2014; Huerta et al., 2008). Studies have shown that cancers have elevated levels of NO in comparison with paired normal tissue. This helps with increased blood flow to tumors as well as maintenance of high growth rate (Heinecke et al., 2014). Expression of nitric oxide synthase (NOS) within tumors is higher than normal cells suggesting that inhibition of this enzyme is a possible mechanism to develop new therapeutic intervention. As a consequence, decreased concentrations of NO might have antiproliferative effects (Tozer et al., 1997). Thus a number of studies have focused on dissipating NO within tumors by using inhibitors of NOS. Due to the decreased levels of NO, tumor growth is slowed. Meanwhile, numerous parallel studies established that increased NO, which is generated by macrophages, within cancers was cytotoxic at elevated concentrations and induced apoptosis (Stuehr and Nathan, 1989; Cui et al., 1994; Farias-Eisner et al., 1994). Due to increased metabolism associated with rapidly dividing cells, tumors also have elevated levels of reactive oxygen species (ROS) including superoxide radical anion. Thus the enhancement of NO within cancers is expected to produce peroxynitrite ($ONOO^-$), a product of rapid combination of NO and superoxide anion radical, that may be highly toxic to cancers (Szabo et al., 2007; Szabó, 2003), while normal cells may be spared due to increased scavenging of NO by hemoglobin as well as low ROS levels. Furthermore, NO synergized with established drugs including doxorubicin (DOX) and cisplatin and contributed to inhibition of drug efflux pumps that decreased intracellular availability of the drug (Ye et al., 2013). These effects appear to be specific toward cancers and not normal cells. The mechanism of this specificity remains to be elucidated. Multidrug resistance (MDR) is the one of the major problems associated with failure of cancer chemotherapy (Gottesman et al., 2002; Szakacs et al., 2006; Wilson et al., 2009). Cancer cells are known to overexpress efflux pumps that reduce the intracellular accumulation of a drug, which leads to MDR (Sarkadi et al., 2006; Simon and Schindler, 1994). P-glycoprotein (P-gp), multidrug resistance related proteins (MRPs), and breast cancer resistance proteins (BCRP-1) are the primary efflux pumps involved in this process (Cole et al., 1992; Uchida et al., 1999; Doyle et al., 1998; Eckford and Sharom, 2009). One of the strategies to reverse MDR is coadministration of anticancer drugs with efflux pump inhibitors. These agents inhibit the efflux pump thus allowing the intracellular accumulation of the anticancer drug, which consequently restores the antitumoral activity in resistance cell lines. These complex molecules belong to the third generation of P-gp inhibitors that have been studied in clinical trials (Abdallah et al., 2015). P-gp inhibitors have some limitations including immunosuppressive effects and cardiovascular effects (Planting et al., 2005; Kuppens et al., 2007; Pusztai et al., 2005). NO has been reported to inhibit efflux pumps and may therefore become an alternative to these P-gp inhibitors. Extensive studies confirm that the antitumor effect of cancer chemotherapy agents was enhanced by the use of NO donors. The major reason is that NO donors generate reactive nitrogen species, which nitrate tyrosine residue of the efflux pumps which leads to increased intracellular drug accumulation and enhanced their anticancer effect against drug-resistant cancer cells (Riganti et al., 2005, 2013). Solid tumors have significant regions where oxygen concentration is diminished, hypoxia (Wilson and Hay, 2011; Shannon et al., 2003; Brown and Wilson, 2004). This condition is rather difficult to treat and hypoxic cells have an increased propensity to develop resistance to DOX, for example. NO can reverse this resistance (Evig et al., 2004; Riganti et al., 2005). The focus of attention thus turned to enhancement of NO within cancers as a therapeutic as well as using NO in conjunction with other clinical drugs. The gaseous nature of NO precluded its facile delivery to cancers. The development of technology to deliver NOS to cancer cells as well as NO donors, which are small molecules that dissociate to generate NO (Sharma et al., 2013; Khodade et al., 2016; Dharmaraja et al., 2014), addressed the issue of controlled generation of NO. These topics have been reviewed extensively elsewhere (Szabo, 2016; Singh and Gupta, 2011; Hirst and Robson, 2010; Bill Cai et al., 2005; Sharma and Chakrapani, 2014). The first part of this manuscript will focus on the synergistic effects of NO with other cancer drugs.

The role of NO in bacteria has been studied extensively. NOS is expressed in bacteria and is referred to in the literature as bacterial NOS (bNOS) (Crane et al., 2010; Sudhamsu and Crane, 2009). Gusarov and Nudler (2005) have reported that NO protects bacteria from oxidative stress induced by the antibiotics and helps in developing resistance. They investigated the effect of NO on the survival of *Bacillus subtilis* by exposing the cells to 10 mM concentration of H_2O_2, which led to cell death. When they pretreated cells with NO (30 μM) 5 seconds prior to H_2O_2 exposure, it was found

NO to protect bacteria 100-fold against H_2O_2-mediated toxicity. A similar effect was observed with the NO donors (SNAP and MAHMA NONOate). No protective effect was observed when cells were treated with NO either simultaneously with or after H_2O_2. Similarly, there was no cytoprotection observed with oxidized product of NO. NO also protects *B. subtilis* against organic peroxides such as *t*-butyl hydroperoxide and cumene hydroperoxide. These results indicate that the NO protects bacteria from oxidative stress. DNA damage is one of the main mechanisms of stress induced by hydroxyl radical formed from H_2O_2 via Fenton reaction. NO can protect bacteria from H_2O_2 by stimulating the activity of a pre-existing H_2O_2 scavenging enzyme such as catalase or NO can suppress the Fenton chemistry. Cytoprotection mechanism of NO in bacteria against H_2O_2 was investigated. They found that NO activates *B. subtilis* catalase enzyme which leads to deactivation of peroxide. These results suggest that the stimulation of catalase by NO is one of the protective mechanisms of NO in bacteria. Next, they checked the role of NO in Fenton chemistry. It was found that DNA damage occurred with 10 mM H_2O_2 and no DNA damage was observed in the presence of cell permeable iron chelator (dipyridyl). Similar protection was observed in case of NO treatment. In order to maintain Fenton process, Fe^{2+} must be continuously reduced from Fe^{3+}. Cys does this process in *Escherichia coli*. In order to verify this, they have used diamide, a specific thioloxidizing reagent. Diamide also shows similar protection. These results suggest that H_2O_2 cytotoxicity was through DNA damage induced by Fenton chemistry and NO protection occurs via the inhibition of Fenton chemistry. But the amount of NO used in this experiment is insufficient to eliminate hydroxyl radical. So the inhibition of Fenton chemistry by NO could be by either scavenging cellular iron or by preventing its reduction. NOS deletion mutant shows more sensitivity toward oxidative damage by enhancing the levels of reduced thiols. Reduced thiols recycle Fe^{2+} from Fe^{3+} and promote Fenton reaction. It was also observed that NO disturbs the recycling of Fe^{2+} by free thiols by inhibiting the thioredoxin system. So the protection mechanism of NO in bacteria from oxidative stress (Crane et al., 2010; Sudhamsu and Crane, 2009) was found to be mediated by suppression of the Fenton reaction, in which NO disturbs Fe^{2+} recycling process. NO also specifically stimulates the catalase activity to detoxify excess H_2O_2 (Fig. 12.6).

NO is produced in bacteria by bNOS to protect bacteria from antibiotic-induced stress (Van Sorge et al., 2013; Gusarov et al., 2009; Macmicking et al., 1997). Inhibition of bNOS resulted in the reduction of NO production, which in turn leads to the conversion of resistant bacteria into more susceptible bacteria, which can be cleared by antibiotic. bNOS is a potential therapeutic target against bacteria by antibiotics induced stress and thus becoming an active area of investigation. A number of studies have been carried out with this goal. For example, Holden et al. (2013, 2015) showed NOS mutant was more sensitive to oxidative stress induced by H_2O_2 or antimicrobial agent acriflavine (ACR) than WT *B. subtilis*. Next they screened library of NOS inhibitors and they found ACR alone showed 23% and only NOS inhibitor 36% of bacterial growth inhibition. Treatment of ACR and NOS inhibitor together showed complete inhibition of bacterial growth. In future coadministration of NOS inhibitor and antibiotic will be a new antimicrobial strategy against bNOS containing bacteria.

There are other reports that suggest that exogenously added NO can have several beneficial effects. Bacteria have adapted to evade stress by forming biofilms. Antibiotics are poorly permeable and thus become ineffective against these pathogens. Biofilms can be dispersed by NO and the resulting disruption exposes bacteria.

Thus the beneficial effects of NO in bacteria appear to outweigh the undesirable ones and this has laid the foundation for newer methodologies to deliver NO to bacteria.

Delivery of NO along with drugs can be achieved by three broad methodologies.

1. *Delivery of NOS gene*: Here, through gene delivery, the intracellular levels of NO can be enhanced. Along with this, a drug can be introduced. The advances toward this goal are discussed herein.
2. *NO donors*: Methodologies to deliver NO using small molecules in conjunction with other drugs will be discussed.
3. *NO-based hybrid drugs*: Here, the drug is covalently linked to an NO donor and upon entry into cells, the conjugate breaks down to generate NO in conjunction with the drug of interest.

DELIVERY OF NOS

Enhancement of intracellular NO can be achieved by increased expression of NOS (Fig. 12.1). Transfection of the relevant gene should result in increased NOS levels. Over the past several decades, significant advances have been made in the area of gene therapy. These topics have been extensively reviewed previously (Szabo, 2016; Singh and Gupta, 2011; Huerta et al., 2008) and in this chapter, the advancement in this area after 2011 is covered. Among the isoforms of NOS, iNOS has the capacity to increase NO concentrations at high nanomolar levels (Knowles and Moncada, 1994). Efforts have been made to encapsulate the plasmid encoding the iNOS protein. Delivery and subsequent expression of this protein will produce NO within cells. Exposure of such cancer cells to cytotoxic agents including DOX revealed the enhanced sensitivity of drug-resistant cells to DOX. Subsequent studies have shown nitration of tyrosine residues in drug efflux pumps results in decreased efflux capacity resulting in larger accumulation of the drug within cells (Riganti et al., 2005; Evig et al., 2004; Ye et al., 2013). Although this methodology is useful, gene delivery is rather cumbersome and its utility may remain limited.

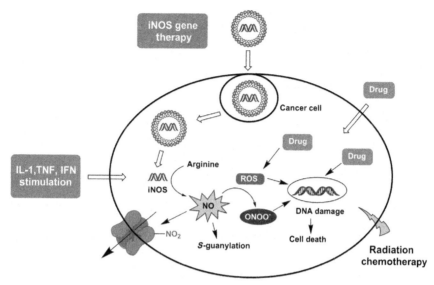

Figure 12.1 Enhancement of intracellular levels of NO by introduction of inducible nitric oxide synthase (iNOS) through gene therapy.

NO AND ANTICANCER DRUG COTREATMENT

Several lines of evidence suggest that the antitumor effect of certain cancer chemotherapy agents enhanced by NO treatment (Fig. 12.2). For example, Wink et al. (1997) reported that the cytotoxic effect of cisplatin was enhanced with pretreatment of NO in Chinese hamster V79 lung fibroblast cells. The effect was different for different NO donors. For example, when pretreatment of DEA/NO and PAPA/NO was carried out, cytotoxicity of cisplatin was enhanced by 180- and 3000-fold, respectively. Whereas pretreatment of cells with GSNO (S-nitrosoglutathione) or SNAP (S-nitroso-N-acetylpenicillamine) did not enhance the cisplatin cytotoxity. Bolus NO, DEA/NO, and PAPA/NO produced more reactive nitrogen species (RNS) than GSNO or SNAP. NO release and the amount of reactive nitrogen species formation are important in the NO-mediated enhancement of cisplatin cytotoxicity. Cook et al. (1997) showed that NO donor (DEA/NO) induces anticancer effect of melphalan in Chinese hamster V79 lung fibroblasts and human breast cancer (MCF-7) cells by 3.6- and 4.3-fold, respectively. Azizzadeh et al. (2001) also found cisplatin cytotoxicity enhancement in a head and neck squamous carcinoma (HNSCC) cell line when cisplatin cotreated with long-acting, but not short-acting NO donors.

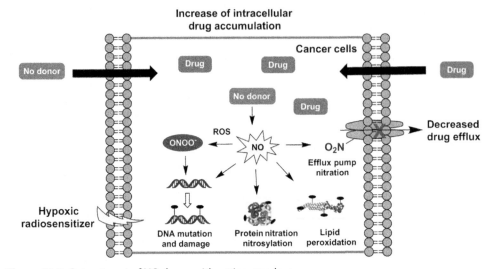

Figure 12.2 Cotreatment of NO donor with anticancer drug.

Evig et al. (2004) achieved the augmentation of NO on DOX by introducing NO donor prior to DOX treatment. Prior depletion of glutathione by incubation of the cells with DL-buthionine-S,R-sulfoximine (BSO) (glutathione synthesis inhibitor) further increased the cytotoxicity of DOX in human breast cancer (MCF-7) cells. Riganti et al. (2005) reported that NO donors, such as sodium nitroprusside (SNP), SNAP, GSNO, were able to increase the intracellular accumulation of DOX in different human cancer cells such as colon cancer (HT29) cells, lung epithelial (A549) cells, and myelogenous leukemic (K562) cells. This result suggested that NO donors inhibit the activity of efflux pumps by nitrating critical tyrosine residues of MRP3 protein, thus allowing the increase of intracellular drug accumulation and enhanced their anticancer effect with different drug-resistant cancer cells. Bratasz et al. (2006) found that the administration of NCX-4016 followed by cisplatin showed a significantly improved toxicity when compared with treatment of cells with NCX-4016 or cisplatin alone against human ovarian cancer cells (HOCCs). De Luca et al. (2011) found that the intracellular accumulation of DOX was enhanced after treatment of S-nitrosoglutathione (GSNO) in drug-resistant breast cancer cell line MCF-7/dx by histone glutathionylation. Curta et al. (2012) showed that simultaneous administration of daunorubicin (DNR) and SNAP caused significant cell death by apoptosis. Combination of these compounds decreased Bcl-2 and survivin, and increased Bax and active-caspase3 expression in DNR resistance K562 cells. Recently, Song et al. (2014) synthesized an NO releasing polymer (TNO_3) and it forms stable micelles in physiological conditions and releases ~90% of NO content in cancer cells for 96

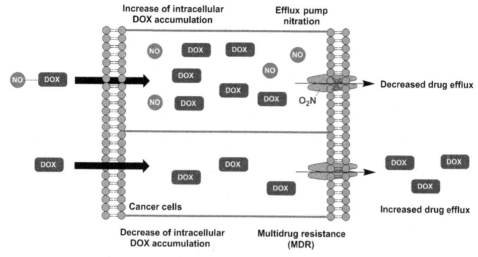

Figure 12.3 Treatment of NO-based hybrid drugs.

hours. TNO_3 enhanced the anticancer effect of chemotherapeutic drug DOX in hepatocarcinoma HepG2 cancer cells. In conclusion, chemosensitizing agents such as NO donor enhance the antitumor effect of chemotherapeutic drugs. However, further investigation is needed for the complete mechanistic interpretation of these antitumor effects.

NO ANTICANCER DRUG HYBRIDS

The broad strategy of NO-based hybrid drugs is shown in Fig. 12.3. DOX is one of the most potent chemotherapy agent used to treat different types of cancers. It acts through DNA intercalation, inhibition of topoisomerase II, and induction of oxidative stress. Chegaev et al. (2011) and Riganti et al. (2013) have synthesized two different types of NO-DOXO hybrids where furoxan and nitrooxy NO donors are conjugated with DOX through an ester linkage (Fig 12.4). Intracellular accumulation of NO-DOXO was found to be more than DOX alone against DOX-resistant human epithelial colon cell line (HT29-dx) with time- and dose-dependent manner. However, Furoxan NO-DOXO shows more accumulation than the nitrooxy NO-DOXO. And the effect was diminished when NO-DOXO was cotreated with c-PTIO (an NO scavenger). This experiment clearly suggests that the intracellular accumulation is because of NO. Cytotoxic effect of NO-DOXO was studied by lactate dehydrogenase (LDH) release assay. NO-DOXO induces more LDH release than DOX alone and the level of LDH was higher for the furoxan NO-DOXO than

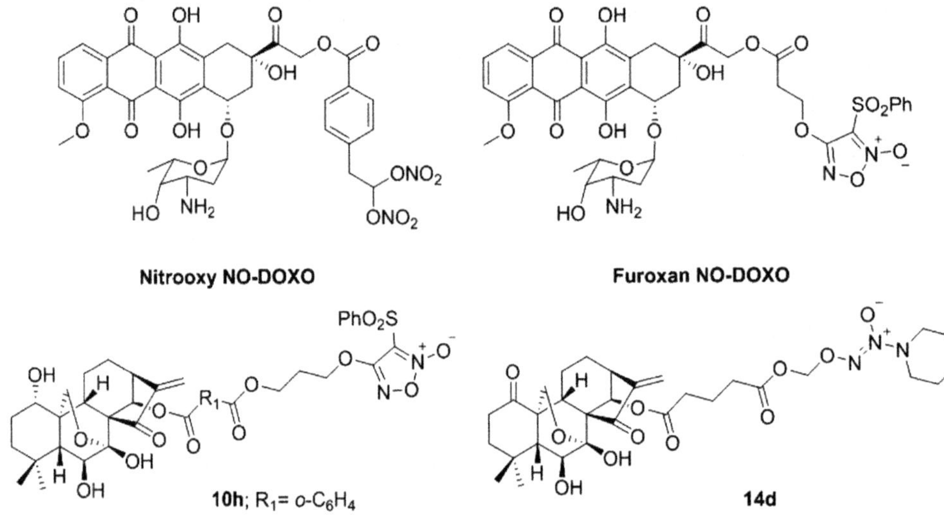

Figure 12.4 Structure of NO-drug hybrids.

for the nitrooxy NO-DOXO. In both the cases, toxicity was diminished in the presence of c-PTIO. There was no significant LDH release when cells were treated with NO donor alone. This indicates that the cytotoxic effect is due to the DOX not due the NO releasing moieties. When NO-DOXO was incubated with HT29-dx cells, tyrosine residues of MRP3 protein were nitrated and no detectable level of nitration of tyrosine occurred in P-gp protein in HT29-dx cells. No nitrotyrosine residues were observed with DOX alone in HT29-dx cells. The amount of nitrotyrosine produced by furoxan NO-DOXO appeared greater than that produced by nitrooxy NO-DOXO. The intracellular accumulation and LDH release of NO-DOXO was well correlated with nitrite release and MRP3 nitration. Furoxan NO-DOXO was found to be more potent than the nitrooxy NO-DOXO.

One of the major problems associated with DOX-based chemotherapy is cardiotoxicity (Minotti et al., 2004; Buzdar et al., 1985). Cardiotoxic studies were performed with both furoxan NO-DOXO and nitrooxy NO-DOXO derivatives against H9c2 cardiomyocytes. Nitrooxy NO-DOXO shows less cardiotoxicity than furoxan NO-DOXO. Next the cytotoxic effect of furoxan NO-DOXO and nitrooxy NO-DOXO derivatives was conducted against nontransformed human colon epithelial CCD-18Co cells. Here, nitrooxy NO-DOXO shows less toxicity than furoxan NO-DOXO. Based on these results, the nitrooxy NO-DOXO derivative was taken for further study. Interestingly, it was found that nitrooxy NO-DOXO was mainly localized in mitochondria and endoplasmic reticulum, whereas it was absent from the Golgi apparatus. Intramitochondrial accumulation of nitrooxy NO-DOXO profile was similar with both HT29 and HT29-dx cells and was higher than DOX alone. Intramitochondrial accumulation of nitrooxy NO-DOXO was diminished in the presence c-PTIO (an NO scavenger). When nitrooxy NO-DOXO was incubated with HT29 and HT29-dx cells, nitrotyrosine residues was observed with MRP-1 and BCRP proteins. That was reduced by treatment of c-PTIO. And the nitrooxy NO-DOXO induces more apoptosis than DOX in HT29-dx cells. Thus it was concluded that NO-DOXO hybrids act by nitrating the MRP efflux pump resulting in higher levels of NO-DOXO and showed higher cytotoxicity than DOX alone. Nitrooxy NO-DOXO shows less cordiotoxicity and was also found to be less toxic against nontransformed human colon epithelial CCD-18Co cells.

Oleanolic acid is a biologically active pentacyclictriterpenoid compound with several promising pharmacological activities, such as hepatoprotective effects, antiinflammatory, antioxidant, and antitumor activities (Sporn et al., 2011; Liby et al., 2007; Liu, 1995). Peptide transporter 1 (PepT1) has been found to be overexpressed by certain tumor cells, including human colon cancer cells (Caco-2) (Nielsen et al., 2001), pancreatic cancer cells (AsPc-1 and Capan-2) (Gonzalez et al., 1998), sarcoma cells (HT-1080) (Nakanishi et al., 1997), prostatic cancer cells (PC-3) (Tai et al., 2013), lung cancer cells (A549) (Endter et al., 2009), and ovarian adenocarcinoma cisplatin-sensitive (2008) and resistant cells (C13) (Negom Kouodom et al., 2012). It has a relatively limited expression in healthy tissues like lung and liver (Endter et al., 2009; Landowski et al., 2005). PepT1 can effectively transport amino acids, peptides, and peptide-like drugs molecules. Oleanolic acid and its derivatives have some limitations such as low solubility and low permeability (Jeong et al., 2007). To overcome these limitations, researchers utilized the differential expression of PepT1 for their design. Series of CDDO–amino acid–NO donor trihybrids (Ai et al., 2015) were synthesized which can be effectively transported to cancer cells by PepT1. Western blot assay supports high levels of PepT1 expression by both HCT-8 and HCT-8/5-FU cells, but in comparison diminished levels in CCD841 cells. Antiproliferative activity of CDDO–amino acid–NO donor trihybrids was studied by MTT assay against HCT-8 and HCT-8/5-FU cells. It was found that the CDDO–amino acid–NO donor trihybrid (**4c**) is the most potent compound and was better than positive controls CDDO-Me and 5-FU in HCT-8 and HCT-8/5-FU cells. CDDO-Me showed similar antitumor activities against both colon cancer and nontumorous colon cells. However **4c** displayed a fivefold decrease in its antiproliferative activity on nontumor CCD841 cells. CDDO–amino acid–NO donor trihybrids composed of CDDO (**5**) and furoxan moiety (**2a**). It was found that the IC_{50} values of CDDO–amino acid–NO donor trihybrids (**4c**) were significantly less than that of individual moieties, **5** and **2a**, and even the combination of **5** and **2a** in HCT-8 and HCT-8/5-FU cells (Fig. 12.5). Interestingly, the most active compounds produce more amount of NO and less NO production from least active compounds with both HCT-8 and HCT-8/5-FU cells. The most active compounds

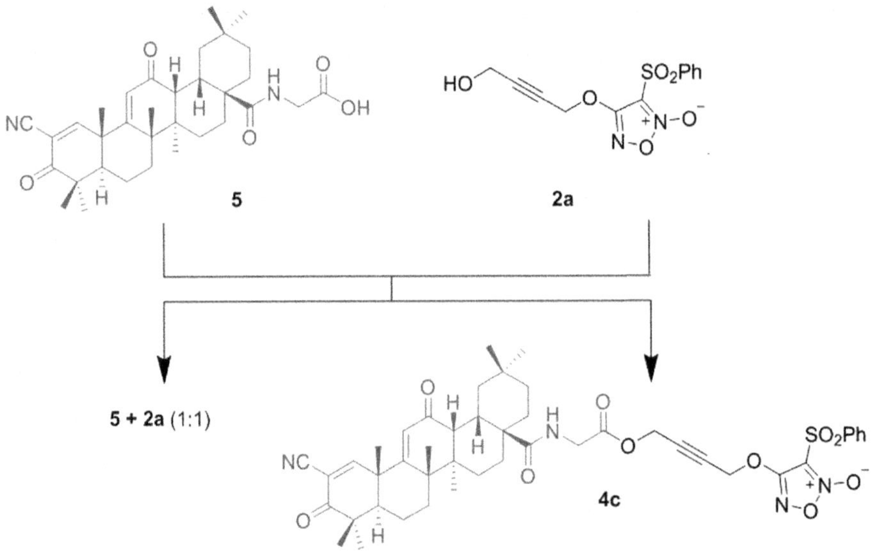

Figure 12.5 Structure of **4c, 5, 2a** and comparison of its anticancer activity.

also produce negligible amounts of NO in nontumor colon CCD841 cells. The intracellular NO release was well correlated with their antitumor activity.

When **4c** was cotreated with different concentrations of c-PTIO (an NO scavenger), its antitumor effect was decreased in a dose-dependent manner. This result suggests that NO plays an important role in the antitumor activity of **4c**. When the compound **4c** was cotreated with Gly-Sar, a high affinity substrate for PepT1, the antitumor effect and NO release from compound **4c** were diminished. These results indicate that the PepT1 plays a crucial role in the intracellular transportation and antitumor activity of **4c**. It was found that compound **4c** increased the levels of nitrated P-gp, MRP-1, and BCRP in a dose-dependent manner, which was partially diminished by pretreatment with c-PTIO in HCT-8/5-FU cells. Thus it was concluded that the antitumor activity of **4c** may be attributed to the synergic effects of CDDO and NO donor moieties as well as PepT1-mediated drug transportation in cancer cells.

Oridonin is a commercially available diterpenoid and shows antiproliferative activity against different cancer cell lines (Ding et al., 2013a, 2013b; Xu et al., 2014). To improve its antiproliferative effect, series of furoxan/oridonin-based hybrids were synthesized (Li et al., 2012). These hybrids showed more potent antiproliferative ability than oridonin and their parent compound with four different cancer cells lines (K562, Bel-7402, MGC-803, and CaEs-17). High levels of NO releasing hybrids show strong inhibitory activity. It was found that the furoxan/oridonin hybrid 10 h (Fig. 12.4) was

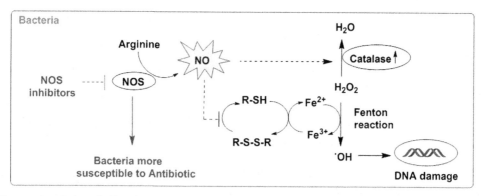

Figure 12.6 Roles of NO in bacteria. NO protects bacteria from oxidative stress by suppressing Cys reduction and simultaneous activation of catalase. Inhibition of NOS results in the reduction of NO production, which makes resistant bacteria more susceptible to antibiotics.

the most potent compound with an IC_{50} value of $0.74\,\mu mol\,L^{-1}$ against Bel-7402 cells, which was stronger than the positive control Taxol ($1.89\,\mu mol\,L^{-1}$). Recently oridonin/NONOate hybrids were designed and screened against different human cancer cell lines (A549, HepG2, HeLa, and K562 cells) (Xu et al., 2016). Oridonin/ NONOate hybrid **14d** (Fig. 12.4) showed potent antitumor activity in all tested cell lines; however, A549 cell appeared to be the most sensitive cell line. And the analogs that released higher amounts of NO had greater antiproliferative activity, while the analogs that produced minor amounts of NO were less active against Bel-7402 cells. NO release from oridonin/NONOate hybrid is well correlated with their antiprolif- erative activity. These results further supported the synergic effect of NO with different anticancer drugs against different cancer cells.

NO-BASED ANTIBACTERIAL AND BIOFILM DISPERSAL AGENTS

Antibiotic resistance is one of the major problems associated with prevention and treatment of an increasing range of infections caused by bacteria. This global health problem affects millions each year and the number is increasing at an alarming rate. In order to overcome this problem, we need new antibacterial strategies includ- ing delivery of NO to pathogens. In addition to its toxicity, NO can disperse bio- films and has been reported to synergize with certain antibiotics. NO is endogenously synthesized by a class of enzymes called NOSs which have three isoforms: nNOS (NOS 1), eNOS (NOS 3). These are continuously expressed and release NO in a cal- cium-dependent manner. iNOS (NOS 2) form, on the other hand, is associated with immune system and produces a burst of NO (nM to μM) for long time (for days) in a calcium-independent manner (Griffith and Stuehr, 1995; Fulton et al., 2001; Thomas et al., 2008). Similarly, superoxide ($O_2.^-$) is also endogenously generated by NADPH oxidase and further single electron reduction produces H_2O_2. Superoxide and hydrogen peroxide are collectively called reactive oxygen species (Kohchi et al., 2009; West et al., 2011; Yang et al., 2013). During immune response both ROS and NO are produced to counter pathogens. It appears that both NO and ROS synergize with each other and show increased damaging effect to pathogen (Fang, 2004). So, the combined antimicrobial effect of NO and ROS is greater than NO or ROS alone as shown in the response to *E. coli*. For example, little bactericidal effect was observed against *E. coli* with treatment of NO or H_2O_2 separately. The effect was enhanced in the presence of cotreatment of NO and H_2O_2, which indicates that although *E. coli* are resistant to NO and H_2O_2 separately, these become highly susceptible to their synergic effect (Pacelli et al., 1995). Bacteria use their iron sulfur cluster for electron transport. NO can diffuse to most parts of the mammalian cells, but is not able to penetrate easily to the interior of the bacteria and ROS is required to interact with the bound iron. So the release of free iron leads to DNA cleavage and bacterial death. These results show synergic effect of NO and ROS and thus suggest that it is a highly potent killing combination for *E. coli*. At the same time, NO protects mammalian cells from ROS-mediated toxicity (Wink et al., 1993). The redox chemistry of NO and ROS in the context of antipathogen activity and immune regulation was discussed previously (Wink et al., 2011). Here we have described synergic effects of NO with existing antibiotics to kill the antibiotic resistant pathogens.

SYNERGIC EFFECT OF NO WITH ANTIBIOTICS

NO has been reported as an antibacterial agent. Controlled and exogenous delivery of NO has been shown to produce promising antimicrobial effects against Gram-positive and Gram-negative bacteria, and antibiotic-resistant strains (Carpenter and Schoenfisch, 2012; Ghaffari et al., 2006). The antimicrobial effect of NO is mainly dependent upon the concentration, release rate, and it has been shown that the antibacterial efficacy of NO is mediated by its oxidation products such as peroxynitrite ($ONOO^-$) and dinitrogen trioxide (N_2O_3) (Fang, 1997; Carpenter and Schoenfisch, 2012; Ghaffari et al., 2006; Hetrick et al., 2008; Jones et al., 2010; Szabo et al., 2007). These reactive species can interact with proteins, DNA, and metabolic enzymes of microbes which lead to cell death. Combination of two broad-spectrum antimicrobial agents will increase antimicrobial effect with reduced drugs concentration at the same time it will reduce the toxic effect to host system. Schoenfisch and coworkers (Privett et al., 2010; Storm et al., 2015) combined NO with another known antimicrobial agent silver sulfadiazine (SSD) and showed the synergic effect of NO and SSD against a range of Gram-positive, Gram-negative, and antibiotic resistance pathogens. Here, a mechanism that partially involves peroxynitrite ($ONOO^-$) has been proposed.

Biofilms are sessile and surface-attached communities of bacteria embedded in a self-produced matrix which acts as a protective membrane for bacteria. Bacteria in biofilms exhibit 10- to 1000-fold higher resistance to various traditional antimicrobial agents compared with their planktonic counterparts (Brooun et al., 2000; Mcdougald et al., 2012). For example, sodium hypochlorite is one of the most effective antibacterial agents and requires a 600-fold increase in concentration to kill biofilm cells of wild-type *Pseudomonas aeruginosa* compared to planktonic cells of the same species (Luppens et al., 2002). Ampicillin has a $2\,\mu g\,mL^{-1}$ minimum inhibitory concentration (MIC) against a β-lactamase negative strain of *Klebsiella pneumonia*, similar strain shows a 2500-fold increased MIC ($5000\,\mu g\,mL^{-1}$) in the biofilm form (Anderl et al., 2000). U.S. National Institutes of Health announced that over 80% of microbial infections in the human body are mediated by biofilms (Davies, 2003). For example, cystic fibrosis (CF) is one of the chronic diseases caused by *P. aeruginosa*. These bacteria show more resistance to host immune defense and frequent antibiotic treatment. An important reason for the persistence of the bacteria is formation of biofilm (Costerton et al., 1995, 1999). The protection mechanism of biofilms makes biofilm-associated infections a major challenge for treatment. Antibiotics have been designed to inhibit the planktonic bacteria. However, the standard antimicrobial treatments typically fail to eradicate biofilms, which lead to chronic infection (Stewart, 2002). In order to address this problem, we need a new antibiofilm strategy, where the molecule has to disperse biofilm into more susceptible planktonic bacteria and which can be cleared by conventional antibiotics. Barraud et al. (2006) showed that NO plays an important role in biofilm dispersal in the model organism *P. aeruginosa*. NO at low concentrations in the micromolar and nanomolar ranges showed a decrease in biofilm biomass and an increase in planktonic biomass. For example, treatment of SNP (500 nM) resulted in a significant increase in the number of planktonic cells. In contrast, at high concentrations in the millimolar ranges, SNP caused an increase in biofilm biomass and a decrease in planktonic biomass was observed. Similar effect was observed with GSNO. The effect was reduced when NO donor cotreated with PTIO (an NO scavenger). Both NO donors shown a reduction in biofilm formation and significantly increased the number of planktonic cells compared to the control. Based on the petri dish biofilm experiment, exposure of 1-day-old biofilms to 500 nM SNP for 24 hours, only a few cells remained attached to the slide and also an increased number of planktonic cells were observed. This result strongly suggested that NO is involved in the biofilm dispersion event. Next various antimicrobial agents against *P. aeruginosa* biofilms and planktonic cells in the presence or absence of the SNP was tested. When biofilms were treated with only antimicrobial agents there is no significant effect was observed. Cotreatment of both SNP and the antimicrobial agents (tobramycin, H_2O_2, sodium dodecyl sulfate) showed almost complete removal of the biofilm. The effect of SNP on *P. aeruginosa* planktonic bacteria was studied; it was found that treatment of 500 nM SNP did not show any reduction in colony forming unit (CFU) in *P. aeruginosa* planktonic cultures compared to CFU in untreated cells. Exposure of planktonic cells to both NO and the antimicrobial agents (tobramycin, H_2O_2, sodium dodecyl sulfate) caused an additional 2-log decrease in CFU counts compared to exposure to the antimicrobial treatment alone. NO-mediated dispersal in *P. aeruginosa* biofilms appears to involve cyclic di-GMP a secondary messenger, the level of which is regulated by phosphodiesterases (PDEs). Low concentration of NO is able to increase the PDE activity and decreased c-di-GMP level in *P. aeruginosa* was identified (Barraud et al., 2009a; Christensen et al., 2013; Chua et al., 2014). NO-mediated biofilm dispersal is not restricted to only *P. aeruginosa* but similar effects were observed with various other species (Barraud et al., 2009b). In future, a compound which can lower the c-di-GMP level in bacteria may be effective drugs against biofilm-based infections. NO donors like SNP spontaneously produce NO in aqueous solution in a nonspecific manner, which leads to host toxicity. To overcome the problems associated with NO delivery, NO donors capable of releasing NO under a specific stimulus are in development. However, controlled and localized generation of NO to biofilm is challenging. Nicolas Barraud et al. (Barraud et al., 2012; Yepuri et al., 2013) designed a prodrug that releases NO by biofilm-specific enzymes (β-lactamase). Cephalosporins with leaving groups at the 3′ position are known to release leaving group during β-lactam ring cleavage by β-lactamase (Faraci and Pratt, 1987). Cephalosporin 3′ position was conjugated with diazeniumdiolate NO donor. Cephalosporin-3′-diazeniumdiolate is stable in buffer in different pH (5, 7, and 9) and it produces NO in a dose-dependent manner with different doses of penicillinase enzyme. NO signal was diminished in the presence of c-PTIO (NO scavenger). This result suggests that the amperometric signal is attributable to NO. Next NO release was studied from prodrug in different bacteria. Prodrug releases significant amount of NO with β-lactamase producing *P. aeruginosa* bacteria than non β-lactamase producing *E. coli*. The prodrug showed a dose-dependent biofilm dispersion response compared with untreated biofilm. Next the effect of antibiotic (tobramycin or ciprofloxacin) on *P. aeruginosa* biofilm was studied and the bactericidal effect was enhanced in the presence of the NO donor compared with controls. Recently Nguyen et al. (2016) developed new dual action polymeric nanoparticles capable of delivering NO and antibiotic simultaneously where NO can disperse the biofilms into an antibiotic susceptible planktonic form, and the antibiotic can clear the bacteria. Gentamicin (antibiotic) and diazeniumdiolate (NO donor) were chosen for dual action. Gentamicin NONOate was conjugated to the biocompatible polymeric nanoparticle via a hydrolysable Schiff base linkage, which allows a slow release of both NO and antibiotic in the middle acidic microenvironment of biofilm (Vroom et al., 1999). Next gentamicin release was tested in different pH; gentamicin release rate was slightly faster in pH 5.5 than at pH 7.4 and 50% of gentamicin release was observed after 17 hours in both pH. Similarly, prolonged NO release was observed from GEN-NO nanoparticles confirmed by amperometric measurement. The signal of NO was reduced in the presence of c-PTIO (NO scavenger) which confirms the signal is because of NO. Based on the Griess assay, GEN-NO nanoparticles produce 75% of NO after 5 hours. The slow release of NO and gentamicin will be helpful for long time dispersion of biofilm (we can avoid a rapid reformation of biofilm) and complete removal of planktonic bacteria. Biofilm dispersal was checked with (1) free NO donor (NONOate), (2) free gentamicin, (3) gentamicin polymer conjugate, and (4) GEN-NO nanoparticles. After 1 hour, GEN-NO nanoparticles (5 µM) showed 83% of biofilm reduction compared with all the control. With only NO donor 30% reduction, free gentamicin 14% reduction was observed and only gentamicin polymer conjugate did not show any biofilm reduction. Next bactericidal effect was investigated with both biofilm and planktonic *P. aeruginosa*. After 1 hour, GEN-NO nanoparticles (10 µM) completely eradicated both biofilm and planktonic cells compared with controls. In conclusion, combination of NO and antibiotic leads to synergistic effects of biofilm dispersal and enhanced bactericidal activity and represents a highly promising strategy to eradicate biofilm and related infections.

CONCLUSION

Numerous advances in the area of concomitant delivery of a drug and NO have been made. Delivery of NOS presents several advantages in the specificity, if transfection can be achieved only for tumors. However, this methodology for NO production may have some limitations in translation to the clinic. Furthermore, the concentration of NO produced may be difficult to control and may present challenges. The use of cotreatment with NO donors and a drug while effective may again suffer for differences in biodistribution of the NO donor and the drug. The NO donor must colocalize in the tumor and must be able to produce NO concurrently with the drug and only in the tumor. This may be challenging unless a drug delivery vehicle is used. This area is ripe for further research. The relative molar ratios of the drug and NO will have to be ascertained and the encapsulation must use this or a similar ratio for maximal efficacy. The last approach of covalent modification presents numerous advantages including delivery of NO at the site of drug action. The challenges would be to localize the conjugate at the tumor site and future work will focus on directed delivery of drug conjugates. Inasmuch as NO and bacteria is concerned, efforts to deliver NO to specific sites remain a major challenge. It is likely that much work will focus on using the biofilm dispersal properties of NO in conjunction with known antibiotics. New materials that can encapsulate NO as well as an antibiotic may have high potential for further development.

REFERENCES

Abdallah, H.M., Al-Abd, A.M., El-Dine, R.S., El-Halawany, A.M., 2015. P-glycoprotein inhibitors of natural origin as potential tumor chemo-sensitizers: a review. J. Adv. Res. 6, 45–62.

Ai, Y., Kang, F., Huang, Z., Xue, X., Lai, Y., Peng, S., et al., 2015. Synthesis of CDDO–amino acid–nitric oxide donor trihybrids as potential antitumor agents against both drug-sensitive and drug-resistant colon cancer. J. Med. Chem. 58, 2452–2464.

Anderl, J.N., Franklin, M.J., Stewart, P.S., 2000. Role of antibiotic penetration limitation in *Klebsiella pneumoniae* biofilm resistance to ampicillin and ciprofloxacin. Antimicrob. Agents Chemother. 44, 1818–1824.

Azizzadeh, B., Yip, H.T., Blackwell, K.E., Horvath, S., Calcaterra, T.C., Buga, G.M., et al., 2001. Nitric oxide improves cisplatin cytotoxicity in head and neck squamous cell carcinoma. Laryngoscope 111, 1896–1900.

Barraud, N., Hassett, D.J., Hwang, S.-H., Rice, S.A., Kjelleberg, S., Webb, J.S., 2006. Involvement of nitric oxide in biofilm dispersal of *Pseudomonas aeruginosa*. J. Bacteriol. 188, 7344–7353.

Barraud, N., Kardak, B.G., Yepuri, N.R., Howlin, R.P., Webb, J.S., Faust, S.N., et al., 2012. Cephalosporin-3′-diazeniumdiolates: targeted NO-donor prodrugs for dispersing bacterial biofilms. Angew. Chem. Int. Ed. 51, 9057–9060.

Barraud, N., Schleheck, D., Klebensberger, J., Webb, J.S., Hassett, D.J., Rice, S.A., et al., 2009a. Nitric oxide signaling in pseudomonas aeruginosa biofilms mediates phosphodiesterase activity, decreased cyclic Di-GMP levels, and enhanced dispersal. J. Bacteriol. 191, 7333–7342.

Barraud, N., Storey, M.V., Moore, Z.P., Webb, J.S., Rice, S.A., Kjelleberg, S., 2009b. Nitric oxide-mediated dispersal in single- and multi-species biofilms of clinically and industrially relevant microorganisms. Microb. Biotechnol. 2, 370–378.

Bill Cai, T., Wang, P.G., Holder, A.A., 2005. NO and NO donors. Nitric Oxide Donors. Wiley-VCH Verlag GmbH & Co. KGaA.

Bratasz, A., Weir, N.M., Parinandi, N.L., Zweier, J.L., Sridhar, R., Ignarro, L.J., et al., 2006. Reversal to cisplatin sensitivity in recurrent human ovarian cancer cells by NCX-4016, a nitro derivative of aspirin. Proc. Natl. Acad. Sci. USA 103, 3914–3919.

Brooun, A., Liu, S., Lewis, K., 2000. A dose–response study of antibiotic resistance in *Pseudomonas aeruginosa* biofilms. Antimicrob. Agents Chemother. 44, 640–646.

Brown, J.M., Wilson, W.R., 2004. Exploiting tumour hypoxia in cancer treatment. Nat. Rev. Cancer 4, 437–447.

Buzdar, A.U., Marcus, C., Blumenschein, G.R., Smith, T.L., 1985. Early and delayed clinical cardiotoxicity of doxorubicin. Cancer 55, 2761–2765.

Carpenter, A.W., Schoenfisch, M.H., 2012. Nitric oxide release: Part II. Therapeutic applications. Chem. Soc. Rev. 41, 3742–3752.

Chegaev, K., Riganti, C., Lazzarato, L., Rolando, B., Guglielmo, S., Campia, I., et al., 2011. Nitric oxide donor doxorubicins accumulate into doxorubicin-resistant human colon cancer cells inducing cytotoxicity. ACS Med. Chem. Lett. 2, 494–497.

Christensen, L.D., Van Gennip, M., Rybtke, M.T., Wu, H., Chiang, W.-C., Alhede, M., et al., 2013. Clearance of *Pseudomonas aeruginosa* foreign-body biofilm infections through reduction of the cyclic di-GMP level in the bacteria. Infect. Immun. 81, 2705–2713.

Chua, S.L., Liu, Y., Yam, J.K.H., Chen, Y., Vejborg, R.M., Tan, B.G.C., et al., 2014. Dispersed cells represent a distinct stage in the transition from bacterial biofilm to planktonic lifestyles. Nat. Commun., 5.

Cole, S., Bhardwaj, G., Gerlach, J., Mackie, J., Grant, C., Almquist, K., et al., 1992. Overexpression of a transporter gene in a multidrug-resistant human lung cancer cell line. Science 258, 1650–1654.

Cook, J.A., Krishna, M.C., Pacelli, R., Degraff, W., Liebmann, J., Mitchell, J.B., et al., 1997. Nitric oxide enhancement of melphalan-induced cytotoxicity. Br. J. Cancer 76, 325–334.

Costerton, J.W., Lewandowski, Z., Caldwell, D.E., Korber, D.R., Lappin-Scott, H.M., 1995. Microbial biofilms. Annu. Rev. Microbiol. 49, 711–745.

Costerton, J.W., Stewart, P.S., Greenberg, E.P., 1999. Bacterial biofilms: a common cause of persistent infections. Science 284, 1318–1322.

Crane, B.R., Sudhamsu, J., Patel, B.A., 2010. Bacterial nitric oxide synthases. Annu. Rev. Biochem. 79, 445–470.

Cui, S., Reichner, J.S., Mateo, R.B., Albina, J.E., 1994. Activated murine macrophages induce apoptosis in tumor cells through nitric oxide-dependent or -independent mechanisms. Cancer Res. 54, 2462–2467.

Curta, J.C., De Moraes, A.C.R., Licínio, M.A., Costa, A., Santos-Silva, M.C., 2012. Effect of nitric oxide on the daunorubicin efflux mechanism in K562 cells. Cell Biol. Int. 36, 529–535.

Davies, D., 2003. Understanding biofilm resistance to antibacterial agents. Nat. Rev. Drug Discov. 2, 114–122.

De Luca, A., Moroni, N., Serafino, A., Primavera, A., Pastore, A., Pedersen, J.Z., et al., 2011. Treatment of doxorubicin-resistant MCF7/Dx cells with nitric oxide causes histone glutathionylation and reversal of drug resistance. Biochem. J. 440, 175–183.

Dharmaraja, A.T., Ravikumar, G., Chakrapani, H., 2014. Arylboronate ester based diazeniumdiolates (BORO/NO), a class of hydrogen peroxide inducible nitric oxide (NO) donors. Org. Lett. 16, 2610–2613.

Ding, C., Zhang, Y., Chen, H., Yang, Z., Wild, C., Chu, L., et al., 2013a. Novel nitrogen-enriched oridonin analogues with thiazole-fused a-ring: protecting group-free synthesis, enhanced anticancer profile, and improved aqueous solubility. J. Med. Chem. 56, 5048–5058.

Ding, C., Zhang, Y., Chen, H., Yang, Z., Wild, C., Ye, N., et al., 2013b. Oridonin ring A-based diverse constructions of enone functionality: identification of novel dienone analogues effective for highly aggressive breast cancer by inducing apoptosis. J. Med. Chem. 56, 8814–8825.

Doyle, L.A., Yang, W., Abruzzo, L.V., Krogmann, T., Gao, Y., Rishi, A.K., et al., 1998. A multidrug resistance transporter from human MCF-7 breast cancer cells. Proc. Natl. Acad. Sci. USA 95, 15665–15670.

Eckford, P.D.W., Sharom, F.J., 2009. ABC efflux pump-based resistance to chemotherapy drugs. Chem. Rev. 109, 2989–3011.

Endter, S., Francombe, D., Ehrhardt, C., Gumbleton, M., 2009. RT-PCR analysis of ABC, SLC and SLCO drug transporters in human lung epithelial cell models. J. Pharm. Pharmacol 61, 583–591.

Evig, C.B., Kelley, E.E., Weydert, C.J., Chu, Y., Buettner, G.R., Patrick Burns, C., 2004. Endogenous production and exogenous exposure to nitric oxide augment doxorubicin cytotoxicity for breast cancer cells but not cardiac myoblasts. Nitric Oxide 10, 119–129.

Fang, F.C., 1997. Perspectives series: host/pathogen interactions. Mechanisms of nitric oxide-related antimicrobial activity.. J. Clin. Invest 99, 2818–2825.

Fang, F.C., 2004. Antimicrobial reactive oxygen and nitrogen species: concepts and controversies. Nat. Rev. Microbiol. 2, 820–832.

Faraci, W.S., Pratt, R.F., 1987. Nucleophilic re-activation of the PC1 beta-lactamase of *Staphylococcus aureus* and of the DD-peptidase of Streptomyces R61 after their inactivation by cephalosporins and cephamycins. Biochem. J. 246, 651–658.

Farias-Eisner, R., Sherman, M.P., Aeberhard, E., Chaudhuri, G., 1994. Nitric oxide is an important mediator for tumoricidal activity in vivo. Proc. Natl. Acad. Sci. USA 91, 9407–9411.

Fulton, D., Gratton, J.P., Sessa, W.C., 2001. Post-translational control of endothelial nitric oxide synthase: why isn't calcium/calmodulin enough? J. Pharmacol. Exp. Ther. 299, 818–824.

Ghaffari, A., Miller, C.C., Mcmullin, B., Ghahary, A., 2006. Potential application of gaseous nitric oxide as a topical antimicrobial agent. Nitric Oxide 14, 21–29.

Gonzalez, D.E., Covitz, K.-M.Y., Sadée, W., Mrsny, R.J., 1998. An oligopeptide transporter is expressed at high levels in the pancreatic carcinoma cell lines AsPc-1 and Capan-2. Cancer Res. 58, 519–525.

Gottesman, M.M., Fojo, T., Bates, S.E., 2002. Multidrug resistance in cancer: role of ATP-dependent transporters. Nat. Rev. Cancer 2, 48–58.

Griffith, O.W., Stuehr, D.J., 1995. Nitric oxide synthases: properties and catalytic mechanism. Annu. Rev. Physiol. 57, 707–736.

Gusarov, I., Nudler, E., 2005. NO-mediated cytoprotection: Instant adaptation to oxidative stress in bacteria. Proc. Natl. Acad. Sci. USA 102, 13855–13860.

Gusarov, I., Shatalin, K., Starodubtseva, M., Nudler, E., 2009. Endogenous nitric oxide protects bacteria against a wide spectrum of antibiotics. Science 325, 1380–1384.

Heinecke, J.L., Ridnour, L.A., Cheng, R.Y.S., Switzer, C.H., Lizardo, M.M., Khanna, C., et al., 2014. Tumor microenvironment-based feed-forward regulation of NOS2 in breast cancer progression. Proc. Natl. Acad. Sci. USA 111, 6323–6328.

Heinrich, T.A., Da Silva, R.S., Miranda, K.M., Switzer, C.H., Wink, D.A., Fukuto, J.M., 2013. Biological nitric oxide signalling: chemistry and terminology. Br. J. Pharmacol. 169, 1417–1429.

Hetrick, E.M., Shin, J.H., Stasko, N.A., Johnson, C.B., Wespe, D.A., Holmuhamedov, E., et al., 2008. Bactericidal efficacy of nitric oxide-releasing silica nanoparticles. ACS Nano. 2, 235–246.

Hirst, D., Robson, T., 2010. Nitric oxide in cancer therapeutics: interaction with cytotoxic chemotherapy. Curr. Pharm. Des. 16, 411–420.

Holden, J.K., Kang, S., Hollingsworth, S.A., Li, H., Lim, N., Chen, S., et al., 2015. Structure-based design of bacterial nitric oxide synthase inhibitors. J. Med. Chem. 58, 994–1004.

Holden, J.K., Li, H., Jing, Q., Kang, S., Richo, J., Silverman, R.B., et al., 2013. Structural and biological studies on bacterial nitric oxide synthase inhibitors. Proc. Natl. Acad. Sci. USA 110, 18127–18131.

Huerta, S., Chilka, S., Bonavida, B., 2008. Nitric oxide donors: novel cancer therapeutics (review). Int. J. Oncol. 33, 909–927.

Jeong, D.W., Kim, Y.H., Kim, H.H., Ji, H.Y., Yoo, S.D., Choi, W.R., et al., 2007. Dose-linear pharmacokinetics of oleanolic acid after intravenous and oral administration in rats. Biopharm. Drug Dispos. 28, 51–57.

Jones, M.L., Ganopolsky, J.G., Labbe, A., Wahl, C., Prakash, S., 2010. Antimicrobial properties of nitric oxide and its application in antimicrobial formulations and medical devices. Appl. Microbiol. Biotechnol. 88, 401–407.

Khodade, V.S., Kulkarni, A., Gupta, A.S., Sengupta, K., Chakrapani, H., 2016. A small molecule for controlled generation of peroxynitrite. Org. Lett. 18, 1274–1277.

Knowles, R.G., Moncada, S., 1994. Nitric oxide synthases in mammals. Biochem. J. 298, 249–258. Pt 2.

Kohchi, C., Inagawa, H., Nishizawa, T., Soma, G., 2009. ROS and innate immunity. Anticancer Res. 29, 817–821.

Kuppens, I.E.L.M., Witteveen, E.O., Jewell, R.C., Radema, S.A., Paul, E.M., Mangum, S.G., et al., 2007. A phase i, randomized, open-label, parallel-cohort, dose-finding study of Elacridar (GF120918) and oral topotecan in cancer patients. Clin. Cancer Res. 13, 3276–3285.

Landowski, C.P., Vig, B.S., Song, X., Amidon, G.L., 2005. Targeted delivery to PEPT1-overexpressing cells: acidic, basic, and secondary floxuridine amino acid ester prodrugs. Mol. Cancer Ther. 4, 659–667.

Li, D.-H., Wang, L., Cai, H., Jiang, B.-W., Zhang, Y.-H., Sun, Y.-J., et al., 2012. Synthesis of novel furozan-based nitric oxide-releasing derivatives of 1-oxo-oridonin with anti-proliferative activity. Chin. J. Nat. Med. 10, 471–476.

Liby, K.T., Yore, M.M., Sporn, M.B., 2007. Triterpenoids and rexinoids as multifunctional agents for the prevention and treatment of cancer. Nat. Rev. Cancer 7, 357–369.

Liu, J., 1995. Pharmacology of oleanolic acid and ursolic acid. J. Ethnopharmacol. 49, 57–68.

Luppens, S.B., Reij, M.W., Van Der Heijden, R.W., Rombouts, F.M., Abee, T., 2002. Development of a standard test to assess the resistance of Staphylococcus aureus biofilm cells to disinfectants. Appl. Environ. Microbiol. 68, 4194–4200.

Macmicking, J.D., North, R.J., Lacourse, R., Mudgett, J.S., Shah, S.K., Nathan, C.F., 1997. Identification of nitric oxide synthase as a protective locus against tuberculosis. Proc. Natl. Acad. Sci. USA 94, 5243–5248.

Mcdougald, D., Rice, S.A., Barraud, N., Steinberg, P.D., Kjelleberg, S., 2012. Should we stay or should we go: mechanisms and ecological consequences for biofilm dispersal. Nat. Rev. Microbiol. 10, 39–50.

Minotti, G., Menna, P., Salvatorelli, E., Cairo, G., Gianni, L., 2004. Anthracyclines: molecular advances and pharmacologic developments in antitumor activity and cardiotoxicity. Pharmacol. Rev. 56, 185–229.

Nakanishi, T., Tamai, I., Sai, Y., Sasaki, T., Tsuji, A., 1997. Carrier-mediated transport of oligopeptides in the human fibrosarcoma cell line HT1080. Cancer Res. 57, 4118–4122.

Negom Kouodom, M., Ronconi, L., Celegato, M., Nardon, C., Marchiò, L., Dou, Q.P., et al., 2012. Toward the selective delivery of chemotherapeutics into tumor cells by targeting peptide transporters: tailored gold-based anticancer peptidomimetics. J. Med. Chem. 55, 2212–2226.

Nguyen, T.-K., Selvanayagam, R., Ho, K.K.K., Chen, R., Kutty, S.K., Rice, S.A., et al., 2016. Co-delivery of nitric oxide and antibiotic using polymeric nanoparticles. Chem. Sci. 7, 1016–1027.

Nielsen, C.U., Andersen, R., Brodin, B., Frokjaer, S., Taub, M.E., Steffansen, B., 2001. Dipeptide model prodrugs for the intestinal oligopeptide transporter. Affinity for and transport via hPepT1 in the human intestinal Caco-2 cell line. J. Control. Release 76, 129–138.

Pacelli, R., Wink, D.A., Cook, J.A., Krishna, M.C., Degraff, W., Friedman, N., et al., 1995. Nitric oxide potentiates hydrogen peroxide-induced killing of Escherichia coli. J. Exp. Med. 182, 1469–1479.

Planting, A.S.T., Sonneveld, P., Van Der Gaast, A., Sparreboom, A., Van Der Burg, M.E.L., Luyten, G.P.M., et al., 2005. A phase I and pharmacologic study of the MDR converter GF120918 in combination with doxorubicin in patients with advanced solid tumors. Cancer Chemother. Pharmacol. 55, 91–99.

Privett, B.J., Deupree, S.M., Backlund, C.J., Rao, K.S., Johnson, C.B., Coneski, P.N., et al., 2010. Synergy of nitric oxide and silver sulfadiazine against gram-negative, gram-positive, and antibiotic-resistant pathogens. Mol. Pharmaceut. 7, 2289–2296.

Pusztai, L., Wagner, P., Ibrahim, N., Rivera, E., Theriault, R., Booser, D., et al., 2005. Phase II study of tariquidar, a selective P-glycoprotein inhibitor, in patients with chemotherapy-resistant, advanced breast carcinoma. Cancer 104, 682–691.

Ridnour, L.A., Thomas, D.D., Switzer, C., Flores-Santana, W., Isenberg, J.S., Ambs, S., et al., 2008. Molecular mechanisms for discrete nitric oxide levels in cancer. Nitric Oxide 19, 73–76.

Riganti, C., Miraglia, E., Viarisio, D., Costamagna, C., Pescarmona, G., Ghigo, D., et al., 2005. Nitric oxide reverts the resistance to doxorubicin in human colon cancer cells by inhibiting the drug efflux. Cancer Res. 65, 516–525.

Riganti, C., Rolando, B., Kopecka, J., Campia, I., Chegaev, K., Lazzarato, L., et al., 2013. Mitochondrial-targeting nitrooxy-doxorubicin: a new approach to overcome drug resistance. Mol. Pharmaceut. 10, 161–174.

Sarkadi, B., Homolya, L., Szakács, G., Váradi, A., 2006. Human multidrug resistance ABCB and ABCG transporters: participation in a chemoimmunity defense system. Physiol. Rev. 86, 1179–1236.

Shannon, A.M., Bouchier-Hayes, D.J., Condron, C.M., Toomey, D., 2003. Tumour hypoxia, chemotherapeutic resistance and hypoxia-related therapies. Cancer Treat. Rev. 29, 297–307.

Sharma, K., Chakrapani, H., 2014. Site-directed delivery of nitric oxide to cancers. Nitric Oxide 43, 8–16.

Sharma, K., Iyer, A., Sengupta, K., Chakrapani, H., 2013. INDQ/NO, a bioreductively activated nitric oxide prodrug. Org. Lett. 15, 2636–2639.

Simon, S.M., Schindler, M., 1994. Cell biological mechanisms of multidrug resistance in tumors. Proc. Natl. Acad. Sci. USA 91, 3497–3504.

Singh, S., Gupta, A.K., 2011. Nitric oxide: role in tumour biology and iNOS/NO-based anticancer therapies. Cancer Chemother. Pharmacol. 67, 1211–1224.

Song, Q., Tan, S., Zhuang, X., Guo, Y., Zhao, Y., Wu, T., et al., 2014. Nitric oxide releasing d-α-tocopheryl polyethylene glycol succinate for enhancing antitumor activity of doxorubicin. Mol. Pharmaceut. 11, 4118–4129.

Sporn, M.B., Liby, K.T., Yore, M.M., Fu, L., Lopchuk, J.M., Gribble, G.W., 2011. New synthetic triterpenoids: potent agents for prevention and treatment of tissue injury caused by inflammatory and oxidative stress. J. Nat. Prod. 74, 537–545.

Stewart, P.S., 2002. Mechanisms of antibiotic resistance in bacterial biofilms. Int. J. Med. Microbiol. 292, 107–113.

Storm, W.L., Johnson, J.A., Worley, B.V., Slomberg, D.L., Schoenfisch, M.H., 2015. Dual action antimicrobial surfaces via combined nitric oxide and silver release. J. Biomed. Mater. Res. A 103, 1974–1984.

Stuehr, D.J., Nathan, C.F., 1989. Nitric oxide. A macrophage product responsible for cytostasis and respiratory inhibition in tumor target cells. J. Exp. Med. 169, 1543–1555.

Sudhamsu, J., Crane, B.R., 2009. Bacterial nitric oxide synthases: what are they good for? Trends Microbiol. 17, 212–218.

Szabó, C., 2003. Multiple pathways of peroxynitrite cytotoxicity. Toxicol. Lett. 140–141, 105–112.

Szabo, C., 2016. Gasotransmitters in cancer: from pathophysiology to experimental therapy. Nat. Rev. Drug Discov. 15, 185–203.

Szabo, C., Ischiropoulos, H., Radi, R., 2007. Peroxynitrite: biochemistry, pathophysiology and development of therapeutics. Nat. Rev. Drug Discov. 6, 662–680.

Szakacs, G., Paterson, J.K., Ludwig, J.A., Booth-Genthe, C., Gottesman, M.M., 2006. Targeting multidrug resistance in cancer. Nat. Rev. Drug Discov. 5, 219–234.

Tai, W., Chen, Z., Cheng, K., 2013. Expression profile and functional activity of peptide transporters in prostate cancer cells. Mol. Pharmaceut. 10, 477–487.

Thomas, D.D., Ridnour, L.A., Isenberg, J.S., Switzer, C.H., Donzelli, S., et al., 2008. The chemical biology of nitric oxide: implications in cellular signaling. Free Radic. Biol. Med. 45, 18–31.

Tozer, G.M., Prise, V.E., Chaplin, D.J., 1997. Inhibition of nitric oxide synthase induces a selective reduction in tumor blood flow that is reversible with L-arginine. Cancer Res. 57, 948–955.

Uchida, S., Shimada, Y., Watanabe, G., Li, Z.G., Hong, T., Miyake, M., et al., 1999. Motility-related protein (MRP-1/CD9) and KAI1/CD82 expression inversely correlate with lymph node metastasis in oesophageal squamous cell carcinoma. Br. J. Cancer 79, 1168–1173.

Van Sorge, N.M., Beasley, F.C., Gusarov, I., Gonzalez, D.J., Von Köckritz-Blickwede, M., Anik, S., et al., 2013. Methicillin-resistant *Staphylococcus aureus* bacterial nitric-oxide synthase affects antibiotic sensitivity and skin abscess development. J. Biol. Chem. 288, 6417–6426.

Vroom, J.M., De Grauw, K.J., Gerritsen, H.C., Bradshaw, D.J., Marsh, P.D., Watson, G.K., et al., 1999. Depth penetration and detection of ph gradients in biofilms by two-photon excitation microscopy. Appl. Environ. Microbiol. 65, 3502–3511.

West, A.P., Shadel, G.S., Ghosh, S., 2011. Mitochondria in innate immune responses. Nat. Rev. Immunol. 11, 389–402.

Wilson, T.R., Johnston, P.G., Longley, D.B., 2009. Anti-apoptotic mechanisms of drug resistance in cancer. Curr. Cancer Drug Targets 9, 307–319.

Wilson, W.R., Hay, M.P., 2011. Targeting hypoxia in cancer therapy. Nat. Rev. Cancer 11, 393–410.

Wink, D.A., Cook, J.A., Christodoulou, D., Krishna, M.C., Pacelli, R., Kim, S., et al., 1997. Nitric oxide and some nitric oxide donor compounds enhance the cytotoxicity of cisplatin. Nitric Oxide 1, 88–94.

Wink, D.A., Hanbauer, I., Krishna, M.C., Degraff, W., Gamson, J., Mitchell, J.B., 1993. Nitric oxide protects against cellular damage and cytotoxicity from reactive oxygen species. Proc. Natl. Acad. Sci. USA 90, 9813–9817.

Wink, D.A., Hines, H.B., Cheng, R.Y.S., Switzer, C.H., Flores-Santana, W., Vitek, M.P., et al., 2011. Nitric oxide and redox mechanisms in the immune response. J. Leukoc. Biol. 89, 873–891.

Xu, S., Pei, L., Wang, C., Zhang, Y.-K., Li, D., Yao, H., et al., 2014. Novel hybrids of natural oridonin-bearing nitrogen mustards as potential anticancer drug candidates. ACS Med. Chem. Lett. 5, 797–802.

Xu, S., Wang, G., Lin, Y., Zhang, Y., Pei, L., Yao, H., et al., 2016. Novel anticancer oridonin derivatives possessing a diazen-1-ium-1,2-diolate nitric oxide donor moiety: design, synthesis, biological evaluation and nitric oxide release studies. Bioorg. Med. Chem. Lett. 26, 2795–2800.

Yang, Y., Bazhin, A.V., Werner, J., Karakhanova, S., 2013. Reactive oxygen species in the immune system. Int. Rev. Immunol. 32, 249–270.

Ye, S., Yang, W., Wang, Y., Ou, W., Ma, Q., Yu, C., et al., 2013. Cationic liposome-mediated nitric oxide synthase gene therapy enhances the antitumor effects of cisplatin in lung cancer. Int. J. Mol. Med. 31, 33–42.

Yepuri, N.R., Barraud, N., Mohammadi, N.S., Kardak, B.G., Kjelleberg, S., Rice, S.A., et al., 2013. Synthesis of cephalosporin-3'-diazeniumdiolates: biofilm dispersing NO-donor prodrugs activated by [small beta]-lactamase. Chem. Commun. 49, 4791–4793.

CHAPTER 13

Hydrogels for Topical Nitric Oxide Delivery

Mathilde Champeau[1], Amedea Barozzi Seabra[2] and Marcelo Ganzarolli de Oliveira[1]

[1]Institute of Chemistry, University of Campinas, UNICAMP, São Paulo, Brazil
[2]Center of Natural and Human Sciences, Universidade Federal do ABC (UFABC), Santo André, São Paulo, Brazil

According to the IUPAC definition, gels are non-fluid polymer or colloidal networks that are expanded by a fluid. In case the liquid that swells the solid material is water, the gel is called a hydrogel. Hydrogels are polymeric networks that can absorb more than 90% of their dry weight in water, while chemical and physical cross-linkings of the polymeric chains make them insoluble in water. The first application of a hydrogel network as biomaterial was carried out in 1954 by Wichterle and Lim (1960), who developed a cross-linked network of poly(2-hydroxyethyl methacrylate) (pHEMA) for use as a soft contact lens. Since this work, hydrogels have been an active field of research and have found applications as biomaterials in tissue engineering and drug delivery (Hoffman, 2001). Due to their softness and capacity to retain large amounts of water, hydrogels have some similarities to living tissues and present good biocompatibility. Their softness also allows their easy adaptation to the shape of the surface on which they are applied. For these reasons, hydrogels have been frequent materials of choice for topical application.

In drug delivery, the two common administration routes that are oral and injection, result in the drug entering the systemic circulation and reaching untargeted organs, thus causing side effects. For this reason, the current trend is to deliver the drug locally at the target site and to control its release rate. Among the local routes of administration, topical drug administration is particularly adapted to treat skin or mucosa disorders such as wounds, vasoconstriction, and bacterial infections. For a convenient application, the drug is generally incorporated into a gel or ointment or it is loaded into a material that is then topically applied. Hydrogels have been used as reservoirs of drugs for topical delivery, because of their above-mentioned properties. Similarly, hydrogels releasing exogenous nitric oxide (NO) have been developed for various topical applications such as vasodilation, wound healing, and antimicrobial treatment, and have been applied on healthy skin, wounded skins, or mucosa. Since gaseous NO cannot be easily stored in materials, different strategies have been invented to design such NO-releasing hydrogels.

Nitric Oxide Donors.

313

This chapter aims at reviewing the different strategies that have been implemented so far to develop NO-releasing hydrogels specifically for topical applications, and the main results that have been obtained in terms of NO release profiles and in vivo applications. The works dealing with NO delivery hydrogels that are not specific for topical applications are not quoted. Hydrogels for topical NO delivery can be classified into four main categories according to the nature of the NO donor and the polymeric matrix:

1. *Hydrogels with dissolved molecular NO donors*: the NO donor is incorporated into the preformed chemically cross-linked hydrogel matrix by absorption from a solution or by mixing the polymeric solution with the NO donor solution, followed by the induction of gelation.

2. *Hydrogels with covalently attached NO donors*: the backbone of the polymeric matrix is functionalized with NO donor moieties.

3. *Hydrogels based on extemporaneous NO release*: two polymeric solutions, or a polymeric solution and an aqueous solution, containing each one a precursor reagent for the production of NO are mixed immediately before application. The reaction between the precursor molecules lead to the final NO-releasing hydrogel.

4. *Supramolecular NO-releasing hydrogels*: the hydrogels are formed by low molecular weight compounds that self-assemble due to noncovalent interactions to form a three-dimensional network. The NO donor is incorporated in the hydrogel during the self-assembling process.

Table 13.1 shows representative examples of NO-releasing hydrogels for each of the above hydrogel categories, along with the identification of the hydrogel and NO donors used, the method employed for triggering the NO release, the NO release profiles and/or doses, the proposed potential applications, and the references. It must be noted that each specific application requires attaining a certain local concentration of exogenous NO to produce the desired biological action. Therefore, controlling the kinetics of NO release is fundamental in all of the proposed applications. As can be seen below, except for the extemporaneous NO release formulations, this control can be achieved by using different strategies, which include the use of light to induce the photochemical release of NO, or the reaction of a precursor NO drug with an enzyme, a reducing species or water. The main advantages and drawbacks of each category of NO donor hydrogels are described in Table 13.2.

HYDROGELS WITH DISSOLVED MOLECULAR NO DONORS

The dissolution of low molecular weight NO donors, such as *S*-nitrosothiols, into a hydrogel matrix is the simplest approach for the NO donor incorporation in a hydrogel. The dispersion of therapeutic amounts of low molecular weight and hydrophilic NO donors is facilitated due to the ability of hydrogels to absorb significant amounts

Table 13.1 Summary of the four main categories of NO-releasing hydrogels for topical applications discussed in the text, with representative examples and identification of the hydrogel and NO donors used, the method employed for triggering the NO release, the reported NO release profiles and/or doses, the proposed potential applications and the corresponding references

Hydrogel categories	Hydrogels	NO donors	Triggers of NO release	NO release profiles and/or doses	Potential applications	References
Dissolved molecular NO donors	pHEMA coated with polyurethane	$[Mn(PaPy_3)(NO)]$ ClO_4 dissolved in the matrix (16 and 32 mM)	Visible light	Continuous increase of NO release over 20 min of irradiation. Steady state reached after 3–5 min	Bactericidal action	Halpenny et al. (2009)
	Carbopol–HPMC–PEG (1.5:1.5:1; 6.7 w/v% solution)	S-nitrosoglutathione	Thermal NO release	First-order exponential growth. The first-order rate constant increases with the initial GSNO charge in the hydrogel	Topical vasodilation	Yoo et al. (2009)
	Pluronic F127	Flutamide derivative (N-(3-aminopropyl)-3-(trifluoromethyl)-4-nitrobenzenamine)	Visible light	NO release is 3.8- to 4-fold higher when F127 forms micelles in comparison with unimers of F127 due to photorearrangement of the flutamide derivative in the core of the micelles	Topical application (not specified)	Taladriz-Blanco and de Oliveira (2014)
	Pluronic F127	S-nitrosoglutathione	Thermal and photochemical (visible light) NO release	Hydrogels containing 50, 100, and 150 mM of GSNO release 5, 6, and 13 $nmol_{NO}\,cm^{-2}$, respectively, after 10 min Irradiation of a hydrogel containing GSNO 13.8 mM increases the rate of NO release from 2.3 $mmol_{NO}\,L^{-1}\,min^{-1}$ to 5.3 $mmol_{NO}\,L^{-1}\,min^{-1}$	Vasodilation and wound healing	Vercelino et al. (2013) Souto et al. (2010) Georgii et al. (2011) Amadeu et al. (2008) Amadeu et al. (2007) Seabra et al. (2007)

(Continued)

Table 13.1 Summary of the four main categories of NO-releasing hydrogels for topical applications discussed in the text, with representative examples and identification of the hydrogel and NO donors used, the method employed for triggering the NO release, the reported NO release profiles and/or doses, the proposed potential applications and the corresponding references (Continued)

Hydrogel categories	Hydrogels	NO donors	Triggers of NO release	NO release profiles and/or doses	Potential applications	References
Covalently attached NO donors	PVA with grafted moieties	Diazeniumdiolate	Water	Over 48 h at pH 6: 1.56 ± 0.08 and $13.0 \pm 0.85\,pmol_{NO}\,cm^{-3}$ released from functionalized PVA containing an NO charge of 0.5 and $5\,mM_{NO}$, respectively	Wound healing	Bohl Masters et al. (2002)
	PVA functionalized with –SNO groups + combined with Pluronic F127	S-nitrosothiol groups grafted to the PVA chain	Thermal NO release	$1200\,pmol_{NO}\,cm^{-2}$ released after 24 h	Wound healing	Schanuel et al. (2015)
	PVA functionalized with –SNO groups	S-nitrosothiol groups grafted to the PVA chain	Thermal NO release	Maximum NO content achieved: $6.2 \pm 1.3\,mmol_{NO}/g_{polymer}$ NO release in PBS at 37°C bursts at 3 min and ends up after 40 min, releasing a totally of $300\,nmol_{NO}\,cm^{-2}$	Topical vasodilation	Marcilli and De Oliveira (2014)

Category	Material	Composition	Mixing method	NO release	Application	Reference
Extemporaneous NO release	Hydroxyethyl cellulose-based hydrogel	1st hydrogel contains sodium nitrite 2nd hydrogel contains ascorbic acid	Mixture of two hydrogels	Not studied	Topical vasodilation	Tucker et al. (1999)
	Agar hydrogel	1st gel contains sodium nitrite and 2nd gel contains ascorbic acid	Mixture of two hydrogels	The real-time NO release was measured through and without a polyester–copolymer membrane permeable to NO: the membrane lowered the peak of NO release by 50% and extended it to 11 min instead of 4 min	Topical vasodilation and microbicidal action	Hardwick et al. (2001)
	Not communicated	1st layer contains sodium nitrite 2nd layer contains ascorbic acid	Two hydrogel layers put into contact	Not described	Wound healing	Edixomed (2016)
Supramolecular hydrogels	Self-assembling peptide	Charged NO donor reacts with a self-assembling peptide	β-Galactosidase	Depends on the amount of enzyme added to the gel varying from 5.2 to 82 $nmol_{NO}$ per 100 mg_{gel}	Wound healing	Gao et al. (2013)

Table 13.2 Summary of the benefits and drawbacks of the four main categories of NO-releasing hydrogels for topical applications discussed in the text

Hydrogel categories	Benefits	Drawbacks
Hydrogels with dissolved molecular NO donors	Easy preparation Possibility of using hydrogel matrices already on the market Easy control of the NO charge	Not adequate for hydrophobic NO donors Leaching of the NO donor before its NO release reaction
Hydrogels with covalently attached NO donors	No leaching of the NO donor Possibility of tuning the NO charge by controlling the extent of the functionalization reaction	Several reaction steps may be necessary. Dialysis may be necessary to remove unreacted reagents
Hydrogels based on extemporaneous NO release	NO release controlled by diffusion of one reagent from one hydrogel/gel through a second hydrogel/gel containing the second reagent	Mixing of the two hydrogels requires special care to assure homogeneity
Supramolecular NO-releasing hydrogels	No leaching of the NO donor if well designed	The NO donor and the supramolecular hydrogel must be specifically tailored to integrate the NO donor into the hydrogel structure

of water. The advantages of the dispersion of NO donors in hydrogels are the simplicity of the methodology and the preservation of the NO donors. In contrast, one disadvantage of this approach is the limitation for the dispersion of NO donors with low water solubilities such as S-nitroso-N-acetylpenicillamine (SNAP) (Hoare and Kohane, 2008). Therefore, the amount and the homogeneity of the NO donors loaded in the hydrogels may be limited.

The use of Pluronic F127, a poly(ethylene oxide)–poly(propylene oxide)–poly(ethylene oxide) triblock copolymer, has been noteworthy in this strategy. Pluronic F127 is FDA approved (Liaw and Lin, 2000) and one of the most employed polymers for topical applications. It has the property of spontaneously assembling into micelles with the increase in temperature, forming a microheterogeneous system where the hydrophobic nuclei of the micelles can dissolve hydrophobic molecules (Fig. 13.1). As a biologically inert matrix, Pluronic F127 has been extensively used for drug delivery in several applications including the coating of burn lesions (Nalbandian et al., 1987) and subcutaneous administrations (Kant et al., 2014). Low molecular weight NO donors have been successfully incorporated into Pluronic F127 hydrogels for topical applications. The advantages of using Pluronic F127 hydrogel as a matrix for NO donors is that it allows easy incorporation of water-soluble NO donors in the liquid cold polymeric solution, which may undergo gelation after topical application

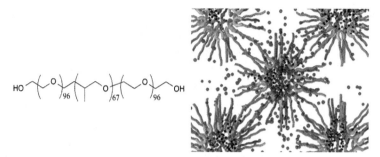

Figure 13.1 Chemical structure of PEO–PPO–PEO and a representation of the packed network of F127 micelles with a hydrophobic NO-releasing drug concentrated in their hydrophobic nuclei.

and heating to the body temperature, allowing a prolonged localized therapeutic action (Shishido et al., 2003; Seabra et al., 2004).

Several articles describe the incorporation of *S*-nitrosothiols (RSNOs) into Pluronic F127 hydrogels and their biological actions in topical applications in animal models and human volunteers. RSNOs are an important class of NO donors that have been employed as experimental drugs for supplying exogenous NO (de Oliveira, 2016). RSNOs act as spontaneous NO donors due to the homolytic S–N bond cleavage with free NO release and the formation of sulfur-bridged dimmers of the RS radicals. This reaction can be promoted by thermal energy and can be catalyzed by copper ions and UV or visible light (Broniowska and Hogg, 2012). It has been shown that Pluronic F127 hydrogels containing *S*-nitrosoglutathione (GSNO) or *S*-nitroso-*N*-acetyl-cysteine (SNAC) release NO spontaneously in the dark and that irradiation of these hydrogels with visible light increases significantly the rate of NO release (Shishido et al., 2003). In addition, it was also reported that Pluronic F127 significantly reduces the rate of thermal and photochemical NO release from the RSNOs in comparison with aqueous RS-NO solutions. This effect was attributed to a cage effect imposed by the hydrogel matrix, which favors the RS-NO radical pair recombination after the homolytic S–N bond cleavage and can be used to modulate the rate of NO delivery.

As an example, Pluronic F127 (25 wt%) containing GSNO (0.3 mol g^{-1}) or SNAC (0.6 mol g^{-1}) were topically applied on the forearm skin of human volunteers and the dermal vasodilation was measured by laser Doppler flowmetry (Seabra et al., 2004). In addition, the diffusion of free NO from the hydrogels was simultaneously measured by cutaneous microdialysis. The results showed a significant increase of dermal blood flow in comparison with the F127 hydrogel vehicle only. Interestingly, the increase in the dermal blood flow was correlated with the amount of nitrite (the oxidation product of NO) in the dermal dialysate, at low and moderate levels of blood flow. However, a reduction of nitrite levels in the dermis was observed at high levels of blood flow. This result was assigned to a faster scavenging of nitrite at higher blood flow level. Overall, these results demonstrate that NO released from topically applied RSNOs-containing Pluronic F127 hydrogels is capable of crossing the lipid-rich structure of the stratum corneum of the intact skin, promoting a local dermal vasodilation.

Similarly, topical application of GSNO-containing Pluronic F127 hydrogel on the foot sole of streptozotocin-induced diabetic rats was also shown to significantly increase dermal blood flow with a maximum peak of blood flow achieved after 30 minutes. This study also showed the absence of side effects associated with the nitration of tyrosine residues of proteins or alterations in heart rate or blood pressure (Seabra et al., 2007). One of the complications of diabetes is impaired micro and macro blood circulation since in this disease, the vasodilatory capacity of the arterioles is compromised by the reduced endogenous production of NO by the endothelial cells of the blood vessels. The above-mentioned results show that topical application of NO donors may potentially reverse this condition at the site of application. Maintenance of normal dermal microcirculation in such topical applications has therefore potential to inhibit the appearance of diabetic ulcers as well as to help in their healing.

Another study has explored the potential application of topical GSNO-containing Pluronic hydrogel to increase the arterial blood flow to the clitoris, in order to treat female sexual dysfunction. Preliminary results with 20 healthy women volunteers, evaluated by Doppler ultrasound, showed a 2.5-fold increase in the systolic and diastolic speeds 15 minutes after application (Souto et al., 2010). Further studies are necessary to better characterize this potential medical application.

The importance of topical administration of hydrogels containing dissolved NO donors has also being demonstrated by investigating their actions on the wound healing process. NO is known to be synthesized by several human skin cells, such as fibroblasts, keratinocytes, Langerhans cells, and melanocytes (Weller, 2003). Many evidences support that NO plays several roles in the wound repair process including cellular proliferation, apoptosis, and collagen deposition (Luo and Chen, 2005). For example, Amadeu et al. (2007) applied Pluronic hydrogel containing GSNO $100\,\mu mol\,L^{-1}$ topically on the excisional wounds of rats during the first 4 days of wounding. Comparison with the control animals, receiving only the vehicle, showed that GSNO-containing hydrogels significantly accelerated the healing process. This result was reflected in an increased collagen fibers deposition and organization in the granulation tissue, an accelerated wound re-epithelialization and wound closure, and an increased amount of fibroblasts. These results suggest that these hydrogel formulations may represent a new strategy to accelerate wound healing with impact on the treatment of chronic wounds. Moreover these results show that the dispersion of GSNO into Pluronic F127 hydrogels allies the advantages of releasing therapeutic amounts of NO in a sustained manner, with the advantages of the hydrogel matrix itself, as a biologically inert and friendly matrix.

It is known that at the inflammatory phase of the wound healing process, NO has a cytotoxic effect, killing microorganisms and controlling the release of growth factors and cytokines, while during the proliferative phase, NO stimulates the synthesis and deposition of collagen fibers. Finally, in the scar tissue formation phase, NO controls the apoptosis of myofibroblasts, leading to the complete re-epithelialization of the granulation tissue. Amadeu et al. (2008) demonstrated that topical application of GSNO-containing Pluronic F127 hydrogel in the inflammatory and proliferative phases significantly enhanced wound contraction and re-epithelialization, compared with topical application of the hydrogel only in the inflammatory or proliferative phases, 5 and 14 days after wounding, respectively, as can be seen in Fig. 13.2.

Other results from the work of Amadeu et al. (2008) showed a superior tissue organization, improved collagen fibers maturation, and decreased levels of inflammatory cells either in the deep or superficial areas of the granulation tissue in the wounds treated with the GSNO-containing hydrogel, applied in both inflammatory and proliferative phases, in comparison with the control treatment. Taking together, these results demonstrated that daily topical application of NO-releasing hydrogel for longer periods is more effective in the acceleration of tissue repair.

In a further work, the beneficial effects of topical GSNO-containing Pluronic F127 (GSNO concentration $200\,\mu mol\,L^{-1}$) in an animal model of chronic wound were evaluated (Georgii et al., 2011). Daily applications during 7 days reduced the amounts of inflammatory cells, enhanced organization of the granulation tissue formation, increased collagen fiber density, hydroxyproline contents, and the number of fibroblasts, while decreased the neovascularization, in comparison with the control group.

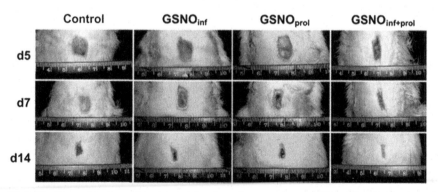

Figure 13.2 Photographs of excisional wounds of rats in control group (Pluronic F127 hydrogel without GSNO), and GSNO-containing hydrogels applied in the inflammatory (GSNO inf), proliferative (GSNO prol), and inflammatory and proliferative (GSNO inf + prol) phases of wound healing; 5 (d5), 7 (d7), and 14 (d14) days after wounding. GSNO concentration: $100\,\mu mol\,L^{-1}$. *Modified from Amadeu T.P., Seabra A.B., de Oliveira M.G. and Costa A.M.A., Nitric oxide donor improves healing if applied on inflammatory and proliferative phase, J. Surg. Res. 149, 2008, 84–93 with permission of Elsevier; Reprinted with permission from Amadeu et al., 2008, John Wiley & Sons.*

These results, in addition to the observed increase in the volume density of myofibroblast cells, in comparison with the control group, indicated a faster wound closure and a superior collagen deposition, opening a perspective for the treatment of chronic wounds such as ischemic and diabetic wounds.

In a more ample study, Vercelino et al. (2013) tested the topical delivery of NO from GSNO-containing Pluronic hydrogels with much higher GSNO concentrations (50–150 mmol L^{-1}) to elicit analgesic action. In this study, the same formulations were applied on the forearm skin of human volunteers to evaluate dermal vasodilation and on the plantar faces of the hind paws of Wistar rats that received an inflammatory stimulus. Laser Doppler measurements showed a dose–response increase in the dermal blood flow and a corresponding dose–response decrease of up to 50% of the hypernociception intensity in the rats. This study thus opened a new perspective for the topical treatment of inflammatory pain with hydrophilic formulation which delivers a very diffusible molecule. This approach may have advantages over the classical topical treatments using hydrophobic nonsteroidal antiinflammatory drugs in hydrophobic vehicles.

More recently, Taladriz-Blanco and de Oliveira (2014) also used the Pluronic F127 hydrogel, but in this case, charged with a different class of NO donor molecule, consisting in the flutamide derivative, N-(3-aminopropyl)-3-(trifluoromethyl)-4-nitrobenzenamine (ATN). Differently from the RSNOs, ATN is thermally stable and releases NO only photochemically with visible light irradiation. As ATN is scarcely soluble in water, the micellar nature of the Pluronic F127 hydrogel allowed its incorporation into the hydrophobic nuclei of the micelles. This study showed that this strategy leads to an enhancement of photochemical NO release from ATN, widening the range of uses of Pluronic F127 hydrogels to encompass the incorporation of hydrophobic NO donors in topical applications.

The strategy of using an NO donor that releases NO exclusively under irradiation was also reported by Halpenny et al. (2009). These authors used several photoactive metal–nitrosyl complexes, which were incorporated into a solution of hydroxyethyl methacrylate (HEMA) monomers. The mixture was polymerized forming a matrix of poly(2-hydroxyethyl methacrylate) (pHEMA), a polymer that is widely used for biomedical applications. Visible light was used to trigger NO release from the metal–nitrosyl complex incorporated in this hydrogel. Discs of this material were coated with a layer of polyurethane (PU) and characterized regarding their microbicidal properties. It was verified that the materials did not release NO sufficiently to cause antimicrobial effects against *Pseudomonas aeruginosa* or *Escherichia coli*. However, in the presence of hydrogen peroxide (H$_2$O$_2$) and methylene blue (MB), as auxiliary bacterial growth attenuators, it was concluded that the combination of NO-releasing pHEMA hydrogel and MB could be used in the fabrication of an antimicrobial wound dressing.

Finally, Yoo et al. (2009) developed a dry GSNO-containing film based on a mixture of Carbopol, hydroxypropyl methylcellulose, and PEG for vaginal application to treat female sexual arousal disorder. This polymeric mixture has been previously used as a gel for controlled delivery. In that case, dry films with mucoadhesiveness were obtained. These films formed gel once absorbing water and further dissolved. The three polymers were first mixed and then GSNO was incorporated into the polymeric solution. After casting, a GSNO-containing film was obtained. The NO release profiles depended on the pH of the medium and on the GSNO charge, which was controlled during GSNO incorporation in the polymeric solution. The released NO was able to activate NO-cyclic guanosine monophosphate (NO-cGMP) signaling pathway of the cells of the vaginal mucosa. The application of a film containing 2 mg of GSNO on the vagina of rats resulted in a blood flow increase up to 210 min, whereas the injection of the same amount of GSNO limited the effect to 20 min. Moreover, the films hindered the burst release observed in case of GSNO injection. The authors underlined that the duration of action was shorter than expected and assigned it to the spontaneous GSNO thermal decomposition, associated with the presence of enzymes that could also trigger GSNO decomposition.

Taking together, the above-cited studies demonstrate that topical application of NO-releasing hydrogels might find important therapeutic applications in the treatment of impaired and diabetic wound healing, in the alleviation of inflammatory pain, and in the promotion of local blood flow.

HYDROGELS WITH COVALENTLY ATTACHED NO DONORS

A special class of NO-releasing hydrogels, where the NO-releasing moieties are covalently bound to the polymer backbone, has been described more recently. These hydrogels are mainly based on a chemical modification of poly(vinyl alcohol) (PVA) which is a hydroxylated polymer with excellent biocompatibility and already used in various biomedical applications (Alves et al., 2011). The first reported modification of PVA to chemically attach NO-releasing groups to its structure was reported by Marcilli and De Oliveira (2014). In this work, the authors obtained chemically cross-linked PVA via esterification with mercaptosuccinic acid (MSA), which contains a sulfhydryl SH group. After this reaction step, the sulfhydryl groups were nitrosated, resulting in a PVA functionalized with SNO groups capable of releasing NO by the same mechanisms observed for low molar weight S-nitrosothiols. The topical application of PVA-SNO films led to a significant increase in the forearm skin blood flow of human volunteers, measured by laser Doppler flowmetry (Fig. 13.3).

A more recent report on the use of PVA hydrogel functionalized with −SNO groups is found in the work of Schanuel et al. (2015), where the authors used the same esterification reaction to attach SNO groups to the PVA backbone. In this case, however, a mixture of MSA (a dicarboxylic cross-linking agent) and thiolactic acid (a sulfhydryl-containing monocarboxylic acid) was used. This mixture led to a less cross-linked PVA matrix, which was therefore capable of undergoing a more extensive

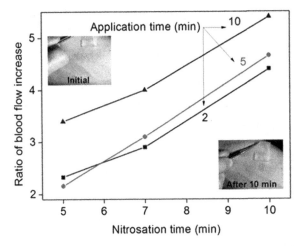

Figure 13.3 Time and dose responses of topical applications of nitric oxide-releasing PVA-SNO films on the forearm skin of the authors. The curves show the ratio of microcirculatory flow increase in response to PVA-SNO films subjected to increasing S-nitrosation times (5, 7, and 10 min) applied for 2, 5, and 10 min. The two inset photographs are representative and show the site on the forearm before application (initial) and the hyperemic response obtained after 10 min of application of the film S-nitrosated for 10 min. *Reprinted with permission from Marcilli, R.H.M., De Oliveira, M.G., 2014. Nitric oxide-releasing poly(vinyl alcohol) film for increasing dermal vasodilation. Colloids Surf. B Biointerf. 116, 643–651, Elsevier.*

swelling with water. This higher swelling capacity may offer an additional advantage to this hydrogel, since it can absorb exudate from lesions in topical applications. This material was tested as a promoter of wound healing in an animal model. In this application, the authors also decided to combine the PVA hydrogel with a bottom layer of Pluronic F127 hydrogel containing or not dissolved GSNO. The combined materials proved to be more effective in accelerating wound healing than the PVA-SNO alone.

Bohl Masters et al. (2002) substituted the pendant hydroxyl groups of PVA by amine groups and was cross-linked by acrylamide cross-linker. The amine groups were reacted with NO gas in water to form diazeniumdiolates moieties. Two hydrogels were prepared containing 0.5 mM and 5 mM of NO, respectively. The two hydrogels released 1.56 ± 0.08 and $13.0 \pm 0.85 \, pmol_{NO} \, cm^{-3}$, respectively, after 48 hours immersion in a solution of pH 6. After 4 days in contact with a human dermal fibroblasts culture, it was observed that the collagen production increased with the concentration of NO in the hydrogel. The effect on wound healing in diabetic mice was investigated. The NO-releasing modified PVA (PVA-NO) and the nonmodified control PVA led to a similar time of wound closure. However, the granulation tissues were thicker in the case where PVA-NO was applied on the wound and the thickness increased with the NO concentration. The mean scar tissue thickness was also higher in animals subjected to PVA-NO with 5 mM of NO compared to nonmodified PVA.

GELS BASED ON EXTEMPORANEOUS NO RELEASE

The class of extemporaneous NO-releasing materials refers to materials prepared at the time of application in order to lead to NO release. In the literature, only one system is found which is based on the reaction between sodium nitrite with ascorbic acid that produces nitrous acid, which is then reduced to free NO. This strategy implies in fact in using two separate hydrogels: one that contains sodium nitrite and another that contains ascorbic acid. At the moment of application, the two hydrogels are mixed or put in contact in order to promote the reduction of the nitrite anion and the release of NO directly at the site of application. This system has been first reported by Tucker et al. (1999) and has been optimized, leading to the development of a product that is currently under clinical trials. Since NO is a vasodilator species, these authors have firstly investigated the effect of a hydrogel based on extemporaneous NO release to treat patients with primary Raynaud's syndrome. This syndrome is characterized by digital vasoconstriction and pain. The authors prepared two hydroxyethyl cellulose-based gels (KY jelly) that contained either sodium nitrite or ascorbic acid (5%, w/v) that could be mixed at the time of attack and topically applied to deliver NO. An equal volume of both hydrogels was then mixed directly on the skin of the forearm or on the finger pulps of healthy patients and of patients with severe Raynaud's syndrome. In order to evaluate the impact of the NO-generating hydrogel, skin microcirculation was measured using the infrared photoplethysmography and laser Doppler flowmetry technics on the sites of application.

In both healthy volunteers and patients with Raynaud's syndrome, the effect of the NO-generating hydrogel tended to raise the local microcirculatory volume and flow. However, the effect was more pronounced on the forearm than on the finger. The authors assigned this difference to the higher baseline of the finger blood flow and to the thicker epidermis of the finger that can hinder the NO diffusion. After removal of the gel, the blood flow decreased but remained higher than the initial baseline after 10 min. However, the authors underlined that the hydrogel was "messy" and not easily applicable. Later, Hardwick et al. (2001) also reported that the acidic nature of this system could entail undesirable secondary effects, particularly on ulcerated skin. In order to solve this drawback, these authors used a hydrophilic membrane permeable to NO, made of polyester–copolymer (Sympatex4 10 μm membrane; Azko Nobel, Wuppertal, Germany) placed between the NO-generating hydrogel and the skin. In this study, the gels of sodium nitrite and ascorbic acid were prepared using agar (0.8%, w/v). The membrane allowed the diffusion of NO and hindered the diffusion of the other molecules. Moreover, the NO release profile after the mixture of the two hydrogels (containing, respectively, 330 mM of reagent) through the membrane was compared to the mixture in the absence of the membrane. The conclusion was that the membrane allowed to reduce the peak of NO release by 50% and to retard it by 7 min. After

20 h, the system was still releasing NO. The vasodilatation test performed on the forearm of healthy volunteers showed that the membrane led to a decrease of 36% in the microvascular flow, compared to the application of the hydrogel alone. Antimicrobial tests were also carried out against *Staphylococcus aureus* and *Escherichia coli*. A range of concentrations of sodium nitrite and ascorbic acid were tested (from 1 to 1000 mM) and the antimicrobial effect was observed for concentrations higher than 50 mM. This hydrogel producing extemporaneous NO has therefore potential to be applied to leg or foot ulcers to promote wound healing.

This NO-generating system has been refined and this research led to the creation of a product by the EdixoMed company (www.edixomed.com, 2016) to target broad medical application such as to treat chronic foot ulcers, venous leg ulcers, and pressure ulcers. This product is composed of two layers of hydrogels, one containing dissolved ascorbic acid and the other one containing sodium nitrite. Once the two layers are put in contact and applied on the wound, NO is released. EdixoMed claims that their dressing is able to absorb exudates, keep an adequate moist environment, be nonadherent on the wound bed, prevent infection, and deliver NO.

Phase II on chronic wound and phase II on acute wound randomized and controlled clinical studies have started in 2015. Their aim is to verify if the NO-generating dressing is able to raise the blood flow in chronic diabetic foot ulcers and to enhance wound healing (www.clinicaltrials.gov Ref. NCT01982565) as well as to improve healing on superficial partial thickness burn wounds and skin graft donor site wounds (www.clinicaltrials.gov Ref. NCT01983085). The preliminary results that have been released showed that 50% of the patients who received the NO-releasing dressing had a wound size lower than 10% of its original size; compared to 16.6% of the patients that were treated by standard procedures. Economical evaluation of the two treatments was performed and showed that the treatment with the NO-generating dressing allowed for decreasing the costs up to £64,000, due to lower adverse health incidence (Stewart et al., 2014).

SUPRAMOLECULAR NO-RELEASING HYDROGELS

Supramolecular hydrogels are formed by low molecular weight compounds that self-assemble due to noncovalent interactions, to create a three-dimensional network. The development of supramolecular hydrogels has been inspired by self-assembling biological systems. When the molecules self-assemble in water, they are called hydrogelators (Du et al., 2015).

Up to now, only one supramolecular hydrogel has been specifically developed for topical delivery of NO. The aim of the authors was to develop a NO molecule with enzyme-responsive properties in order to mimic biological systems that control NO concentration (Gao et al., 2013). Since the dispersion of such free molecule in the hydrogel matrix could

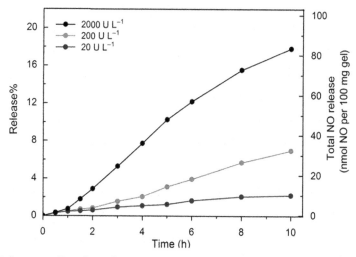

Figure 13.4 Release profile of NO from the supramolecular hydrogels containing 0.5 wt% of NO donor with the addition of different concentrations of β-galactosidase. *Reproduced from Gao, J., Zheng, W., Zhang, J., Guan, D., Yang, Z., Kong, D., et al., 2013. Enzyme-controllable delivery of nitric oxide from a molecular hydrogel. Chem. Commun. 49, 9173–9175, with permission from The Royal Society of Chemistry.*

entail its leaching during application, the authors developed an original strategy to provide a localized NO delivery. A β-galactosidase-responsive caged NO molecule was reacted with a short peptide able to self-assemble. The formation of the supramolecular hydrogel was verified by rheometry. The addition of the enzyme β-galactosidase allowed for removing a protective group from the NO donor and hence for releasing NO. The quantity of NO released was dependent on the amount of β-galactosidase added on the hydrogel (Fig. 13.4) and proceeds continuously at least during 10 h after the addition of enzyme. Moreover, no NO release was observed in absence of the enzyme.

The wound healing potential of this system was evaluated in mice. The daily addition of β-galactosidase on the NO-releasing supramolecular hydrogel during 4 days led to a significant decrease of the wound area (81.9% after 7 days, compared to 64.6% when only the hydrogel was applied). Moreover, the topical application of the hydrogel with β-galactosidase addition, allowed for higher angiogenesis on the center of the wound and on the border as well as to higher neovascularization. Interestingly, no significant difference in the wound area reduction was observed between the wounds treated with PBS and the one on which the caged NO donor was delivered, followed by a daily addition of enzyme. The authors accounted this result for the quick diffusion of the free NO donor into the body of the animal, highlighting the advantage of developing a formulation in which the NO donor cannot leach.

Another self-assembling hydrogel of poly-β-cyclodextrin and modified dextran has recently been developed for NO delivery (Kandoth et al., 2013). An NO photodonor was designed to be incorporated in the β-cyclodextrin cavities during the assembling, in order to avoid its leaching. Even if the targeted application does not specifically focus on topical applications, such use can be envisaged.

The fact that NO-releasing supramolecular hydrogels have been sparsely developed must rely on the necessity to tailor a specific NO donor that can be inserted in the self-assembling system. For example, the synthesis route developed by Gao et al. (2013) to prepare the previously described β-galactosidase-responsive molecular hydrogel, required an extensive preparation procedure of the caged NO donor and of the peptide, followed by a click reaction of the two molecules to form the hydrogelator, the purification of hydrogelator and, finally, the formation of the hydrogel.

CONCLUSION

From the four main categories of NO-releasing hydrogels that can be devised for topical applications, the one based on dissolved molecular NO donors is by far the most feasible, since it involves basically a dissolution process. This strategy, however, implies the use of NO donors with high water solubility, which limits them to mono or polypeptides containing NO-releasing moieties, such as −SNO and −NNO, and to some metal−nitrosyl complexes. The main drawbacks in these cases are the leaching of the intact NO donors to the target tissues and, more important, the low stability of some NO donors that, while assuring spontaneous NO release, demands special storage conditions for an acceptable shelf life. Hydrogels with covalently attached NO donors arise in this setting as an alternative to avoid the leaching of loose NO donors and allow the chemical attachment of a large variety of functional NO-releasing groups. This strategy brings a great versatility to this class of hydrogels, but one must keep in mind that it may also bring different levels of complexity, regarding the reaction and purification steps involved. Thus only a few options may in fact be feasible for scaling up. The hydrogels classified here as based on extemporaneous NO release are also one of the simplest approaches to obtain an NO-releasing hydrogel. However, so far, they remain exclusively based on the mixture of sodium nitrite and acidified ascorbate. Therefore, some advance in this proposal is surely needed and can be expected for the near future. From simplicity to complexity, one arrives at the supramolecular NO-releasing hydrogels, which evoke an elegant chemistry and also demand elegant solutions to be able to dispute a place in the market of forthcoming medical products for topical NO delivery.

REFERENCES

Alves, M.H., Jensen, B.E.B., Smith, A.A.A., Zelikin, A.N., 2011. Poly(vinyl alcohol) physical hydrogels: new vista on a long serving biomaterial. Macromol. Biosci. 11, 1293–1313.
Amadeu, T.P., Seabra, A.B., de Oliveira, M.G., Costa, A.M.A., 2007. S-nitrosoglutathione-containing hydrogel accelerates rat cutaneous wound repair. J. Eur. Acad. Dermatol. Venereol. 21, 629–637.

Amadeu, T.P., Seabra, A.B., de Oliveira, M.G., Costa, A.M.A., 2008. Nitric oxide donor improves healing if applied on inflammatory and proliferative phase. J. Surg. Res. 149, 84–93.

Bohl Masters, K.S., Leibovich, S.J., Belem, P., West, J.L., Poole-Warren, L.A., 2002. Effects of nitric oxide releasing poly(vinyl alcohol) hydrogel dressings on dermal wound healing in diabetic mice. Wound Repair Regen 10, 286–294.

Broniowska, K.A., Hogg, N., 2012. The chemical biology of S-nitrosothiols. Antioxid. Redox Signal. 17, 969–980.

Clinical Pilot Study: Nitric oxide generating gel dressing in patients with diabetic foot ulcers. ClinicalTrials. gov Identifier: NCT01982565. https://clinicaltrials.gov/ct2/show/study/NCT01982565 (accessed 04.09.16).

Clinical Pilot Study: Nitric oxide generating gel dressing in patients with diabetic foot ulcers. ClinicalTrials.gov Identifier: NCT01983085 https://clinicaltrials.gov/ct2/show/study/NCT01983085 (accessed 04.09.16).

De Oliveira, M.G., 2016. S-nitrosothiols as platforms for topical nitric oxide delivery. Basic Clin. Pharmacol. Toxicol. 119, 49–56.

Du, X., Zhou, J., Shi, J., Xu, B., 2015. Supramolecular hydrogelators and hydrogels: from soft matter to molecular biomaterials. Chem. Rev. 115, 13165–13307.

Edixomed, 2016. http://www.edixomed.com/ (accessed 04.09.16).

Gao, J., Zheng, W., Zhang, J., Guan, D., Yang, Z., Kong, D., et al., 2013. Enzyme-controllable delivery of nitric oxide from a molecular hydrogel. Chem. Commun. 49, 9173–9175.

Georgii, J.L., Amadeu, T.P., Seabra, A.B., de Oliveira, M.G., Costa, A.M.A., 2011. Topical S-nitrosoglutathione-releasing hydrogel improves healing of rat ischaemic wounds. J. Tissue Eng. Regen. Med. 5, 612–619.

Halpenny, G.M., Steinhardt, R.C., Okialda, K.A., Mascharak, P.K., 2009. Characterization of pHEMA-based hydrogels that exhibit light-induced bactericidal effect via release of NO. J. Mater. Sci: Mater. Med 20, 2353–2360.

Hardwick, J.B.J., Tucker, A.T., Wilks, M., Johnston, A., Benjamin, N., 2001. A novel method for the delivery of nitric oxide therapy to the skin of human subjects using a semi-permeable membrane. Clin. Sci. 100, 395–400.

Hoare, T.R., Kohane, D.S., 2008. Hydrogels in drug delivery: progress and challenges. Polymer 49, 1993–2007.

Hoffman, A.S., 2001. Hydrogels for biomedical applications. Ann. N.Y. Acad. Sci. 944, 62–73.

Kandoth, N., Mosinger, J., Gref, R., Sortino, S., 2013. A NO photoreleasing supramolecular hydrogel with bactericidal action. J. Mater. Chem. B 1, 3458–3463.

Kant, V., Gopal, A., Kumar, D., Gopalkrishnan, A., Pathak, N.N., Kurade, N.P., et al., 2014. Topical pluronic F-127 gel application enhances cutaneous wound healing in rats. Acta Histochem. 116, 5–13.

Liaw, J., Lin, Y.-C., 2000. Evaluation of poly(ethylene oxide)–poly(propylene oxide)–poly(ethylene oxide) (PEO–PPO–PEO) gels as a release vehicle for percutaneous fentanyl. J. Control. Release 68, 273–282.

Luo, J.D., Chen, A.F., 2005. Nitric oxide: a newly discovered function on wound healing. Acta Pharmacol. Sin. 26, 259–264.

Marcilli, R.H.M., De Oliveira, M.G., 2014. Nitric oxide-releasing poly(vinyl alcohol) film for increasing dermal vasodilation. Colloids Surf. B Biointerf. 116, 643–651.

Nalbandian, R.M., Henry, R.L., Balko, K.W., Adams, D.V., Neuman, N.R., 1987. Pluronic F-127 gel preparation as an artificial skin in the treatment of third-degree burns in pigs. J. Biomed. Mater. Res. 21, 1135–1148.

Schanuel, F.S., Raggio Santos, K.S., Monte-Alto-Costa, A., De Oliveira, M.G., 2015. Combined nitric oxide-releasing poly(vinyl alcohol) film/F127 hydrogel for accelerating wound healing. Colloids Surf. B Biointerf. 130, 182–191.

Seabra, A.B., Fitzpatrick, A., Paul, J., de Oliveira, M.G., Weller, R., 2004. Topically applied S-nitrosothiol-containing hydrogels as experimental and pharmacological nitric oxide donors in human skin. Br. J. Dermatol. 151, 977–983.

Seabra, A.B., Pankotai, E., Fehér, M., Somlai, L.K., Bíró, L., Szabó, C., et al., 2007. S-nitrosoglutathione-containing hydrogel increases dermal blood flow in streptozotocin-induced diabetic rats. Br. J. Dermatol. 156, 814–818.

Shishido, S.M., Seabra, A.B., Loh, W., de Oliveira, M.G., 2003. Thermal and photochemical nitric oxide release from *S*-nitrosothiols incorporated in Pluronic F127gel: potential uses for local and controlled nitric oxide release. Biomaterials 24, 3543–3553.

Souto, S., Palma, P., Riccetto, C., Seabra, A.B., de Oliveira, M.G., Palma, T., et al., 2010. Impact of topic administration of nitric oxide donor gel in the clitoridian blood flow, assessed by Doppler ultra-sound. Acta. Urol. Esp. 34, 708–712.

Stewart, J., McCardle, J., Young, M., Edmonds, M., Kennon, B., Daintree, C., et al., 2014. Poster ProNOx1: improved healing and outcomes for diabetic foot wounds with a nitric oxide generating dressing. In: Wounds UK Annual Conference, Harrogate, UK.

Taladriz-Blanco, T., de Oliveira, M.G., 2014. Enhanced photochemical nitric oxide release from a flutamide derivative incorporated in Pluronic F127 micelles. J. Photochem. Photobiol. A 293, 65–71.

Tucker, A.T., Pearson, R.M., Cooke, E.D., Benjamin, N., 1999. Effect of nitric-oxide-generating system on microcirculatory blood flow in skin of patients with severe Raynaud's syndrome: a randomised trial. Lancet. 354, 1670–1675.

Vercelino, R., Cunha, T.M., Ferreira, E.S., Cunha, F.Q., Ferreira, S., de Oliveira, M.G., 2013. Skin vasodilation and analgesic effect of a topical nitric oxide-releasing hydrogel. J. Mater. Sci: Mater. Med. 24, 2157–2169.

Weller, R., 2003. Nitric oxide: a key mediator in cutaneous physiology. Clin. Exp. Dermatol. 5, 511–514.

Wichterle, O., Lim, D., 1960. Hydrophilic gels for biological use. Nature 185, 117–118.

Yoo, J.W., Acharya, G., Lee, C.H., 2009. In vivo evaluation of vaginal films for mucosal delivery of nitric oxide. Biomaterials 30, 3978–3985.

INDEX

Note: Page numbers followed by "*f*" and "*t*" refer to figures and tables, respectively

A

A549 cells. *See* Lung epithelial cells (A549 cells)
Acetylcholinesterase inhibitors, 145–147
Acetylcholine (ACh), 122–123, 145–147, 245–246
Acid tolerance response (ATR), 178
Acidification, 178
Acidified nitrate, 178
Acinetobacter baumannii, 230
ACR. *See* Antimicrobial agent acriflavine (ACR)
Acute chest syndrome (ACS), 60
Acute inflammation, 78–79
Acute lymphoblastic leukemia (ALL), 99–100
    cell lines, 88
Acute respiratory distress syndrome (ARDS), 60
Acylated homoserine lactones (AHL), 184
AD. *See* Alzheimer's disease (AD)
Adenosine TriPhosphate (ATP), 267–269
Aedes aegypti, 44
Aedes albopictus, 43
Aedes genus, 42–43
AH. *See* Arterial hypertension (AH)
AHAP3. *See* N-(6-aminohexyl)
        aminopropyltrimethoxysilane (AHAP3)
AHL. *See* Acylated homoserine lactones (AHL)
AlbSNO. *See* S-nitrosoalbumin (AlbSNO)
Aldehyde dehydrogenase (ALDH-2), 178, 247
Alkylamines, 205–207
ALL. *See* Acute lymphoblastic leukemia (ALL)
ALS. *See* Amyotrophic lateral sclerosis (ALS)
Alternative pathway (AP), 193–194
Alzheimer's disease (AD), 145–147, 265, 270–275,
        276*f*–277*f*
    NODs in, 282–283
AMA1. *See* Apical membrane antigen 1 (AMA1)
American trypanosomiasis, 35–36
Amino dicarboxylic acid, 133
4-Amino-5-methylamino-2', 7'-difluorofluorescein
        (DAF-FM), 17
3-Aminopropyltrimethoxysilane, 205–207
4-Aminopyridine (4-ampy), 20–21

Aminosilane, 208
Amodiaquine, 27
4-ampy. *See* 4-Aminopyridine (4-ampy)
Amyl nitrite vapors, 178
Amyloid beta (Aβ), 145–147, 270, 274
    amyloid beta-peptide, 145–147
    production, 282
Amyloid precursor protein (APP), 270
Amyotrophic lateral sclerosis (ALS), 145
Anaerobic
    growth, 173
    pathways, 171–172
    processes, 173
    respiration, 183–184
Anaerobic regulation of arginine deiminase and
        nitrate reduction (ANR), 171–172
Androgen receptor (AR), 99
Angina, 178
Angiogenesis, 58
    GSNO influences angiogenesis in rats, 162–163
Anionic liposomes, 232–233
Anopheles albimanus, 44
ANR. *See* Anaerobic regulation of arginine
        deiminase and nitrate reduction (ANR)
Anti-*Burkholderia mallei* activity, 174
Antibacterial agent in bacteria, 223–225
Antibacterial properties, 175–176
Antibiofilm agent, NO as, 173–175
Antibiotic(s)
    NO hybrid antimicrobials based on, 182–183,
        183*f*
    resistance, 303–304
    synergic effect of NO with, 304–307
Anticancer drugs, 182, 293–294
    hybrids, 299–303
    NO and anticancer drug cotreatment, 297–299,
        298*f*
Antidengue virus activity, 42–43
Antifungal agents, 182–183
Antihypertensive activity, 182

Antimalarial drugs, 28–29
Antimicrobial
    actions of NO, 46
    mechanisms, 174
    properties, 178, 180–183
    treatments, 175
Antimicrobial agent acriflavine (ACR), 295–296
Antimicrobial agents
    diazeniumdiolates, 179–181, 179f
    dual action NO
        antimicrobials based on QS inhibitors,
            182–183
        donors as antimicrobials, 181–184
    nitrates, 177t, 178
    nitric oxide, 169
    NO, 173–175
        as antimicrobial and antibiofilm agent,
            173–175
        biological effects, 170–171
        biosynthesis, 169–170, 170f
        donors as antimicrobials, 176–181
        gas, 175–176
        hybrid antimicrobials based on antibiotics,
            182–183, 183f
        inactivation chemistry, 170
        regulation in microbes, 171–173, 172f
    S-NO, 181
Antioxidant response element (ARE), 91
Antipathogenic response, 75–76
AOM. See Azoxymethane (AOM)
AP. See Alternative pathway (AP)
Apical membrane antigen 1 (AMA1), 31
Apoptosis, 58
APP. See Amyloid precursor protein (APP)
AR. See Androgen receptor (AR)
ARDS. See Acute respiratory distress syndrome
        (ARDS)
ARE. See Antioxidant response element (ARE)
Arginase, 58
Arginine, 58–59, 76
Artemisinin-combined therapies, 27
Arterial hypertension (AH), 248–251
Asialoglycoprotein receptors (ASGPR), 100
Aspirin, 96, 182
"Aspirin-like" compounds, 81–84, 84f
Astrocytes cells, 279
Atherosclerosis, 57
ATN. See N-(3-aminopropyl)-3-(trifluoromethyl)-
        4-nitrobenzenamine (ATN)

Atovaquone, 27
ATP. See Adenosine TriPhosphate (ATP)
ATP-sensitive K+ channels (KATP), 256
ATR. See Acid tolerance response (ATR)
Autoinducers, 173
Azoxymethane (AOM), 89
Aβ. See Amyloid beta (Aβ)

B
B16F10 tumor cell line, 20–21, 22f
Baby hamster kidney (BHK-21), 46
BACE. See Beta-site APP cleaving enzyme
        (BACE)
Bacillus anthracis nitric oxide synthase (bNOS),
        224–225
Bacillus subtilis, 173
Bacteria, 171, 296
    antimicrobial effects against Gram-positive and
        Gram-negative, 304
    by bNOS, NO production in, 295–296
    infection, 178
    life cycle, 171
    NO in, 294–296, 303f
    in self-produced matrix, 305–307
Bacterial NOS (bNOS), 294–296
Bacteriocidal activity, 175–176
BBB. See Blood brain barrier (BBB)
BCF. See Forskolin (BCF)
BCRP-1. See Breast cancer resistance proteins
        (BCRP-1)
BdlA chemotaxis protein, 223–224
BDNF. See Brain-derived neurotrophic factor
        (BDNF)
Benznidazole, 36–37
Benzofuroxans, 177f
β-catenin signaling, 89
β-galactosidase-responsive caged NO molecule,
        326–327
β-lactam unit, 182–183
β-lactamase penicillinase, 182–183
Beta-site APP cleaving enzyme (BACE), 270
BHK-21. See Baby hamster kidney (BHK-21)
Bilharzia. See Schistosomiasis
Binuclear ruthenium system, 14
Biochemical pathway, 170–171
Biofilm(s), 67, 171, 173–174, 305–307. See also
        Implanted sensors
    application of gNO, 221
    bactericidal effect, 222

delivery of NO via small molecular weight
      donors, 234–236
dispersal agents, 303–304
dual function of NO, 223–225
eradication, 233–234
life cycle, 183–184
NO, 221
      signaling on dispersal, 225f
NO donors
      control by, 222f
      controlling of clinical interest, 226–234
      NO delivery using nanoparticles, 227–234
      NO liposomes, 232–233
      NO releasing dendrimers, 233–234
      NO releasing polysaccharides, 229–232
nontraditional antibiotic agents, 223
reduction, 231
Biofouling, 229–230, 235–236
Biological
      actions of spacer in NO-ASA, 93
      effects of NO, 170–171
      functions, 169, 176
      pathways, 171
      response to implanted sensors, 192–196
            to polymers in blood, 193–194
            to polymers in subcutaneous tissue, 195–196
Biomedical devices, NO releasing, 60–61
Biomolecules, 20–21, 171
Biosynthesis of NO, 169–170, 170f
2, 2'-Bipyridine (bpy), 20–21
3, 3-bis(aminoethil)-1-hydroxy-2-oxo-1-triazene
      (NOC-18), 149–150
      Hydra regeneration studies on NO-cGMP
            pathway with, 149–150
      locust studies on NO-cGMP with, 151–152
(Bis[amino[(2-aminophenyl)thio]methylene]
      butanedinitrile), 99
Blastocystis hominis, 25
      infection, 44–45
      NO donors and, 44
Blood
      biological response to polymers in, 193–194
      vessels, 58
Blood brain barrier (BBB), 272
bNOS. See Bacillus anthracis nitric oxide synthase
      (bNOS)Bacterial NOS (bNOS)
BOP. See N-nitrosobis(2-oxopropyl)amine (BOP)
bpy. See 2, 2'-Bipyridine (bpy)
Brain, 171

Brain damage, NO in, 265–275
      AD, 270–275, 276f–277f
      ischemia–reperfusion injury, 267–270, 271f
      TBI, 265–267, 268f
Brain inflammation. See also Hypertension
      neuroinflammation and NO in damage brain,
            265–275
      NO in brain physiology, 263–264
      NODs as neuroprotective agents in
            in AD, 282–283
            in cerebral ischemia–reperfusion injury, 275–280
            in TBI, 280–282
Brain-derived neurotrophic factor (BDNF), 157,
      162, 281
Breast cancer resistance proteins (BCRP-1),
      293–294
1-Bromoacetyl-3, 3-dinitroazetidine (ABDNAZ).
      See RRx-001
Brugia malayi, 45
BSO. See Buthionine-(S,R)-sulfoxime (BSO)
BTMOS. See Isobutyltrimethoxysilane (BTMOS)
Buthionine-(S,R)-sulfoxime (BSO), 103–104,
      297–299
(Z)-1-[N-Butyl-N-[6-(N-methylammoniohexyl)
      amino]]-diazen-1-ium-1, 2-diolate
      (DBHD/N₂O₂), 198, 204–205
BxPC-3 pancreatic cancer, 89–90

C

c-di-GMP. See Cyclic diguanosine monophosphate
      (c-di-GMP)
C-Nitrosoamines, 177f
c-PTIO, 302
C-Rel, 266–267
C57Bl/6 mice, 30
CAA. See Cerebral amyloid angiopathy (CAA)
Calpains, 281
Cancer
      cell lines, 302–303
      cells to cytotoxic agents, 296
      cellular actions of NO, 76–79
      chemotherapy agents, 297–299
      colon cancer cells, 297–299
      diazeniumdiolate-based NO-releasing
            compounds, 94–97, 94f
      HNO-NSAIDs, 97–100
      NO-ASA
            biological actions of spacer in, 93
            molecular targets in cancer, 88–93

Cancer (*Continued*)
  NO-NSAIDs
    cancer prevention with, 86–93
    and rationale for development, 80–81
    structural features of nitrate NO-NSAIDs, 81–84, 82*f*
  NO-releasing coxibs, 84–86, 85*f*
  NONO-NSAIDs, 95–97, 95*f*
  NOSH-NSAIDs, 100–102, 101*f*, 103*f*
  NSAIDS
    protecting against cancer, 79
    side effects associated with, 79–80
  relationship with NO and, 293–294
  RRx-001 aerospace compound for cancer treatment, 102–105
  therapy, 1–2
Capsular relaxation, 57
Carbon monoxide, 221
Carboplatin (CPT), 98
Carcinogenic air pollutant, 169
Cardiopulmonary bypass (CPB), 60
Cardiovascular diseases, 182, 248, 257
Catalytic generation from endogenous RSNOs, 213–215
CBF. *See* Cerebral blood flow (CBF)
CCI model. *See* Controlled cortical impact model (CCI model)
*cd*₁-dNir. *See* Cytochrome nitrite reductase (*cd*₁-dNir)
Ceftazidime, 175–176
Cell kinetics, 88
Cell toxicity, 153–155
Cellular actions of NO, 76–79
  cancer as inflammatory disease, 76–79
Cellular imaging, 17–21
  fluorescence microscopy images of B16F10 cells, 18*f*
  liposome delivery, 19–21
  NO detection and ROS/RNS generation, 17*f*
Cellular inflammation, 58
Cellular systems, 170
Cellulose nanocrystals (CNC), 231–232
Central atom, 1
Central nervous system (CNS), 141–142, 263
  inflammation, 263
Cephalosporin, 182–183
Cephalosporin-3′-diazeniumdiolate, 182–183
Cephalosporin-Furoxan, 183*f*
Cerebral amyloid angiopathy (CAA), 272

Cerebral blood flow (CBF), 265
Cerebral hypoperfusion, 272
Cerebral ischemia–reperfusion injury, NODs in, 275–280
Cerebral malaria (CM), 27
CF. *See* Cystic fibrosis (CF)
cfu. *See* Colony forming units (cfu)
cGK-I. *See* cGMP-dependent kinase (cGK-I)
cGMP. *See* Cyclic guanosine monophosphate (cGMP)
cGMP-dependent kinase (cGK-I), 178, 252
Chagas´ disease, 35–36
Chemokines, 78–79
Chemopreventive agents, 79
Chemotherapeutic drugs, 37
Chemotherapy, 32
Chitin, 67–68
Chitosan, 67–70
  chitosan/TPP nanoparticles, 39
  copper chitosan, 70
  oligosaccharides, 230
  topical copper chitosan particles, 70–71
Chloroquine, 27
Ciliary neurotrophic factor (CNTF), 157
Cinaciguat, 133
Ciprofloxacin, 182–183
cis-[Ru(bpy)₂(SO₃) (NO)]PF-6-9 (FONO), 132
cis-[RuCl(bpy)₂(NO)](PF₆)₂, 1–2
Cisplatin, 1–2
L-Citruline, 76
CL. *See* Cutaneous leishmaniasis (CL)
CM. *See* Cerebral malaria (CM)
CNC. *See* Cellulose nanocrystals (CNC)
cNOS. *See* Constitutive NOS (cNOS)
CNS. *See* Central nervous system (CNS)
CNTF. *See* Ciliary neurotrophic factor (CNTF)
Coagulation, 193
Colon cancer cells (HT29 cells), 297–299
Colony forming units (cfu), 42
Colorectal adenomas, 79
Colorectal cancer risk, 76
"Complex", in the context of chemistry, 1
Compliment system, 193–194
Constitutive enzymes, 169–170
Constitutive NOS (cNOS), 55–56, 76
Controlled cortical impact model (CCI model), 160
Converted chitin, 68

Copper (Cu), 56–57
  Cu(I), 56–57
  Cu(II), 56–57
  particles, 214–215
Copper chitosan, 70
  catalytic cycle, 71*f*
Copper-containing nitrite reductase (Cu-dNir), 172–173
Copper-nitrosyl (Cu(I)-NO$^+$), 56–57
Corpora cavernosa, 122–125
  relaxation responses, 132
Covalently attached NO donors, hydrogels with, 314, 323–324
COX inhibition. *See* Cyclooxygenase inhibition (COX inhibition)
COX-2 expression, 275, 279–280
COX-inhibiting nitric oxide donators (CINODs). *See* Nitric oxide-releasing NSAIDs (NO-NSAIDs)
coxibs. *See* Cyclooxygenase-2 inhibitors (coxibs)
CPB. *See* Cardiopulmonary bypass (CPB)
CPT. *See* Carboplatin (CPT)
CREB. *See* Cyclic AMP-responsive-element-binding protein (CREB)
Cross-tolerance phenomenon, 250
Cu-dNir. *See* Copper-containing nitrite reductase (Cu-dNir)
Cu(II) in dibenzo[e,k]-2, 3, 8, 9-tetraphenyl-1, 4, 7, 10-tetraaza-cyclododeca-1, 3, 7, 9-tetraene (Cu(II)-DTTCT), 214
Cutaneous leishmaniasis (CL), 32
CV. *See* Cyclic voltammograms (CV)
Cyanide ion (CN$^-$), 1–2, 129
Cyclic AMP-responsive-element-binding protein (CREB), 144
Cyclic diguanosine monophosphate (c-di-GMP), 174, 223–224
Cyclic guanosine monophosphate (cGMP), 28, 57, 76–78, 123–124, 141–142, 144, 170–171, 243–244, 264
  pathway, 141–142, 142*f*
Cyclic voltammograms (CV), 5
Cyclicnucleotide-gated ion channels, 170–171
Cyclohexane nitrate, 257
Cyclooxygenase inhibition (COX inhibition), 80–81
Cyclooxygenase-2 (COX-2), 266
Cyclooxygenase-2 inhibitors (coxibs), 84–86
Cyclosporine A, 127–129
Cysteine and *S*-nitrosocysteine (CysNO), 213–214

Cysteine proteases, 46
Cysteine thiol groups, 171
Cystic fibrosis (CF), 175, 305–307
Cytochrome nitrite reductase (cd_1-dNir), 172–173
Cytokines, 56, 78–79
Cytostatic effects, 173
Cytotoxic effects, 171, 173
  of NO-DOXO, 299–300
Cytotoxicity, 32–33
  mechanisms of NO donor agents, 17–21
    fluorescence microscopy images of B16F10 cells, 18*f*
    liposome delivery, 19–21
    NO detection and ROS/RNS generation, 17*f*

D
DA. *See* Dopamine (DA)
DAF-FM. *See* 4-Amino-5-methylamino-2', 7'-difluorofluorescein (DAF-FM)
Daunorubicin (DNR), 297–299
DBHD/N$_2$O$_2$. *See* (Z)-1-[N-Butyl-N-[6-(N-methylammoniohexyl)amino]]-diazen-1-ium-1, 2-diolate (DBHD/N$_2$O$_2$)
DEA NONOate. *See* Diethylamine diazeniumdiolate (DEA NONOate)
DEA NONOate-cephalosporin prodrug (DEACP), 182–183, 183*f*
DEA/NO-aspirin. *See* O^2-(acetylsalicyloyloxyme thyl)-1-(N,N-diethylamino)-diazen-1-ium-1, 2-diolate (DEA/NO-aspirin)
DEACP. *See* DEA NONOate-cephalosporin prodrug (DEACP)
Deinococcu radiodurans, 224–225
Dendrimers, 66
  NO releasing, 233–234
Dengue fever (DF), 25, 42–44
  NO donors and, 43–44
Denitrification, 56
  activity, 173
  enzymes, 171–172
  pathway, 56
  process, 171–172
Dermal vasodilation, 319–320, 322
DETA/NO. *See* Diethylenetriamine NONOate NO adduct (DETA/NO)
DF. *See* Dengue fever (DF)
Diabetics, 59

Diazeniumdiolates, 64, 65*f*, 96–98, 176, 177*f*,
    179–181, 179*f*, 305–307
  diazeniumdiolate-based NO-releasing
      compounds, 94–97, 94*f*
  as NO donors for sensor coatings, 198–210
    diazeniumdiolated *N*-(6-aminohexyl)-
        3-minopropyltrimethoxysilane, 201*f*
    hybrid sol–gel/PU glucose sensor, 209*f*
    intravascular oxygen sensors, 206*f*
    NO releasing oxygen sensor, 203*f*
    sensor configuration, 204*f*
  protected diazeniumdiolates, 179*f*
  spermine, 180–181
Dichotomous role, 75–76
Diethylamine diazeniumdiolate (DEA NONOate),
    182–183
Diethylenetriamine NONOate NO adduct
    (DETA/NO), 42, 45, 180–181
Dihydroartemisinin scaffold, 28–29
Dinitrogen trioxide (N_2O_3), 26, 46
Dinitrosyl iron complex (DNIC), 76–78
Dioctyl sebacate (DOS), 198
Dipropylenetriamine NONOate (*DPTA/NO*), 29
Disinfectants, 67
Dissimilatory nitrate respiration regulator (DNR
    regulator), 171–172
Dissolved molecular NO donors, hydrogels with,
    314–323
  chemical structure of PEO–PPO–PEO, 319*f*
  dissolution of low molecular weight NO donors,
      314–318
  low molecular weight NO donors, 318–319
  Pluronic F127 hydrogel without GSNO, 321*f*
  topical administration of hydrogels, 320
  topical application of NO-releasing hydrogels,
      323
DMHD/N_2O_2. *See* (*Z*)-1-[*N*-Methyl-*N*-[6-
    (*N*-methylammoniohexyl)amino]]-diazen-
    1-ium-1, 2-diolate (DMHD/N_2O_2)
DNA
  alkyl transferases, 174
  base damage, 58
  fragmentation, 75–76
  repair enzymes, 174
  replication, 174
DNIC. *See* Dinitrosyl iron complex (DNIC)
DNR. *See* Daunorubicin (DNR)
DNR regulator. *See* Dissimilatory nitrate
    respiration regulator (DNR regulator)

Dodecyl-modified dendrimers, 233
Donor carriers, 65–67
Dopamine (DA), 142–143
DOS. *See* Dioctyl sebacate (DOS)
DOX. *See* Doxorubicin (DOX)Drug doxorubicin
    (DOX)
Doxorubicin (DOX), 33–34, 299–300
DPTA/NO. *See* Dipropylenetriamine NONOate
    (*DPTA/NO*)
Drug
  carriers, 65
  compartmentalization, 18
  delivery, 313
  development, new insights on, 253–257
  efflux, 293–294, 296
Drug doxorubicin (DOX), 297–299
"Dynamite", 245

E

EC. *See* Endothelial cells (EC)
ECMO. *See* Extracorporeal membrane oxygenation
    (ECMO)
ED. *See* 1, 2-Epoxy-9-decene (ED)Erectile
    dysfunction (ED)
EDL. *See* Extensor digitorum longus (EDL)
EDRF. *See* Endothelium-derived relaxing factor
    (EDRF)
EGDE. *See* Ethylene glycol diglycidyl ether (EGDE)
EGFR. *See* Epidermal growth factor receptor
    (EGFR)
EHBPs. *See* Esterprotected hydroxy benzyl
    phosphates (EHBPs)
Electrospinning technique, 33–34
Elephantiasis tropica. *See* Lymphatic filariasis
Endogenous RSNOs, catalytic generation from,
    213–215
Endothelial cells (EC), 121, 122*f*
Endothelial dysfunction, 132, 248–249, 272
Endothelial inflammatory activation, 121
Endothelial nitric oxide synthase (eNOS), 26,
    55–56, 60, 62–63, 76, 122, 124*f*, 169–170,
    243, 263–264, 303–304
Endothelial-NO deficiency, 272–273
Endothelium, 57
Endothelium-derived relaxing factor (EDRF), 55,
    141–142, 245–246
Endotoxins, 56
eNOS. *See* Endothelial nitric oxide synthase
    (eNOS)

Enteropathogens, 178
Enzymatic bioactivation, 247
Enzymatic mechanism, 178
Enzymatic process, 56–57
Epidermal growth factor receptor (EGFR),
    104–105
1, 2-Epoxy-9-decene (ED), 233–234
EPS. *See* Extracellular polymeric substances (EPS)
ER. *See* Estrogen receptor (ER)
Eradication, 230
Erectile dysfunction (ED), 121–122
    and prevalence, 126
Erectile function, NO donors for, 127, 128*t*
Erectile physiology, NO in, 123–126
    NO/cGMP signaling cascade, 125*f*
Erection, 124–125, 131
ESBL. *See* Extended-spectrum β-lactamase
    (ESBL)
Escherichia coli, 173, 230
Esterprotected hydroxy benzyl phosphates
    (EHBPs), 93
Estrogen receptor (ER), 97–98
Ethylene glycol diglycidyl ether (EGDE), 69–70
Ethyltrimethoxysilane (ETMOS3), 208
Eukaryotes, 221
Eukaryotic cells, 180–181
Exogenous calmodulin, 55–56
Exogenous gNO, 175–176
Exogenous NO, 171
Exogenous SNAC on motor functional recovery in
    rats, 158–159
Extemporaneous NO release
    gels based on, 325–326
    hydrogels based on, 314
Extended-spectrum β-lactamase (ESBL), 180–181
Extensor digitorum longus (EDL), 158–159
Extracellular polymeric substances (EPS), 226
Extracorporeal membrane oxygenation (ECMO),
    60

F
F344 rats, 75–76
FAD. *See* Flavine adenine dinucleotide (FAD)
FAL. *See* Formaldoxime (FAL)
FAM. *See* Formamidoxime (FAM)
FBGCs. *See* Foreign body giant cells (FBGCs)
Fenton reaction, 58
Ferric uptake regulator (Fur), 173
Fibrin network, 193

Fibrinogen, 193
Fick's Law, 231–232
Flavine adenine dinucleotide (FAD), 55–56
Flavine mononucleotide (FMN), 55–56
Flavonoids, 42–43
Flurbiprofen benzyl nitrate, 89
5-Flurouracil (5-FU), 87–88
FMN. *See* Flavine mononucleotide (FMN)
Focal cerebral ischemia, 277–278
FONO. *See cis*-[Ru(bpy)$_2$(SO$_3$) (NO)]PF-6-9
    (FONO)
Foreign body giant cells (FBGCs), 195–196
Formaldoxime (FAL), 131
Formamidoxime (FAM), 131
Forskolin (BCF), 157
Framingham Heart study, 76
Free radical gas, 169
Free radical NO (NO^0), 58
Free radicals, 76–78
FT-IR spectra, 69
5-FU. *See* 5-Flurouracil (5-FU)
Fur. *See* Ferric uptake regulator (Fur)
Furoxans, 28, 177*f*

G
Ganglion cell layer (GCL), 152–153
GAP-43. *See* Growth associated protein-43
    (GAP-43)
GAPDH. *See* Glyceraldehyde 3-phosphate
    dehydrogenase (GAPDH)
Gaseous NO (gNO), 175–176, 221
Gastric bactericidal functions, 178
Gastrointestinal system (GI system), 80–81, 178
    homeostasis, 84–86
GCL. *See* Ganglion cell layer (GCL)
Gels, 313
Genes, 171–172
    deletion of, 173
    expression, 173
Genetic regulator in bacteria, 223–225
Gentamicin, 305–307
GFP-transfected MDA-MB-231 cells, 97–98
GI system. *See* Gastrointestinal system
    (GI system)
Gibbs free energy, 3
D-Glucosamine, 68
Glucose oxidase (GOx), 208–209
Glutamatergic neurons, 142–143
Glutathione (GSH), 94, 161

Glutathione *S*-transferases (GSTs), 91, 98
Glyceraldehyde 3-phosphate dehydrogenase
    (GAPDH), 38–39, 97–98
Glycerin, 253, 253*f*
Glyceryl trinitrate (GTN), 130, 178, 179*f*, 245–246
Glyceryl trinitrate-induced apoptosis, 75–76
Glycoprotein Ib (GPIb), 193
gNO. *See* Gaseous NO (gNO)
Goldfish optic nerve regeneration studied with
    NOR-2 and SNAP, 152–155, 154*f*
GOx. *See* Glucose oxidase (GOx)
GPIb. *See* Glycoprotein Ib (GPIb)
Gram-negative bacteria, 175, 184, 225
Gram-negative pathogen, 181
Gram-positive bacteria, 225
Gram-positive pathogen, 181
Gram-positive *Streptococcus mutans*, 233
Growth associated protein-43 (GAP-43), 144
GSH. *See* Glutathione (GSH)
GSK-3β activity, 283
GSNO, 280–283
    influences angiogenesis in rats, 162–163
GSTs. *See* Glutathione *S*-transferases (GSTs)
GTN. *See* Glyceryl trinitrate (GTN)
Guanosine triphosphate (GTP), 28, 170–171,
    196–197
Guanylate cyclase, 178

H

H&E. *See* Hematoxylin and eosin (H&E)
Halofantrine, 27
HCC cells. *See* Hepatocellular carcinoma cells
    (HCC cells)
HD. *See* Huntington's disease (HD)
Head and neck squamous carcinoma cell (HNSCC
    cell), 297–299
HED. *See* Human equivalent doses (HED)
Helicobacter pylori, 183
2–HEMA. *See* O²–acetoxymethyl–1–[*N*–(2–
    hydroxyethyl)–*N*–methylamino]diazen–1–
    ium–1, 2–diolate (2–HEMA)
HEMA monomers. *See* Hydroxyethyl methacrylate
    monomers (HEMA monomers)
Hematoxylin and eosin (H&E), 162
Heme oxygenase-1 (HO-1), 29, 144
Hemenitric oxide/oxygen (H-NOX), 174
    H-NOX-diguanylate cyclase/phosphodiesterase
        protein, 174
Hepatocellular carcinoma cells (HCC cells), 100
Heterodimeric enzyme, 170–171

HIF-1. *See* Hypoxia-inducible factor-1 (HIF-1)
High blood pressure, 257
HIV. *See* Human immunodeficiency virus (HIV)
HMGCoA. *See* 3-Hydroxy-3-methyl-glutaryl-
    CoA (HMGCoA)
4-HNE. *See* 4-Hydroxynonenal (4-HNE)
HNO-NSAIDs, 97–100
    JS-K and PABA/NO, 98–100
H-NOX. *See* Hemenitric oxide/oxygen (H-NOX)
HNSCC cell. *See* Head and neck squamous
    carcinoma cell (HNSCC cell)
HO-1. *See* Heme oxygenase-1 (HO-1)
HOCCs. *See* Human ovarian cancer cells
    (HOCCs)
Host defense, 26, 40, 44
HPU. *See* Hydrophilic polyurethane (HPU)
HT29 cells. *See* Colon cancer cells (HT29 cells)
Human brain injuries, 160
Human cavernosal smooth muscle cells, 129
Human colon cancer cell lines, 87–88
Human equivalent doses (HED), 87–88
Human immunodeficiency virus (HIV), 41
Human ovarian cancer cells (HOCCs), 297–299
Huntington's disease (HD), 145, 147–148
HUVECs, 97–98
Hybrid drugs, 182
Hydra regeneration studies on NO-cGMP pathway
    with NOC-18, 149–150
Hydra vulgaris, 149–150
Hydrogelators, 326
Hydrogels, 231–232, 313
    with covalently attached NO donors, 323–324,
        324*f*
    with dissolved molecular NO donors, 314–323
        chemical structure of PEO–PPO–PEO, 319*f*
        dissolution of low molecular weight NO
            donors, 314–318
        low molecular weight NO donors, 318–319
        Pluronic F127 hydrogel without GSNO, 321*f*
        topical administration of hydrogels, 320
        topical application of NO-releasing hydrogels,
            323
    gels based on extemporaneous NO release,
        325–326
    as reservoirs of drugs for topical delivery, 313
    supramolecular NO–releasing hydrogels, 326–
        328, 327*f*
    for topical NO delivery, 314
Hydrogen peroxide (H₂O₂), 26, 129, 208–209, 322
Hydrogen sulfide (H₂S), 100–101, 221

Hydrolysable Schiff base linkage, 305–307
Hydrophilic polyurethane (HPU), 208
3-Hydroxy-3-methyl-glutaryl-CoA (HMGCoA), 182
(8-[2-Hydroxy-3-(isopropylamino)propoxy]-3-chromanol, 3-nitrate), 155–156
Hydroxybutyl(butyl) nitrosamine model (OH-BBN model), 87–88
Hydroxyethyl methacrylate monomers (HEMA monomers), 322
[(*E*)-Hydroxyimino]-5-nitro-3-hexenamide (NOR-3), 29
Hydroxyl radicals (OH•), 26, 174
Hydroxylamines, 177*f*
4-Hydroxynonenal (4-HNE), 160
Hypertension. *See also* Brain inflammation
   AH, 248–251
   historical perspective, 245–246
   new insights on drug development, 253–257
   nitrate tolerance, 251–253
   organic nitrates, 246–251
   physiological NO production, 244*f*
Hypoperfusion, 266
Hypoxia-inducible factor-1 (HIF-1), 162–163

I

IBSs. *See* Inflammatory bowel diseases (IBSs)
Ibuprofen, 96
ICUs. *See* Intensive care units (ICUs)
IHC. *See* Immunohistochemistry (IHC)
IL. *See* Interleukin (IL)
Immunohistochemistry (IHC), 160
Implanted sensors, 191–192. *See also* Biofilm(s)
   biological response to, 192–196
      to polymers in blood, 193–194
      to polymers in subcutaneous tissue, 195–196
   NO, 196–197
   NO donors and generators, 197–215
      catalytic generation from endogenous RSNOs, 213–215
      development of diazeniumdiolates as, 198–210
      RSNOs, 210–212
In vivo sensor, 195–196
Indomethacin, 96
Inducible nitric oxide synthase (iNOS), 26, 55–56, 75–76, 122, 169–170, 243, 263–264, 270, 303–304
INF-γ cytokine. *See* Interferon gamma cytokine (INF-γ cytokine)
Inflammatory bowel diseases (IBSs), 56

Inflammatory disease, cancer as, 76–79
Inhalable PLGA microparticles, 42
Inhibitory mechanism, 38–39
Inorganic nitroso compounds, 65*f*
iNOS. *See* Inducible nitric oxide synthase (iNOS)
Inositol 1, 4, 5-triphosphate receptor (IP3 receptor), 243–244
Intensive care units (ICUs), 191–192
Interferon gamma cytokine (INF-γ cytokine), 30
Interleukin (IL), 270
   IL-1β, 270
   IL-6, 267
   IL-10, 36
Intestinal cells, 44
Intracellular calcium, 55–56
Intravascular glucose/lactate sensors, 207–208
Intravascular oxygen sensors, 206*f*
Ion-selective electrode (ISE), 198
IP3 receptor. *See* Inositol 1, 4, 5-triphosphate receptor (IP3 receptor)
IPA/NO-aspirin. *See* O^2-(acetylsalicyloyloxymethyl)-1-(*N*-isopropylamino)-diazen-1-ium-1, 2-diolate (IPA/NO-aspirin)
I/R injury. *See* Ischemia/reperfusion injury (I/R injury)
Iron, 56
Iron-limiting conditions, 173
Ischemia/reperfusion injury (I/R injury), 158–159, 267–270, 271*f*
   GSNO effects after I/R injury in rats, 159–162
ISDN. *See* Isosorbide dinitrate (ISDN)
ISE. *See* Ion-selective electrode (ISE)
ISMN. *See* Isosorbide mononitrate (ISMN)
Isobutyltrimethoxysilane (BTMOS), 227, 229
Isoenzymes, 170–171
Isoniazid, 41
Isosorbide dinitrate (ISDN), 34, 130, 178, 179*f*, 246
Isosorbide mononitrate (ISMN), 42, 178, 179*f*, 232–233, 246
Isosorbite dinitrate, 75–76
Isosorbite mononitrate, 75–76

J

Japanese encephalitis virus (JEV), 46
JS-K. *See* O^2-(2, 4-Dinitrophenyl)1-[(4-thoxycarbonyl)piperazin-1-yl]diazen-1-ium-1, 2-diolate (JS-K)

K

K562 cells. *See* Myelogenous leukemic cells (K562 cells)
KATP. *See* ATP-sensitive K^+ channels (KATP)
235-kDa rhoptry protein (Py235), 31
Ketoconazole–NO donor hybrid drugs, 183
KTpClPB. *See* Potassium tertakis(4-chlorophenyl) borate (KTpClPB)

L

Lactate dehydrogenase (LDH), 299–300
Laser Doppler measurements, 322
LDH. *See* Lactate dehydrogenase (LDH)
LDL. *See* Low-density lipoprotein (LDL)
Leishmania species, 32, 181
Leishmaniasis, 32–35
    NO and, 32
    NO donors and, 32–35, 35*f*
Lewis acid, 1
LFB. *See* Luxol fast blue (LFB)
Lipid peroxidation, 174
Liposome delivery, 19–21, 19*f*, 20*f*
    overlap of fluorescence microscopy on B16F10 cells, 22*f*
    in vitro dark toxicity, 21*f*
Liposomes, 66
    NO, 232–233
Litomosoides sigmodontis, 45
LLNO1. *See Trans*-[Ru(NH$_3$)$_4$ (caffeine)(NO)]C1$_3$ (LLNO1)
Locust studies on NO-cGMP with NOC-18 and SNP, 151–152
Long-term depression (LTD), 143
Long-term potentiation (LTP), 143
Low-density lipoprotein (LDL), 125
Lung epithelial cells (A549 cells), 297–299
Luxol fast blue (LFB), 162
Lymphatic filariasis, 45
Lymphatic vessel dysfunction, 61

M

m-NO-ASA, 91
MAC. *See* Membrane attack complex (MAC)
Macrophages, 59, 195–196
    M1, 59
    M2, 59
MAHMA NONOate, 226, 231–232, 234–235, 235*f*

Malaria, 25, 27–32
    NO and, 27–28
    NO donors and, 28–31, 31*f*
MALES study. *See* Men's Attitudes to Life Events and Sexuality study (MALES study)
Manganese SuperOxide Dismutase (MnSOD), 266
MAPK. *See* Mitogen-activated protein kinase (MAPK)
Massachusetts Male Aging Study (MMAS), 126
Mast cells (MCs), 195
Matrix metalloproteinase 9 (MMP9), 147
Maximum relaxing effect (ME), 253–254, 254*t*
MB. *See* Methylene blue (MB)
MCAO. *See* Middle cerebral artery occlusion (MCAO)
MCF-7 breast cancer cell lines, 89–90
MCs. *See* Mast cells (MCs)
MDR-tuberculosis, 41
ME. *See* Maximum relaxing effect (ME)
2-ME. *See* 2-Methoxyestradiol (2-ME)
Medicinal inorganic chemistry, 1–2
Mefloquine, 27
Melanoma (M21), 103–104
Membrane attack complex (MAC), 193–194
Membrane–bound cytochrome *bc* complex, 172–173
Men's Attitudes to Life Events and Sexuality study (MALES study), 126
3-Mercapto-1, 2-propanediol, 227
Mercaptopropionic acid (MPA), 20–21
Mercaptosuccinic acid (MSA), 39, 227, 323
Merozoites surfaces protein 1 (MSP 1), 31
Messenger RNA (mRNA), 280
Metabolite NO, 178
Metal ligand charge transfer band (MLCT), 12–13
Metal-based drugs, 1–2
Metal–NO complexes, 177*f*
Methemoglobin, 58
2-Methoxyestradiol (2-ME), 162–163
(*Z*)-1-[*N*-Methyl-*N*-[6-(*N*-methylammoniohexyl) amino]]-diazen-1-ium-1, 2-diolate (DMHD/N$_2$O$_2$), 198
Methylene blue (MB), 322
Methylprednisolone (MP), 158–159
Metronidazole, 183
MIC. *See* Minimum inhibitory concentration (MIC)

MIC 50%. *See* Minimum inhibitory concentration 50% (MIC 50%)

Micelles, 66

Miconazole, 180–181

Microarray studies, 223–224

Microbes, NO regulation in, 171–173, 172*f*

Microbial species, 56

Microglial cells, 279

Middle cerebral artery occlusion (MCAO), 162–163

Minimum inhibitory concentration (MIC), 305–307

Minimum inhibitory concentration 50% (MIC 50%), 180–181

Mitochondria, 18–19

Mitochondrial energy production, 169–170

Mitochondrial NOS (mNOS), 169–170

Mitochondrial permeability transition pore (MPTP), 277–278

Mitogen-activated protein kinase (MAPK), 88, 90–91

ML. *See* Mucosal leishmaniasis (ML)

MLCK. *See* Myosin light chain kinase (MLCK)

MLCP. *See* Myosin light chain phosphatase (MLCP)

MLCT. *See* Metal ligand charge transfer band (MLCT)

MLV. *See* Multilamellar vesicles (MLV)

MMAS. *See* Massachusetts Male Aging Study (MMAS)

MMP9. *See* Matrix metalloproteinase 9 (MMP9)

mNOS. *See* Mitochondrial NOS (mNOS)

MnSOD. *See* Manganese SuperOxide Dismutase (MnSOD)

mNSS. *See* Modified neurological severity scores (mNSS)

Modified neurological severity scores (mNSS), 163

Molecular mass, 169

Molecular targets of NO-ASA in cancer, 88–93

Molsidomine, 234–235, 235*f*

MOM-protected diazeniumdiolate, 179*f*, 180

3-Morpholino-sydnonimine (SIN-1), 39, 65*f*, 160

3-Morpholinosydnonimine stimulates guanylate cyclase, 130

MP. *See* Methylprednisolone (MP)

MPA. *See* Mercaptopropionic acid (MPA)

MPTP. *See* Mitochondrial permeability transition pore (MPTP)

mRNA. *See* Messenger RNA (mRNA)

MRPs. *See* Multidrug resistance related proteins (MRPs)

MS. *See* Multiple sclerosis (MS)

MSA. *See* Mercaptosuccinic acid (MSA)

MSP 1. *See* Merozoites surfaces protein 1 (MSP 1)

Mucosal leishmaniasis (ML), 32

Multidrug resistance (MDR), 293–294

Multidrug resistance related proteins (MRPs), 293–294

Multidrug resistant (MDR), 41

Multilamellar vesicles (MLV), 232–233

Multiple sclerosis (MS), 145

Mycobacterium tuberculosis, 41

Myelogenous leukemic cells (K562 cells), 297–299

Myosin light chain kinase (MLCK), 249

Myosin light chain phosphatase (MLCP), 243–244

N

N-(2-aminoethyl)-3-aminopropyltrimethoxysilane, 205–207

N-(2-aminoethyl)-3-aminopropyltrimethoxysilane/ isobutyltrimethoxysilane xerogel, 210

N-(3-aminopropyl)-3-(trifluoromethyl)-4-nitrobenzenamine (ATN), 322

N-(6-aminohexyl)-3-aminopropyltrimethoxysilane, 201–202, 205–207

N-(6-aminohexyl)aminopropyltrimethoxysilane (AHAP3), 208, 229

(Na₂ [Fe(CN))₅(NO)])). *See* Sodium nitroprusside (SNP)

NAC. *See* *N*-acetylcysteine (NAC)

NAc. *See* Nucleus accumbens (NAc)

N-acetyl-ᴅ-glucosamine, 68

N-acetylcysteine (NAC), 103–104

NAD(P)H:quinone oxireductase (NQO), 91

NADPH. *See* Nicotinamide adenine dinucleotide phosphate (NADPH)

NADPH disphosphorase (NADPHd), 152 staining, 152

NADPHd. *See* NADPH disphosphorase (NADPHd)

ʟ-NAME. *See* *N*-nitro-ʟ-arginine methyl ester (ʟ-NAME)

N-aminohexyl-*N*-aminopropyltrimethoxysilane, 227

NANC cells. *See* Nonadrenergic, noncholinergic cells (NANC cells)

Nanocarriers, 66
Nanoparticles, NO delivery using, 227–234
Nar. *See* Nitrate reductase (Nar)
N-bound diazeniumdiolates, 179–180
NCX-4016, 65*f*
NDBP. *See* 2-Nitrate-1, 3-dibuthoxypropan (NDBP)
N-diazen-1-ium-1, 2 diolate functional group, 95
N-diazeniumdiolates, 179*f*, 230
NE. *See* Noradrenaline (NE)
Negative logarithm of the EC50 (pD2), 253–254, 254*t*
Neglected diseases, 25–26
    NO donors, 26
        B. hominis infection, 44–45
        DF, 42–44
        leishmaniasis, 32–35
        lymphatic filariasis, 45
        malaria, 27–32
        NO action mechanisms against pathogens, 46, 47*f*
        other neglected diseases, 45–46
        schistosomiasis, 40–41
        trypanosomiasis, 35–40
        tuberculosis, 41–42
Nerve regeneration. *See also* Brain
    AD, 145–147
    examples of NO donors evaluation, 148–151, 148*f*
    goldfish optic nerve regeneration studied with NOR-2 and SNAP, 152–155, 154*f*
    HD, 147–148
    locust studies on NO-cGMP with NOC-18 and SNP, 151–152
    neuronal growth and synaptic plasticity, 143
    neuroprotectant, 143–144, 145*f*
    NO
        in neurodegeneration, 145
        in neuroregeneration, 141–143
    PD, 147
    rats
        exogenous SNAC on motor functional recovery in, 158–159
        GSNO effects after I/R injury in, 159–162
        GSNO influences angiogenesis in rats, 162–163
        reinnervation after penile nerve crush studies with SNP in, 157–158

regenerating RGCs of cats with nipradilol, 155–157
NET. *See* Neuroendocrine tumors (NET)
Neurobehavioral dysfunctions, 159–160
Neurodegeneration, 141. *See also* Neuroregeneration
    NO in, 145
Neuroendocrine tumors (NET), 104–105
Neuroinflammation, 265–275. *See also* Brain
    AD, 270–275, 276*f*–277*f*
    ischemia–reperfusion injury, 267–270, 271*f*
    TBI, 265–267, 268*f*
Neuronal growth, 143
Neuronal nitric oxide synthase (nNOS), 55–56, 76, 122, 169–170, 243, 263–264, 303–304
Neuronal peroxynitrite production in TBI, 281
Neuronal survival, 57
Neuronal transmissions, 57
Neuroprotectant, 143–144, 145*f*
    NO-induced cGMP immunoreactivity, 144*f*
Neuroprotection, 269
Neuroprotective agents in brain inflammation NODs
    in AD, 282–283
    in cerebral ischemia–reperfusion injury, 275–280
    in TBI, 280–282
Neuroregeneration, 141. *See also* Neurodegeneration
    NO in, 141–143
Neurotoxic effects of NO, 145, 146*f*
Neurotoxicity, 269
Neurotransmission, 141–143
Neutrophils, 195, 197
NF-κB. *See* Nuclear Factor kappa B (NF-κB)
NG-monomethyl-L-arginine (L-NMMA), 62
NG-nitro-L-arginine (L-NNA), 62
N-Hydroxyguanidines, 176, 177*f*
7-NI. *See* 7-NitroIndazole (7-NI)
Nicotinamide adenine dinucleotide phosphate (NADPH), 55–56, 243, 263
Nifurtimox, 36
Nipradilol, regenerating RGCs of cats with, 155–157
NIRs. *See* Nitrite reductases (NIRs)
NISSL staining, 163
Nitrate (NO_3^-), 172–173, 176, 177*t*, 178, 179*f*, 223–224

Nitrate reduction pathway, 56
Nitrate tolerance, 95, 251–253
Nitrate NO-NSAIDs, structural features of,
    81–84, 82*f*
  cancer prevention with NO-NSAIDs, 86–93
  second-generation NO-NSAIDs, 81–84
Nitrate reductase (Nar), 172–173, 178
2-Nitrate-1, 3-dibuthoxypropan (NDBP), 250–
    251, 254
Nitrergic nerve fibers, 123
Nitric oxide (NO), 1–2, 55, 75–76, 77*f*, 121, 141,
    169, 191–192, 196–197, 221, 243–244, 263,
    293–294, 313
  antibacterial and biofilm dispersal agents,
    303–304
  with antibiotics, synergic effect of, 304–307
  and anticancer drug cotreatment, 297–299, 298*f*
  anticancer drug hybrids, 299–303
  as antimicrobial and antibiofilm agent,
    173–175
  antimicrobials based on QS inhibitors, dual
    action, 182–183
  in bacteria, 294–296, 303*f*
  biological effects, 170–171
  biosynthesis, 169–170, 170*f*
  in body, 57–67
    agents for regulating NO, 61–64
    complications from attenuation of NO
      production, 59–60
    donor carriers, 65–67
    NO donors, 64–65
    NO releasing biomedical devices, 60–61
    in wound healing, 58–59
  in brain damage, 265–275
    AD, 270–275, 276*f*–277*f*
    ischemia–reperfusion injury, 267–270, 271*f*
    TBI, 265–267, 268*f*
  in brain physiology, 263–264
  cellular actions, 76–79
  delivery, 296
    of NOS, 296
  derivative species physical and chemical
    properties, 2–4
    back-bonding representation, 4*f*
    bond length, vibrational energy, and reduction
      potential of, 4*t*
    cis-[RuCl(bpy)$_2$(NO)] electrochemical
      reduction mechanism, 9*f*
  dimerization reaction of NO, 3*f*
  electron configuration, 3*f*
  NO oxidation reaction steps, 4*f*
  diazeniumdiolates, 179–181, 179*f*
  dual action NO
    antimicrobials based on QS inhibitors,
      182–183
    donors as antimicrobials, 181–184
  effects, 296
  enhancement of intracellular levels, 297*f*
  in erectile physiology, 123–126
  gas as antimicrobial agent, 175–176
  in human health, 55
  hybrid antimicrobials based on antibiotics,
    182–183, 183*f*
  inactivation chemistry, 170
  leishmaniasis and, 32
  malaria and, 27–28
  in neurodegeneration, 145
  in neuroregeneration, 141–143
  nitrates, 177*t*, 178
  nitrosyl ruthenium complexes as NO delivery
    agents, 2–5
    NO derivative species physical and chemical
      properties, 2–4
    by reduction process, 5–11
  regulation in microbes, 171–173, 172*f*
  releasing oxygen sensor, 203*f*
  ruthenium-nitrogen oxide derivatives as NO
    sources, 5
  S-NO, 181
  sol–gel film, 227
  structure of NO-drug hybrids, 300*f*
  treatment of NO-based hybrid drugs, 299*f*
  trypanosomiasis and, 36
Nitric oxide analyzer (NOA), 71
Nitric oxide synthases (NOSs), 55–56, 76, 77*f*,
    89, 122, 149–150, 169, 243, 263, 293–295,
    303–304
  delivery of, 296
  isoforms, 122–123
Nitric oxide-releasing NSAIDs (NO-NSAIDs),
    80–81
  second-generation NO-NSAIDs, 81–84
Nitrite, nitric oxide reductase regulator
    (NNR regulator), 171–172
Nitrite (NO$_2^-$), 172–173, 178
  binding, 56–57

Nitrite reductases (NIRs), 56–57, 172–173
  nitrite binds to metal center, 57*f*
Nitroaspirin, 65*f*
Nitrogen dioxide (NO₂), 46
Nitroglycerin, 64, 65*f*
Nitroglycerine (NTG), 75–76, 243–244
7-NitroIndazole (7-NI), 269
Nitrooxy NO-DOXO, 300–301
Nitrooxyl (–ONO²), 28
Nitrosamines, 131–132
"Nitrosative stress", 145
Nitrosonium cation (NO⁺), 58
Nitrosonium ion (NO⁺), 126–127, 129
Nitrosyl (NO⁺), 1–3
Nitrosyl ruthenium complexes, 1–2, 17
  cellular imaging, 17–21
  central atom, 1
  as NO delivery agents, 2–5
    chemical structure, 6*t*
    nitrosyl ruthenium complex with hydroxide
      ion reaction, 7*f*
    pH effect on metal ligand charge transfer
      band, 7*f*
    by reduction process, 5–11
  photo-nitric oxide release promoting by
    ruthenium complexes, 11–16
  ruthenium-nitrogen oxide derivatives as NO
    sources, 5
  spectral profile change
    with pH, 7*f*
    with time, 8*f*
  synergistic effect between ruthenium complex
    and nitric oxide, 9
    cell viability, 11*f*, 11*t*
    cytosolic concentration of nitric oxide, 10*f*
Nitrosyl-heme complex, 170–171
3-Nitrotyrosine (3-NT), 76–78, 160
Nitrovasodilators, 75–76
Nitroxyl (HNO/NO⁻), 3, 58, 94, 126–127
NMDAR. *See* N-methyl-D-aspartate receptors
  (NMDAR)
N-methyl-3-aminopropyltrimethoxysilane,
  205–207
N-methyl-D-aspartate receptors (NMDAR),
  141–144, 171
L-NMMA. *See* NG-monomethyl-L-arginine
  (L-NMMA)
N-nitro-L-arginine methyl ester (L-NAME), 62,
  143, 149–150

N-Nitrosoamines, 177*f*
N-nitrosobis(2-oxopropyl)amine (BOP), 87–88
L-NNA. *See* NG-nitro-L-arginine (L-NNA)
nNOS. *See* Neuronal nitric oxide synthase (nNOS)
NNR regulator. *See* Nitrite, nitric oxide reductase
  regulator (NNR regulator)
NnrR. *See* Nitrite, nitric oxide reductase regulator
  (NNR regulator)
NO donor–amodiaquine conjugates (NO–AQ), 28
NO donors (NODs), 64–65, 121, 126–127, 141,
  275–277, 314
  in AD, 282–283
  as antimicrobials, 176–181
  biofilms
    control by, 222*f*
    controlling of clinical interest, 226–234
    NO delivery using nanoparticles, 227–234
    NO liposomes, 232–233
    NO releasing dendrimers, 233–234
    NO releasing polysaccharides, 229–232
  in cerebral ischemia–reperfusion injury,
    275–280
  dual action NO donors as antimicrobials,
    181–184
  for erectile function, 127, 128*t*
  evaluation for nerve regeneration, 148–151, 148*f*,
    150*f*
    Hydra regeneration studies on NO-cGMP
      pathway with NOC-18, 149–150
  and generators to sensors, 197–215
    catalytic generation from endogenous
      RSNOs, 213–215
    development of diazeniumdiolates as, 198–210
    RSNOs, 210–212
  as neuroprotective agents in brain inflammation
    with polymeric materials, 199*f*
  in TBI, 280–282
  to treating neglected diseases, 26
    B. hominis infection, 44–45
    leishmaniasis, 32–35, 35*f*
    lymphatic filariasis, 45
    malaria, 28–31, 31*f*
    neglected diseases, 45–46
    NO action mechanisms against pathogens,
      46, 47*f*
    schistosomiasis, 40, 41*f*
    trypanosomiasis, 37–39, 38*f*
    tuberculosis, 42
NO reductase (Nor), 172–173

NO-ASA
    biological actions of spacer in, 93
    molecular targets, 88–93
        cell kinetics, 88
        COX-2, 89–90
        MAPK, 90–91
        modulation of proinflammatory cytokines, 92
        NF-κB, 88
        nitric oxide synthase, 89
        oxidative stress, 91–92
        PPARδ, 90
        S-nitrosylation, 92
        Wnt/β-Catenin signaling, 89
        xenobiotic metabolizing enzymes, 91
NO-based compounds as cancer therapeutics, 80
NO-celecoxib, 84–86
NO-cGMP. See NO-cyclic guanosine
        monophosphate (NO-cGMP)
NO-cGMP pathway, 152–153
    Hydra regeneration studies on, 149–150
    locust studies on NO-cGMP with NOC-18 and
        SNP, 151–152
NO-cGMP transduction pathway, 143
NO-cyclic guanosine monophosphate
        (NO-cGMP), 322–323
NO-CYS. See S-nitrosocysteine (SNOC)
NO-donating statins, 134
NO-DOXO, 299–300
NO-NSAIDs. See Nitric oxide-releasing NSAIDs
        (NO-NSAIDs)
NO-releasing coxibs, 84–86, 85f
NO-releasing hydrogels, 314, 315t–317t, 318t
NO-releasing modified PVA (PVA-NO), 324
NO-responsive regulatory networks, 171
NO-responsive transcriptional activators, 171–172
NO-rofecoxib, 84–86
NO-sGC-cGMP pathway, 133
NO-valdecoxib, 84–86
NO-xerogels, 228–229
NO⁰. See Free radical NO (NO⁰)
NOA. See Nitric oxide analyzer (NOA)
NO-AQ. See NO donor–amodiaquine conjugates
        (NO-AQ)
NOC-18. See 3, 3-bis(aminoethil)-1-hydroxy-2-
        oxo-1-triazene (NOC-18)
NOC5, 31
NODs. See NO donors (NODs)
Nonadrenergic, noncholinergic cells (NANC cells),
        122–123

Nonenzymatic activation of NTG, 247
Nonenzymatic mechanism, 178
NONO-NSAIDs, 95–97, 95f
NONOates. See Diazeniumdiolates
Nonsmall cell lung cancer cell lines (NSCLC cell
        lines), 99
Nonsteroidal anti inflammatory drugs (NSAIDs),
        78–79
    protect against cancer, 79
    side effects associated with, 79–80
        NO-based compounds as cancer therapeutics,
            80
Nonsteroidal antiinflammatory drugs (NSAIDs),
        182
Nontoxic concentrations, 175
Nontraditional antibiotic agents, 223
Nor. See NO reductase (Nor)
NOR-2, goldfish optic nerve regeneration studied
        with, 152–155
NOR-3. See [(E)-Hydroxyimino]-5-nitro-3-
        hexenamide (NOR-3)
Noradrenaline (NE), 142–143
NOS1. See Neuronal nitric oxide synthase
        (nNOS)
NOS2. See Inducible nitric oxide synthase
        (iNOS)
NOS3. See Endothelial nitric oxide synthase
        (eNOS)
NOSH-NSAIDs, 100–102, 101f, 103f
NOSs. See Nitric oxide synthases (NOSs)
NQO. See NAD(P)H:quinone oxireductase
        (NQO)
NRF-2. See Nuclear factor erythroid 2-related
        factor 2 (NRF-2)
NSAIDs. See Nonsteroidal anti inflammatory drugs
        (NSAIDs)Nonsteroidal antiinflammatory
        drugs (NSAIDs)
NSCLC cell lines. See Nonsmall cell lung cancer
        cell lines (NSCLC cell lines)
3-NT. See 3-Nitrotyrosine (3-NT)
NTG. See Nitroglycerine (NTG)
Nuclear factor erythroid 2-related factor 2 (NRF-
        2), 29
Nuclear Factor kappa B (NF-κB), 266–267,
        278–279
    NF-κB–DNA interaction, 88
    signaling, 88
Nucleophilic aromatic substitution reactions, 180
Nucleus accumbens (NAc), 142–143

O

O^2-(2, 4-Dinitrophenyl)1-[(4-thoxycarbonyl)
   piperazin-1-yl]diazen-1-ium-1, 2-diolate
   (JS-K), 94, 94f, 98–100
O^2–acetoxymethyl–1–[N–(2–hydroxyethyl)–N–
   methylamino]diazen–1–ium–1, 2–diolate
   (2–HEMA), 96
O^2-(acetylsalicyloyloxymethyl)-1-(N,N-
   diethylamino)-diazen-1-ium-1, 2-diolate
   (DEA/NO-aspirin), 97–98
O^2-(acetylsalicyloyloxymethyl)-1-(N-
   isopropylamino)-diazen-1-ium-1, 2-diolate
   (IPA/NO-aspirin), 97–98
O^2-[2, 4-dinitro-5-(N-methyl-N-4-
   carboxyphenylamino)phenyl]1-N,N-
   dimethylamino)diazen-1-ium-1, 2-diolate
   (PABA/NO), 98–100
O^2-acetoxymethyl-protected (PROLI/NO), 96
O^2-arylated diazeniumdiolate, 179f, 180
O^2-diazeniumdiolate, 179f, 180
O^2-diazeniumdiolates, 180
O^2-unsubstituted N-diazen-1-ium-1, 2-diolate. See
   Diazeniumdiolates
OA. See Oleanolic acid (OA)
Octyl-modified dendrimers, 233
ODQ. See (1H-[1, 2, 4]Oxadiazolo[4, 3-a]
   quinoxalin-1-one) (ODQ)
OGG1. See 8-Oxo-deoxyguanosine glycosylase
   (OGG1)
OH-BBN model. See Hydroxybutyl(butyl)
   nitrosamine model (OH-BBN model)
Oleanolic acid (OA), 100, 301–302
o-nitrophenyloctyl ether (o-NPOE), 214
Optic nerve (OpN), 155–156
Oral PDE-5 inhibitors, 130
Organic nitrates, 130, 176, 177f, 245–251
Organoselenium species, 214
Oridonin, 302–303
OS. See Oxidative stress (OS)
1, 2, 5-Oxadiazole 2-oxide. See Furoxan
   (1H-[1, 2, 4]Oxadiazolo[4, 3-a]quinoxalin-1-one)
   (ODQ), 99, 149–150
Oxaliplatin, 87–88
Oxidation, 58
Oxidative stress (OS), 91–92, 125–126, 264, 269
   hypothesis, 252–253
Oximes, 131, 177f
8-Oxo-deoxyguanosine glycosylase (OGG1), 99

Oxygen (O_2), 55–56, 170
   sensors, 208
Oxyhemoglobin, 58

P

$P.$ $berghei$ ANKA (PbA), 28
PABA/NO. See O^2-[2, 4-dinitro-5-(N-methyl-
   N-4-carboxyphenylamino)phenyl]1-N,
   N-dimethylamino)diazen-1-ium-1,
   2-diolate (PABA/NO)
PAMAM. See Poly(amidoamine) (PAMAM)
Pancreatic carcinoma (PANC-1), 103–104
Parasitic diseases, 29, 45–46
Parasympathetic nervous system (PSNS), 157–158
Parkinson's disease (PD), 145, 147
Pathogens, NO action mechanisms against, 46, 47f
PbA. See $P.$ $berghei$ ANKA (PbA)
PBIT. See S,S'-1, 4-phenylene-bis(1, 2-ethanediyl)
   bis-isothiourea (PBIT)
PCa. See Prostate cancer (PCa)
PD. See Parkinson's disease (PD)
pD2. See Negative logarithm of the EC50 (pD2)
PDE-5 inhibitors, 132
PDEs. See Phosphodiesterases (PDEs)
PDI. See Protein-disulfide isomerase (PDI)
PDMS. See Polydimethylsiloxane (PDMS)
PDT. See Photodynamic therapy (PDT)
PEG. See Polyethyleneglycol (PEG)
Pelouze's salt, 179
Pendant hydroxyl groups of PVA, 324
Penile detumescence, 122–123
Penile erectile function
   EC, 122f
   ED and prevalence, 126
   nitrosamines, 131–132
   NO donors, 126–127
      for erectile function, 127, 128t
   NO in erectile physiology, 123–126
   NO-donating statins, 134
   NOS isoforms, 122–123
   organic nitrates, 130
   oximes, 131
   PDE-5 inhibitors, 132
   S-nitrosothiols, 131
   SGC stimulators and activators, 133–134
   SNAP, 129
   SNP, 129
   sodium nitrite, 127–129

spermine NONOate, 131
sydnonimines, 130
Penile nerve crush studies with SNP in rats,
    reinnervation after, 157–158
Pentaerythritol tetranitrate (PETN), 246–247
PEO–PPO–PEO, chemical structure of, 319*f*
PepT1. *See* Peptide transporter 1 (PepT1)
Peptide thiol (SH), 181
Peptide transporter 1 (PepT1), 301–302
Peripheral nerve injury, 144
Peripheral nervous system (PNS), 141–142
Peroxiredoxin 1 (PRX1), 99
Peroxisome proliferator-activated receptor δ
    (PPARδ), 90
Peroxynitrite (ONOO⁻), 26, 46, 58, 76–78, 125,
    145, 159–160, 170, 263–264
PET. *See* Polyethylene terephthalate (PET)
PETN. *See* Pentaerythritol tetranitrate (PETN)
PGE2. *See* ProstaGlandin E2 (PGE2)
PGI₂. *See* Prostacyclin (PGI₂)
P-glycoprotein (P-gp), 293–294
P-gp. *See* P-glycoprotein (P-gp)
PGs. *See* Prostaglandins (PGs)
pHEMA. *See* Poly(2-hydroxyethyl methyl acrylate)
    (pHEMA)
Phosphodiesterases (PDEs), 305–307
Photo-induced electron transfer, 14, 14*f*
Photo-nitric oxide release promoting by
    ruthenium complexes, 11–16
    effects of RuNO²⁺:Rupz²⁺, 16*f*
    as NO deliver agents, 12*f*
    NO release and cellular mechanisms, 13*f*
    photo-induced electron transfer in, 14*f*
    photochemical pathway NO release, 13*f*
    supramolecular formation of ruthenium
        complexes, 15*f*
    visible light irradiation effects, 16*f*
Photochemical experiments, 1–2
Photodynamic therapy (PDT), 17
Photoluminescence (PL), 20–21
Photosensitization processes, 18, 20–21
Photosensitizer drugs (PS drugs), 18
Phthalocyanine–ruthenium compound, 14–15
Physiological processes, 170
PKG. *See* Protein kinase G (PKG)
PL. *See* Photoluminescence (PL)
Plasmodium falciparum, 27
Plasmodium vivax, 29
Platinum compounds, 1–2

PLGA. *See* Poly(lactide-*co*-glycolide) (PLGA)
Pluronic F127, 318–320
PMBN. *See* Polymyxin B nonapeptide (PMBN)
PMMA. *See* Polymethylmethacrylate (PMMA)
p-NO-ASA isomers, 88
PNPE. *See* Poly(nitrosated) polyester (PNPE)
PNS. *See* Peripheral nervous system (PNS)
PO. *See* Propylene oxide (PO)
Poly-β-cyclodextrin, 327–328
Poly(2-hydroxyethyl methyl acrylate) (pHEMA),
    214, 313, 322
Poly(amidoamine) (PAMAM), 233–234
Poly(ethylene oxide)-poly(propylene oxide)-poly
    (ethylene oxide) triblock copolymer,
    318–319
Poly(lactide-*co*-glycolide) (PLGA), 33–34, 207–208,
    229
Poly(nitrosated) polyester (PNPE), 227
Poly(sulfhydrylated polyester) (PSPE), 227
Poly(vinyl alcohol) (PVA), 323
Poly(vinyl chloride) (PVC), 228
Poly(vinylpyrrolidone) (PVP), 210
Polydimethylsiloxane (PDMS), 199–200
Polyesterification reaction, 227
Polyethylene terephthalate (PET), 66, 213–214
Polyethyleneglycol (PEG), 230
Polymeric materials, 192
    NO donors with, 199*f*
Polymers, 66
    biological response
        in blood, 193–194
        in subcutaneous tissue, 195–196
    polymeric matrix, 314
Polymethylmethacrylate (PMMA), 212, 227
Polymyxin B nonapeptide (PMBN), 180–181
Polysaccharides, NO releasing, 229–232
Polyurethane (PU), 198, 208–209, 227, 322
Potassium tertakis(4-chlorophenyl)borate
    (KTpClPB), 198
PPARδ. *See* Peroxisome proliferator-activated
    receptor δ (PPARδ)
Praziquantel (PZQ), 40
Primaquine, 27
Proinflammatory cytokines, 92, 173, 275
Prokaryotic cells, 180–181
PROLI NONOate, 234–235, 235*f*
PROLI/NO. *See* O²-acetoxymethyl-protected
    (PROLI/NO)
Proline-based *N*-diazeniumdiolate, 179–180, 179*f*

Propylene oxide (PO), 233–234
Prostacyclin (PGI$_2$), 84–86
ProstaGlandin E2 (PGE2), 275
Prostaglandins (PGs), 80–81
Prostate cancer (PCa), 99
Protected diazeniumdiolates, 179f
Protein kinase C activation, 121
Protein kinase G (PKG), 76–78, 123–124, 141–142,
        144, 149–150, 243–244, 264
Protein β-catenin, 99–100
Protein-disulfide isomerase (PDI), 147
Proteins, 145–147
Proteus mirabilis, 228
Protonation, 180
PRX1. *See* Peroxiredoxin 1 (PRX1)
PS drugs. *See* Photosensitizer drugs (PS drugs)
Pseudomonas aeruginosa, 171–173, 175–176,
        183–184, 223–224, 230–231
Pseudotolerance, 252
PSNS. *See* Parasympathetic nervous system (PSNS)
PSPE. *See* Poly(sulfhydrylated polyester) (PSPE)
PU. *See* Polyurethane (PU)
Purified chitin, 68
PVA. *See* Poly(vinyl alcohol) (PVA)
PVA-NO. *See* NO-releasing modified PVA
        (PVA-NO)
PVC. *See* Poly(vinyl chloride) (PVC)
PVP. *See* Poly(vinylpyrrolidone) (PVP)
Py235. *See* 235-kDa rhoptry protein (Py235)
(Pyrazoyl)benzenesulfonamides, 84–86
PZQ. *See* Praziquantel (PZQ)

Q

Quality of life (QOL), 270
Quantum dots (QDs), 20–21
Quinone methide intermediate (QM
        intermediate), 93
Quorum sensing (QS), 173
    dual action NO antimicrobials based on QS
        inhibitors, 182–183

R

Rats
    exogenous SNAC on motor functional recovery
        in, 158–159
    GSNO
        effects after I/R injury in, 159–162
        influences angiogenesis in, 162–163
    reinnervation after penile nerve crush studies
        with SNP in, 157–158

Raynaud's syndrome, 325–326
RBCs. *See* Red blood cells (RBCs)
Reactive nitrogen and oxygen intermediates
        (RNOS), 174
Reactive nitrogen species (RNS), 26, 99, 144, 171,
        297–299
Reactive oxygen and nitrogen species (RONS), 3,
        91–92
Reactive oxygen species (ROS), 26, 99, 122–123,
        145–147, 252, 264, 266, 293–294, 303–304
Red blood cells (RBCs), 27
Redox-sensitive enzyme, 133
Reperfusion-induced ROS, 269
Research and Training in Tropical Diseases (TDR),
        25–26
Residues to form *S*-nitrosothiols (RSNO), 39,
        76–78
Retinal ganglion cells (RGCs), 152–153
    regeneration of cats with nipradilol, 155–157
RGCs. *See* Retinal ganglion cells (RGCs)
Rhodamine B isothiocyanate (RITC), 230
    confocal fluorescence images of, 231f
Rhodnius prolixus, 39
Rifampicin, 41
RITC. *See* Rhodamine B isothiocyanate (RITC)
RNOS. *See* Reactive nitrogen and oxygen
        intermediates (RNOS)
RNS. *See* Reactive nitrogen species (RNS)
RONS. *See* Reactive oxygen and nitrogen species
        (RONS)
ROS. *See* Reactive oxygen species (ROS)
RRx-001, 102–103
    aerospace compound for cancer treatment,
        102–105
RSNO. *See* Residues to form *S*-nitrosothiols
        (RSNO)
[Ru(bdqi)(terpy)(NO)]Cl$_3$, 9
[Ru(phthalocyanine)(NO$_2$)(NO)] complex, 13–14
Ruthenium complexes, photo-nitric oxide release
        promoting by, 11–16
    effects of RuNO^{2+}:Rupz^{2+}, 16f
    as NO deliver agents, 12f
    NO release and cellular mechanisms, 13f
    photo-induced electron transfer in, 14f
    photochemical pathway NO release, 13f
    supramolecular formation of ruthenium
        complexes, 15f
    visible light irradiation effects, 16f
Ruthenium-nitrogen oxide derivatives as NO
        sources, 5

S

SAG. *See* Standard antileishmanial sodium stibogluconate (SAG)
Salmonella, 223–224
"Sandwich", 245–246, 246f
Sarcoplasmic reticulum (SR), 243–244
SCC VII. *See* Squamous cell carcinoma (SCC VII)
SCD. *See* Sickle cell disease (SCD)
Schistosoma genus, 40
Schistosomiasis, 40–41
  NO donors and, 40, 41f
Sciatic functional index (SFI), 158–159
SCYX-7158, 36
SD rats. *See* Sprague–Dawley rats (SD rats)
Second-generation NO-NSAIDs, 81–84
Serotonin reuptake transporter (SERT), 151
Severe malaria, 27
SFI. *See* Sciatic functional index (SFI)
sGC. *See* Soluble guanylate cyclase (sGC)
SH. *See* Peptide thiol (SH)
Shewanella oneidensis, 224
SHRs. *See* Spontaneously hypertensive rats (SHRs)
Sickle cell disease (SCD), 60
Silver sulfadiazine (SSD), 304
SIN-1. *See* 3-Morpholino-sydnonimine (SIN-1)
Singlet oxygen, 17
siRNA. *See* Small interfering RNS (siRNA)
SLN. *See* Solid lipid nanoparticles (SLN)
Small interfering RNS (siRNA), 155
Small molecular weight donors, delivery of NO via, 234–236
SMAR. *See* Superior mesenteric artery rings (SMAR)
SNAC. *See* S-nitroso-N-acetylcysteine (SNAC)
SNAP. *See* S-nitroacetylpenicillamine (SNAP)
SNAP-PDMS. *See* S-nitroso-N-acetyl-D-penicillamine to PDMS (SNAP-PDMS)
S-nitroacetylpenicillamine (SNAP), 33, 43, 127, 129, 153–155, 179f, 181, 275–277, 297–299, 314–318
  goldfish optic nerve regeneration studied with, 152–155
S-nitrosated dextran-cysteamine, 230
S-nitrosation, 174
S-nitroso proteins, 181
S-nitroso-N-acetyl-D-penicillamine to PDMS (SNAP-PDMS), 212
S-nitroso-N-acetyl-DL-penicillamine, 211–212

S-nitroso-N-acetylcysteine (SNAC), 32–33, 158–159, 179f, 181, 211–212, 319
S-nitroso-penicillamine (SPA), 214
S-nitroso-protein (SNO-protein), 92
S-nitrosoalbumin (AlbSNO), 213–214
S-nitrosoamines, 177f
S-nitrosocysteine (SNOC), 129, 282
S-nitrosoglutathione (GSNO), 29, 64, 131, 159–160, 179f, 181, 214, 264, 275–277, 297–299, 319
  effects after I/R injury in rats, 159–162, 161f
S-nitrosothiol-modified silica/chitosan core–shell nanoparticles, 231–232
S-nitrosothiols (S-NO), 64, 65f, 131, 179f, 181, 210–212, 314–319
  catalytic generation from endogenous, 213–215
S-nitrosylation, 92, 171
S-nitrothiols, 176
S-NO. *See* S-nitrosothiols (S-NO)
SNO-protein. *See* S-nitroso-protein (SNO-protein)
SNOC. *See* S-nitrosocysteine (SNOC)
SNP. *See* Sodium nitroprusside (SNP)
Sodium nitrite (NaNO$_2$), 44, 127–129
Sodium nitroprusside (SNP), 1–2, 29, 125, 127, 129, 151–152, 234–235, 235f, 245–246, 275–277, 297–299, 305–307
  reinnervation after penile nerve crush studies with SNP in rats, 157–158
Sol–gel particle, 208–209
  hybrid sol–gel/PU glucose sensor, 209f
Sol gels, 227
Solid lipid nanoparticles (SLN), 9
Soluble guanylate cyclase (sGC), 57, 76–78, 121, 141–142, 144, 170–171, 196–197, 243
  stimulators and activators, 133–134
SPA. *See* S-nitroso-penicillamine (SPA)
Spermate derivatives, 179–180, 179f
Spermine NONOate, 131, 277
Spontaneously hypertensive rats (SHRs), 84–86, 249–250, 254–255
Sporadic AD, 282
Sprague–Dawley rats (SD rats), 143
Squamous cell carcinoma (SCC VII), 103–104
SR. *See* Sarcoplasmic reticulum (SR)
S,S'-1, 4-phenylene-bis(1, 2-ethanediyl)bis-isothiourea (PBIT), 89
SSD. *See* Silver sulfadiazine (SSD)
Standard antileishmanial sodium stibogluconate (SAG), 34

Staphylococcus aureus, 175–176, 230
Stroke, 162–163
Stronglyoides stercoralis, 45–46
Subcutaneous tissue, biological response to polymers in, 195–196
Superior mesenteric artery rings (SMAR), 253–254
Superoxide (O_2^-), 26, 46, 125–126, 171
Supramolecular NO-releasing hydrogels, 314, 326–328, 327f
Supramolecular systems, 14–15
Surface-bound C3b, 193–194
Sydnonimines, 65f, 130, 177f
Sympathetic system neurotransmitter noradrenaline, 122–123
Synaptic plasticity, 143
Synaptophysin, 162
Synergic effect of NO with antibiotics, 304–307
Synergistic effect between ruthenium complex and nitric oxide, 9

T
Tau hyperphosphorylation, 283
TBI. *See* Traumatic brain injury (TBI)
TDDA. *See* Tridoceylamine (TDDA)
TDR. *See* Research and Training in Tropical Diseases (TDR)
Temozolomide (TMZ), 98
Tetrahydrobiopterin (BH_4), 134, 243
TGD. *See* Tumor growth delay (TGD)
Thioredoxin-1 (Trx-1), 91–92
Thrombosis, 193
TMZ. *See* Temozolomide (TMZ)
TNF-α. *See* Tumor Necrosis Factor-α (TNF-α)
TNG. *See* Trinitroglycerin (TNG)
Tobramycin, 175–176, 182–183
Topical application
   of GSNO-containing Pluronic F127 hydrogel, 320, 321f
   NO-donating devices
      chitosan, 67–70
      copper chitosan, 70
      NIRs, 56–57
      NO in body, 57–67
      NOSs, 55–56
      topical applicators, 67–71
      topical copper chitosan particles, 70–71

Topical copper chitosan particles, 70–71
TPP. *See* Tripolyphosphate (TPP)
Trans-[$Ru(NH_3)_4$ (caffeine)(NO)]C1$_3$ (LLNO1), 132
Trans-[RuCl([15]aneN$_4$)(NO)]Cl$_2$, 1–2
Trans-[RuCl([15]aneN$_4$)NO]$^{2+}$, 37
Transcriptional activator, 171–172
Transforming growth factor beta, 36
Transition metals, 76–78, 170
Traumatic brain injury (TBI), 159–160, 265–267, 268f
   NODs in, 280–282
Tridoceylamine (TDDA), 198
Trinitroglycerin (TNG), 34–35
Tripolyphosphate (TPP), 39
TrkB. *See* Tyrosine receptor kinase B (TrkB)
Troposphere, 169
True tolerance, 252
Trx-1. *See* Thioredoxin-1 (Trx-1)
Trypanosoma brucei gambiense, 35–36
Trypanosoma brucei rhodesiense, 35–36
Trypanosoma cruzi, 35–36
Trypanosomiasis, 35–40
   NO and, 36
   NO donors and, 37–39, 38f
Tuberculosis, 25, 41–42
   NO donors and tuberculosis, 42
Tumor growth delay (TGD), 103–104
Tumor Necrosis Factor-α (TNF-α), 267
Tyrosine hydroxylase, 147
Tyrosine receptor kinase B (TrkB), 162, 271f, 281

U
Ubiquitin–proteasome system, 147
Ultrafast short-burst NO donors, 179–180
Unilamellar vesicles (ULV), 232–233
Urinary infectious bacteria, 178
Urinary system, 178
Urinary tract infection, 178

V
Valinomycin (VAL), 198
Vascular
   grafts, 61
   relaxation, 75–76
Vascular cellular adhesion molecule-1 (VCAM-1), 96–97

Vascular endothelial growth factor (VEGF), 58,
    162–163, 195
Vascular smooth muscle cell (VSMC), 122*f*, 243
Vasculature, 58
Vasodilation, 57, 313
Vasodilator-stimulated phosphoprotein (VASP), 59
Vasorelaxant effect
    by cyclohexane nitrate, 256*f*
    by NDBP, 255*f*
VASP. *See* Vasodilator-stimulated phosphoprotein
    (VASP)
VCAM-1. *See* Vascular cellular adhesion
    molecule-1 (VCAM-1)
Vector-borne zoonotic disease, 32
VEGF. *See* Vascular endothelial growth factor
    (VEGF)
Vero cell line, 37–38
Versatile biopolymer, 67–70
Vibrio harveyi, 183–184
Vinyl protected prodrug V-PYRRO, 94

Visceral leishmaniasis (VL), 32
Visible light irradiation, 14–15
Vitamin E, 134
VL. *See* Visceral leishmaniasis (VL)
von Willebrand's Factor (vWF), 193
VSMC. *See* Vascular smooth muscle cell
    (VSMC)
vWF. *See* von Willebrand's Factor (vWF)

W
Water-soluble NO, 230
Whole brain radiotherapy (WBRT), 104–105
Wnt signaling, 89
World Health Organization (WHO), 27, 41, 248
Wound healing, 313, 320–321, 327
    NO in, 58–59
    process, 58

X
Xenobiotic metabolizing enzymes, 91